Glencoe Science

Reading Essentials *for* BIOLOGY

The Dynamics of Life

An Interactive Student Textbook

bdol.glencoe.com

New York, New York Columbus, Ohio Chicago, Illinois Peoria, Illinois Woodland Hills, California

The McGraw·Hill Companies

Send all inquiries to:
Glencoe/McGraw-Hill
8787 Orion Place
Columbus, OH 43240-4027

ISBN 0-07-870181-3

Printed in the United States of America.

6 7 8 9 10 11   009   10 09 08 07

Contents

To the Student

Reading Essentials for Biology is designed to help you read, learn, and understand biology. Biology content is presented by sections in a total of 39 chapters. Each section is divided into *Before You Read*, *Read to Learn*, and *After You Read.*

Before You Read helps you organize your thoughts by drawing from prior knowledge, asking questions about information you would like to better understand, or finding clues in the text about the topics that will be covered.

In *Read to Learn*, the text focuses on key biology concepts. Key terms are reinforced and redefined. *Read to Learn* contains margin features to help you understand, organize, and reinforce new information. As you read, a ✓ at the end of a paragraph provides a visual clue for answering the *Reading Check* question in the margin.

After You Read presents a Mini Glossary featuring the key terms from the section and an activity using the terms. Additional activities help you organize, summarize, and analyze the content in the *Read to Learn* section.

Name Date Class

Section 1.1

What is biology?

Before You Read

When you hear the word *biology*, what do you think of? In the space below, make a list of some of the topics you think you will learn about as you study biology.

Read to Learn

Create a Quiz After you have read this section, create a five-question quiz based on what you have learned. Then, partner with another student and exchange quizzes. After taking the quizzes, review your answers together.

The Science of Biology

Biology is the study of life. When you study biology you will learn about all of the different kinds of living things. You will learn where they live, what they are like, how they depend on each other, and how they behave. ✓

One of the main ideas in biology is that living things depend on each other. They are part of what is called the balance of nature. All living things interact with other living things and with the world they live in. Without these relationships, nothing would survive.

Humans need plants and animals to supply us with food and raw materials, such as wood, oil, and cotton. Plants provide the oxygen in our air.

Biologists Study the Diversity of Life

The study of one living thing always involves studying other living things. Knowing how human life depends on nature and other living things is the only way that humans can expect to understand how to keep Earth healthy. With this knowledge, researchers can find ways to prevent diseases. Scientists can also find ways to help save living things that are in danger of becoming extinct and solve other problems.

✓ Reading Check

1. What is biology?

Characteristics of Living Things

Sometimes it is difficult to tell the difference between living and nonliving things. At times, nonliving things have one or more of the characteristics of life, but it is necessary to have all of the characteristics of life to be considered living. Things that have all of the characteristics of life are known as **organisms.** All organisms are made of one or more cells. Each cell contains the genetic material DNA that has the information needed to control the life processes of the organism.

What are the characteristics of life?

One of the first things biologists look for when they are searching for characteristics of life is structure, or **organization.** Whether an organism is made of a single cell or billions of cells, all of its parts work together in an orderly living system.

Another important characteristic of life is reproduction. **Reproduction** is the ability of an organism to make more of the same type of organism. The new organisms that are made are called offspring. Although reproduction is not needed for the survival of an individual organism, it must occur for the continuation of the organism's species. A **species** (SPEE sheez) consists of a group of organisms that can mate with each other and produce offspring that are able to reproduce. For example, there are many species of crocodiles including the American crocodile, the Australian freshwater crocodile, and the saltwater crocodile. American crocodiles reproduce only American crocodiles. Without reproduction, the species would die out. ✓

2. What is a species?

____

____

____

Another characteristic of life is that growth and development must take place. An organism begins life as a single cell. As time passes, it grows and develops. As growth and development take place, the organism takes on the characteristics of its species. **Growth** results in the formation of new structures and an increase in the amount of living material. **Development** refers to the changes that occur in each organism's life.

Section 1.1 **What is biology?,** *continued*

One more characteristic of life is the ability to adjust to surroundings, or the **environment.** Anything in the environment—air, water, temperature, weather, other organisms—that causes the organism to react is called a **stimulus** (plural, stimuli). The organism's reaction to the stimulus is called a **response.** An organism also has the ability to control its internal environment in order to maintain conditions suitable for survival. For example, an organism must make constant adjustments to maintain the right amount of water and minerals in its cells. This ability is called **homeostasis** (hoh mee oh STAY sus). Without the ability to adjust to internal changes, an organism would die. ✓

Reading Check

3. Name three or more things that make up the environment.

____

____

____

____

How do organisms respond to change?

Organisms use energy to grow, develop, respond to stimuli, and maintain homeostasis. **Energy** is the ability to cause change. Organisms get their energy from food.

Any behavior, structure, or internal process that allows an organism to make changes in response to environmental factors and live long enough to reproduce is called an **adaptation** (a dap TAY shun). For example, the leaves of many desert plants have a thick, waxy coating. This is an adaptation that helps these plants conserve water. Having large eyes is an adaptation that lets owls see well at night. The gradual change in a species over time due to adaptations is called **evolution** (e vuh LEW shun).

Think it Over

4. Conclude Some trees drop their leaves in the fall in response to: (Circle your choice.)

a. higher temperatures
b. lower temperatures
c. more daylight hours

After You Read

Mini Glossary

adaptation (a dap TAY shun): any structure, behavior, or internal process that enables an organism to respond to environmental factors and survive to produce offspring

biology: the study of life that seeks to provide an understanding of the natural world

development: the changes that take place during an organism's life; one of the characteristics of all living things

energy: the ability to cause change

environment: the surroundings to which an organism must adjust; includes air, water, weather, temperature, organisms, and other factors

evolution (e vuh LEW shun): gradual change in a species through adaptations over time

growth: changes in an organism resulting in an increase in the amount of living material and the formation of new structures; one of the characteristics of all living things

homeostasis (hoh mee oh STAY sus): an organism's ability to control its internal environment to maintain conditions suitable for survival

organism: anything that possesses all the characteristics of life

organization: the orderly structure of cells in an organism; one of the characteristics of all living things

Section 1.1 What is biology?, *continued*

reproduction: the production of offspring; a characteristic of all living things

response: an organism's reaction to a change in its environment

species (SPEE sheez): a group of organisms capable of mating with each other and producing offspring who can also reproduce

stimulus: anything in the environment that causes an organism to react

1. Review the terms and their definitions in the Mini Glossary above. Then on the lines below, write a sentence for each of the following words: **adaptation, evolution,** and **homeostasis.**

2. Use the web diagram below to help you review what you have read about organisms. List the four characteristics biologists use to recognize living things.

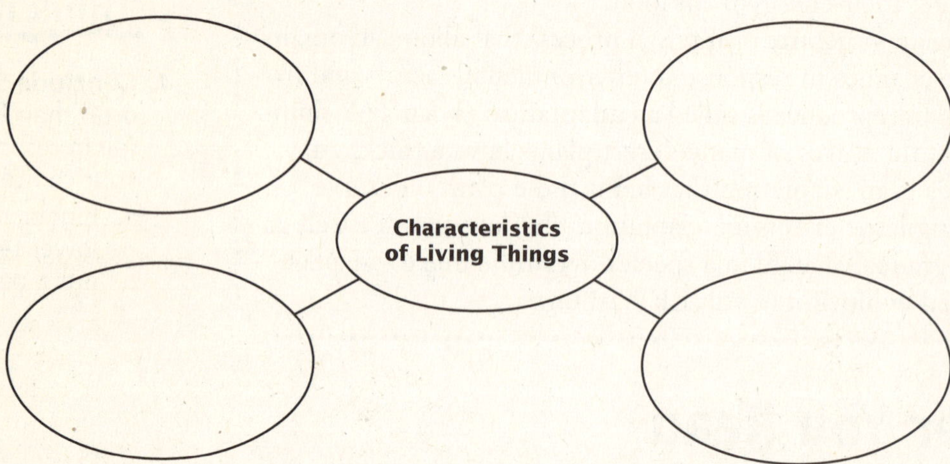

3. Give two examples of ways in which humans depend on other living things.

Visit the Glencoe Science Web site at **science.glencoe.com** to find your biology book and learn more about what biology is.

Name Date Class

Section 1.2 The Methods of Biology

Before You Read

Find the key terms **hypothesis** and **experiment** in the Read to Learn section below. Highlight and read the words and their definitions. Then, think of an experiment you might like to conduct. On the lines provided, explain the question you hope to answer with your experiment.

Read to Learn

Observing and Hypothesizing

Even though biologists and other scientists study many different types of things, they all use the same basic steps. The common steps they use to do research and answer questions are called **scientific methods.** Scientists often figure out questions to ask and answer just by observing the world around them.

What is a hypothesis?

Forming a **hypothesis** (hi PAHTH us sus) is a research method scientists use often. A hypothesis is an explanation for a question or problem that can be tested. For example, imagine that the number of birds in an area decreased after snakes came into the area. A scientist might make the hypothesis that the snakes were the reason the number of birds decreased. ✓

A scientist who forms a hypothesis must be certain that it can be tested. Before testing a hypothesis, scientists make observations and do research. The results of the experiment will help the scientist answer whether or not the hypothesis is supported.

Experimenting

To a scientist, an **experiment** is a test of a hypothesis by collecting information under controlled conditions.

Make Flash Cards Making flash cards is a good way to learn chapter material. For each paragraph, think of a question your teacher might ask on a test. Write the question on one side of the flash card. Then write the answer on the other side. Quiz yourself until you know the answers.

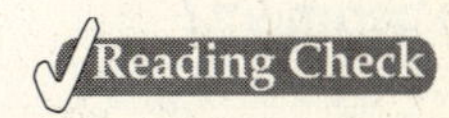

1. What is a hypothesis?

Reading Check

2. What are the two groups in a controlled experiment?

____

____

What is a controlled experiment?

Controlled experiments involve two groups—the control group and the experimental or test group. The **control** is the part of an experiment that represents the standard conditions. In other words, the control receives no experimental treatment. The experimental group is the test group that receives experimental treatment.

For instance, imagine an experiment to learn how fertilizer affects plant growth. Fertilizer would be used in the experimental group but not in the control group. All other conditions—soil, light, and water—would be the same for both groups.

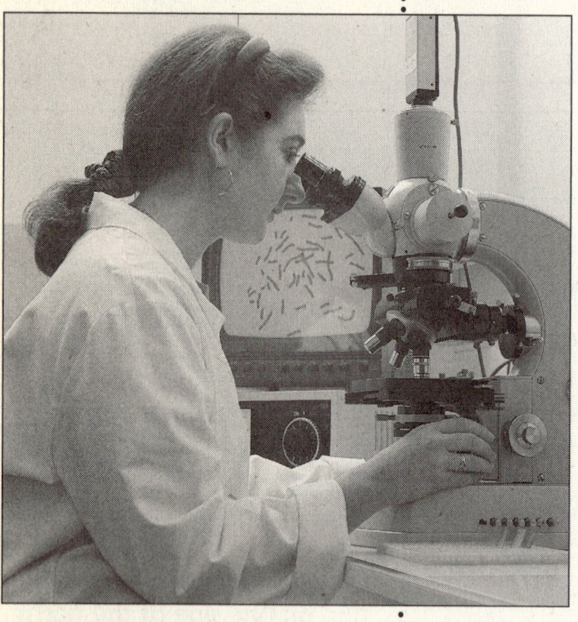

In this experiment, using fertilizer is the independent variable. The **independent variable** is the one condition in an experiment that is tested. How much the plants grow is the dependent variable. The **dependent variable** is the condition that changes because of a change in the independent variable.

Safety is another important factor that scientists think about when carrying out investigations and experiments. It is important to know about dangers that may exist from doing an experiment before you begin it. Anyone doing an experiment has a responsibility to follow safety procedures. They must keep themselves and others out of danger.

Reading Check

3. Who is responsible for making sure that safety procedures are followed when conducting an experiment?

____

____

____

How are theories formed?

The information gathered from experiments is called **data.** A scientist carefully reviews or analyzes experimental results to decide if the data supports the hypothesis. Scientists repeat their experiments in order to gather more data. Data are considered reliable only when repeating the experiment several times produces similar results.

Scientists also compare the results of their experiments with the results of other studies. They research published information in scientific journals and computer databases. It is important to have details of an experiment presented in scientific journals and databases so scientists can compare their results with those of similar studies. It lets other scientists test the results by repeating the experiment. If many scientists get the same results, it helps support the hypothesis. A hypothesis that is supported by many different investigations and observations becomes a **theory.**

After You Read

Mini Glossary

control: in an experiment, the standard against which results are compared

data: information gathered from an experiment

dependent variable: the condition in an experiment that results from the changes made to the independent variable

experiment: an investigation that tests a hypothesis by collecting information under controlled conditions

hypothesis (hi PAHTH us sus): an explanation for a question or problem that can be tested

independent variable: in an experiment, the condition that is tested because it affects the outcome of the experiment

scientific methods: common steps that scientists use to do research and answer questions

theory: an explanation of a natural phenomenon or event that is supported by a large body of scientific evidence obtained from many different investigations and observations

1. Review the terms and their definitions in the Mini Glossary above. Write a sentence using at least two of the terms.

____

____

2. Use the pyramid diagram to help you review what you have read. Arrange the steps used in scientific research in the order that they usually take place. Place the letter next to each step in the right order in the pyramid.

   a. Conduct experiments
   b. Form a hypothesis
   c. Observe and identify a problem to solve
   d. Study results data to see if hypothesis is supported

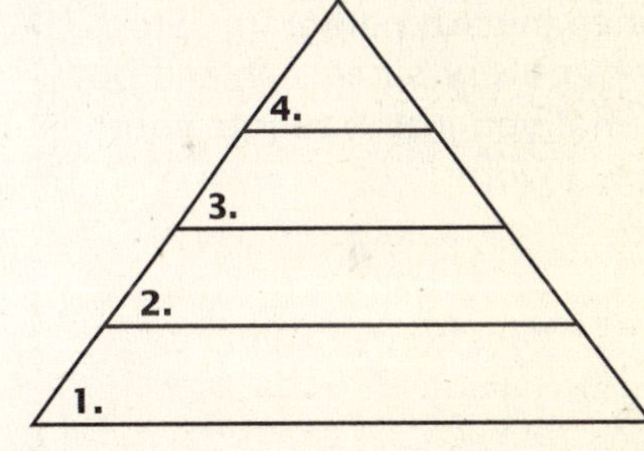

3. Choose one of the question headings in the Read to Learn section. Write the question in the space below. Then, write your answer to that question on the lines that follow.

Question:

Answer:

____

____

Visit the Glencoe Science Web site at **science.glencoe.com** to find your book and learn more about the methods of biology.

Name Date Class

Section 1.3 The Nature of Biology

Before You Read

In this section, you will learn how biologists use scientific research to study the natural world and make important discoveries. Think of any important discoveries in the field of biology that you already know about. Give as many examples as you can on the lines below.

____

____

____

Read to Learn

STUDY COACH

In Your Own Words Highlight the main idea in each paragraph. Stop after every paragraph and put what you just read into your own words.

1. What are the two main types of scientific research?

____

____

Kinds of Information

Scientific information can usually be broken down into two main types—quantitative or qualitative. In quantitative research, results are compared by using numbers. Imagine an experiment to see when different materials begin to melt. The temperature at which wax, iron, and glass each begins to melt is different. Temperatures often are measured in degrees, which are numbers on a scale. These temperatures are a type of quantitative data in quantitative research. ✓

Qualitative research is based on observation. It is also called descriptive research because it describes scientists' observations when they do their research. If a scientist wanted to figure out how a beaver builds a dam, numbers would not be very helpful. The scientist would observe the beaver and see how the dam is built. Then the scientist would describe, in detail, all the steps the beaver takes to build the dam.

Science and Society

Scientific research often provides society with important information. What we learn from scientific research cannot be defined as good or bad. Ethics must play a role in deciding how the information will be used. **Ethics** are the moral principles and values held by humans. Ethics are how we decide what is right or wrong, good or bad. Suppose scientists develop a new vaccine to cure a disease, but they can only produce 1000 doses each year. Ethics help society decide who should receive those doses. Society as a whole must take responsibility for making sure that scientific discoveries are used in an ethical way.

Section 1.3 The Nature of Biology, *continued*

Some scientific study is done only to learn new things. This type of science is called pure science. Pure science is not done so that the results can be used for a specific need. The research is filed away for later use.

Science that solves a problem is technology. **Technology** (tek NAHL uh jee) means using scientific research to meet society's needs or solve its problems. Technology has helped reduce the amount of manual labor needed to make and raise crops. It has also helped cut down on environmental pollution.

After You Read

Mini Glossary

ethics: the moral principles and values held by society

technology (tek NAHL uh jee): the application of scientific research to society's needs and problems.

1. Review the terms and their definitions in the Mini Glossary above. Then, write the definitions of both terms in your own words on the lines below.

____

____

2. Use the partially completed outline below to help you review what you have read. Fill in the blanks to provide additional information.

I. Scientific Research

A. Quantitative Research

1. What is it? ______________________________

2. Give an example: ______________________________

B. Qualitative Research

1. What is it? ______________________________

2. Give an example: ______________________________

3. Give an example of how technology has helped human life and the world around us.

____

____

Visit the Glencoe Science Web site at **science.glencoe.com** to find your book and learn more about the nature of biology.

Name Date Class

Section 2.1

Organisms and Their Environment

Before You Read

This section discusses organisms and their environment. All of us come into contact with a variety of organisms every day. On the lines below, list all of the organisms you can think of that you come into contact with during a typical week.

Read to Learn

STUDY COACH

Create a Quiz After you have read this section, create a quiz based on what you have learned. After you have written the questions, be sure to answer them.

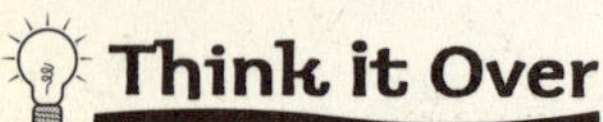

1. **Describe** What animals share your world?

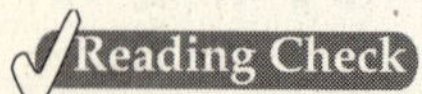

2. What do ecologists study?

Sharing the World

Every day you share your environment with many organisms. These can be as small as houseflies or mosquitoes. They can be dust mites that you cannot even see. Larger animals include dogs, raccoons, and deer. You need to know about your environment. The reason is simple: Your environment affects you and every other organism in it.

The study of plants and animals, where they live, what they eat, and what eats them, is called natural history. Natural history tells us about the health of the world we live in.

What is ecology?

The branch of biology that developed from natural history is known as ecology. **Ecology** is the study of relationships between organisms and their environment.

Ecologists use both qualitative and quantitative research. They gather qualitative information by observing organisms. They gather quantitative data by making measurements and doing experiments. Ecologists study organisms both in the lab and where the organisms naturally live. ✓

The Biosphere

The **biosphere** (BI uh sfihr) is the portion of Earth that supports living things. It includes the air, land, and water where organisms can be found. ✓

The biosphere supports a wide variety of organisms in a wide range of conditions. Climates, soils, plants, and animals can be very different in different parts of the world. All over the world, though, living things are affected by both the physical, nonliving environment and by other living things.

Reading Check

3. What is the biosphere?

___

___

___

How is the environment organized?

The nonliving parts of the environment are called **abiotic** (ay bi AH tihk) **factors.** Some examples are temperature, moisture, light, and soil. Ecology includes the study of abiotic factors because they are part of an organism's life. To truly know about moles, for example, ecologists must learn the type of soil moles dig their tunnels in. To get a complete picture of the lives of trout, it is important to know the type of river bottom where they lay their eggs. ✓

Living things also are affected by biotic factors. **Biotic** (by AH tihk) **factors** are all the living organisms in an environment. Even goldfish in a bowl are affected by fishes, plants, or other organisms that share their bowl. All organisms depend on others directly or indirectly for food, shelter, reproduction, or protection.

Reading Check

4. Name four examples of abiotic factors.

___

___

___

___

Levels of Organization

Ecologists study individual organisms. They study relationships among organisms of the same species and connections among organisms of different species. They also study the effects of abiotic factors on species that live together. To make it easier to examine all of these biotic and abiotic interactions, ecologists have organized the living world into levels. The levels are the organism by itself, populations, communities, and ecosystems.

Section 2.1 Organisms and Their Environment, *continued*

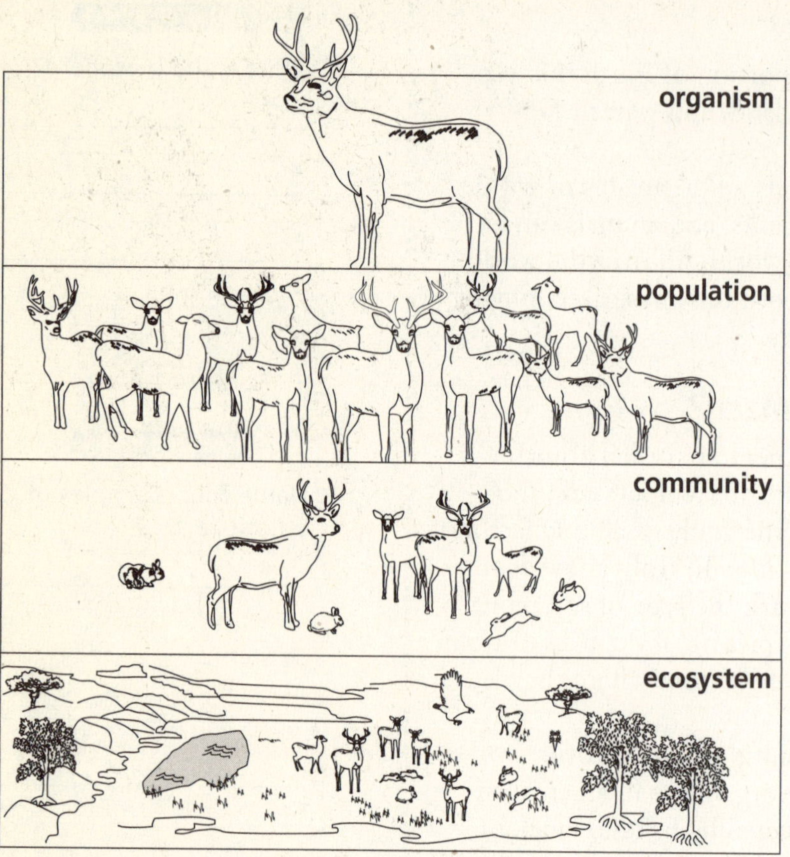

What is a population?

A **population** is a group of organisms that belongs to the same species. Population members breed with each other and live in the same area at the same time. How organisms in a population share the things that they need in their environment is important. It may determine how far apart the organisms live and how large the population becomes. Members of the same population may compete with each other for food, water, mates, or other resources.

Some species have adaptations that reduce competition within a population. For instance, frogs have a life cycle in which the young tadpoles and adult frogs look very different and have different diets. Tadpoles eat algae and frogs eat insects; therefore, they are not competing with each other for food.

How do communities interact?

No species lives entirely alone. Every population shares its environment with other populations. This creates what is called a biological community. A **biological community** is made up of different populations in a certain area at a certain time.

In a biological community, changes in one population may cause changes in other populations. For instance, if the number of mouse-eating hawks in a community increases slightly, the number of mice in that community will decrease slightly. Other changes can be more extreme. For example, one population may grow so large that it threatens the food supply of another population.

In a healthy forest community, there are many populations that depend on each other. These might include birds eating insects, squirrels eating nuts from trees, mushrooms growing from decaying leaves or bark, and raccoons fishing in a stream. While these populations are connected to each other, they are all affected by abiotic factors. These relationships between different populations and their surroundings create an ecosystem. An **ecosystem** is made of all of the different populations in a biological community and the community's abiotic factors. ✓

There are two major kinds of ecosystems—terrestrial and aquatic. Terrestrial ecosystems are those located on land. Examples include forests, fields, and a rotting log. Aquatic ecosystems are found in both freshwater and salt water. Freshwater ecosystems include ponds, lakes, and streams. Oceans are a type of saltwater, or marine, ecosystem.

Organisms in Ecosystems

Different types of organisms make their homes in different places. Some species of birds live in only one type of forest. In these areas, they find food, avoid enemies, and reproduce. Prairie dogs make their homes underground in grasslands. The place where an organism lives out its life is known as a **habitat.** ✓

What place does a species have in its habitat?

Though several species may share a habitat, the food, shelter, and other needed items in that habitat are often used in different ways by each species. For example, if you turn over a log, you may find a community of millipedes, centipedes, insects, slugs, and earthworms. At first it might seem that the members of this community are competing for the same food because they all live in the same habitat, but each population feeds in different ways, on different things, and at different times. Each species has its own niche. A **niche** (neesh) is all strategies and adaptations a species uses in its environment. It is how the species meets its specific needs for food and shelter. It is how and where the species survives and reproduces. A species' niche includes all its interactions with the biotic and abiotic parts of its habitat.

Two species cannot exist for long in the same community if they both have the same niche. There is too much competition. In the end, one species will gain control over the resources in the community. The other species will either die out in that area, move somewhere else, or change in some way to fill another niche.

Reading Check

5. What is an ecosystem made of?

____

____

____

Reading Check

6. What is a habitat?

____

____

____

Think it Over

7. Infer Which of the following does a polar bear use to survive in its habitat? (Circle your choice.)

a. burrowing instinct to dig deep in the soil

b. thick coat to protect it from the cold

Symbiosis

People once thought that animals in the same environment fought each other for survival. In reality, most species survive because of the relationships they have with other species. A relationship in which there is a close and permanent association between organisms of different species is called **symbiosis** (sihm bee OH sus). Symbiosis means living together. There are three major kinds of symbiosis—mutualism, commensalism, and parasitism.

Reading Check

8. What is the name of the relationship in which both species benefit?

____

____

____

Mutualism **Mutualism** (MYEW chuh wuh lih zum) is a relationship between two species that live together in which both species benefit. The relationship between ants and an acacia (uh KAY shuh) tree is a good example of mutualism. The ants protect the tree by attacking any animal that tries to feed on the tree. The tree provides nectar as a food for the ants. The tree also provides a home for the ants. In an experiment, ecologists removed the ants from some acacia trees. Results showed that the trees with ants grew faster and lived longer than the trees with no ants. ✓

Commensalism **Commensalism** (kuh MEN suh lih zum) is a relationship in which only one species benefits and the other species is not harmed or helped. For example, mosses sometimes grow on the branches of trees. This does not help or hurt the trees, but the mosses get a good habitat.

Parasitism **Parasitism** (PER uh suh tih zum) is a relationship in which a member of one species benefits at the expense of another species. For instance, when a tick lives on a dog, it is good for the tick but bad for the dog. The tick gets food and a home, but the dog could get sick. The tick is a parasite. A parasite is the organism that benefits from the relationship. The dog is a host. The host is the organism that is harmed by the relationship.

What relationship do predators and prey have?

Another type of relationship is that between a predator and its prey. Predators are organisms that seek out and eat other organisms. The organisms that are eaten are called prey. Predators are found in all ecosystems. Some eat animals and plants. Some eat only animals. Lions and birds that eat insects are predators.

After You Read

Mini Glossary

abiotic (ay bi AH tihk) factors: nonliving parts of an organism's environment; air currents, temperature, moisture, light, and soil are examples

biological community: a community made up of interacting populations in a certain area at a certain time

biosphere (BI uh sfihr): portion of Earth that supports life; extends from high in the atmosphere to the bottom of the oceans

biotic (by AH tihk) factors: all the living organisms that inhabit an environment

commensalism (kuh MEN suh lih zum): symbiotic relationship in which one species benefits and the other species is neither harmed nor benefits

ecology: scientific study of interactions between organisms and their environments

ecosystem: interactions among populations in a community and the community's physical surroundings, or abiotic factors

habitat: place where an organism lives out its life

mutualism (MYEW chuh wuh lih zum): symbiotic relationship in which both species benefit

niche (neesh): all strategies and adaptations a species uses in its environment; includes all biotic and abiotic interactions as an organism meets its needs for survival and reproduction

parasitism (PER uh suh tih zum): symbiotic relationship in which one organism benefits at the expense of another

population: group of organisms of the same species that interbreeds and lives in the same place at the same time

symbiosis (sihm bee OH sus): permanent, close association between two or more organisms of different species

1. Review the terms and their definitions in the Mini Glossary above. Circle the three terms that identify specific types of relationships between organisms that live in the same ecosystem. On the lines below, give an example of each type of relationship.

___

___

___

2. Use the pyramid diagram below to help you review what you have read. List the four levels that ecologists have organized the living world into. Start with the least complex level at the bottom and work your way up.

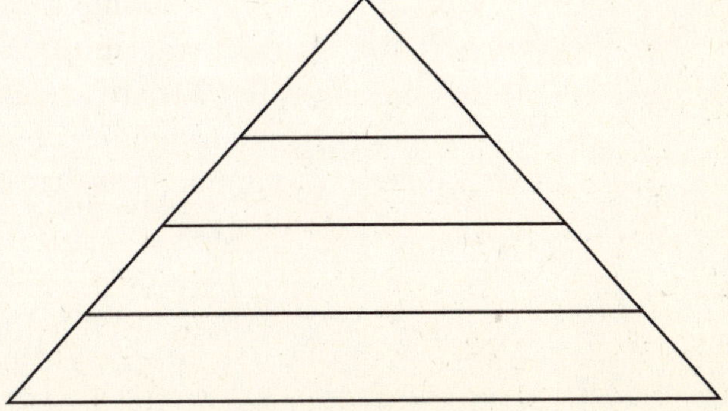

Section 2.1 Organisms and Their Environment, *continued*

3. In Column 1 are some new concepts you learned about in this section. Column 2 gives one example of each concept. Use the line next to each concept to put the letter of the example that matches it.

New Concept Column 1	Example Column 2
_____ 1. abiotic factor	a. an owl eating a mouse
_____ 2. habitat	b. a rain forest
_____ 3. predator-prey relationship	c. rain
_____ 4. biological community	d. a tick on a cat
_____ 5. parasitism	e. millipedes, centipedes, insects, slugs, and earthworms under a log

Visit the Glencoe Science Web site at **science.glencoe.com** to find your biology book and learn more about organisms and their environment.

Section 2.2 Nutrition and Energy Flow

Before You Read

This section discusses how organisms interact with their environment to get the food and energy they need. On the lines below, explain how a pet's food might be different if it had to live in the wild and get its own food.

Read to Learn

How Organisms Obtain Energy

One of the most important things about a species' niche is how the species gets its energy. Ecologists study the flow of energy through communities to discover nutritional relationships between organisms.

Autotrophs The ultimate source of energy for all life is the sun. Plants use the sun's energy to make food. This process is called photosynthesis (foh toh SIN thuh suhs). This makes plants autotrophs. **Autotrophs** (AW tuh trohfs), or producers, are organisms that use light energy or energy stored in chemical compounds to make energy-rich compounds. Grass, trees, and other plants are the most familiar autotrophs, but some one-celled organisms, such as green algae, also make their own food.

Heterotrophs Some organisms cannot make their own food. They must eat other organisms to get their food and energy. These organisms are called consumers, or **heterotrophs** (HE tuh ruh trohfs). Some heterotrophs, such as rabbits, feed only on autotrophs. Other heterotrophs, such as lions, feed only on other heterotrophs. Still other heterotrophs, such as bears and humans, feed on both autotrophs and heterotrophs.

Decomposers There are other organisms called **decomposers.** They break down the complex compounds of dead and decaying plants and animals. They change these compounds into simpler forms that they can use for fuel. Some protozoans, many bacteria, and most fungi are decomposers. ✓

STUDY COACH

Make an Outline Make an outline of the information you learn about in this section. Use the headings in the reading as a starting point. Include the boldface terms.

✓ Reading Check

1. What do decomposers do?

Flow of Matter and Energy in Ecosystems

When you eat food, such as an apple, you consume matter. Matter, in the form of carbon, nitrogen, and other elements, flows through the levels of an ecosystem from producers to consumers. Scientists call this flow of matter *cycling*. The apple is more than matter, though. It also contains some energy from sunlight. This energy was trapped in the apple as a result of photosynthesis. As you cycle the matter in the apple by eating it, some trapped energy is transferred from one level of the ecosystem to the next. At each level, a certain amount of energy is also transferred to the environment as heat.

What are food chains?

Ecologists study feeding relationships and symbiotic relationships to learn how matter and energy flow in ecosystems. These scientists sometimes use a simple model called a food chain. **Food chains** show how matter and energy move through an ecosystem. In a food chain, nutrients and energy move from autotrophs to heterotrophs to, in the end, decomposers. A food chain is drawn using arrows. The arrows show the direction in which energy is transferred. An example of a simple food chain in a forest ecosystem is shown below.

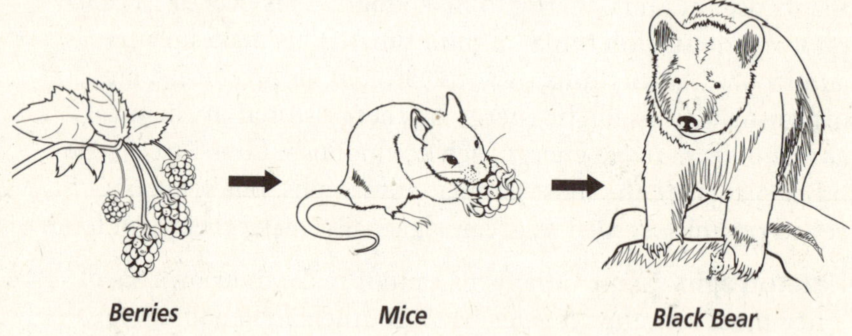

Most food chains are made up of two, three, or four transfers, or steps. Each organism in a food chain represents a feeding step, or **trophic** (TROH fihk) **level,** in the transfer of energy and matter. The amount of energy in the last transfer is only a small part of what was available at the first transfer. At each transfer, some of the energy is given off as heat.

Think it Over

2. **Interpret** When energy is transferred between trophic levels, what is always given off to the environment? (Circle your choice.)
   a. photosynthesis
   b. heat
   c. water

What is a food web?

A food chain shows only one possible path for the transfer of matter and energy through an ecosystem. Many other paths may exist because many different species can be on each trophic level. For instance, in the food chain example on the previous page, there are many animals in the forest other than mice that eat berries. Also, many different kinds of organisms eat more than one type of food. This means that a single species may feed at several different trophic levels. The black bear, for instance, does not eat only mice. It also eats berries. There also are other animals in the forest that eat berries and mice. For these reasons, ecologists also use food webs. **Food webs** are models that show all possible feeding relationships at each trophic level in a community. A food web is a more realistic model than a food chain because most organisms depend on more than one type of organism for food.

How does energy flow through an ecosystem?

Food chains and food webs deal with both matter and energy. When ecologists want to focus only on energy, they use another type of model—an ecological pyramid. An ecological pyramid shows how energy flows through an ecosystem. There are different types of ecological pyramids. Each pyramid has the autotrophs, or first trophic layer, at the bottom. Higher trophic layers are then layered on top of one another. ✓

The pyramid of energy shows that the amount of available energy becomes less from one trophic level to the next. The total energy transfer from one trophic level to the next is only about ten percent because organisms do not use all the food energy in the trophic level below them. An organism uses energy to do all the things necessary for life. Organisms use energy to move, to interact with their environment, and to digest their food. They also use energy to build body tissue. Some of this energy is given off as heat. The law of conservation of energy states that energy is neither lost nor gained. Even though some of the energy transferred at each trophic level enters the environment as heat, it is still energy. It is just in a different form.

A pyramid of numbers shows the number of organisms eaten by the level above it. In most cases, the number of organisms decreases at each higher trophic level.

✓ Reading Check

3. What type of organisms appear at the bottom of the ecological pyramid?

____

____

____

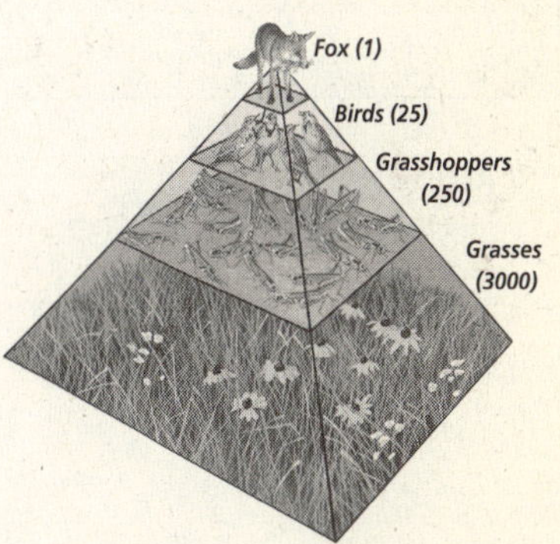

Reading Check

4. What is biomass?

Biomass is the total weight of living matter at each trophic level. A pyramid of biomass shows the total dry weight of living material at each trophic level.

Cycles in Nature

Matter, in the form of food, moves through every organism. In this way, matter is found at every trophic level. Matter is never made or destroyed. It just changes form as it cycles through the different trophic levels. There is the same amount of matter today as there was when life on Earth began.

What is the water cycle?

Water also cycles through different stages. It is always moving between the atmosphere and Earth. For instance, when you leave a glass of water out for a few days, some of it seems to disappear. It has evaporated or changed into water vapor in the air. Similarly, water from lakes and oceans evaporates. At some point, this water vapor condenses, or comes together, and makes clouds. After even more condensation, drops of water form. This water then falls back to Earth as rain, ice, or snow.

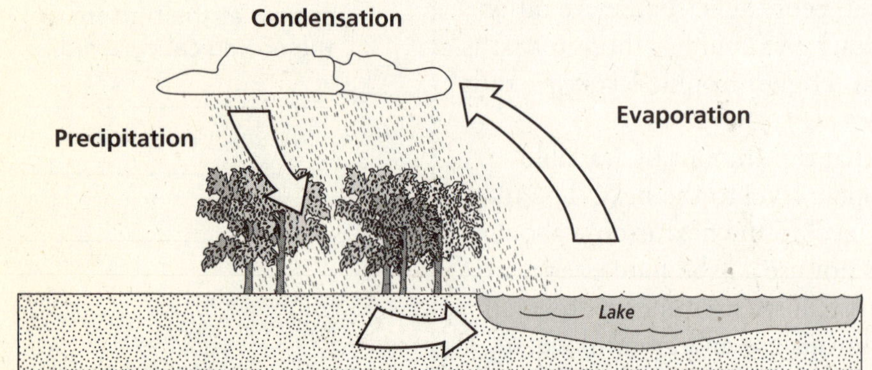

In the water cycle, water is constantly moving between the atmosphere and Earth.

Plants and animals need water to live. When the water falls to Earth, plants and animals use it. Plants pull water from the ground and lose water from their leaves. Losing water this way puts water vapor back into the air. This continues the water cycle. Animals also take in water. They breathe out water vapor in every breath. When animals urinate, they return water to the environment. This water then continues in the cycle.

What is the carbon cycle?

Carbon has its own cycle. All life on Earth is based on carbon, and all living organisms need carbon.

The carbon cycle starts with autotrophs. In photosynthesis, autotrophs use the sun's energy to change carbon dioxide gas into energy-rich forms of carbon.

Autotrophs use this carbon for growth and energy. Heterotrophs then feed on autotrophs or feed on other animals that have already fed on autotrophs. The heterotrophs then use the carbon for growth and energy. As autotrophs and heterotrophs use this carbon, they release carbon dioxide into the air. The carbon cycle continues very slowly. How rapidly it cycles depends upon whether the carbon is found in soil, leaves, roots, in oil or coal, in animal fossils, or in calcium carbonate reserves.

What is the nitrogen cycle?

Nitrogen is another element important to living things. Although 78 percent of air is nitrogen, plants cannot use this form well. There are bacteria, though, that change the nitrogen from air to a form plants can better use. This form is found in the soil. Plants use this nitrogen to make proteins. Animals eat the plants and change the plant proteins into animal proteins. These proteins are used in building muscle and blood cells. Urine is an animal waste that lets animals get rid of nitrogen they do not need. This urine returns nitrogen to the soil. When organisms die and decay, nitrogen returns to the soil. Plants then reuse this nitrogen. Soil bacteria also act on these dead organisms and put nitrogen back into the air. In this way, nitrogen is always cycling through the system. ✓

Reading Check

5. What type of organisms change nitrogen in the air into a form plants can better use?

____

____

____

What is the phosphorous cycle?

Phosphorus also cycles through ecosystems. It is another element that all organisms need. It cycles in two ways. In the short-term cycle, plants get phosphorus from the soil. Animals get phosphorus from eating plants. When these animals die, their decaying bodies release phosphorus back into the soil to be used again.

In the long-term cycle, materials containing phosphorus are washed into rivers and oceans. As millions of years pass, the phosphorus becomes locked in rocks. Millions of years later, as the environment changes, some of the rock is no longer covered. As this rock wears away, the phosphorus is released back into the environment.

After You Read

Mini Glossary

autotrophs (AW tuh trohfs): organisms that use energy from the sun or energy stored in chemical compounds to make their own nutrients

biomass: the total mass or weight of all living matter in a given area

decomposers: organisms, such as fungi and bacteria, which break down and absorb nutrients from dead organisms

food chain: a simple model that shows how matter and energy move through an ecosystem

food web: a model that shows all the possible feeding relationships at each trophic level in a community

heterotrophs (HE tuh ruh trohfs): organisms that cannot make their own food and must feed on other organisms for energy and nutrients

trophic (TROH fihk) level: an organism that represents a feeding step in the movement of energy and materials through an ecosystem

1. Review the terms and their definitions in the Mini Glossary above. Define **autotrophs** and **heterotrophs** in your own words. Give an example of an autotroph and a heterotroph.

____

____

____

2. Use the table below to help you review what you have read. For each of the organisms shown in the food chain, choose three facts from the list below the table that are true. Then write the facts in the table under the correct organism.

Grass →	Rabbit →	Wolf

performs photosynthesis
heterotroph
eats other heterotrophs
on highest trophic level
provides energy for other heterotrophs
heterotroph
on lowest trophic level
autotroph
eats autotrophs

Section
2.2 Nutrition and Energy Flow, *continued*

3. Next to each of the statements below, place a T if the statement is true or an F if the statement is false.

_____ a. Although 78 percent of air is nitrogen, plants cannot use this form well.

_____ b. Heterotrophs can get the energy they need directly from the sun.

_____ c. During part of its cycle, phosphorus falls back to Earth as rain, ice, or snow.

_____ d. All life on Earth is based on carbon.

Visit the Glencoe Science Web site at **science.glencoe.com** to find your biology book and learn more about nutrition and energy flow.

Section 3.1 Communities

Before You Read

This section discusses organisms and how they are affected by where they live. Think about the kinds of plants and animals you see around you. On the lines below, list as many types of plants and animals you can think of that live in your community. Think of an organism that would have trouble surviving where you live. Write its name below, too.

____

____

____

Read to Learn

Mark the Text **Identify the Main Point** Highlight the main point in each paragraph. Highlight in a different color an example that helps explain the main point.

1. Anything that limits an organism's ability to live in an environment is called a

____

____

____

Life in a Community

A community is made up of populations of organisms that live in a common area. For example, weeds, beetles, worms, and bacteria might live in someone's lawn. In that same lawn, there would also be moisture and soil. This lawn community is made up of populations of organisms and abiotic factors. They all contribute to the life of the lawn. The factors these populations share create their environment.

Different combinations of living and nonliving factors interact in different places around the world. They form communities where conditions may be good for one form of life but not for another.

What is a limiting factor?

Anything that limits an organism's ability to live in a particular environment is known as a **limiting factor.** For example, in most mountain areas, there is a timberline. Trees cannot grow above the timberline. It is too high, too cold, too windy, and the soil is too thin. ✓

Something that limits one population in a community can have an indirect effect on another life form. A lack of water could cause grass to die in an area. If an animal group living in that area depends on grass for food, its population could decrease as a result.

What is the range of tolerance?

The ability of living things to survive the changes in their environment is called **tolerance.** An organism reaches its limits of tolerance when it gets too much or too little of an environmental factor. For example, corn plants need warm, sunny weather and a regular supply of water. If these conditions do not exist over a long period of time, the plants may survive, but they will not produce a crop.

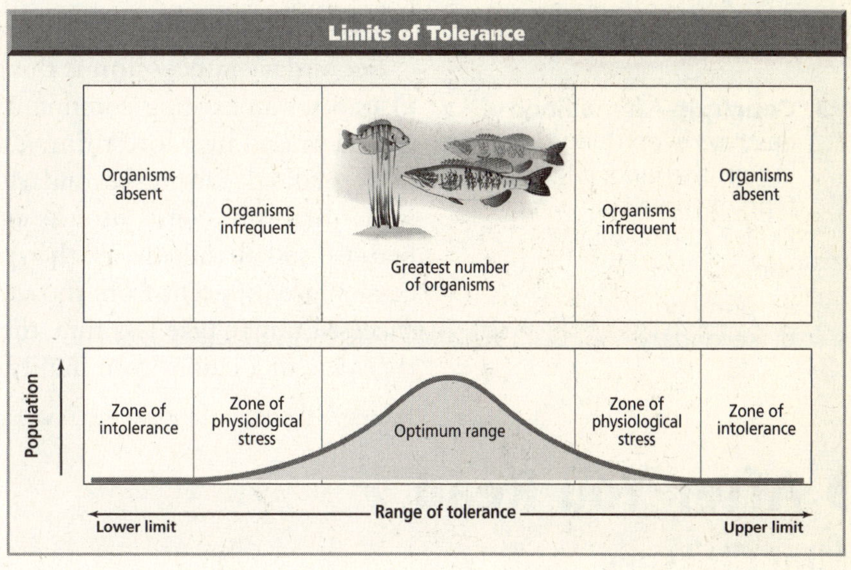

The figure above shows the range of tolerance for organisms. Notice that in the zones of intolerance, the corn plants would not grow. In the zones of stress, the crop would be poor. The corn plants grow the best in the optimum range.

Succession: Changes over Time

Succession (suk SE shun) is the process of gradual, natural change and species replacement that takes place in the communities of an ecosystem over time. There are two types of succession—primary and secondary. ✓

Reading Check

2. What is succession?

____

____

____

What is primary succession?

Primary succession takes place on land where there are no living organisms. For example, when lava flows from a volcano, it destroys everything around it. When it cools, land forms, but there are no living organisms in the new land. The first species to live in such an area is called a pioneer species. Decaying lichens, along with bits of sediment in cracks and crevices of rock, make up the first stage of soil development. Gradually, other life forms take hold. After some time, primary succession slows down and the community becomes stable. Pioneer species eventually die. Once little or no change occurs, the community is called a **climax community.** A climax community can last for hundreds of years.

Section 3.1 **Communities,** *continued*

Think it Over

3. Conclude After a flood destroys everything growing on the land, which type of succession is most likely? (Circle your choice.)

a. primary
b. secondary

What is secondary succession?

Secondary succession is the pattern of changes that takes place after an existing community is destroyed. The destruction can be caused by a forest fire or when a field is plowed over and not replanted. During secondary succession, as in primary succession, organisms come into the area and change gradually. But, because soil already exists, the species involved in secondary succession are different from those in primary succession. Secondary succession may take less time than primary succession to reach the stage of a climax community.

After You Read

Mini Glossary

climax community: a stable community that undergoes little or no change

limiting factor: any biotic or abiotic factor that restricts the existence, numbers, reproduction, or distribution of organisms

primary succession: colonization of barren land by pioneer organisms

secondary succession: sequence of changes that take place after a community is disrupted by natural disasters or human actions

succession (suk SE shun): the orderly, natural changes that take place in the communities of an ecosystem

tolerance: the ability of an organism to withstand changes in biotic and abiotic environmental factors

1. Review the terms and their definitions in the Mini Glossary above. Highlight the three terms that deal with changes in an ecosystem. Then circle the term that refers to a situation in which there are few or no changes.

2. Use the flowchart below to help you review what takes place after a volcanic eruption. Fill in the blank box.

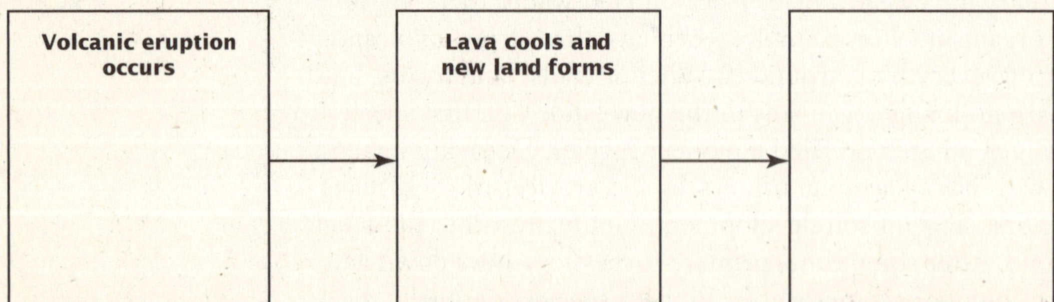
Volcanic eruption occurs
Lava cools and new land forms

Section

3.1 Communities, *continued*

3. Column 1 lists new concepts that you learned about in this section. Column 2 gives an example of each concept. Draw a line from each concept to its correct example.

Column 1	Column 2
1. limiting factors	a. the first organisms to grow on a new patch of cooled, hardened lava
2. pioneer species	b. weeds and wildflowers beginning to grow in a field after a corn crop is harvested
3. tolerance	c. ability of mosquitoes to survive in very different conditions all over the world
4. climax community	d. cold temperatures and high winds that prevent tree growth in mountain areas
5. secondary succession	e. an old forest that has not had any fire damage in over 200 years

Visit the Glencoe Science Web site at **science.glencoe.com** to find your biology book and learn more about communities.

Section 3.2 Biomes

Before You Read

Deserts are very different from rain forests. Each is home to species that can find everything they need to survive in their unique environment. This section discusses areas with their own specific sets of characteristics. Highlight each of the key terms that introduces one of these types of areas. Then use a different color to highlight important facts about each.

Read to Learn

Make Flash Cards Make flash cards to help you learn section material. Think of a quiz question for each paragraph. Write the question on one side of the flash card and the answer on the other side. Keep quizzing yourself until you know all of the answers.

What is a biome?

A **biome** is a large group of ecosystems that shares the same type of climax community. All the ecosystems within the biome have similar climates and organisms. Biomes located on land are called terrestrial biomes. Biomes located in bodies of water are known as aquatic biomes.

Aquatic Biomes

Approximately 75 percent of Earth's surface is covered with water. Most of that water is salty. Salt water is found in oceans, seas, and some inland lakes. Freshwater is found in rivers, streams, ponds, and most lakes.

The oceans contain a large amount of biomass, or living material. Much of this biomass is made up of tiny organisms that humans cannot see. Large marine animals depend on these organisms for food.

What is an estuary?

An **estuary** (ES chuh wer ee) is a coastal body of water partly surrounded by land that forms where rivers meet the ocean. Freshwater and salt water come together in an estuary. The amount of salt in an estuary depends on how much freshwater the river brings into the estuary. ✓

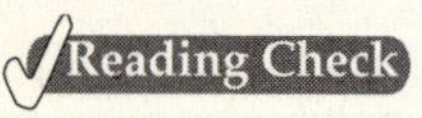

1. What is an estuary?

____

____

____

Grasses that can grow in salt water can become very thick in an estuary. Their stems and roots trap food material for small organisms like snails, crabs, and shrimp. These organisms feed on the trapped, decaying materials. The nutrients in the food pass through the food chain when these smaller organisms are eaten by larger predators, including birds.

What are the effects of tides?

Each day, the gravitational pull of the sun and moon causes ocean tides to rise and fall. The area of shoreline that lies between the high and low tide lines is called the **intertidal zone.**

Many animals that live in the intertidal zone, such as snails and sea stars, have suctionlike adaptations. These allow the animals to hold onto rocks when wave action is strong. Other animals make their own strong glue that helps them stay in place. Clams, worms, snails, and crabs survive by burying themselves into the sand.

Reading Check

2. Why are plankton important?

____

____

____

Where do most marine organisms live?

Most of the organisms that live in the marine (saltwater) biome are **plankton.** Plankton are tiny organisms that float in the waters of the photic zone. The **photic zone** is the area of the ocean that is shallow enough for sunlight to penetrate. Plankton include autotrophs (organisms that make their own nutrients), diatoms, eggs, and very young marine animals. Plankton are important because they form the base of the entire aquatic food chain. This means that every aquatic animal either eats plankton or eats an animal that eats plankton.

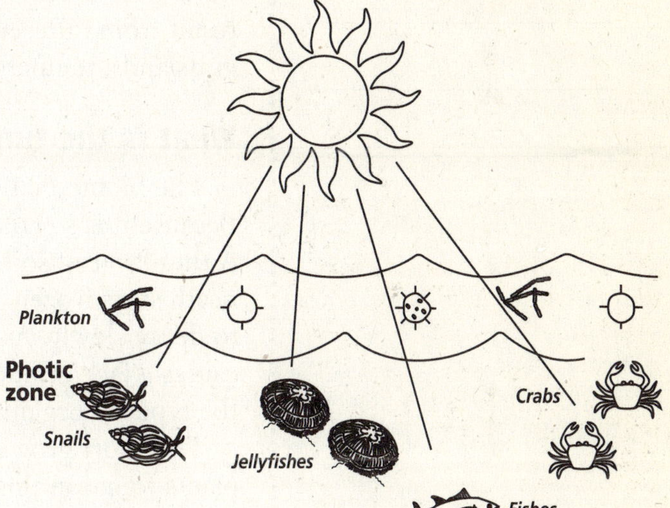

What are freshwater biomes?

Bodies of freshwater are another kind of biome. Lakes and ponds serve as home to many organisms. Plants grow around the shorelines and into the water. The shallow waters where these plants grow are home to tadpoles, aquatic insects, worms, certain fishes, and many other living things. All the life forms are part of the local food chain.

In deeper waters, it is colder and there are fewer species. Dead organisms drift to the bottom. There, bacteria break down the organisms and recycle the nutrients. Organisms decay more slowly at the bottom of a deep lake than in shallow water.

Terrestrial Biomes

Terrestrial biomes vary greatly. At the north pole, the weather is very cold and there are no plants. As you move south, the weather gets warmer and there is a change in the size, number, and kinds of plants that cover the ground. As you continue south, the temperatures rise and you encounter forests. Still farther south are grasslands and deserts, with high summer temperatures and little rainfall. Near the equator, you find lush growth and much rainfall.

Reading Check

3. What five abiotic factors does climate include?

____

____

____

____

____

How does climate affect biomes?

Climate is a group of abiotic factors that influences the kind of climax communities that develop in an area. Climate includes wind, cloud cover, temperature, humidity, and the amount of rain and snow an area receives. The most common terrestrial biomes that result from differences in climate are tundras, taigas, deserts, grasslands, temperate forests, and tropical rain forests. ✓

What is the tundra biome?

The biome that circles the north pole is called the **tundra.** Because it is so cold, only a few grasses and small plants grow. A thin layer of soil may thaw in the summer, but the soil underneath stays frozen. The cold causes any organisms that die there to decay slowly. As a result, nutrients are recycled slowly. This causes a lack of nutrients in the soil. The lack of nutrients limits the types of organisms the tundra can support.

The short growing season in the tundra limits the types of plants found in this biome, but the plants that do grow there live a long time and are resistant to drought and cold. These include grasses, dwarf shrubs, and cushion plants. Animals that live on the tundra include arctic foxes, weasels, lemmings, hares, snowy owls, hawks, musk oxen, caribou, and reindeer.

What is the taiga biome?

South of the north pole is the **taiga** (TI guh). The climate here is warmer and wetter than the tundra and forests of trees, such as fir and spruce, are found. Animals such as elk, deer, moose, squirrels, voles, weasels, and a variety of birds call the taiga home. The taiga stretches across much of Canada, northern Europe, and Asia.

What is the desert biome?

The driest biome is the desert biome. A **desert** is an arid region with little to no plant life. The plants that do grow there are well adapted to these dry areas. In fact, many desert plants need little rainfall. Their leaves, stems, and coatings conserve water. Many also have spines, thorns, or poisons to protect against plant eaters.

Many small desert mammals are plant eaters that hide during the heat of the day and search for food at night. The kangaroo rat is a plant eater that does not have to drink water. Instead, these rodents get water from the plants they eat. Coyotes, hawks, and owls are meat eaters. They feed on snakes, lizards, and small desert mammals. Scorpions and some snakes use their venom to capture their prey.

What is the grassland biome?

Grasslands, or prairies, are large communities covered with rich soil, grasses, and other grasslike plants. Grasslands most often exist in climates that have a dry season. In this type of biome, there is not enough water to support forests. Many of the grasses die off each winter, but the roots of the grasses survive and enlarge every year. This forms a continuous underground mat called sod.

Grasslands are known as the breadbaskets of the world because of the many types of grains that can grow there. Many other plants grow well there too, including wildflowers. At some times of the year, the grasslands are populated by herds of grazing animals including deer and elk. Other prairie animals include jackrabbits and prairie dogs. Foxes and ferrets prey on prairie dogs. Many species of insects, birds, and reptiles also make the grasslands their home.

Think it Over

4. **Infer** Because of the many types of grains that can grow in these areas, they are called the breadbaskets of the world. (Circle your choice.)
   a. tundras
   b. grasslands
   c. tropical rain forests

What is the rain forest biome?

Rain forests are home to more types of life than any other biome. There are two types of rain forests—the temperate rain forest and the tropical rain forest. **Tropical rain forests** have warm temperatures, wet weather, and lush plant growth. They are located near the equator. The hot climate of these rain forests allows organisms to decay quickly. Plants must quickly absorb these nutrients before they are carried away from the soil by rain. Temperate rain forests are less common, but they also have large amounts of moisture. ✓

Reading Check

5. What are the two types of rain forests?

____

____

What is the temperate forest biome?

Temperate or **deciduous forests** have precipitation that ranges from 70 to 150 cm annually. The soil has a rich humus top layer and a deeper layer of clay. If mineral nutrients in the humus are not immediately absorbed by roots, they might be washed into the clay and lost from the food web for many years. The forests are thick with broad-leaved trees that lose their leaves every year. Some of these trees are maple, oak, and elm. Animals that live in this biome include mice, rabbits, bears, and many different birds.

After You Read

Mini Glossary

biome: a group of ecosystems with the same climax communities; biomes on land are called terrestrial biomes, those in water are called aquatic biomes

desert: an arid region with sparse to almost no plant life; the driest biome

estuary (ES chu wer ee): coastal body of water, partially surrounded by land, where salt water and freshwater mix

grassland: biome of large communities covered with rich soil, grasses, and similar plants

intertidal zone: the area of shoreline that lies between the high and low tide lines

photic zone: portion of the marine biome that is shallow enough for sunlight to penetrate

plankton: small organisms that drift and float in the waters of the photic zone

taiga (TI guh): the biome just south of the tundra; characterized by a boreal or northern coniferous forest composed of fir, hemlock, and spruce trees and acidic, mineral-poor soils

temperate/deciduous forest: biome composed of forests of broad-leaved hardwood trees that lose their foliage annually; receives 70–150 cm of precipitation annually

tropical rain forest: biome located near the equator; has warm temperatures, lots of rain, and lush plant growth

tundra: the biome that circles the north pole; treeless land with long summer days and short periods of winter sunlight

1. Review the terms and their definitions in the Mini Glossary above. Use one of these terms that describes a biome in a complete sentence. Write your sentence in the space below.

____

____

____

____

Section
3.2 Biomes, *continued*

2. Fill in the blanks below with the following words to make correct statements about the material you read in this section: **desert, tundra, plankton, photic zone, rain forest.**

a. The ________________ biome is home to more types of life than any other biome.

b. The ________________ is so cold that very little life exists there.

c. The driest biome is the ________________ biome.

d. The part of the marine biome that is shallow enough for sunlight to penetrate is called the ________________.

e. The base of the entire marine biome food chain is formed by ________________.

Visit the Glencoe Science Web site at **science.glencoe.com** to find your biology book and learn more about biomes.

Name Date Class

Section 4.1 Population Dynamics

Before You Read

This section discusses populations—how they grow and why they stop growing. The factors that affect how populations grow and why they stop growing are called population dynamics. What populations of living things can you observe where you live? Think about populations of plants and animals near your home. List them below. Have those populations ever increased or decreased? Write down reasons for changes in those populations.

Read to Learn

Mark the Text **Identify Growth Factors** Highlight the factors that affect the growth of populations.

Principles of Population Growth

Every organism is a member of a population. A population is a group of organisms of the same species that live in a specific area. You may have a population of grass in your backyard. There also may be a population of ants there. You may even have a population of bacteria on your bathroom door handle.

Populations grow, die, increase, and decrease. Factors that affect the growth rate of populations include food, space, disease, and predators. Scientists have discovered that populations of organisms tend to grow in the same way. Populations start by growing slowly. In the beginning, there are just a few organisms that reproduce. Soon the rate of population growth increases because there are more organisms reproducing.

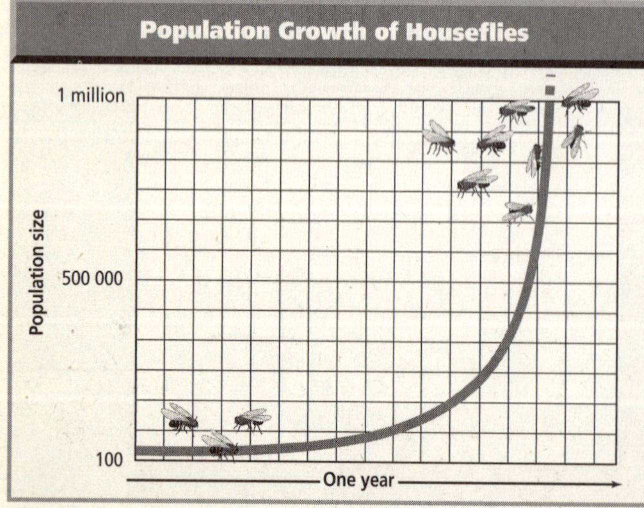

How do populations grow?

Scientists use graphs to show how populations grow. The graph of a growing population looks like a J-shaped curve. The flat part of the J shows slow growth. The part of the J that rises shows rapid growth. The J-shaped curve is a picture of exponential growth. **Exponential growth** means that as a population gets larger, it grows at a faster rate. For example, you may have one dandelion in your backyard. A few weeks later you may have twenty dandelions. A few weeks after that you may have hundreds of dandelions. The reason is exponential growth. One dandelion produces many seeds. Twenty dandelions produce hundreds of seeds.

Can populations grow and never stop growing? No, there are factors that slow down or limit growth. These are called limiting factors. The effects of limiting factors also can be shown on a graph. The J-shaped curve begins to look like an S-shaped curve as population growth slows or levels off.

When populations run out of food or space, growth starts to slow down. Population growth also slows when disease or predators attack populations. Later, you will read about other reasons why population growth slows. ✓

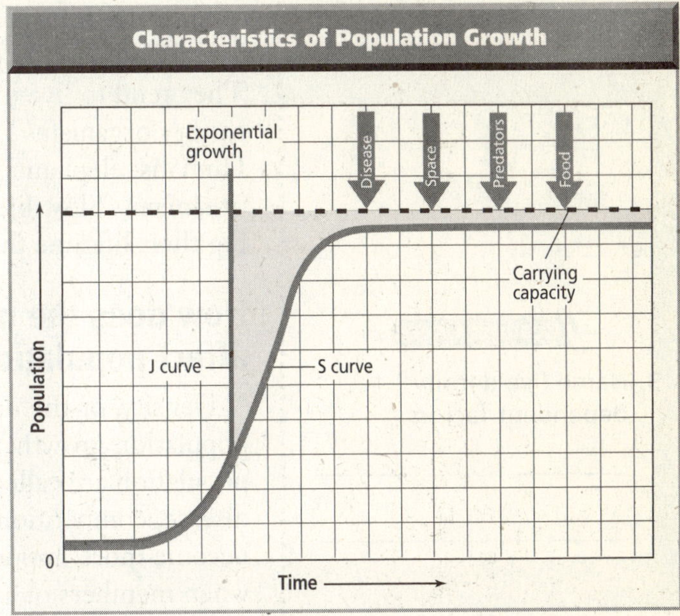

What is an environment's carrying capacity?

The **carrying capacity** of an environment is the number of organisms of one species that can be supported for an unlimited amount of time. Until carrying capacity is reached, there are more births than deaths in the population. If a population grows larger than the carrying capacity of the environment, there will be more deaths than births. The number of organisms in a population is sometimes more than the environment can support and sometimes less than the environment can support.

1. What limits a population's growth?

______

______

______

Reproduction Patterns

An organism's population growth is shaped by its reproductive pattern, or **life-history pattern.** If an organism has a rapid life-history pattern, it will reproduce early in life. It also will produce many offspring in a short period of time. A mosquito is an example of a rapid life-history organism. Rapid life-history organisms have a small body size and a short life span. They live in environments that change extensively. Their populations grow quickly but decline when the environment becomes unfavorable. As soon as conditions are favorable, the small population will grow rapidly.

Think it Over

2. Analyze What type of body size do rapid life-history organisms have? (Circle your choice.)

a. small
b. large

Slow life-history patterns belong to large species that have long lives. They usually grow into adults slowly and reproduce slowly. They tend to live in more stable environments. Often, slow life-history organisms will keep their populations at carrying capacity. Humans, elephants, and trees are examples of slow life-history organisms. Slow life-history organisms have far fewer young during their lifetime than do rapid life-history organisms.

Reading Check

3. Name five density-dependent factors.

____

____

____

____

____

How does the number of organisms in an area affect population growth?

Density, or the number of organisms in a given area, affects population growth. Factors that are related to the density of the population are called **density-dependent factors.** They include disease, competition, predators, parasites, and food. These factors become more important as the population increases. For example, when members of a population live far apart, disease spreads slowly. When the members live close together, disease spreads quickly. This is true for both plants and animals. It is true for populations of people. Some scientists think that the presence of HIV/AIDS in populations is a limiting factor in the growth of those populations. ✓

Density-independent factors affect populations, no matter how large or small. Density-independent factors include volcanic eruptions, temperature, storms, floods, drought, chemical pesticides, and major habitat disruption. Imagine a pond containing a population of fish. If a drought caused the pond to dry up, then that population of fish would die. It would not matter if there were 10 or 100 fish in the pond. Drought is a density-independent factor. Human populations also can be affected by density-independent factors. Rivers sometimes overflow their banks after a heavy rain. If a town is flooded, it does not matter how many people live in the town; everyone feels the effects of the flood.

Think it Over

4. Analyze An example of a density-independent factor is (Circle your choice.)

a. a frost that destroys tomato plants.

b. a fungus that spreads from plant to plant.

Organism Interactions Limit Population Size

Populations also are limited by contact with other organisms in a community. The relationship between predator and prey is a good example. A cat controls the population of mice around a house. A swarm of locusts eats and destroys acres of lettuce on a farm. Bears eat the salmon swimming up a river. All of these are examples of interactions between organisms in a community. Sometimes predators can have a large impact on a population. For example, when brown tree snakes were brought to the island of Guam, they had no natural predators on the island. The snake population increased unchecked. As a result, bird populations have been almost completely destroyed by the snakes. When predators consume prey on a large enough scale, the size of the prey population can be greatly reduced. ✓

Sometimes populations of predators and their prey experience changes in their numbers over time. Scientists observe cycles of population increases and decreases. Some are quite regular and predictable. An example is the interaction of the snowshoe hare and the Canadian lynx. The lynx, a member of the cat family, eats the snowshoe hare. As you can see, predator populations affect the size of prey populations.

Reading Check

5. Why did the brown tree snake population on Guam increase unchecked?

____

____

____

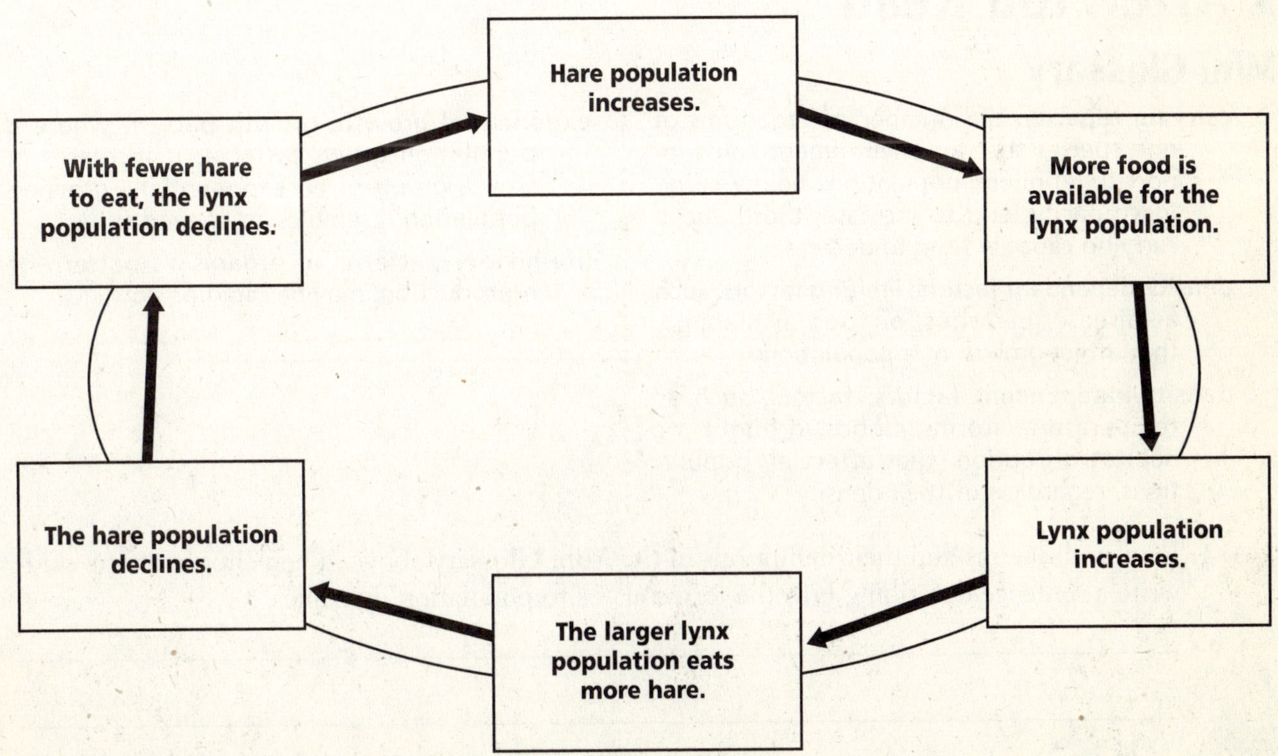

Usually the lynx population catches the young, old, injured, or sick members of the hare population. This makes more resources available for the remaining healthy members of the hare population.

Is there competition among members of the same population?

The lynx and hare are members of different populations, but members of the same population also interact with each other. Populations can increase so that members are competing for food, water, and territory. Competition is a density-dependent factor. When only a few individuals need the available resources, there is no problem. When the population becomes so large that demand for resources is greater than the supply of resources, the population size will decrease.

Sometimes populations become crowded and organisms begin to show signs of stress. Individual animals may become aggressive. They may stop caring for their young or even lose their ability to bear young. Stress also makes animals more at risk for disease. All of these stress symptoms are limiting factors for growth. They keep populations below carrying capacity.

After You Read

Mini Glossary

carrying capacity: the number of organisms of one species that an environment can support indefinitely; populations below carrying capacity tend to increase; those above carrying capacity tend to decrease

density-dependent factors: limiting factors, such as disease, parasites, or food availability, that affect growth of a population

density-independent factors: factors, such as temperature, storms, floods, drought, or habitat disruption, that affect all populations, regardless of their density

exponential growth: growth pattern where a population grows faster as it increases in size; a graph of an exponentially growing population resembles a J-shaped curve

life-history pattern: an organism's pattern of reproduction; may be rapid or slow

1. Review the terms and their definitions in the Mini Glossary above. Then choose a term and write a sentence describing how the term relates to population dynamics.

2. The web diagram below identifies three factors that affect the growth of populations. Write two examples of each factor in the diagram.

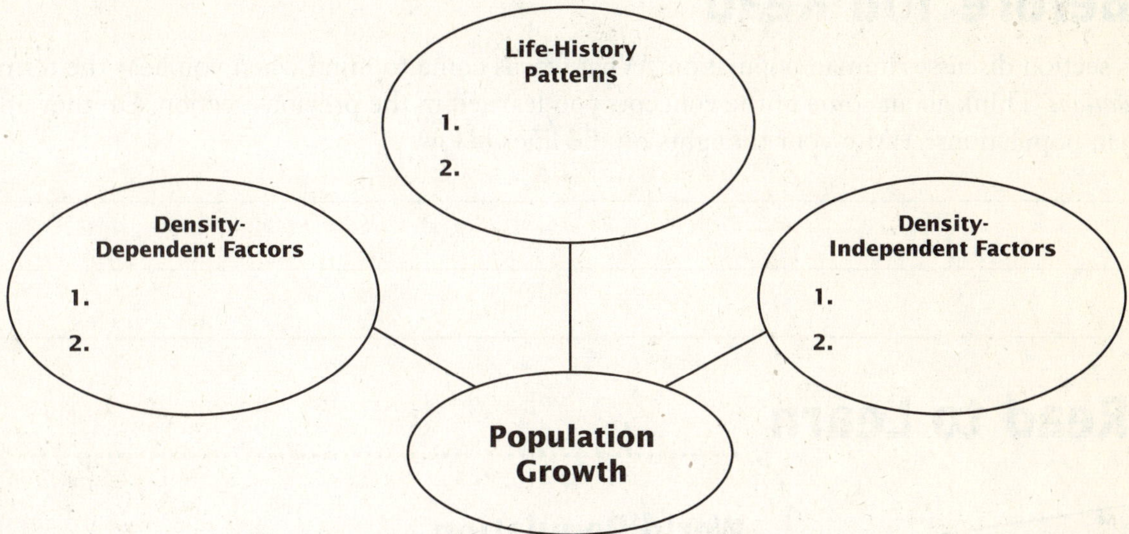

Visit the Glencoe Science Web site at **science.glencoe.com** to find your biology book and learn more about population dynamics.

Section 4.2 Human Population

Before You Read

This section discusses human population. What words come to mind when you hear the term *human population?* Think about some of the concepts you learned in the previous section. Do they apply to human populations? Write your thoughts on the lines below.

Read to Learn

Mark the Text **Identify the Main Point** Skim the section and highlight the main idea of each paragraph.

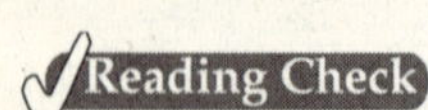

1. How are humans different from other populations?

World Population

A census, which counts all the people who live in a country, is taken every ten years in the United States. It also collects information about where people live and their economic condition. The United Nations keeps similar information for each country in the world. The study of information about human populations is called **demography** (de MAH gra fee). Information about populations includes size, density, distribution, movement, birthrates, and death rates.

Human population has grown at a rapid rate in recent years. Scientists estimate that it took from the time of the first humans to 1800 for the world human population to reach 1 billion. In 1930, there were 2 billion people. In 1999, there were 6 billion people. If this rate continues, scientists estimate that in 2050 there will be 9 billion people on Earth.

What factors affect the growth of human populations?

Remember that populations can keep growing as long as they have enough resources. Human populations are different from other populations because humans can consciously change their environment. Humans can grow their own resources by farming and raising farm animals. Humans can control limiting factors such as disease. Many illnesses that killed people in the past can now be treated with medicine. People live longer and they are able to produce more children. The children grow up and they produce more children, causing the population to grow.

How can you determine the growth rate of a population?

To determine the growth rate of a population, you need to consider different factors. Those factors are illustrated in the chart below.

Factors that Add to a Population	Factors that Subtract from a Population
Births	Deaths
People moving into a population	People moving out of a population

The **birthrate** is the number of live births per 1000 people in a given year. The **death rate** is the number of deaths per 1000 people in a given year. Keeping track of people who move into and out of a population is not always possible. As a result, it is easier to use the birthrate and death rate to calculate the Population Growth Rate (PGR). The following formula is used to calculate PGR:

Birthrate – Death Rate = Population Growth Rate

When the birthrate of a population is the same as the death rate, then the PGR is zero. It is important to remember, though, that if the PGR is zero, it does not mean that the population is remaining exactly the same. It means that new people are entering the population (through birth) at the same rate that people are leaving the population (through death). The population is changing, but it is stable. Your school population will help you understand this idea. If new students arrive at the same rate that students graduate, the population of your school remains the same even though the students that make up the population change each year.

What is the effect of a positive growth rate?

When the PGR is above zero, or positive, it means that more people are entering the population than are leaving. The population is growing. The PGR is positive when there are more births than deaths in the population. A PGR also can be less than zero, or negative. This happens when there are more deaths than births.

Doubling time is the time needed for a population to double in size. Doubling only happens with growing populations. A negative growth rate means that a population will not double. A slow growth rate means that it will take a long time for the population to double. Doubling time can be calculated for a population.

Think it Over

2. Interpret If the population growth rate is positive it means there are (Circle your choice.)

a. more births than deaths.

b. more deaths than births.

What is the age structure of a population?

You may have filled out a survey that asked if you were between the ages of 10 and 14 or between the ages of 15 and 19. Your answer tells where you are in the age structure of the population. **Age structure** refers to the number of people at each different age level. Scientists are interested in the ages of the population, and how many people are males and how many are females. This information tells the scientists if the population is stable, growing, or becoming smaller.

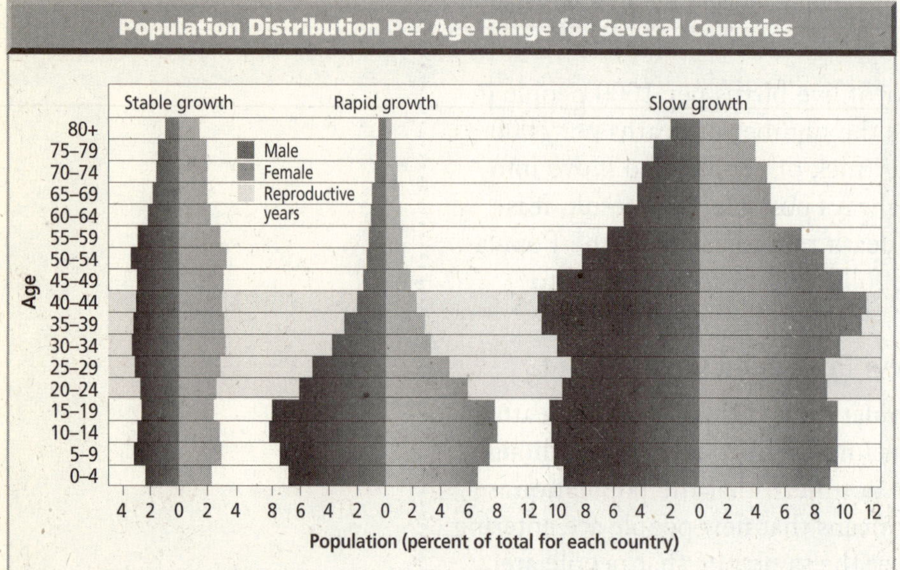

Scientists put age information into a graph called an age structure graph. The graph shows at a glance the ages of the largest group in the population. When a large amount of the population is made up of children, then the population is experiencing rapid growth. When there are more adults than children, the population is declining. When the amount of people in the different age levels is about equal, the population is stable.

Think it Over

3. Interpret a Graph In the age structure graph above, which country has the smallest percentage of people in the 80+ age group? (Circle your choice.)

a. stable growth country
b. rapid growth country
c. slow growth country

Does environment affect population growth?

The needs of human populations are different all over the world. Some are concerned with basic needs such as food and water. Some need to maintain the healthy conditions they already have. Sometimes populations grow too fast. In these populations, there is not enough food, water, or living space for everyone. Maybe the population creates more waste than can be handled. Sometimes the conditions of the environment cause disease to spread, or pollution affects the water. These things affect the stability of human populations. The environment plays an important part in the growth of populations.

After You Read

Mini Glossary

age structure: proportions of a population that are at different age levels

birthrate: number of live births per 1000 population in a given year

death rate: number of deaths per 1000 population in a given year

demography (de MAH gra fee): study of population characteristics such as growth rate, age structure, and geographic distribution

doubling time: time needed for a population to double in size

1. Review the terms and their definitions in the Mini Glossary above. Circle the four key terms that scientists use to gather important information about a population.
2. Use the table below to review what you have read. Place each of the characteristics from the list that follows the table under the correct heading.

Increasing Population	Decreasing Population

Will not have doubling time
Positive growth rate
More adults than children
More children than adults
Will have doubling time
Negative growth rate

Visit the Glencoe Science Web site at **science.glencoe.com** to find your biology book and learn more about human population.

Section 5.1 Vanishing Species

Before You Read

Dinosaurs are probably the most familiar organisms that are extinct, or no longer exist. Many plants and animals that are alive today are in danger of dying out. Think of one animal and one plant that would change the way you live if they became extinct. Write the names of those organisms on the lines below and tell why you think each is so important.

____

____

____

Read to Learn

Mark the Text **Identify Threats to Biodiversity** Highlight all the threats to biodiversity that you read about in this section.

1. Where is the most biodiversity found?

____

____

____

Biological Diversity

Biological diversity is also called biodiversity. **Biodiversity** refers to all the different species in a specific area. The most common way to measure biodiversity is to count all the species in a certain area. For example, an area of farmland might grow one type of corn. It might also contain hundreds of species of insects and several species of birds and other animals. The same size area of tropical rain forest might include hundreds of different types of plants, thousands of different insects, and hundreds of different types of birds. The rain forest has a greater biodiversity than the cornfield because it has a greater number of species in the same area of space. Areas of the world that have the highest biodiversity are those found in warm, tropical places near the equator. Such places include rain forests, coral reefs, and tropical lakes. ✓

What do biodiversity studies tell us?

Studying Biodiversity For many years, ecologists have studied the plants and animals on islands to learn more about biodiversity. In one study, scientists chose some small islands and counted the spiders and insects on the islands. Then they removed all species from the islands except for the trees. This is what they found. (1) Insects and spiders returned first. (2) The farther the island was from the mainland, the longer it took for the species to move back. (3) Eventually, the islands had about the same number of species as they had in the beginning. However, some of the species were different. They also found that for the most part, the larger the island, the more species that live on it.

Biodiversity research is not easy. Some scientists work for weeks at a time hundreds of meters off the ground in trees in rain forests. Others work counting all the species that live in coral reefs. Still others work in labs studying the DNA of members of different populations. They look to see how or if these populations are changing.

Importance of Biodiversity

All living things are interdependent. This means they depend on other living things to stay alive. For example, animals need plants to survive. Most animals either eat plants or eat other animals that eat plants. Also, many plants need animals to survive. Some plant species, for instance, need bees or other animals to take their pollen from one plant to another so that the plants can reproduce. This is just one example of how plants need animals to survive. ✓

Reading Check

2. What does it mean to say that all living things are interdependent?

____

____

____

How do species depend on each other?

Species that live together in the same area often depend on each other. They form symbiotic relationships. Symbiotic relationships, you will remember, are close, permanent, dependent relationships between two or more organisms of different species. Because of this dependence, when one species permanently disappears, it affects the species that remain. An organism suffers when a plant or animal it feeds upon is permanently removed from a food chain or food web. A population may exceed the area's carrying capacity if its predators are removed.

Biodiversity Brings Stability Biodiversity can bring stability to an ecosystem. Compare the following ecosystems.

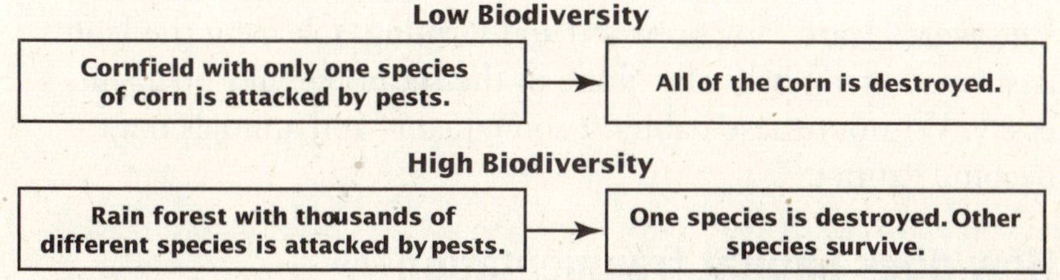

Low Biodiversity
Cornfield with only one species of corn is attacked by pests.
All of the corn is destroyed.
High Biodiversity
Rain forest with thousands of different species is attacked by pests.
One species is destroyed. Other species survive.

Cornfields contain only one type of plant. Therefore, biodiversity is low and the ecosystem may not remain stable if there is a change in species. Rain forests have many types of plants. They have high biodiversity, which means the ecosystem could remain stable even if there is a change in species.

Importance to People Humans also depend on other organisms. People rely on plants and animals for food. People also rely on plants for wood, cotton, and many types of medical drugs. These are just a few examples of how people depend on animals and plants.

If biodiversity continues, people will always have a supply of living things. One day, drugs to cure cancer or HIV might even be found in some of these living things.

Loss of Biodiversity

Biodiversity can be lost when species become extinct. **Extinction** (ek STINGK shun) is the disappearance of a species when the last of its members dies. A certain amount of natural extinction goes on all the time. But now it seems that more than usual is occurring. This could be due to a difference between human needs and available resources.

A species is considered to be an **endangered species** when its numbers become so low that extinction is possible. When a species is likely to become endangered, it is called a **threatened species.** African elephants are one example of a threatened species.

Threats to Biodiversity

The complex interactions among species make each ecosystem unique. Within each ecosystem every species is usually well adapted to its specific habitat (the place where a species lives). This means that changes to a species' habitat can threaten the species with extinction. In fact, habitat loss is one of the biggest reasons for decline in biodiversity.

Habitat Loss One example of habitat loss is in the Amazonian rain forest. In the 1970s and 1980s, thousands of hectares of land were cleared for firewood and farming. Clearing the land destroyed many habitats. None of them can be built up again easily. Without these habitats, some plants and animals may become extinct.

Reading Check

3. What is habitat fragmentation?

____

____

____

How does habitat fragmentation affect biodiversity?

Habitat Fragmentation Another threat to biodiversity is habitat fragmentation. **Habitat fragmentation** is the separation of wilderness areas from other wilderness areas. The fragmented areas are similar to islands. They are cut off from other habitats. ✓

Fragmentation can cause species diversity to drop. This happens when some species leave an area that has become unsuitable for them. Then, other species that depend on these species also leave or die out. When species leave a fragmented area or die out, overall species diversity declines.

Habitat fragmentation can cause members of the same species to become separated from each other. Some members get stuck in one fragmented area while others get stuck in another. Then the two populations cannot interact. This means that the members of one population cannot breed with members of the other population. Genes from one population cannot mix with the genes from another. This is known as genetic isolation.

Edge and Size The edge of a habitat is where one habitat meets another. A forest meeting a field or a road cutting through a forest are two examples of edges. Conditions along the edges are different than they are in the middle of a habitat. These different conditions are called **edge effects.** Because of the different conditions, different organisms may live at the edge of a habitat than in the middle of a habitat.

How does pollution affect the environment?

Habitat Degradation Habitat degradation is another threat to biodiversity. **Habitat degradation** is damage to a habitat caused by air, water, or land pollution. Air pollution can cause breathing problems. It can also irritate the eyes and nose. **Acid precipitation**—rain, snow, sleet, and fog with low pH values—has damaged some forests, lakes, statues, and buildings. Acid precipitation forms when pollutants in the air combine with water vapor in the air. When these acidic droplets hit Earth, the moisture takes nutrients out of the soil. Without the nutrients, many plants become sick and die.

The Sun gives off waves, called ultraviolet waves, which can damage living things. Earth has an area in its atmosphere, between 15 km and 35 km altitude, called the **ozone layer,** which absorbs some of these ultraviolet waves before they reach Earth. Pollution has damaged the ozone layer. Today more ultraviolet waves reach the Earth than in the past. Over some parts of Antarctica, the amount of ozone overhead is reduced by as much as 60 percent during the Antarctic spring. This seasonal reduction of ozone is caused by chemicals such as chlorofluorocarbons (CFCs), which are produced by humans.

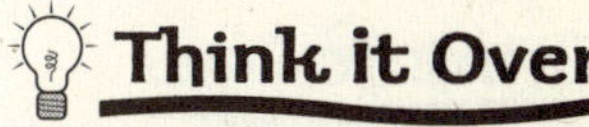

4. Interpret Which diagram demonstrates habitat fragmentation? (Circle your choice.)

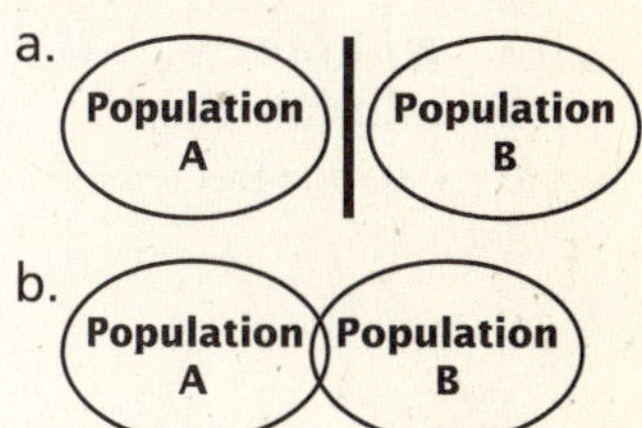

Reading Check

5. Why do more ultraviolet waves reach Earth today?

____

____

____

Think it Over

6. Conclude Which of the following does not threaten biodiversity? (Circle your choice.)

a. acid rain
b. rain forests
c. exotic species

Water Pollution Water pollution degrades, or damages, habitats in streams, rivers, lakes, and oceans. Many different kinds of pollutants can harm the living things in these habitats. Examples include detergents, fertilizers, and industrial chemicals that end up in streams and rivers.

Land Pollution Land pollution comes in many forms. One form is chemicals used to kill plant pests. Many years ago DDT was often sprayed on food crops to control insects. Birds that ate the insects, plants, or fish exposed to DDT had high levels of DDT in their bodies. The DDT passed from the birds to the predators that ate them. Some of the predators, such as the bald eagle, laid eggs with very thin shells because of the DDT in their bodies. The thin shells cracked easily and many of the eagle chicks died. The use of DDT was banned in the United States in 1972.

Exotic Species Sometimes people bring new plants or animals to an ecosystem where these organisms have not lived before. In other words, these species are not native to the area. Such species are called **exotic species.** When they begin to live in a new area, they can upset the biodiversity in that ecosystem. For example, many years ago, goats were taken to Catalina Island off the coast of California. There had never been goats on the island. As the goats multiplied, they ate more and more of the plants on the island. Eventually, 48 kinds of plants that used to be on the island were gone. They disappeared because the goats had eaten all of them. When exotic species are taken to a new area they can multiply quickly because they do not have natural predators in the new area.

After You Read

Mini Glossary

acid precipitation: rain, snow, sleet, or fog with a low pH, which causes damage to forests, lakes, statues, and buildings

biodiversity: the variety of species in a particular area

edge effects: different environmental conditions that occur along the boundaries of a habitat

endangered species: a species in which the number of individuals becomes so low that extinction is possible

exotic species: species that are not native to an area

extinction: the disappearance of a species when the last of its members dies

habitat degradation: damage to a habitat by air, water, and land pollution

habitat fragmentation: the separation of wilderness areas from other wilderness areas

ozone layer: a region of Earth's atmosphere between 15 km and 35 km altitude

threatened species: a species that is likely to become endangered

Section 5.1 Vanishing Species, *continued*

1. Review the terms and their definitions in the Mini Glossary on page 48. Circle the terms that refer to things that pose a threat to the biodiversity of an area.

2. Use the outline to help you review this section. Use the question clues below and the headings in Read to Learn to fill in the blanks.

I. Biological Diversity

A. (What is it?) ______________________

B. (How is it measured?) ______________________

C. (Where is the highest biodiversity?) ______________________

II. Importance of Biodiversity

A. (What is true of all living things?) ______________________

____

B. (What can biodiversity bring to an ecosystem?) ______________________

____

C. (What benefit might preserving biodiversity bring humans in the future?) ______________________

____

III. ______________________

A. Extinction can occur

B. Species can become endangered or threatened

IV. ______________________

A. Habitat fragmentation

B. Habitat degradation

3. In Column 1 are some new concepts you learned about in this section. Column 2 gives one example of each concept. Write the letter from Column 2 on the line next to the concept that matches the example.

Column 1	Column 2
____ 1. habitat fragmentation	a. all the species in an area of rain forest
____ 2. biodiversity	b. DDT
____ 3. extinct species	c. acid precipitation
____ 4. land pollution	d. a road splitting a forest in half
____ 5. habitat degradation	e. triceratops dinosaurs

Visit the Glencoe Science Web site at **science.glencoe.com** to find your biology book and learn more about vanishing species.

Section 5.2 Conservation of Biodiversity

Before You Read

This section discusses ways to keep plants and animals and their environments safe and healthy. On the lines below, list things you do to keep a favorite plant or animal safe and healthy.

Read to Learn

Mark the Text **Restate the Main Point** Highlight the main point in each paragraph. State each main point in your own words.

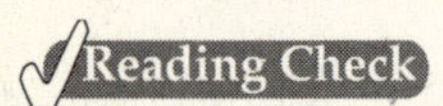

1. What is conservation biology?

Conservation Biology

Conservation biology is the study and carrying out of ways to protect biodiversity. Conservation biology develops ways to conserve, or save, species and natural resources. **Natural resources** are the parts of the environment that are needed by or useful to living organisms. Sunlight, water, air, and plants and animals are all natural resources. These must be thought of when any conservation activity is planned. ✓

Protecting Species and Habitats In the United States and in many other countries, laws have been made to protect endangered and threatened species.

Habitats, the place where a species lives, must be protected, too. Many habitats have been protected by the creation of natural preserves and national parks. The United States has many national parks. Some examples are Yellowstone, Crater Lake, and Big Bend National Parks.

Habitat Corridors In some areas, habitat corridors are created to connect wilderness areas that are separated from one another. **Habitat corridors** are protected strips of land that allow organisms to move from one wilderness area to another. Research has shown that these can help both animal and plant species to remain strong and be able to reproduce.

How do sustainable use programs protect ecosystems and benefit people?

Plants and animals that live in national parks must still be protected from harm. National parks in the United States and other countries hire rangers to help protect all species in the parks. Some parks allow sustainable use of the natural resources in the parks. **Sustainable use** lets people use the natural resources in ways that will benefit them and maintain the ecosystem. For example, in some rain forests, people are allowed to harvest crops like Brazil nuts that grow naturally. The people are able to sell the nuts to make money while the ecosystem is maintained.

How do reintroduction programs work?

Reintroduction programs can help save organisms that are in danger. **Reintroduction programs** take members of an endangered species and breed and raise them in protected habitats. When many of the organisms have been raised, they are released back into the area where they would naturally live. ✓

Ferrets were reintroduced to their habitat this way. Ferrets feed on prairie dogs. At one time, the land where prairie dogs lived was being reduced by rural land use. This was destroying the habitat of the prairie dogs. As a result, the ferrets were losing their food supply and becoming endangered. A number of ferrets were captured and a breeding program was begun. Many black-footed ferrets have been reintroduced to the wild.

An organism that is kept by humans is said to be in **captivity.** Some species exist only in captivity in zoos or other special places. When endangered species are in captivity, scientists try to increase their number. Then they try to reintroduce the species into the habitat where they would live naturally. It is easier to reintroduce plants into the wild than animals. Some animals raised in captivity may lose the behaviors needed to survive in the wild.

✓ Reading Check

2. What are reintroduction programs?

____

____

____

After You Read

Mini Glossary

captivity: when people keep members of a species in zoos or other conservation facilities

conservation biology: field of biology that studies methods and carrying out ways to protect biodiversity

habitat corridors: natural strips of land that allow the migration of organisms from one wilderness area to another

natural resources: those things in the environment that are useful or needed for living organisms

Section 5.2 Conservation of Biodiversity, *continued*

reintroduction programs: programs that take members of an endangered species and breed and raise them in protected habitats; when many of the organisms have been raised, they are released back into the area where they would naturally live

sustainable use: philosophy that promotes letting people use resources in wilderness areas in ways that will benefit them and maintain the ecosystem

1. Review the terms and their definitions in the Mini Glossary above. Use one of the terms in a sentence that supports conservation efforts in your community.

____

____

____

2. Use the diagram below to help you review this section. In the top two boxes, list one method used to protect the natural resources. In the bottom three boxes, explain what the methods listed attempt to accomplish.

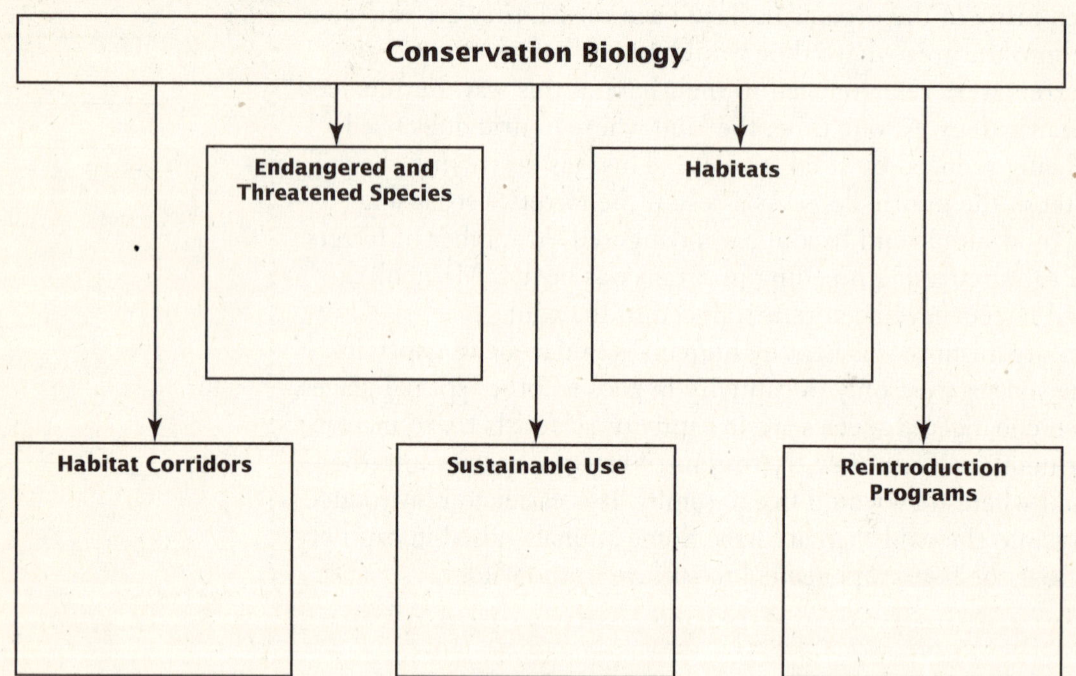

Visit the Glencoe Science Web site at **science.glencoe.com** to find your biology book and learn more about conservation of biodiversity.

Name Date Class

Section 6.1 Atoms and Their Interactions

Before You Read

In this section you will learn about the smallest particles that make up everything you can see, feel, or touch. Think about what living and nonliving things have in common. On the lines below, write a sentence explaining your ideas.

Read to Learn

Elements

Everything—whether it is a rock, frog, or flower—is made of things called elements. An **element** is a substance that cannot be broken down into simpler chemical substances. Suppose you found a nugget of pure gold. You could grind it into a billion bits of powder and every tiny bit would still be gold. You could treat the gold with every chemical there is, but the gold would never break down into simpler substances. That is because gold is an element.

Mark the Text **Identify Definitions** Highlight the definition of each word that appears in bold.

Natural Elements in Living Things Only about 25 of the elements on Earth are needed by living organisms. Four of these elements—carbon, hydrogen, oxygen, and nitrogen—together make up more than 96 percent of the mass of a human body. Mass is the measure of matter a body contains. Other elements are needed by organisms but only in very small amounts. These elements are called trace elements. Plants get trace elements by taking them in through their roots. Animals get trace elements from the foods they eat.

Reading Check

1. What is an element?

Each element has an abbreviation made of one or two letters. The abbreviation is called a symbol. For example, the symbol C stands for carbon. H stands for hydrogen. O stands for oxygen. N stands for nitrogen. As mentioned above, these elements are important to the human body.

Atoms: The Building Blocks of Elements

All elements are made of atoms. **Atoms** are small particles that are the basic building blocks of all matter.

What is the structure of an atom?

Reading Check

2. What are the two types of charged particles in an atom?

____

____

All atoms have the same basic structure. The center of the atom is the **nucleus** (NEW klee us; the plural of nucleus is nuclei). All nuclei contain positively charged particles called protons, which give the nuclei a positive charge. Most nuclei also contain neutrons. These are particles with no charge. The space around the nucleus contains negatively charged particles called electrons. Electrons are held in this space by their attraction to the positively charged nucleus. The space in which electrons stay is divided into energy levels. The first level can hold only two electrons. The second level can hold no more than eight electrons. The third level can hold up to 18 electrons.

Boron Atom

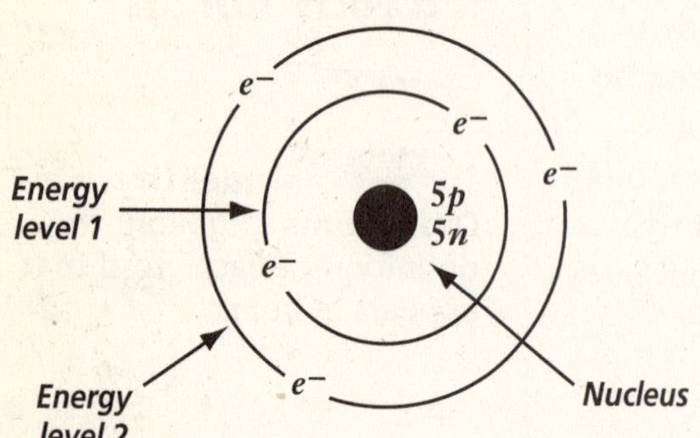

Each atom has the same number of electrons as it has protons. The atom as a whole has no charge. The positive charges of the electrons cancel out the negative charges of the protons. The figure to the left shows an example of a boron atom. Its nucleus contains five protons and five neutrons.

Each element has distinct characteristics that result from the number of protons in the nuclei of that element. For instance, the elements iron, gold, and oxygen each have a different number of protons in their atoms' nuclei.

Isotopes of an Element

Atoms of the same element always have the same number of protons. Sometimes, though, they can have different numbers of neutrons. Atoms of the same element that have different numbers of neutrons are called **isotopes** (I suh tohps). Different isotopes of the same element act differently. Scientists have developed some useful ways to use different isotopes in medicine.

Compounds and Bonding

A **compound** is a substance made of atoms of two or more elements. These atoms are chemically combined. Table salt is a compound of the elements sodium and chlorine. If an electrical current is passed through melted salt, the salt breaks down into these two elements. The element chlorine is a poison. When it is combined with sodium to form salt, the chlorine is harmless. The properties of a compound are different from the properties of the individual elements that make it up.

Reading Check

3. What is a compound?

____

____

____

Why do atoms combine?

Atoms chemically combine with other atoms to form compounds only when the result is more stable than the individual atoms. An atom becomes more stable when the energy level farthest from the nucleus is filled with the maximum number of electrons it can hold. For most atoms, this means when the second level has eight electrons. One way to fill energy levels is to share electrons with other atoms. This is what hydrogen atoms do. When two atoms share electrons, the force that holds the atoms together is called a **covalent** (koh VAY lunt) **bond.**

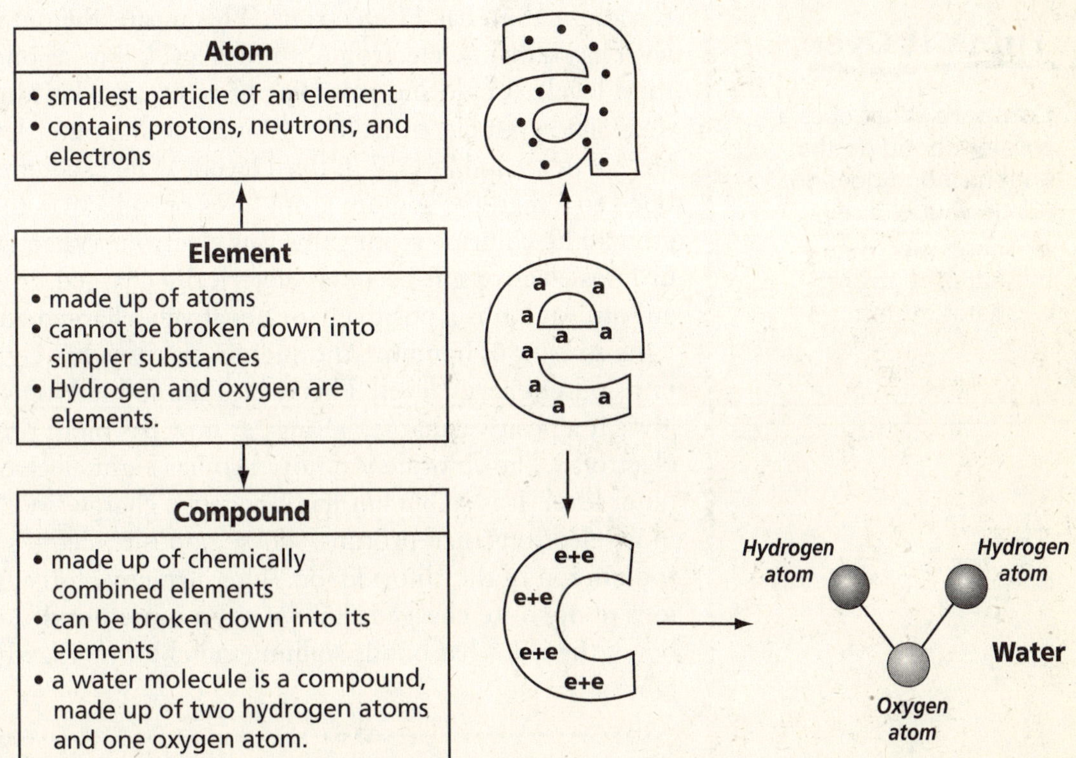

Atom
• smallest particle of an element
• contains protons, neutrons, and electrons
Element
• made up of atoms
• cannot be broken down into simpler substances
• Hydrogen and oxygen are elements.
Compound
• made up of chemically combined elements
• can be broken down into its elements
• a water molecule is a compound, made up of two hydrogen atoms and one oxygen atom.
a a a a a a a a a
e+e e+e e+e e+e
Hydrogen atom
Hydrogen atom
Water
Oxygen atom

Reading Check

4. What is a group of atoms held together by covalent bonds called?

Hydrogen atoms have only one energy level with one electron. To be full, this level would need two electrons. For this reason, a hydrogen atom will share its one electron with another hydrogen atom. This makes both atoms more stable. The two shared electrons move in the space around the nuclei of both atoms. The positively charged nuclei attract the negatively charged electrons that are shared. This attraction holds the two atoms together. When a group of atoms is held together by covalent bonds, the group is called a **molecule.** A molecule has no charge, positive or negative.

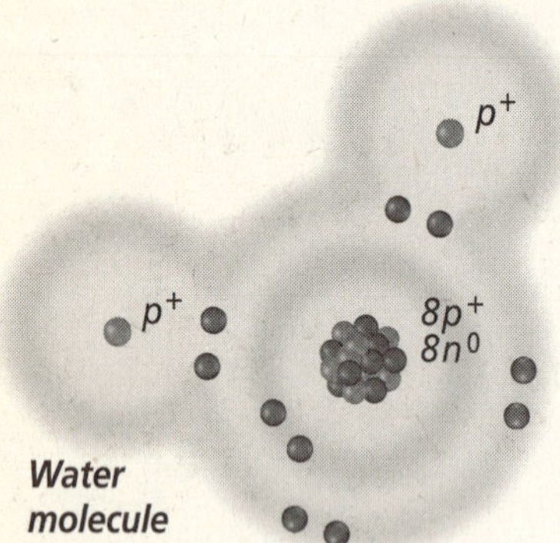

Water molecule

How do covalent bonds form?

Most compounds in organisms have covalent bonds. Water is a great example. In a molecule of water, there are two hydrogen atoms and one oxygen atom. Each hydrogen atom shares its one electron. The oxygen atom shares the six electrons it has in its outer energy level. By sharing eight electrons, all three atoms become stable.

How do ionic bonds form?

Not all atoms bond together by sharing electrons. A sodium atom and a chlorine atom bond in another way to make table salt. A sodium atom has 11 electrons. This means the outer energy level has only one electron (2 in first level, 8 in second level, 1 in third level). A chlorine atom has 17 electrons. This means that there are seven electrons in a chlorine atom's outer level (2 in first level, 8 in second level, 7 in third level). When sodium and chlorine bond together, the sodium atom loses one electron to the chlorine atom. The chlorine atom gains this electron. When an atom gains or loses electrons, it becomes electrically charged. It is then called an **ion.** An ion is a positively or negatively charged atom.

As a result of bonding, the sodium ion now has eight electrons in its outer energy level. This makes the sodium ion stable, but gives it a positive charge, because it now has more protons than electrons. The chloride ion now also has eight electrons in its outer level. It is stable but has a negative charge since it now has more electrons than protons. These opposite charges attract the sodium ion to the chloride ion. This attractive force between two ions of opposite charge is known as an **ionic bond.** An ionic bond, then, is what bonds sodium and chlorine together to make table salt.

Think it Over

5. Compare What does a covalent bond do that an ionic bond does not? (Circle your choice.)

a. shares electrons
b. shares neutrons
c. shares atoms

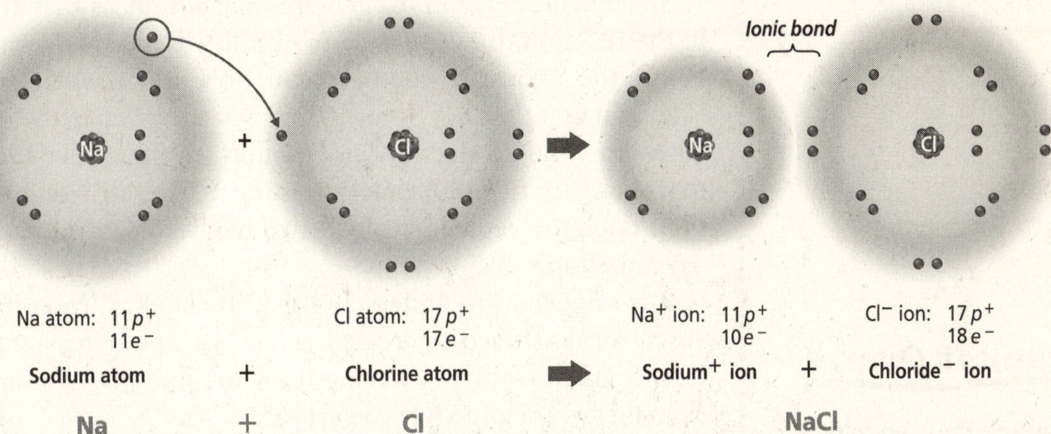

Chemical Reactions

The forming or breaking of bonds is what causes substances to combine in different ways to make other substances. Chemical reactions take place when bonds are formed and broken. In organisms, chemical reactions occur inside cells. All of the chemical reactions that take place within an organism are known as that organism's **metabolism.** These reactions break down and build substances that are important for the organism to function properly. ✓

How do you write chemical equations?

Scientists describe what happens in chemical reactions by using chemical equations (i KWA zhunz). These are a combination of element symbols and numbers. They explain how many substances are involved in a reaction. They also show how these substances join together or come apart. For example, hydrogen and oxygen bond together, or react, to form water. Here is the equation that describes this:

$$2H_2 + O_2 \rightarrow 2H_2O$$

H is the symbol for hydrogen. O is the symbol for oxygen. But the numbers are important, too. The subscript numbers (the smaller numbers after a symbol) show how many atoms of each element are in a substance. The number before a substance tells how many molecules are in the substance. If there is no number before a substance, this means that there is only one molecule of it. ✓

So, in the sample equation, $2H_2$ means that there are two molecules of hydrogen. Each molecule is made of two hydrogen atoms. O_2 means that there is one molecule of oxygen, and the molecule is made up of two oxygen atoms.

Reading Check

6. What causes substances to combine in different ways to make other substances?

____

____

____

Reading Check

7. Explain the meaning of each of the following symbols in a chemical equation.

Number before an element:

____

____

Subscript number after an element:

____

____

8. Analyze Pouring milk over cereal creates a solution. (Circle your choice. Then, on the lines provided, explain your answer.)

a. True
b. False

____

____

____

____

Since hydrogen and oxygen are the substances that are involved in the reaction, they are called reactants. Reactants always appear before the arrow in a chemical equation. The result of a reaction is known as a product. It always comes after the arrow in the equation. The product in this reaction is water. $2H_2O$ means that there are two molecules of water that result from this reaction. Each molecule is made of two hydrogen atoms and one oxygen atom.

It is important to understand that in chemical reactions, atoms are never destroyed or created. They are simply rearranged. Equations show this by having the same number of atoms of each element before and after the arrow.

Mixtures and Solutions

Mixtures When elements combine chemically to form a compound, the elements no longer have their original properties. Sometimes, though, substances mix together but do not chemically combine. This makes a mixture. A **mixture** is a combination of substances in which the individual substances keep their own properties. Stirring sand and sugar together in a bowl, for example, makes a mixture. Neither the sand nor the sugar changes. They do not combine chemically.

Solutions A solution is a special type of mixture. In a **solution,** one or more substances (solutes) dissolve in another substance (solvent). Equal amounts of the solute are then found throughout the solvent. For example, stirring a pack of powdered drink mix in some water makes a solution. The powder in the drink mix dissolves and is the solute. The water is the solvent. If you want the drink to taste right, you must know how much solute (drink mix powder) to dissolve in your solvent (water). The measure of the amount of solute dissolved in a solvent is called concentration. The more solute that is dissolved in a given amount of solvent, the greater the concentration.

Acids and Bases Chemical reactions take place only when conditions are just right. A reaction might need a certain temperature or a certain amount of energy. It might depend on the right concentration of a substance in a solution. Chemical reactions in organisms also depend on the pH of the environment inside the organism. The **pH** measures how acidic or basic a solution is. A scale from below 0 to above 14 is used to measure pH. Substances with a pH below 7 are acidic. An **acid** is any substance that forms hydrogen ions (H^+) in water. (Remember, ions have either a

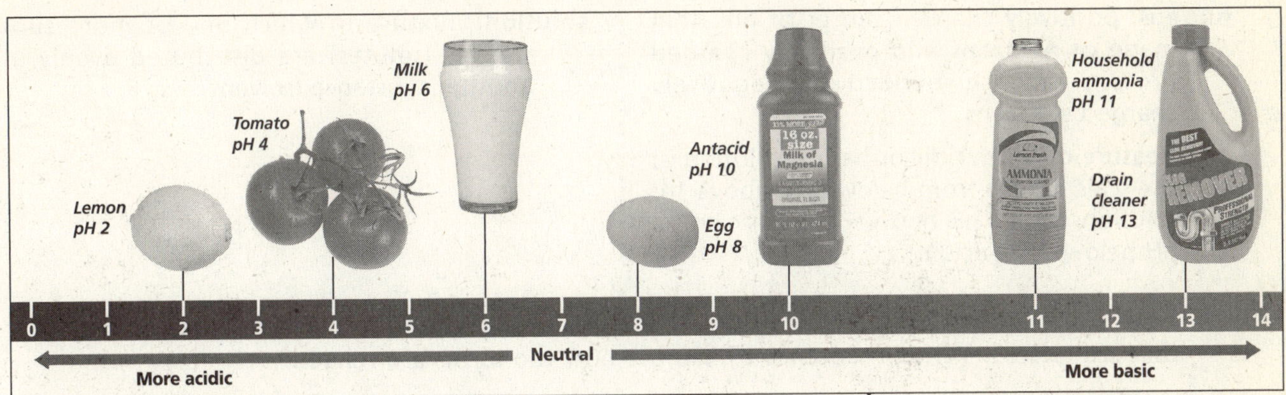

positive or negative charge. That is what the + or − means after an element's symbol.) For example, when hydrogen chloride (HCl) is added to water, hydrogen ions (H^+) and chloride ions (Cl^-) are formed. That is why hydrogen chloride in a solution with water is called hydrochloric acid. This acidic solution contains many, many hydrogen ions (H^+). The solution's pH is below 7.

Substances with a pH above 7 are basic. A **base** is any substance that forms hydroxide ions (OH^-) in water. For example, if sodium hydroxide (NaOH) is dissolved in water, it forms sodium ions (Na^+) and hydroxide ions (OH^-). This basic solution contains many, many hydroxide ions (OH^-) and has a pH above 7.

If a substance is not acidic or basic, it is neutral. Neutral substances have a pH of 7. The figure above shows common acids and bases.

9. What is the pH of a neutral substance?

____

____

After You Read

Mini Glossary

acid: any substance that forms hydrogen ions (H^+) in water. Acids have a pH below 7.

atom: smallest particle of an element; basic building block of all matter

base: any substance that forms hydroxide ions (OH^-) in water. Bases have a pH above 7.

compound: substance composed of atoms of two or more elements that are chemically combined

covalent (koh VAY lunt) bond: chemical bond formed when two atoms share electrons

elements: substances that cannot be broken down into simpler chemical substances

ion: atom or group of atoms that gain or lose electrons; has an electrical charge

ionic bond: chemical bond formed by the attractive forces between two ions of opposite charge

isotopes: atoms of the same element that have different numbers of neutrons in the nucleus

metabolism: all of the chemical reactions that occur within an organism

mixture: combination of substances in which individual components keep their own properties

molecule: group of atoms held together by covalent bonds; has no overall charge

Section 6.1 Atoms and Their Interactions, *continued*

nucleus: positively charged center of an atom made of neutrons and positively charged protons, and surrounded by negatively charged electrons

pH: measure of how acidic or basic a solution is; the scale ranges from below 0 to above 14; solutions with a pH above 7 are basic and a pH below 7 are acidic

solution: mixture in which one or more substances (solutes) are distributed evenly in another substance (solvent)

1. Review the terms and their definitions in the Mini Glossary. Then choose two of the terms that are related. Write a sentence explaining how the terms are related.

____

____

____

2. Draw a line from each term in Column A to the best explanation of the term in Column B.

Column A	Column B
1. compound	a. substance with a pH of 7
2. element	b. an organism's metabolism
3. covalent bond	c. forms hydroxide ions in water
4. chemical reactions	d. forms hydrogen ions in water
5. mixture	e. what holds hydrogen and oxygen together in a water molecule
6. solution	f. drink mix dissolved in water
7. acid	g. sugar and sand stirred in a bowl
8. neutral	h. what holds sodium and chloride together in table salt
9. base	i. gold
10. ionic bond	j. table salt

Visit the Glencoe Science Web site at **science.glencoe.com** to find your biology book and learn more about atoms and their interactions.

Section 6.2 Water and Diffusion

Before You Read

This section tells about the importance of water in our bodies and in other living organisms. How important is water in your life? Could you go a day without using water? On the lines below, list the ways water is important in your life.

____

____

____

____

Read to Learn

Water and Its Importance

When was the last time you had a drink of water? Water is perhaps the most important compound in living organisms. In fact, water makes up 70 to 95 percent of most organisms.

STUDY COACH

Identify the Main Point Skim the section and highlight the main idea of each paragraph.

What are the facts about water?

You learned in the last section that when two atoms share electrons, the force that holds them together is called a covalent bond. This group of atoms held together by a covalent bond forms a molecule. Some molecules do not share the electrons equally. They form a polar bond. A molecule with a polar bond is called a **polar molecule.** A polar molecule has a positive end and a negative end. For example, the electrons in a water molecule spend more time near the oxygen nucleus than they do near the hydrogen nuclei. This makes water a polar molecule. Part A of the figure to the right shows a water molecule. Part B of the figure shows that a water molecule is a polar molecule.

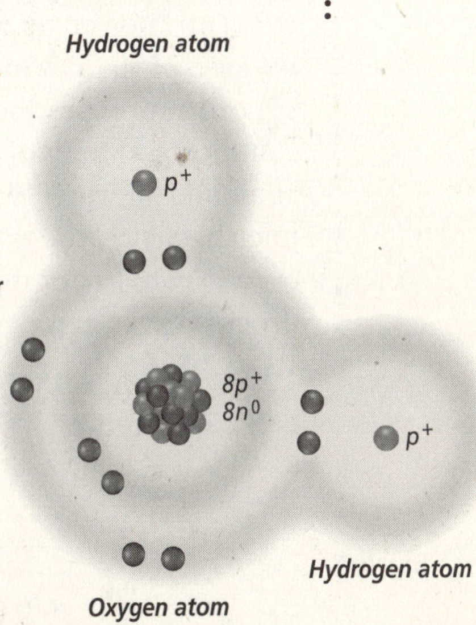

A In a covalent bond between hydrogen and oxygen, the electrons spend more time near the oxygen nucleus than near the hydrogen nucleus.

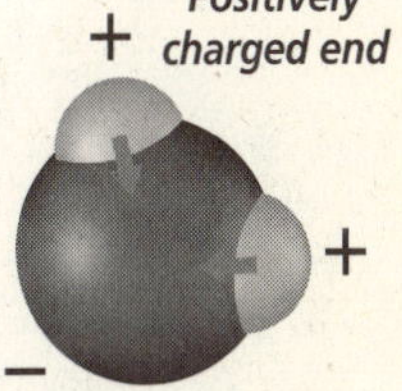

B Since oxygen attracts the electrons more than hydrogen does, the oxygen end of a water molecule is slightly negative and the hydrogen end is slightly negative.

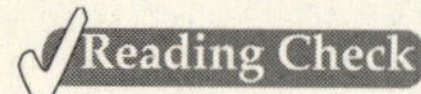

1. What allows water to dissolve salt and sugar?

Polar water molecules attract ions. You learned in the last section that ions are positively and negatively charged atoms. Because of this attraction, water can dissolve many ionic compounds such as salt. It also can dissolve many other polar molecules such as sugar.

Water molecules also attract other water molecules. The positively charged hydrogen atoms of one water molecule attract the negatively charged oxygen atoms of another water molecule. When water molecules bond with other water molecules, they form a weak bond called a **hydrogen bond.** Hydrogen bonds are important because they hold molecules, such as proteins, together.

Because water is a polar molecule, it is able to creep up thin tubes, such as those found in plants. This allows plants to get water from the ground.

Water has a number of other special characteristics. Water resists temperature changes. It takes more heat to raise the temperature of water than it does to raise the temperature of most other liquids. Water also loses a lot of heat when it cools. Water expands when it freezes. As a result, ice is less dense than liquid water. This is why ice floats when it forms in water.

Diffusion

Atoms and molecules of gases, liquids, and some solids move randomly. The molecules, or particles, have energy of motion. They constantly move in straight lines until they collide with other particles. When one particle hits into another particle, both rebound off each other and move in different directions.

Diffusion happens when particles move from an area of higher concentration to an area of lower concentration. Diffusion results because of the random movement of particles. An area of higher concentration has more particles in it than an area of lower concentration. Think of concentrated orange juice. If placed in a container of water, which has a lower concentration of orange juice particles, the orange juice will begin to spread into the water even if you do not mix it. This is diffusion. Diffusion is a slow process because it depends on the random movement of particles.

What affects the speed of diffusion?

There are three factors that affect how quickly particles diffuse: concentration, temperature, and pressure. Concentration is the main factor in controlling the speed of diffusion. The more concentrated a substance, the more quickly it diffuses. When there are more randomly moving particles, there are more collisions among them. More collisions increase the chance that the particles will bump each other into areas where no other particles exist. If temperature increases, the particles' motion speeds up (think of boiling water). This speeds up the diffusion process. Increasing pressure will also speed up particle motion. This, too, speeds up diffusion.

Result of Diffusion Diffusion is complete when all the particles in a mixture become evenly distributed or mixed. At this point, the particles are still moving, but the concentration of the particles will not change. This is called **dynamic equilibrium.** The figure below shows how diffusion occurs. The last illustration in the figure shows dynamic equilibrium.

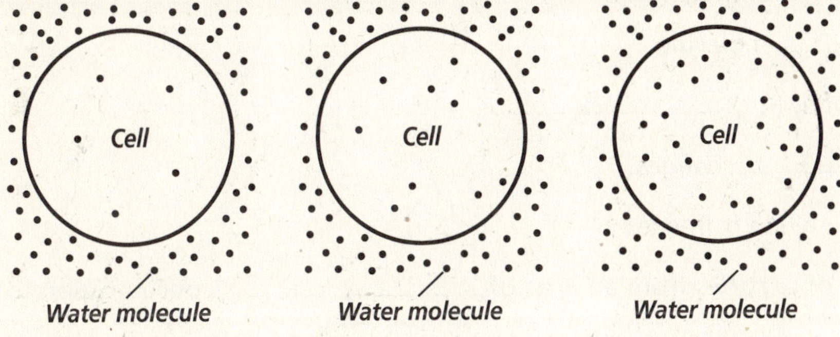

Diffusion in Living Systems Diffusion is one of the ways cells move substances in and out of the cell. This is also illustrated in the figure above. Notice in the first illustration that the concentration is higher outside the cell. In the second illustration, the concentration level in the cell is increasing. In the last illustration, the concentration level is equal on the inside and outside of the cell. In other words, there is dynamic equilibrium.

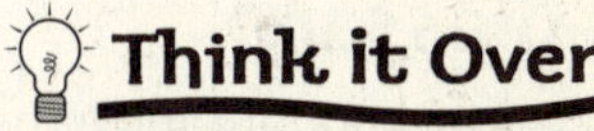

2. Apply Which will diffuse more quickly? (Circle your choice.)

a. concentrated orange juice in cold water
b. concentrated orange juice in warm water

3. Why do cells use diffusion?

____

____

After You Read

Mini Glossary

diffusion: random movement of particles from an area of higher concentration to an area of lower concentration

dynamic equilibrium: result of diffusion where there is continuous movement of particles but no overall change in concentration

hydrogen bond: weak chemical bond formed by the attraction of positively charged hydrogen atoms to other negatively charged atoms

polar molecule: molecule with an unequal distribution of charge, resulting in the molecule having a positive end and a negative end

1. Review the terms and their definitions in the Mini Glossary above. Select one term and write a definition of it in your own words.

____

____

____

2. Complete the sentences about water and diffusion with information you learned from your reading.

a. Water is a ________________ molecule.

b. Water can creep up thin tubes in ________________.

c. Water resists ________________ changes.

d. Water ________________ when it freezes.

e. Diffusion is the movement of particles from an area of ________________ concentration to an area of ________________ concentration.

f. The three factors that affect the rate of diffusion are ________________, ________________, and ________________.

g. Diffusion allows cells to move substances ________________ and ________________ of the cell.

Visit the Glencoe Science Web site at **science.glencoe.com** to find your biology book and learn more about water and diffusion.

Section 6.3 Life Substances

Before You Read

This section explains the chemical construction of many of the substances that make up our bodies and the bodies of other living organisms. What do you think your body is made of? On the lines below, list substances that you think make up your body.

Read to Learn

The Role of Carbon in Organisms

Carbon is one of the substances found in living organisms. Carbon atoms can form covalent bonds with other carbon atoms and with many other elements. When a carbon atom bonds by sharing one electron, it forms a single bond. When it bonds by sharing two electrons, it forms a double bond. When a carbon atom bonds by sharing three electrons, it forms a triple bond. The figure below illustrates the three types of bonds.

Molecular Chains As one carbon atom bonds to another and then that one bonds to another, they form straight chains, branched chains, or rings. These chains and rings can contain almost any number of carbon atoms and can include atoms of other elements as well. The chains and rings are called carbon compounds.

Carbon compounds sometimes contain only one or two carbon atoms. But some carbon compounds contain tens, hundreds, or thousands of carbon atoms. These large compounds are called biomolecules.

State the Main Ideas As you read this section, stop after every few paragraphs and put what you have just read into your own words.

Mark the Text Highlight the main idea in each paragraph.

Single Bond

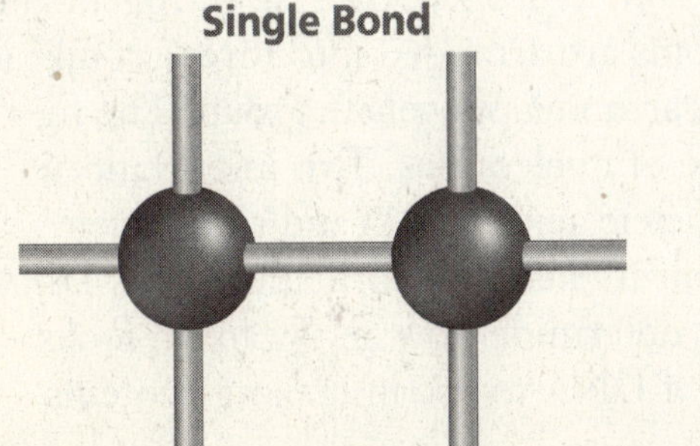

Double Bond

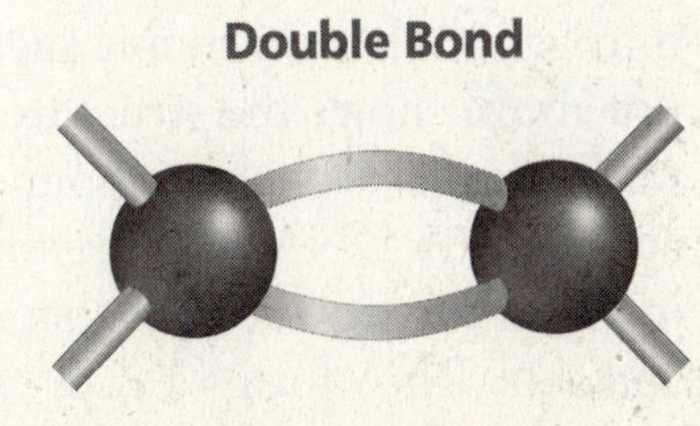

Triple Bond

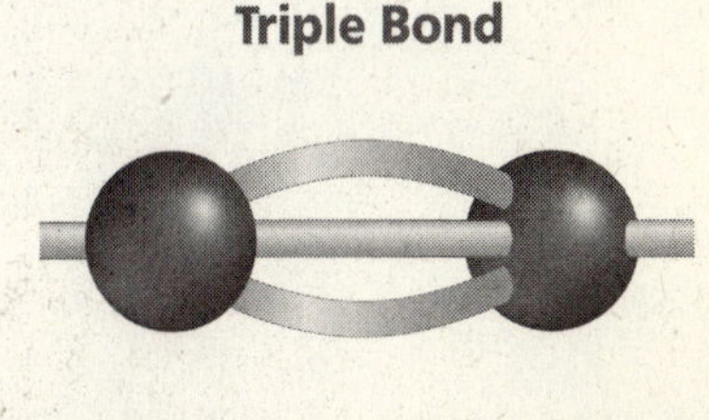

Section **6.3** **Life Substances,** *continued*

Reading Check

1. What three elements are carbohydrates made of?

______

______

______

What are examples of biomolecules?

Carbohydrates are one type of biomolecule. **Carbohydrates** are organic compounds made of carbon, hydrogen, and oxygen. They are used by cells to store and release energy. Starch and sugars are examples of carbohydrates.

Lipids are another type of biomolecule. **Lipids** are large and are made mostly of carbon and hydrogen, with a small amount of oxygen. Fats, oils, waxes, and steroids are all lipids. Lipids do not dissolve in water because their molecules are not attracted by water molecules. Water molecules do not attract lipids because lipids are nonpolar molecules. Lipids are used by cells for energy storage, insulation, and protective coatings, such as in membranes.

Another type of biomolecule is protein. **Proteins** are necessary for all life because they provide structure for tissues and organs and carry out cell metabolism (you learned in Section 6.1 that metabolism is all of the chemical reactions that occur within an organism). They provide the body with the ability to move muscles. They also are needed to transport oxygen in the bloodstream. Proteins are large and complex and are made up of carbon, hydrogen, oxygen, nitrogen, and sometimes sulfur.

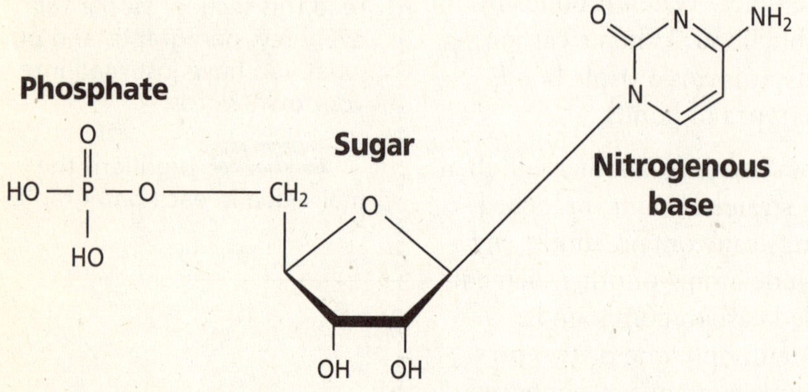

Think it Over

2. **Analyze** DNA is an example of a (Circle your choice.)
   a. nucleic acid.
   b. biomolecule.
   c. carbon compound.
   d. all of the above.

Enzymes are a particular type of protein. **Enzymes** change the speed of chemical reactions within the body. In some cases, enzymes speed up a reaction that would ordinarily take more time. For example, enzymes speed up the digestion of food.

A **nucleic** (noo KLAY ihk) **acid** is a biomolecule that stores cellular information in the form of a code. Nucleic acids are important compounds necessary for life. They are made of smaller units called nucleotides.

Nucleotides consist of carbon, hydrogen, oxygen, nitrogen, and phosphorus atoms. These atoms are arranged into three groups: a nitrogenous base, a simple sugar, and a phosphate group. The figure above shows the structure of nucleotides. Two important nucleic acids are deoxyribonucleic acid (DNA) and ribonucleic acid (RNA). DNA is an organism's master information code. DNA includes the instructions that determine how an organism looks and acts. RNA forms a copy of DNA to use in making proteins.

After You Read

Mini Glossary

carbohydrates: organic compounds used by cells to store and release energy; composed of carbon, hydrogen, and oxygen

enzyme: type of protein found in all living things that changes the speed of chemical reactions

lipids: large organic compounds made mostly of carbon and hydrogen with a small amount of oxygen; examples are fats, oils, waxes, and steroids; are insoluble in water and used by cells for energy storage, insulation, and protective coatings, such as in membranes

nucleic (noo KLAY ihk) acid: complex biomolecules, such as RNA and DNA, that store cellular information in cells in the form of a code

nucleotides: subunits of nucleic acid formed from a simple sugar, a phosphate group, and a nitrogenous base

proteins: large, complex biomolecules essential to all life composed of carbon, hydrogen, oxygen, nitrogen, and sometimes sulfur; provide structure for tissues and organs and help carry out cell metabolism

1. Review the terms and their definitions in the Mini Glossary above. Choose two terms that are related to each other. On the lines below, tell how these terms are related.

2. Complete the diagram with information you learned from reading the section.

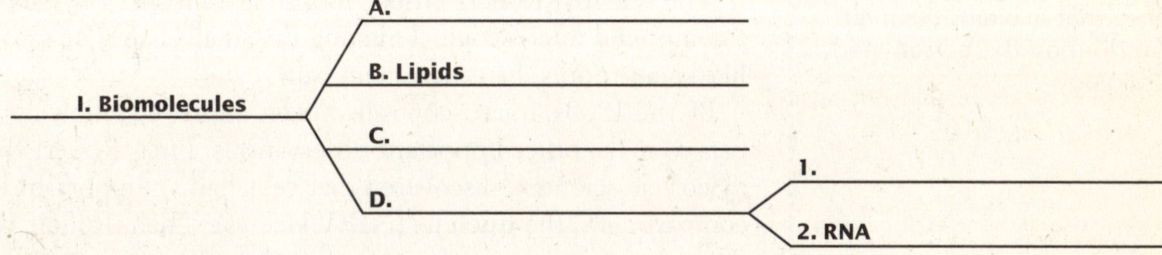

Visit the Glencoe Science Web site at **science.glencoe.com** to find your biology book and learn more about life substances.

Name Date Class

Section 7.1

The Discovery of Cells

Before You Read

This section introduces cells. Skim the reading below and find two important facts about cells. Write those two facts in the space below.

Read to Learn

Mark the Text **Identify Scientists** Underline each scientist's name introduced in this section. Say the name aloud. Then highlight the sentence that explains the main contribution the person made to biology.

The History of the Cell Theory

The invention of microscopes made it possible for scientists to view and study cells. **Cells** are the basic units of living organisms. In the 1600s, Anton van Leeuwenhoek (LAY vun hook) used a single lens microscope to view bacteria, which until then could not be seen. Later, **compound light microscopes** used several lenses and could magnify objects up to 1500 times their original size.

The scientist Robert Hooke looked at thin slices of cork under a compound microscope. Thinking the small shapes he saw looked like small rooms, he called them cells.

By the 1800s, microscopes had been improved, allowing scientists to make other important observations. First, Robert Brown, a Scottish scientist, discovered that cells had an important inner compartment, the **nucleus** (NEW klee us). Then, Rudolf Virchow figured out that the nucleus controls the cell's activities. Later, two German biologists, Matthias Schleiden and Theodor Schwann, did their own experiments and learned that all living things are made of one or more cells.

What is cell theory?

The experiments of Schleiden, Schwann, and other scientists led to the development of what is called the **cell theory.** It is one of the fundamental ideas of the science of biology. The three main parts of the cell theory are summarized below:

1. All living things are made of one or more cells.
2. Cells are the basic units of structure and function in living things.
3. All cells come from other cells.

How do microscopes help scientists learn about cells?

In the 1930s and 1940s, microscopes were improved. **Electron microscopes** allowed scientists to magnify an object up to 500 000 times using a beam of electrons instead of a beam of light. A scanning electron microscope (SEM) lets scientists see a cell's three-dimensional shape. A transmission electron microscope (TEM) lets scientists see the structures inside a cell.

Microscopes are continually being improved so scientists can gather more information about cells.

Two Basic Cell Types

Using microscopes, scientists saw that all cells contain small structures called **organelles.** Each organelle has a specific function in the cell. Some cell organelles are held together by a membrane, but others are not.

Scientists group cells into two categories—cells that have membrane-bound organelles and cells that do not. Cells that do not contain membrane-bound organelles are called **prokaryotes** (pro kar ee AWTS). Unicellular organisms, such as bacteria, are prokaryotes.

If the cell has organelles that are held together by a membrane, the cell is called a **eukaryote** (yew kar ee AWT). Most cells you can think of are eukaryotic. These include most of the multicellular organisms you know. Having membrane-bound organelles is an advantage for eukaryotic cells because chemical reactions in different parts of the cell can happen at the same time.

Eukaryotic cells have a central organelle called a nucleus that controls all of the cell's activities. Prokaryotes do not have an organized nucleus. Instead, they have loose strands of DNA.

Reading Check

1. What are the three main ideas of cell theory?

____

____

____

Think it Over

2. Compare Which cells are more complex? (Circle your choice.)

a. prokaryotic
b. eukaryotic

Reading Check

3. What does a nucleus do?

____

____

____

After You Read

Mini Glossary

cell: the basic unit of all living things

cell theory: theory that states all organisms are made of one or more cells; the cell is the basic unit of organisms; and all cells come from preexisting cells

compound light microscope: microscope using a series of light and lenses to magnify objects

electron microscope: microscope using a beam of electrons instead of lenses to magnify objects

eukaryote (yew kar ee AWT): unicellular or multicellular organisms (like yeast, plants, and animals) that contain a nucleus and membrane-bound organelles

nucleus (NEW klee us): the cell organelle that controls the cell's activities and contains DNA

organelle: membrane-bound structures with particular functions within some cells

prokaryote (pro kar ee AWT): unicellular organisms (like bacteria) that lack membrane-bound organelles

1. Circle two terms from the Mini Glossary above that are related to each other. On the lines below, tell how these terms are related.

____

____

2. Use the Venn diagram below to help you review what you have read. List what makes prokaryotic and eukaryotic cells different. Then list their common characteristics in the middle.

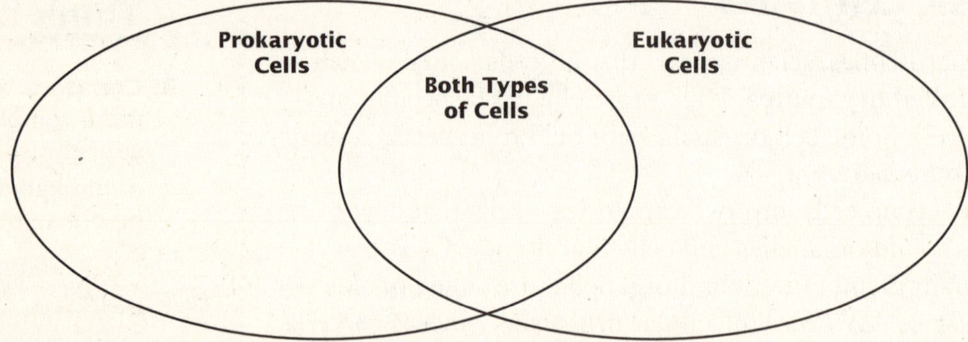

3. Write the three main ideas of the cell theory in the spaces below.

The Cell Theory

Visit the Glencoe Science Web site at **science.glencoe.com** to find your biology book and learn more about the discovery of cells.

Name Date Class

Section 7.2 The Plasma Membrane

Before You Read

A window screen in your home allows air to pass through while keeping insects out. In this section you will learn about a cell structure that has the same basic function: allowing some things to pass through while keeping other out. On the lines below, list some things that you think would be allowed to pass into a cell, some that would be allowed to pass out of the cell, and some that would be kept out of the cell.

Read to Learn

Maintaining a Balance

The **plasma membrane** is the flexible boundary of a cell that separates a cell from its surroundings. It allows nutrients to enter the cell and waste to be removed. Keeping this healthy balance within a cell is called homeostasis. ✓

To maintain homeostasis, the plasma membrane allows some molecules into the cell and keeps others out. This is called **selective permeability.** Some molecules are allowed in at any time. Other molecules are only admitted at certain times and in limited amounts. Others are not allowed in at all.

Reading Check

1. What is the function of the plasma membrane?

Structure of the Plasma Membrane

Earlier you learned that lipids are large molecules made up of glycerol and three fatty acids. A **phospholipid** (fahs foh LIH pid) is made up of glycerol, two fatty acids, and a phosphate group. The plasma membrane is made up of two layers of phospholipids arranged back to back in what is called a phospholipid bilayer.

The phosphate group is an important part of the plasma membrane. The group is found in the head of the phospholipid molecule. The head is polar. The tails of fatty acid chains hang from the head. The tails are nonpolar. The two phospholipid layers are arranged so the polar heads are facing out, and the nonpolar tails are facing in.

Think it Over

2. Which part of a phospholipid molecule contains the phosphate group? Which is nonpolar?

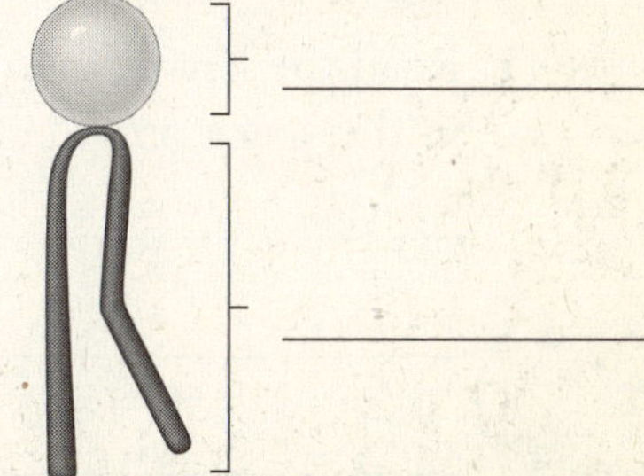

Plasma membrane

Phospholipids

Surface protein

Phospholipids

Transport protein

Reading Check

3. Name three elements that float in the plasma membrane.

What is the fluid mosaic model?

Water, which all living things need, is also polar. The polar phosphate heads on the surface of the membrane allow the polar water to interact with the membrane. Because they are nonpolar, fatty acid tails avoid water. This makes a barrier that is water-soluble on the outside of the membrane, but water-insoluble inside the membrane. This prevents water-soluble molecules from easily moving through the plasma membrane.

This organization of the plasma membrane is called the **fluid mosaic model.** It is fluid because the phospholipids are not fixed in one place, but float in the membrane. Protein molecules also float with the phospholipids. The model is called a mosaic because of the patterns the proteins create on the membrane's surface.

Specific proteins called **transport proteins** work to regulate which molecules are allowed to enter and which are allowed to leave the cell. Other proteins help cells identify chemical signals. Proteins on the inner surface of the membrane help support the cell structure.

Cholesterol also floats on the surface of the membrane. It helps keep the fatty acid tails from sticking together. Cholesterol is an important part of a healthy diet partly because of the role it plays in the plasma membrane.

After You Read

Mini Glossary

fluid mosaic model: structural model of the plasma membrane where phospholipids and proteins float within the surface of the membrane

phospholipid (fahs foh LIH pid): a large molecule with a glycerol backbone, two fatty acid chains, and a phosphate group

plasma membrane: the flexible boundary of a cell

selective permeability: a process in which a membrane allows some molecules to pass through while keeping others out

transport proteins: proteins that move needed substances or waste materials through the plasma membrane into or out of the cell

1. Read the key terms and definitions in the Mini Glossary above. Circle one key term. Then, in the space provided, write the definition of the term in your own words.

Section 7.2 The Plasma Membrane, *continued*

2. Use the partially completed outline below to help you review what you have read. Fill in the blanks where missing information is needed.

I. Parts of a phospholipid molecule

A. 2 fatty acids

B. 1 ______

C. 1 ______

II. Fluid mosaic model of plasma membrane

A. Called fluid because ______

B. Called mosaic because ______

3. Choose one of the following headings in the Read to Learn section: **Maintaining a Balance** or **Structure of a Plasma Membrane.**

Turn the heading into a question. Write the question in the space below. Then write your answer to that question on the lines that follow.

Question:

Answer:

______

______

______

Visit the Glencoe Science Web site at **science.glencoe.com** to find your book and learn more about the plasma membrane.

Section 7.3 Eukaryotic Cell Structure

Before You Read

In this section you will learn about the various parts of eukaryotic cells. For cells to function correctly, each part does its job. Think of a group that you belong to. On the lines below, list the members of the group. Consider how well the group functions when all members work together.

____

____

Read to Learn

Mark the Text **Restate the Main Point** Highlight the main idea in each paragraph. As you read this section, stop after every few paragraphs and put what you have just read into your own words.

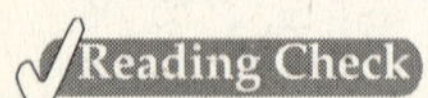

1. Which organelle in the nucleus makes ribosomes?

____

____

____

Cellular Boundaries

As you have already learned, the plasma membrane acts as the flexible boundary of the cell. In plant cells, fungi, bacteria, and some protists, there is an additional boundary—the cell wall. The **cell wall** is a rigid wall outside the plasma membrane. It gives extra support and protection to the cell.

The Nucleus and Cell Control

The nucleus directs the activity of a cell's organelles. It also contains directions to make proteins. Strands of DNA called **chromatin** are located within the nucleus. The chromatin tells the cell what kinds of proteins to make. When a cell divides, the chromatin condenses and becomes chromosomes.

How do ribosomes make protein?

The **nucleolus** inside the nucleus is an important organelle. It makes ribosomes. **Ribosomes** are where the cell makes protein according to the DNA directions. Ribosomes are not bound by a membrane. They are made up of RNA and protein. ✓

To make protein, ribosomes move out of the nucleus and into a thick fluid in the cell called **cytoplasm.** The nuclear envelope is the boundary between the nucleus and the cytoplasm. It is a membrane of two phospholipid bilayers. Ribosomes and copies of the DNA pass through this nuclear envelope on the way to the cytoplasm.

Assembly, Transport, and Storage

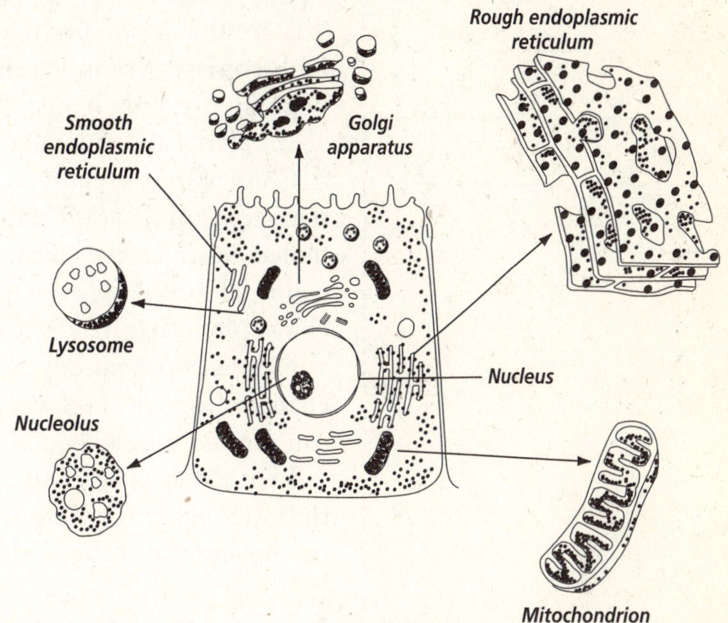

A eukaryotic cell contains the endoplasmic reticulum (ER). The ER is a series of tightly folded membranes floating in the cytoplasm.

What is the purpose of the ER?

The **endoplasmic reticulum** is the site of cellular chemical reactions. There are two types of ER. The areas where ribosomes in the cytoplasm attach to the ER's surface are called rough endoplasmic reticulum. The ribosomes' job is to make proteins. Each protein the ER makes has a specific job. Other ribosomes float freely in the cytoplasm and carry out specific tasks.

The surface of the ER that does not have ribosomes attached is called the smooth endoplasmic reticulum. The smooth ER also performs chemical activities. For example, it makes and stores lipids. ✓

What tasks are performed by other organelles?

Once the proteins are created, they move to another organelle, the Golgi (GAWL jee) apparatus. The **Golgi apparatus** sorts proteins and then packs them into structures called vesicles. The vesicles are then sent to where they are needed.

Another important part of the cell is the vacuole. A **vacuole** is a sac inside a membrane. Inside the vacuole, materials needed by the cell are temporarily stored. Some vacuoles store food and enzymes. Others store waste products until they can be removed from the cell.

Lysosomes have the job of removing the waste. They are organelles filled with digestive enzymes. Lysosomes join a vacuole and send their enzymes into it. The enzymes digest the contents.

STUDY COACH

Mark the Text **Identify the Parts** Place a check mark beside each part of the eukaryotic cell in the diagram above after you have read about it.

✓ Reading Check

2. What are the two types of endoplasmic reticulum?

____

____

Reading Check

3. What do chloroplasts do?

____

____

____

Energy Transformers

Cells need energy to do their work. Chloroplasts and mitochondria are the organelles that provide the energy that cells need. **Chloroplasts** are in green plants and some protists. They capture light energy and change it into chemical energy. **Mitochondria** are in plants and animals. They also transform energy for cells.

Chloroplasts are a type of plastid. **Plastids** are plant organelles that store things. Some store food, some store pigments—molecules that give color. In fact, plastids are named for their color. Chloroplasts contain **chlorophyll,** a green pigment. It is chlorophyll that makes leaves and stems green.

Plastids store energy in sugar molecule bonds. Mitochondria change the sugar molecule bonds into other bonds that organelles can use more quickly and easily to get their energy.

Like chloroplasts, mitochondria have an outer membrane and a tightly folded inner membrane. Molecules for storing energy are on the surface of the folds.

Organelles for Support and Locomotion

Inside the cell is the **cytoskeleton,** which supports the organelles in the cytoplasm. The cytoskeleton can be taken apart in one place and put back together in another. When this is done it changes the cell's shape.

Reading Check

4. What is the job of microtubules and microfilaments?

____

____

____

How does the cytoskeleton help cells move?

The cytoskeleton is a network of microtubules and microfilaments. **Microtubules** are tiny cylinders of protein. **Microfilaments** are even smaller solid protein fibers. They support the organelles and help materials move around the cell. Animal and protist cells also contain pairs of centrioles. Centrioles are made of microtubules. They are very important in cell division.

Two other organelles made of microtubules are cilia (SIH lee uh) and flagella (fluh JEL uh). These organelles help the cell to move and to feed. **Cilia** are short projections that look like hairs. They move in a wavelike motion. A cell has many cilia. **Flagella** are longer projections that move in a whip-like motion. A cell usually has one or two flagella. In one-celled organisms the cilia and flagella are the most important ways of moving.

After You Read

Mini Glossary

cell wall: rigid wall outside the plasma membrane for additional support and protection

chlorophyll: green pigment that traps light energy and gives leaves and stems their green color

chloroplast: organelles that capture light energy and convert it to chemical energy

chromatin: strands of DNA containing directions for making proteins

cilia: short, hair-like projections on a cell's surface, with an oar-like motion

cytoplasm: clear, jelly-like fluid inside a cell

cytoskeleton: cell support structure within the cytoplasm

endoplasmic reticulum: site of cellular chemical reactions

flagella: longer projections on a cell's surface, with a whip-like motion

Golgi apparatus: an organelle that sorts proteins into packages and packs them into vesicles

lysosome: organelles that remove waste from the cell

microfilament: tiny, solid protein fibers that are part of the cytoskeleton

microtubule: thin hollow cylinders made of protein that are part of the cytoskeleton

mitochondria: organelles in plants and animals that transform energy for the cell

nucleolus: organelle within the nucleus that makes ribosomes

plastid: a plant organelle used for storage

ribosome: site where DNA makes protein

vacuole: membrane-bound compartments for temporary storage of materials

1. Read the key terms and definitions in the Mini Glossary above. Highlight the name of an organelle that is found in plants but not animals. Then, in the space provided, explain what the organelle does.

____

____

2. Choose two structures that work closely together in the cell to perform an important task or tasks. In your own words, describe in the space below what the structures do and how they do it. Be sure to tell how the work of one structure relates to the work of the other.

____

____

____

____

____

____

____

Section 7.3 Eukaryotic Cell Structure, *continued*

3. Use the chart to help you review what you have learned. Identify the structure and specific task for the various parts of a eukaryotic cell.

Eukaryotic Cell Structure		
Cell Part	**Structure**	**Task**
cell wall		
chromatin		
nucleus		
nucleolus		
ribosomes		
cytoplasm		
nuclear envelope		
endoplasmic reticulum		
Golgi apparatus		
vacuole		
lysosome		

Visit the Glencoe Science Web site at **science.glencoe.com** to find your book and learn more about eukaryotic cell structure.

Name Date Class

Section 8.1 Cellular Transport

Before You Read

This section is about cell transport. You will learn how substances move in and out of cells. Think about the ways you have seen things being moved from place to place. When are things easily transported? When it is more difficult, what types of equipment might be used to help move things along? Write your thoughts on the lines below.

when the water moves it can easily move.

Read to Learn

Osmosis: Diffusion of Water

The plasma membrane of a cell is called a selectively permeable membrane. This means only certain particles, or molecules, are let in and out of a cell.

Water is the only substance the plasma membrane always allows in or out. The movement of water from an area of higher concentration to an area of lower concentration, or diffusion, across the plasma membrane is called **osmosis** (ahs MOH sus). This water flow through the membrane helps create homeostasis. Homeostasis is the regulation of the cell's internal environment. ✓

How does osmosis affect cells?

Most cells undergo osmosis because they are floating in water solutions. When a cell is floating in an isotonic solution, the water moves in and out of the cells at the same rate. In an **isotonic solution,** dissolved substances inside and outside of the cell have the same concentration. Because the amount of water moving in and out of the cells is the same, the cells keep their normal shape.

When a cell is floating in a hypotonic solution, more water enters the cell through osmosis than leaves it. In a **hypotonic solution,** the concentration of dissolved substances is lower outside of the cell than inside. The cell swells because of the extra water that enters. As a result, the cell's internal pressure increases.

In a hypertonic solution, water moves out of the cell during osmosis. When a cell is floating in a **hypertonic solution,** the concentration of dissolved substances outside of the cell is higher

Mark the Text **Read for Understanding** As you read this section, highlight any sentence that you reread. Reread any sentences you highlighted.

✓ **Reading Check**

1. What is the term for the flow of water across the plasma membrane?

Osmosis

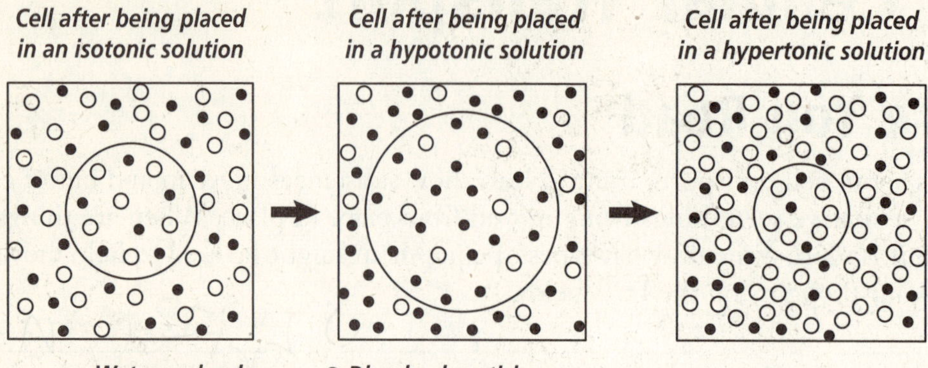

than inside of the cell. As water leaves the cell, the cell shrinks, and the pressure inside of the cell decreases.

2. What is the movement called when a cell does not use energy to move particles across the cell's membrane?

passive transport.

Passive Transport

Some molecules, like water, can pass through the plasma membrane by simple diffusion. When the cell uses no energy to move such particles, the movement is called **passive transport.** ✓

Special proteins in the plasma membrane move materials across the membrane. These are called transport proteins. When transport proteins help the passive transport of materials, the process is called **facilitated diffusion.**

Active Transport

In **active transport,** a transport protein called a carrier protein helps move particles across the membrane against a force. That force is called a concentration gradient. A concentration gradient develops when there are more molecules on one side of a membrane than the other. Because the transport protein has to work hard against the concentration gradient, energy is needed for the carrier protein to move the particles. The transport of substances across cell membranes is what helps cells maintain homeostasis.

Transport of Large Particles

Endocytosis (en doh si TOH sus) is the process in which a cell takes in material from its surroundings and then releases the material inside the cell. During endocytosis, a cell surrounds and takes in material from its environment. This material does not pass directly through the membrane. Instead, it is engulfed and enclosed by a portion of the cell's plasma membrane. It then breaks away and moves to the inside of the cell.

Exocytosis is the reverse process of endocytosis. It is the process of a cell taking material from inside itself and secreting or expelling it from the cell. The material can be wastes or other chemicals. The figure below illustrates endocytosis and exocytosis.

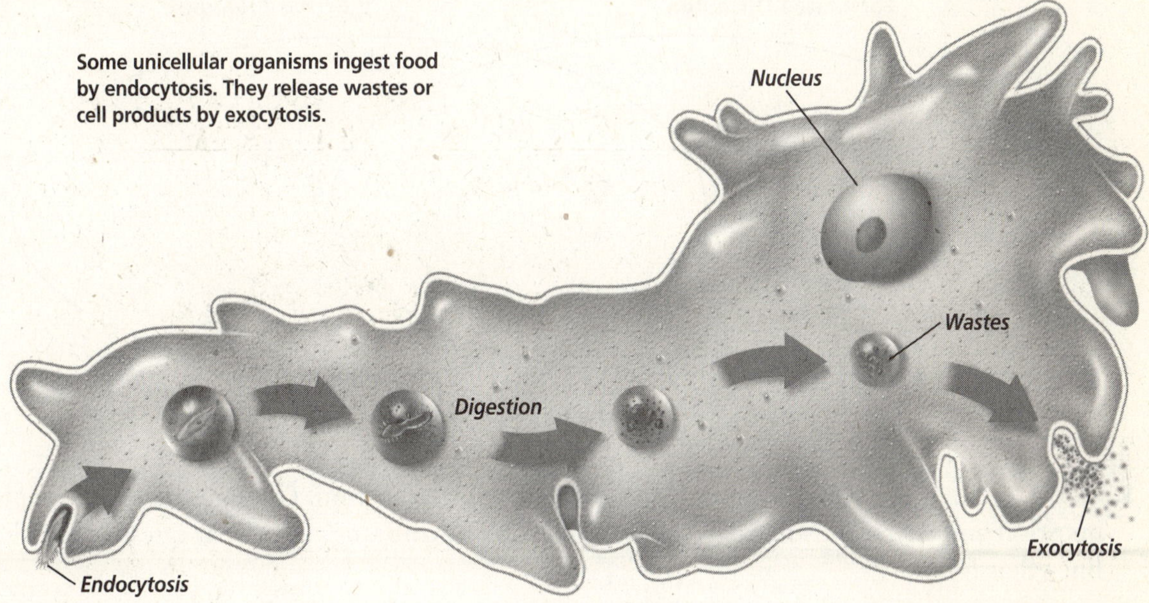

Some unicellular organisms ingest food by endocytosis. They release wastes or cell products by exocytosis.

After You Read

Mini Glossary

active transport: energy-needing process by which cells transport materials across the cell membrane against a concentration gradient

endocytosis (en doh si TOH sus): active transport process where a cell engulfs materials with a portion of the cell's plasma membrane and releases the contents within the cell

exocytosis: active transport process by which materials are secreted or expelled from a cell

facilitated diffusion: passive transport of materials across a plasma membrane by transport proteins embedded in the plasma membrane

hypertonic solution: in cells, solution in which the concentration of dissolved substances outside the cell is higher than the concentration inside the cell; causes a cell to shrink as water leaves

hypotonic solution: in cells, solution in which the concentration of dissolved substances is lower in the solution outside the cell than the concentration inside the cell; causes a cell to swell and possibly burst as water enters the cell

isotonic solution: in cells, solution in which the concentration of dissolved substances in the solution is the same as the concentration of dissolved substances inside a cell

osmosis (ahs MOH sus): diffusion of water across a selectively permeable membrane depending on the concentration of solutes on either side of the membrane

passive transport: movement of particles across cell membranes by diffusion or osmosis; the cell uses no energy to move particles across the membrane

Section 8.1 Cellular Transport, *continued*

1. Highlight two terms in the Mini Glossary on page 81 that identify specific types of active cell transport.
2. Use the Venn diagram to help you review what you have read about cell transport.

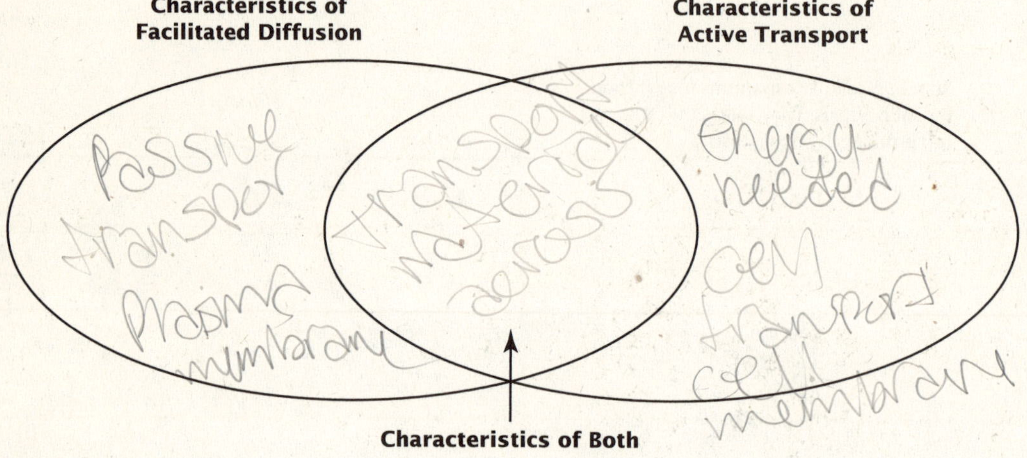

3. Choose one of the main headings in the Read to Learn section. Change the heading into a question and write it in the space below. Then write your answer to that question on the lines that follow.

Question: How does osmosis affect cells?

Answer: they are floating in the water solution

Visit the Glencoe Science Web site at **science.glencoe.com** to find your book and learn more about cellular transport.

Section 8.2 Cell Growth and Reproduction

Before You Read

In this section you will learn about the way cells grow and divide. Have you ever watched someone trim a bush or a tree? What would have happened if the bush or tree had not been trimmed? Write down a few of your thoughts on the lines below.

____

____

____

Read to Learn

Cell Size Limitations

As you have learned, the plasma membrane lets nutrients into the cell and allows wastes to leave. Inside the cell, nutrients and wastes move by diffusion. Because a cell's size can slow the rate of diffusion, cells have to have a way of limiting their growth. Fortunately, cells divide before they become too big and unable to function well. Cell division also has other purposes.

Mark the Text **Identify the Main Idea** As you read this section, stop after every paragraph and put what you read into your own words. Highlight the main idea in each paragraph.

Cell Reproduction

When cells divide, new cells are produced from one cell. The two cells that are produced are identical to the original cell. Just before cells divide, several short, stringy structures appear in the nucleus. These structures are called chromosomes. ✓

Reading Check

1. What structures appear in the nucleus shortly before cell division?

____

____

____

What do chromosomes do?

Chromosomes (KROH muh sohmz) contain DNA and are the carriers of the genetic material that is copied and passed from generation to generation. For most of a cell's lifetime, chromosomes exist as something called chromatin (KROH muh tihn). **Chromatin** is long, stringy strands of DNA. Without the proper amount of DNA, the cell cannot survive. Therefore the chromosomes must be accurately passed on to new cells.

The Cell Cycle

The **cell cycle** is the time of growth and division of a cell. A cell's life cycle is divided into two periods. There is a period of

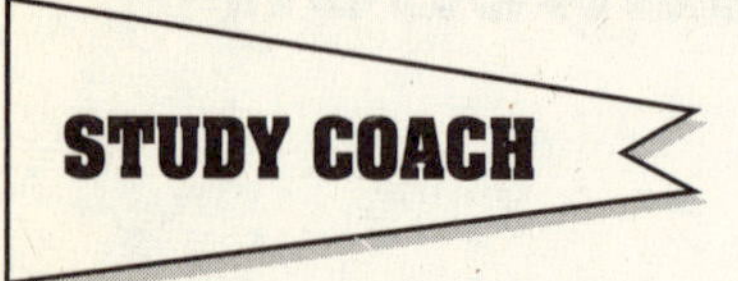

Mark the Text **Analyze the Diagram** As you read about mitosis, highlight each stage in the diagram below.

growth called interphase. There is also a period of nuclear division called mitosis.

What is interphase?

During **interphase** a cell grows in size, carries on metabolism, duplicates chromosomes, and prepares for division. Interphase is the busiest phase of the cell cycle.

What is mitosis?

Mitosis (mi TOH sus) follows interphase. It is the process of nuclear division in which two daughter cells form. Each of these daughter cells contains a complete set of chromosomes that are identical to those of the parent cell.

What are the phases of mitosis?

There are four phases of mitosis. Each phase merges into the next phase. The four phases are prophase, metaphase, anaphase, and telophase as shown in the illustration to the left.

Prophase is the first and longest phase. During prophase the chromatin coils up to form chromosomes. Each duplicated chromosome is made up of two identical halves, called **sister chromatids. Centromeres** (SEN truh meers) hold the sister chromatids together.

During **metaphase,** the second phase of mitosis, the doubled chromosomes are pulled to the center of the cell. **Anaphase** is the third phase of mitosis. During this phase, the centromeres of the sister chromatids split apart. This separates the sister chromatids from each other. In **telophase,** the last phase of mitosis, the chromatids move to opposite sides of the cell. Two nuclei are formed—one on each side of the cell. Finally, a new double membrane begins to form between the two new nuclei.

Stages of Mitosis

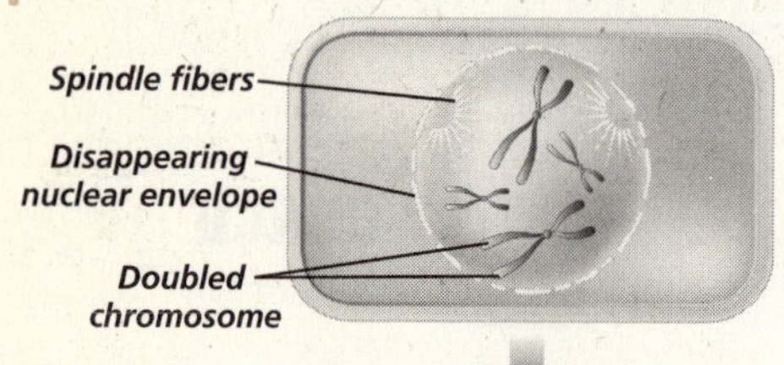

A Prophase The chromatin coils to form chromosomes.

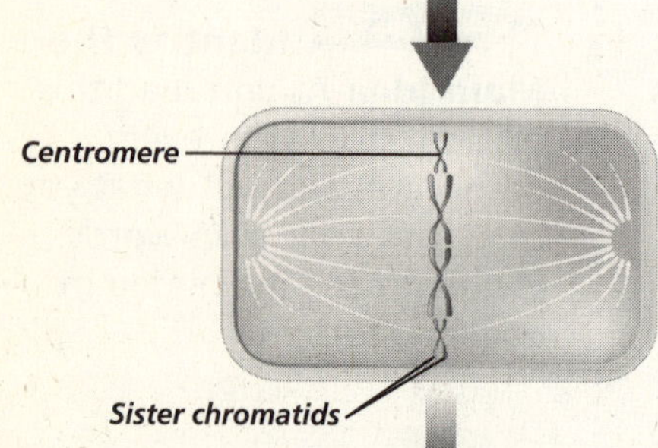

B Metaphase The chromosomes move to the center of the cell.

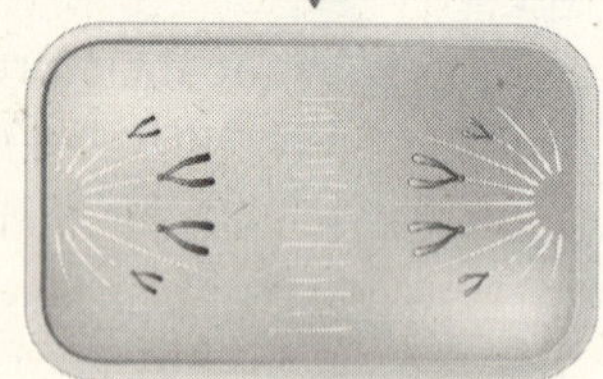

C Anaphase Centromeres split and sister chromatids are pulled to the opposite sides of the cell.

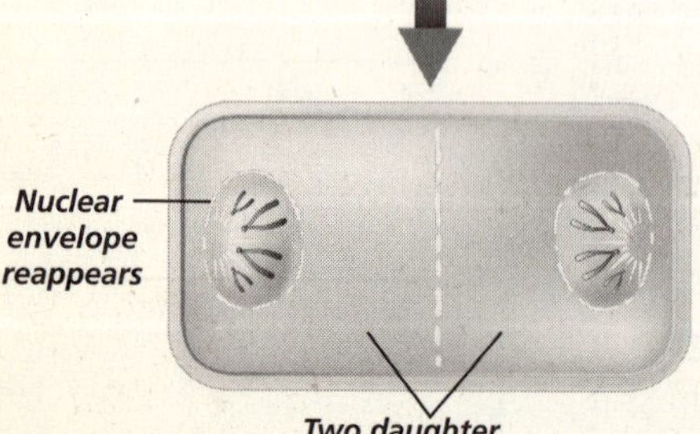

D Telophase Two new nuclei are formed and a double membrane begins to form between them.

Cytokinesis

Following telophase, the cell's cytoplasm divides and separates into two new identical cells. This is called **cytokinesis** (si toh kih NEE sus).

Results of Mitosis

When mitosis is complete, one-celled organisms remain as single cells. The organism simply multiplied into two organisms. These daughter cells eventually will repeat the same cell cycle as the parent cell and will grow and divide. In larger organisms, cell growth and reproduction result in groups of cells that work together as **tissue** to perform a certain function. Tissues organize in combinations to form **organs.** Organs perform specific complex tasks within the organism. Multiple organs that work together form an **organ system,** such as the digestive system. The stomach is one organ in the digestive system. It functions to digest food. It is important to remember that no matter how complex the organ system or organism becomes, the cell is still the most basic unit of that organization. ✓

Reading Check

2. What do tissues organize to form?

____

____

____

After You Read

Mini Glossary

anaphase: the third phase of mitosis in which the centromeres split and the sister chromatids of each chromosome are pulled apart

cell cycle: continuous sequence of growth (interphase) and division (mitosis) in a cell

centromere (SEN truh meer): cell structure that joins two sister chromatids of a chromosome

chromatin (KROH muh tihn): long strands of DNA found in the eukaryotic cell nucleus; coils up to form chromosomes

chromosomes (KROH muh sohmz): cell structures that contain DNA and carry the genetic material that is copied and passed from generation to generation of cells

cytokinesis (si toh kih NEE sus): cell process following mitosis in which the cell's cytoplasm divides and separates into new identical cells

interphase: cell growth period where a cell increases in size, carries on metabolism, and duplicates chromosomes prior to division

metaphase: short second phase of mitosis where doubled chromosomes move to the center of the cell

mitosis (mi TOH sus): period of nuclear cell division in which two daughter cells are formed, each containing a complete set of chromosomes

organ: group of two or more tissues organized to perform complex activities within an organism

organ system: multiple organs that work together to perform a specific life function

prophase: first and longest phase of mitosis where chromatin coils into visible chromosomes

Section 8.2 Cell Growth and Reproduction, *continued*

sister chromatid: identical half of duplicated parent chromosome formed during the prophase stage of mitosis; the halves are held together by a centromere

telophase: final phase of mitosis during which new cells prepare for their own independent existence

tissue: groups of cells that work together to perform a specific function

1. Circle the terms from the Mini Glossary that identify the phases of mitosis. Then, in the space provided, list them in the order they occur.

   1. ______________________ 3. ______________________

   2. ______________________ 4. ______________________

2. Use the diagram below to help you review the cell cycle by providing two facts for each period.

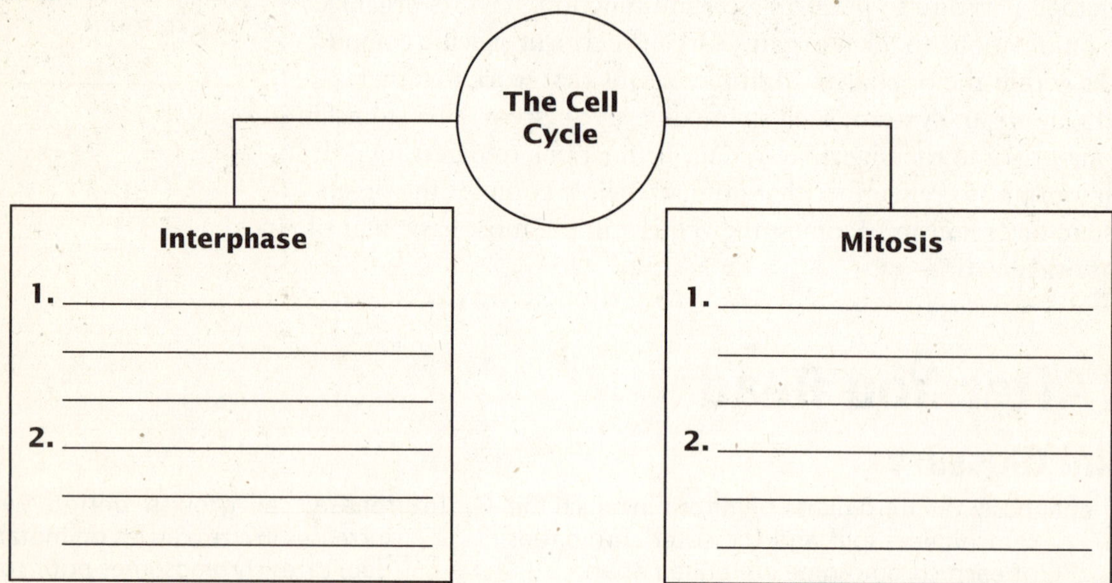

3. Fill in the blanks with the following terms: **tissues, sister chromatids, cell cycle, mitosis,** and **chromosomes.**

   a. Two identical halves of a duplicated parent chromosome are called ______________.

   b. The process of cell division is called ______________.

   c. ______________ are the carriers of genetic material that is copied and passed from generation to generation.

   d. The ______________ is the period of growth and division of a cell.

   e. ______________ are groups of cells that work together to perform specific functions.

Visit the Glencoe Science Web site at **science.glencoe.com** to find your biology book and learn more about cell growth and reproduction.

Name Date Class

Section 8.3 Control of the Cell Cycle

Before You Read

This section is about cancer and what happens when cells grow out of control. Experts agree that a healthful diet can reduce the risk of cancer. On the lines below list foods that you think belong in a healthful diet.

____

____

____

Read to Learn

Normal Control of the Cell Cycle

The cell cycle is controlled by conditions both inside the cell and in the cell's environment. When something goes wrong with normal cell conditions, cells lose control of the cell cycle.

Cancer is a growth that occurs when uncontrolled cell division takes place. This loss of control may be caused by changes in enzyme production inside the cell or by some outside factor, such as air or water pollution. An enzyme, you will remember, is a type of protein found in all living things that changes the rate of chemical reactions. ✓

Summarize As you read this section, highlight the main point in each paragraph. Then, write a short paragraph summarizing what you have learned.

Cancer: A Mistake in the Cell Cycle

Scientists think that cancer is caused by changes to one or more of the genes that produce the substances that help to control the cell cycle. A **gene** is a part of DNA that controls the production of a protein. Cancer occurs when something causes the damaged genes to go into action. When that happens, cancerous cells form masses of tissue called tumors. Tumors keep normal cells from getting needed nutrients. This causes damage to organs. Cancer cells can spread throughout the body to other organs as well.

✓ Reading Check

1. What occurs when uncontrolled cell division takes place?

____

____

____

What are the causes of cancer?

Cancer is the second leading cause of death in the United States. Only heart disease leads to more deaths each year. But the causes of cancer are hard to pinpoint. This is because both environmental and genetic factors play a part. We do know that many environmental factors damage the genes that control the cell cycle. Environmental factors include cigarette smoke, air and water pollution, and exposure to ultraviolet radiation from the sun. Cancer may also be caused by genetic factors such as viral infections that damage the genes.

Think it Over

2. **Infer** Which snack is more healthful? (Circle your choice.)
   a. an apple
   b. a chocolate candy bar

How can cancer be prevented?

A healthful lifestyle reduces the risk of cancer. Experts agree that diets low in fat and high in fiber content can reduce the risk for many kinds of cancer. Fruits, vegetables, and grain products are good food choices because they are low in fat and high in fiber. Other healthful choices such as daily exercise and not using tobacco also are known to lower the risk of cancer.

After You Read

Mini Glossary

cancer: uncontrolled cell division that may be caused by environmental factors and/or changes in enzyme production in the cell cycle

gene: segment of DNA that controls protein production and the cell cycle

1. Write a sentence using both of the terms in the Mini Glossary above.

____

____

2. Place the following events in the appropriate box based on the order in which they occur:
Uncontrolled cell division occurs.
Cancerous cells form tumors that deprive normal cells of nutrients.
Cancer occurs.
Environmental factors or viral infections damage genes.

1.	2.
3.	4.

Visit the Glencoe Science Web site at **science.glencoe.com** to find your biology book and learn more about control of the cell cycle.

Section 9.1 The Need for Energy

Before You Read

This section tells about the need our bodies have for energy. Think about the many ways your body uses energy. When do you need energy? Write a list of these times. After you read this section, add any other times that you learned about.

____

____

____

____

Read to Learn

Create a Quiz After you have read this section, create a five-question quiz based on what you have learned. After you have written the questions, be sure to answer them.

Cell Energy

We need energy to live. In fact, all living organisms need energy to live. Plants and other green organisms are able to make energy from sunlight and store it to use later. Some organisms, like animals, cannot make their own energy. They must eat other organisms to get the energy they need.

Many of the things our bodies do, called cell processes, need energy. Some cell processes are muscles contracting during exercise, your heart pumping, and cell division. Your brain also needs energy to do its work.

How do our cells get energy?

After exercising, your body needs a quick source of energy. Perhaps you eat a granola bar to satisfy the need. The cells in our bodies often need a quick source of energy. There is a molecule in your cells called **adenosine triphosphate** (uh DEH nuh seen • tri FAHS fayt), or **ATP** for short which provides quick energy for cells when they need it. ✓

Reading Check

1. What does ATP do?

____

____

____

Forming and Breaking Down ATP

ATP has an adenosine molecule, a ribose sugar, and three phosphate groups held together by chemical bonds. When one of the chemical bonds is broken, one of the phosphate groups is released. Energy is also released. This quick release of energy is then available for a cell to use.

Section 9.1 The Need for Energy, *continued*

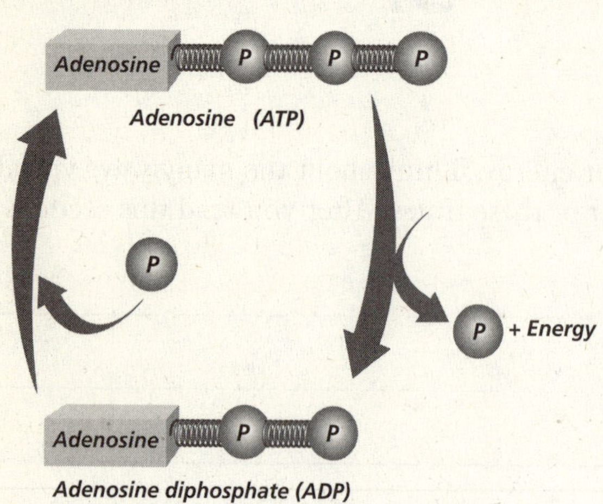

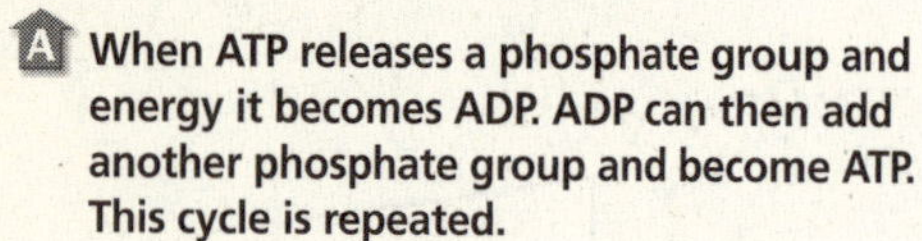

A When ATP releases a phosphate group and energy it becomes ADP. ADP can then add another phosphate group and become ATP. This cycle is repeated.

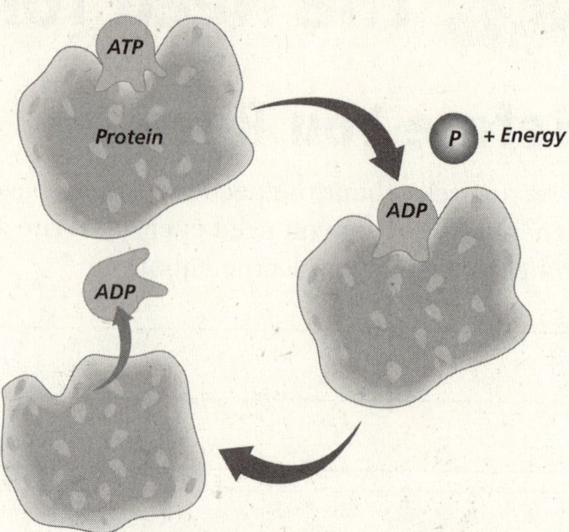

B A protein binds to ATP and breaks the chemical bond, releasing energy and a phosphate group. ATP becomes ADP and is released from the protein. The energy is used by the cell. This cycle is repeated.

When one of the phosphate groups is released, ATP becomes **adenosine diphosphate,** or **ADP.** ADP has only two phosphate groups. ADP can add another phosphate group and become ATP again. The cycle of the formation and breakdown of ATP creates a source of energy. The figure above shows the cycle of ATP.

How do cells get the energy they need from ATP?

Many proteins have a special place where ATP can bind itself. When ATP releases its energy by breaking the phosphate bond, the cell uses the energy. After releasing the energy, ATP becomes ADP and is released from the protein. As you have learned, at this point ADP can bind with another phosphate group and form ATP again. This cycle is repeated, providing a renewable source of energy for the cell.

Uses of Cell Energy

Cells use the energy they receive from ATP in many ways. Some cells make new molecules with the energy. Other cells use the energy to build membranes and cell organelles. Some cells use energy to maintain homeostasis, which is the regulation of their internal environment. Kidneys use energy to eliminate wastes from the bloodstream. At the same time the kidneys are eliminating wastes, they are using energy to keep needed substances in the bloodstream.

Think it Over

2. **Analyze** A cell's internal environment is kept stable through (Circle your choice.)
   a. ATP.
   b. homeostasis.
   c. ADP.

After You Read

Mini Glossary

ADP (adenosine diphosphate): molecule formed from the releasing of a phosphate group from ATP; results in a release of energy that is used for biological reactions

ATP (adenosine triphosphate): energy-storing molecule in cells composed of an adenosine molecule, a ribose sugar, and three phosphate groups; energy is stored in the molecule's chemical bonds and can be used quickly and easily by cells

1. On the lines below, tell how the two terms in the Mini Glossary above are related.

____

____

____

2. Use the diagram to help you review what you have read. Fill in the boxes to show the complete formation and breakdown cycle of ATP.

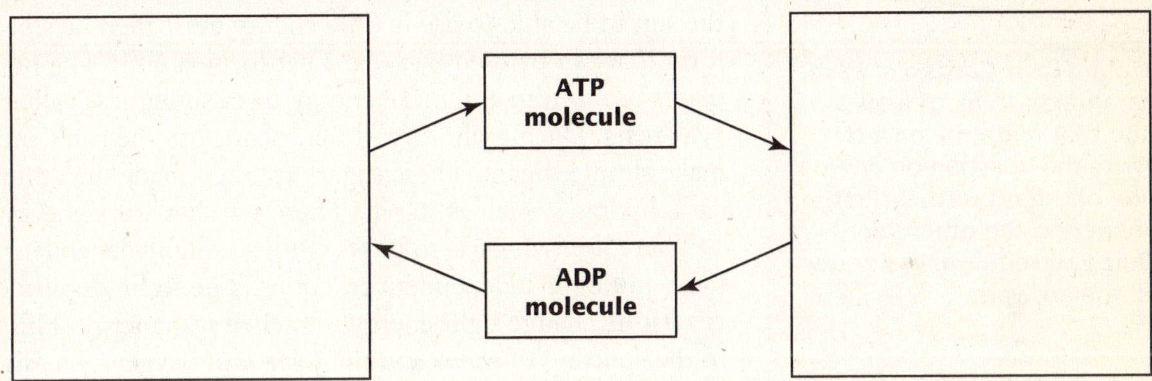

3. Fill in the missing number in the statements that describe the formation and breakdown of ATP.

ATP contains _____ phosphate group(s).	_____ phosphate group(s) is released.	ADP contains _____ phosphate group(s).	ADP adds _____ phosphate group(s) to become ATP.

Visit the Glencoe Science Web site at **science.glencoe.com** to find your biology book and learn more about the need for energy.

Section 9.2 Photosynthesis: Trapping the Sun's Energy

Before You Read

This section tells how plants trap sunlight to make energy. The process, called photosynthesis, is one of the key terms for this section. Highlight each of the key terms in the Read to Learn section. Then use a different color to highlight the definition of each key term.

Read to Learn

Make Flash Cards For each paragraph, think of a question that might be on a test. Write the question on one side of a flash card. Write the answer on the other side. Quiz yourself until you know all the answers.

1. What are the two parts of photosynthesis?

____

____

Trapping Energy from Sunlight

Plants and other green organisms must trap light energy from the sun to be able to use it. The energy must then be stored in a form that can be used by cells. That form is ATP. The process plants use to trap and make energy from sunlight is called **photosynthesis.** During photosynthesis, plants use the sun's energy to make simple sugars. These sugars are then made into complex carbohydrates, such as starch. These starches store energy.

There are two parts to photosynthesis: light-dependent reactions and light-independent reactions. The **light-dependent reactions** change light energy into chemical energy. This results in the splitting of water and the release of oxygen. The **light-independent reactions** produce simple sugars. ✓

Where does photosynthesis take place?

The chloroplast is the part of the plant's leaf where photosynthesis takes place. The chloroplast contains **pigments,** which are molecules that take in specific wavelengths of sunlight. Wavelengths of sunlight transfer energy. The most common pigment in the chloroplast is chlorophyll.

Chlorophyll is a plant pigment that absorbs most wavelengths of sunlight except green. Because it cannot take in the green wavelength, it reflects the green. This is what makes leaves look green. In the fall, leaves stop producing chlorophyll, so other pigments become visible. This gives leaves a wide variety of colors.

Light-Dependent Reactions

The first phase of photosynthesis needs sunlight. The sunlight strikes the chlorophyll in the plant's leaves and the energy from the sunlight is transferred to electrons in the chlorophyll. The electrons move from the chlorophyll to an electron transport chain. An **electron transport chain** is a line of proteins embedded in a membrane along which the electrons are passed down. Each protein in the line passes the electron to the next protein. As the electrons pass along this line, they lose some of their energy. If you filled a bucket with water and passed it along a line of people very quickly, some of the water would spill out. This is similar to how the electrons lose their energy.

Reading Check

2. What happens during the Calvin cycle?

___

___

___

What happens to the lost energy?

The energy the electrons lose can be used to form ATP from ADP, which you learned about in the last section. Energy that is not used to form ATP can be stored for use in the light-independent reactions. The energy is stored in an electron carrier called NADPH, shown at the right. NADPH carries the energy to the light-independent reactions.

Light-Dependent Reactions

Light energy transfers to chlorophyll.

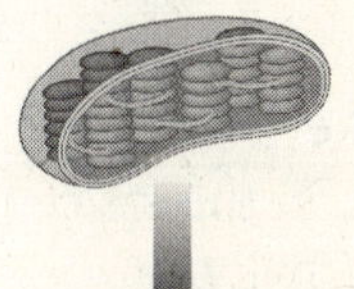

Chlorophyll passes energy down through the electron transport chain.

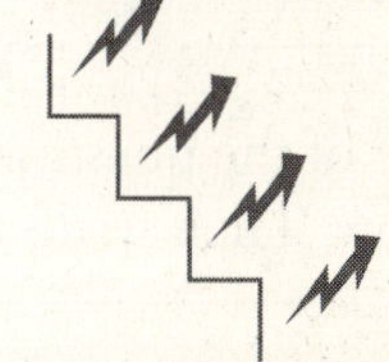

Energized electrons provide energy that

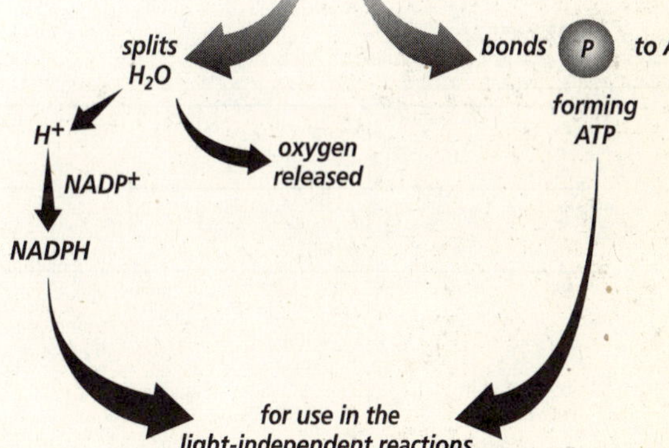

Light-Independent Reactions

The second phase of photosynthesis does not need sunlight. It takes place in the chloroplast and is called the Calvin cycle. The **Calvin cycle** uses carbon dioxide to form sugars. The sugars then become stored energy. It is a large task for a cell to make sugars from carbon dioxide. The ATP and NADPH produced in the light-dependent reactions are used. The chloroplast breaks down this large task into very small steps. The end result is energy that is stored in the plant as sugars. Organisms that eat plants use these sugars to give them energy. The energy is used in cellular respiration, which you will learn about in the next section. ✓

After You Read

Mini Glossary

Calvin cycle: series of reactions during the light-independent phase of photosynthesis in which sugars are formed from carbon dioxide using ATP and NADPH from the light-dependent reactions

chlorophyll: light-absorbing pigment in plants and some other green organisms that is required for photosynthesis; absorbs most wavelengths of light except for green

electron transport chain: series of proteins embedded in a membrane along which energized electrons are transported; as electrons are passed from molecule to molecule, energy is released

light-dependent reactions: phase of photosynthesis where light energy is converted to chemical energy in the form of ATP; results in the splitting of water and the release of oxygen

light-independent reactions: phase of photosynthesis where energy from light-dependent reactions is used to produce sugar and additional ATP molecules

photosynthesis: process by which plants and other green organisms trap energy from sunlight with chlorophyll and use this energy to convert carbon dioxide and water into simple sugars

pigments: molecules that absorb specific wavelengths of sunlight

1. Read the key terms and their definitions in the Mini Glossary above. Circle one of the key terms and, in the space provided, define it in your own words.

____

____

____

2. Select one of the question headings in the Read to Learn section. Write the question in the space below. Then write your answer to that question on the lines that follow.

Question:

Answer:

____

____

____

3. Complete the concept diagram below using the following terms: **ATP, Calvin cycle, chlorophyll, electron transport chain, light-dependent reactions, light-independent reactions,** and **NADPH.**

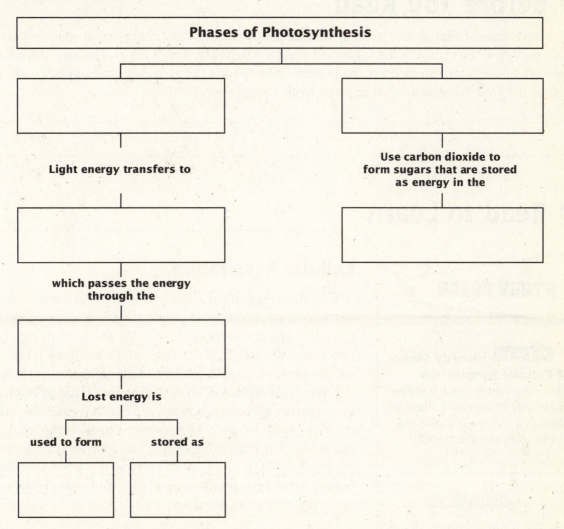

Phases of Photosynthesis

Light energy transfers to

which passes the energy through the

Lost energy is

used to form

stored as

Use carbon dioxide to form sugars that are stored as energy in the

Visit the Glencoe Science Web site at **science.glencoe.com** to find your biology book and learn more about photosynthesis.

Name Date Class

Section 9.3 Getting Energy to Make ATP

Before You Read

You have learned that during photosynthesis, plants use the sun's energy to produce sugars that provide energy. Plants convert the sugars into carbohydrates to store energy. Humans cannot directly convert sunlight into energy. Yet the human body can use the energy stored in plants. On the lines below, list your favorite foods that come from a plant source.

____

____

Read to Learn

Mark the Text **Identify Stages of Cellular Respiration** Circle the name of each stage of cellular respiration. Place a number (1, 2, or 3) in the margin to indicate the stage.

1. What are the three stages of cellular respiration?

____

____

____

Cellular Respiration

Cellular respiration is the process in which mitochondria break down food molecules to produce ATP. You will remember that mitochondria are organelles in plants and animals that transform energy for the cell. Cellular respiration has three stages: glycolysis, the citric acid cycle, and the electron transport chain. ✓

Glycolysis (gli KAH lih sis) is the first stage of cellular respiration. During glycolysis, a series of chemical reactions take place in the cytoplasm of a cell. The reactions break down glucose, which is a six-carbon sugar compound. The glucose is broken down into pyruvic (pie RUE vik) acid. Pyruvic acid is a three-carbon compound. Glycolysis produces two ATP molecules for every glucose molecule that is broken down.

What happens after glycolysis?

After glycolysis is complete, the pyruvic acid molecules move into the mitochondria.

The second stage of cellular respiration is called the citric acid cycle. The **citric acid cycle,** also called the Krebs cycle, is a series of chemical reactions similar to the Calvin cycle, which you learned about in the last section. Like the Calvin cycle, the molecule used in the first reaction of the citric acid cycle is also one of the end products. For every turn of the citric acid cycle, one molecule of ATP and two molecules of carbon dioxide are produced.

The electron transport chain is the third stage of cellular respiration. The electron transport chain in cellular respiration is like the electron transport chains in the first phase of photosynthesis. The electrons are passed from protein to protein in the electron transport chain and lose energy as they move down the chain. Some of this energy is used to make ATP. Some is used by an enzyme to pump H^+ ions into the center of the mitochondrion. As a result, the mitochondrion inner membrane becomes positively charged because of the high concentration of positively charged hydrogen ions. The gradient of H^+ ions that results across the inner membrane of the mitochondrion provides the energy for ATP production.

The final electron acceptor at the bottom of the chain is oxygen, which reacts with four hydrogen ions ($4H^+$) and four electrons to form two molecules of water (H_2O). This is why oxygen is so important to our bodies. Without oxygen, the proteins in the electron transport chain cannot pass along the electrons. If a protein cannot pass along an electron to oxygen, it cannot accept another electron. Very quickly, the entire chain becomes blocked and ATP production stops.

Fermentation

Sometimes your cells do not get oxygen for a short time. This can happen during heavy exercise. When this happens, an **anaerobic** (a nuh RO bik) process called fermentation takes place. Anaerobic processes are those that do not need oxygen. ✓

Fermentation follows glycolysis. It interrupts the process of cellular respiration. It allows the cells to continue making ATP until oxygen is available again.

Reading Check

2. When might your body perform fermentation processes?

____

____

____

How does lactic acid fermentation work?

You know that under anaerobic conditions, the electron transport chain backs up because oxygen is not present as the final electron acceptor. **Lactic acid fermentation** is a process that supplies energy when oxygen is not available. During lactic acid fermentation, pyruvic acid uses NADH to form lactic acid and NAD^+, which is then used in glycolysis. This allows two ATP molecules to be formed for each glucose molecule. The lactic acid is transferred from muscle cells, where it is produced during strenuous exercise, to the liver, which converts it back to pyruvic acid. The lactic acid that builds up in muscle cells results in muscle fatigue.

What is alcoholic fermentation?

Alcoholic fermentation is a process used by some types of bacteria and yeast cells to convert pyruvic acid into carbon dioxide (CO_2) and ethyl alcohol. Bread gives us an example of alcoholic fermentation. The yeast cells in bread give off CO_2, which makes bubbles in the dough. These bubbles make the bread dough rise.

Comparing Photosynthesis and Cellular Respiration

Reading Check

3. List two ways photosynthesis and cellular respiration are similar.

____

____

____

____

As you remember, photosynthesis produces food molecules and cellular respiration breaks down the food molecules. These two processes are very different, but they are also similar in some ways.

Both photosynthesis and cellular respiration use electron carriers and a cycle of chemical reactions to form ATP. Both also use electron transport chains in their processes.

Even though these two processes use some similar tools, the results are very different. Photosynthesis makes high-energy carbohydrates and oxygen. Cellular respiration breaks down carbohydrates to make ATP and compounds that provide less energy. Also, one of the end products of cellular respiration is CO_2, which is one of the beginning products for photosynthesis. The oxygen produced during photosynthesis is a critical molecule necessary for cellular respiration. The table below compares photosynthesis and cellular respiration.

Photosynthesis	Cellular Respiration
Food synthesized	Food broken down
Energy from sun stored in glucose	Energy of glucose released
Carbon dioxide taken in	Carbon dioxide given off
Oxygen given off	Oxygen taken in
Produces sugars from PGAL	Produces CO_2 and H_2O
Requires light	Does not require light
Occurs only in presence of chlorophyll	Occurs in all living cells

After You Read

Mini Glossary

alcoholic fermentation: anaerobic process in which cells convert pyruvic acid into carbon dioxide and ethyl alcohol; carried out by many bacteria and fungi such as yeasts

anaerobic (a nuh RO bik): chemical reactions that do not require oxygen

cellular respiration: chemical process where mitochondria break down food molecules to produce ATP; the three stages of cellular respiration are glycolysis, the citric acid cycle, and the electron transport cycle

citric acid cycle: in cellular respiration, a series of chemical reactions that break down glucose and produce ATP; energizes electron carriers that pass the energized electrons on to the electron transport chain

glycolysis: in cellular respiration, a series of anaerobic chemical reactions in the cytoplasm that break down glucose into pyruvic acid; forms a net profit of two ATP molecules

lactic acid fermentation: a series of anaerobic chemical reactions in which pyruvic acid uses NADH to form lactic acid and NAD^+, which is then used in glycolysis; supplies energy when oxygen for aerobic respiration is scarce

1. Read the key terms and definitions in the Mini Glossary above. List two terms that are anaerobic chemical reactions.

____

____

2. Use the partially completed outline below to help you review what you have read about cellular respiration. Fill in the blanks where missing information is needed.

I. Glycolysis

A. Chemical reactions break down ______________________.

B. ______________________ is formed.

II. Citric acid cycle

A. Series of ______________________

B. ______________________ used in first reaction is also available at end of cycle.

C. Also called ______________________.

III. Electron transport chain

A. ______________________ are passed from protein to protein.

B. ______________________ lose energy.

C. ______________________ formed.

Visit the Glencoe Science Web site at **science.glencoe.com** to find your biology book and learn more about getting energy to make ATP.

Section 10.1 Mendel's Laws of Heredity

Before You Read

Gregor Mendel was a mathematician and a monk. Mendel used his skills in math to understand why some characteristics are passed from parent to offspring and other characteristics are not. On the lines below, tell why you think having math skills is important in life.

Read to Learn

Mark the Text **Identify Concepts** Highlight each question head in this section. Then use a different color to highlight the answers to the questions.

1. What is the difference between self-pollination and cross-pollination?

Why Mendel Succeeded

Gregor Mendel, an Austrian monk, discovered important facts about heredity. **Heredity** is the passing on of characteristics from parents to offspring. These characteristics are called **traits.** Mendel was the first person to predict which traits would be passed from parents to offspring. The study of heredity is called **genetics.**

Mendel used garden peas for his experiments. Garden peas produce male and female sex cells called **gametes. Fertilization** occurs when the male sex cell unites, or joins, with the female sex cell. The united gametes form a new fertilized cell called a **zygote** (ZI goht). The zygote becomes part of a seed.

In garden peas, as with most flowers, the male sex cells are the grains of pollen. When pollen is transferred from the male reproductive organ to the female reproductive organ, it is called **pollination.** Garden peas are self-pollinators. That means the pollen from a flower pollinates the female sex cells within that same flower. The seeds that develop carry the traits of that plant. This was important for Mendel. When he wanted to have the gametes of different plants unite, Mendel opened the petals of a flower and removed the male reproductive organs on one plant, and dusted the female reproductive organ with the pollen from a different plant. This is cross-pollination. The seeds that develop from cross-pollination have traits of the two different plants. ✓

How did Mendel proceed?

To get accurate results, Mendel needed to carefully control his experiment. He studied one trait at a time to control variables. He decided to study how height in pea plants is passed from parent to offspring. He used plants that were true breeding. That meant the plants always passed the same trait from parent to offspring. He took pollen from a true-breeding tall pea plant and cross-pollinated a true-breeding short pea plant.

Mendel's Monohybrid Crosses

Crossing a tall pea plant with a short pea plant produced offspring called hybrids. A **hybrid** is the offspring of parents that have different forms of a trait, such as tall and short height. The first hybrids that Mendel produced are known as monohybrid crosses. *Mono* means one. Since the parent plants that Mendel used differed from each other by only one trait—height—the offspring are called monohybrids. ✓

Reading Check

2. What is a monohybrid?

____

____

____

What were Mendel's results?

The results of Mendel's experiment were interesting. Mendel cross-pollinated a six-foot tall pea plant with a pea plant less than two feet tall. When he planted the seeds from this cross, all of the offspring grew as tall as the taller parent plant. The short trait did not appear at all.

Mendel allowed the offspring, known as the first generation, to self-pollinate. Mendel planted the seeds from the first generation. There were more than 1000 plants in the second generation. Three-fourths of the plants were as tall as the tall parent plant. One-fourth of the plants were as tall as the short parent plant. The short trait had reappeared. The ratio of tall to short plants in the second generation was three tall plants for every one short plant.

How do you identify the generations?

In genetics, abbreviations are used for the generations. The original parents are known as the P_1 generation. *P* stands for "parent." The offspring of the parents are called the F_1 generation. The *F* stands for "filial," which means son or daughter. When you cross two F_1 plants with each other their offspring are the F_2 generation, the second filial generation. ✓

Reading Check

3. List the abbreviations for each generation and tell what they stand for.

____

____

____

Section

10.1 Mendel's Laws of Heredity, *continued*

Mendel did similar monohybrid crosses with seven pairs of traits. He used traits such as whether a seed was wrinkled or round, yellow or green. In every case he found that one trait of a pair did not appear in the F_1 generation. Then the trait reappeared in one fourth of the F_2 plants. For example, when he crossed a plant that produced round seeds with a plant that produced wrinkled seeds, all of the offspring (F_1) had round seeds. But when the plants of the F_1 generation self-pollinated, one fourth of their offspring (F_2) had wrinkled seeds.

From these results Mendel determined that each organism has two factors that control each of its traits. Today we know that these factors are genes. Genes exist in alternative forms, such as, tall and short, round and wrinkled. The alternative forms are called **alleles** (uh LEELZ). Mendel's pea plants had two alleles for height. A plant could have two alleles for tallness, two alleles for shortness, or one allele for tallness and one allele for shortness. The organism receives or inherits one allele from the female parent and one allele from the male parent.

Reading Check

4. If a plant has a dominant allele for height and a recessive allele for height, which allele will determine the height of the plant?

____

____

____

What is dominance?

But why did the offspring of a short plant crossed with a tall plant all grow into tall plants? Mendel called the observed trait **dominant** and the trait that disappeared **recessive.** A dominant allele will mask, or cover up, a recessive allele. The allele for tall plants is dominant to the allele for short plants. The plants that had a tall and short allele were tall because the tall allele is dominant and the short allele is recessive. The plants with two alleles for tallness were tall. The plants with two alleles for shortness were short. In the F_1 generation of plants each plant had one tall allele and one short allele. That is why the offspring of a tall and short plant were all tall.

When the results of crosses are written down, scientists use the same letter for different alleles of the same trait. An uppercase letter is used for the dominant allele, and a lowercase letter is used for a recessive allele. So in writing down the results of Mendel's experiment, *T* is used for the dominant allele for tallness, while *t* is used for the recessive allele for shortness. The dominant allele is always written first.

Mendel took the facts that he learned from his experiments to create rules or laws to explain heredity. The first of his laws is called the **law of segregation.** This law says that every organism has two alleles of each gene and when gametes are produced the alleles separate. Each gamete receives one of these alleles. During fertilization, these gametes randomly pair to produce four combinations of alleles.

Reading Check

5. Explain the difference between phenotype and genotype.

____

____

____

Phenotypes and Genotypes

You cannot always tell by looking at an organism what genes it might pass on. Sometimes tall plants crossed with each other produce both tall and short offspring. Sometimes a short plant and a tall plant produce all tall offspring.

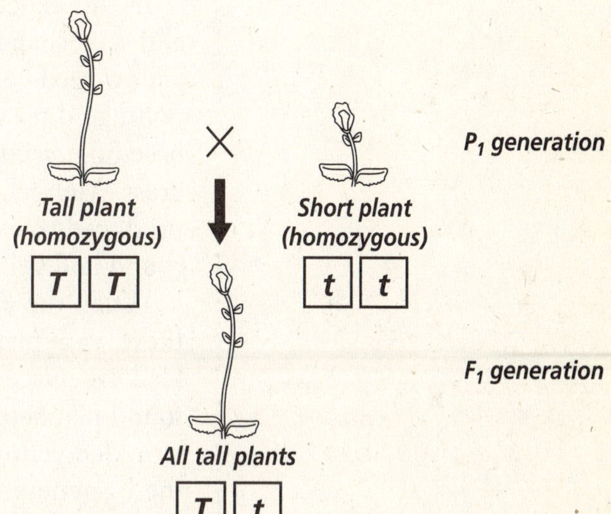

Two organisms can look alike but have different allele combinations. The way an organism looks and behaves is called its **phenotype** (FEE noh tipe). The phenotype of a tall plant is tall. The plant could have an allele combination of *TT* or *Tt.* The allele combination of an organism is called the **genotype** (JEE noh tipe). The genotype cannot always be determined even if you know the phenotype. ✓

What are homozygous and heterozygous alleles?

An organism is **homozygous** (hoh moh ZI gus) for a trait if the two alleles for the trait are the same. So a plant with two alleles for tallness (*TT*) would be homozygous for the trait of height. Remember that since tallness is dominant, a *TT* plant is homozygous dominant for height. A short plant always has two alleles for shortness (*tt*). Therefore, a short plant would be homozygous recessive for height.

What if the two alleles are not the same? An organism is **heterozygous** (heh tuh roh ZI gus) for a trait if its two alleles for the trait are different from each other. A tall plant that has one allele for tallness and another allele for shortness (*Tt*) is heterozygous for the trait of height. ✓

Reading Check

6. Explain the difference between homozygous and heterozygous.

____

____

____

Mendel's Dihybrid Crosses

In Mendel's first set of experiments he used monohybrid crosses. He used plants that differed from each other in only one trait. Later, Mendel used pea plants that differed from each other in two traits. A cross involving two different traits is called a dihybrid cross. *Di* means two. Mendel wanted to know whether, in a dihybrid cross, the two traits would stay together in the next generation, or whether they would be passed on independently of each other.

In this experiment Mendel observed the traits for seed color and seed shape. He knew from previous experiments that the yellow seed color and round seed shape were dominant. He cross-pollinated true-breeding round, yellow seeds (*RRYY*) with true-breeding wrinkled, green seeds (*rryy*). This created a dihybrid cross. Mendel discovered that in the F_1 generation all the plants produced round yellow seeds. That was not surprising since round and yellow are dominant traits.

When the F_1 plants self-pollinated and produced offspring there were plants with round yellow seeds and plants with wrinkled green seeds. That was not surprising either. But Mendel also found plants producing two other seed types, round green, and wrinkled yellow. When Mendel sorted and counted the plants of the F_2 generation he found a ratio of phenotypes—9 round yellow, 3 round green, 3 wrinkled yellow, 1 wrinkled green.

What is the law of independent assortment?

To explain his results, Mendel created his second law of heredity, known as the **law of independent assortment.** This law states that genes for different traits are inherited independently of each other. In a dihybrid cross you can see both of Mendel's laws at work. The plants in the F_1 generation of the dihybrid cross had the genotype *RrYy*. When a plant with this genotype produces gametes, the alleles *R* and *r* will separate from each other. That is the law of segregation at work. The *R* and *r* alleles will also separate from the *Y* and *y* alleles. That is the law of independent assortment at work. The alleles will then recombine in four different ways. If the alleles for seed shape and color were inherited together, only two kinds of seeds would have been produced. Instead, four different kinds of seeds were produced. ✓

Reading Check

7. What does the law of independent assortment state?

____

____

____

Punnett Squares

Reginald Punnett, an English biologist, came up with a way to predict the proportions of possible genotypes in offspring. It is called a Punnett square. If you know the genotypes of the parents you can use a Punnett square to predict the possible genotypes of their offspring. For example, in Mendel's original experiment, the F_1 generation had the genotype *Tt.* That means that half the gametes for each plant would contain the *T* allele, and the other half would contain the *t* allele. A Punnett square can show the possible combinations for offspring with parents with this genotype.

A Punnett square for a single trait is two boxes tall and two boxes wide. The genotype of one parent is listed across the top. The genotype for the other parent is listed along the left side. Look at the illustration to see the possible combinations. A Punnett square can also be created for dihybrid crosses. It would be larger, four boxes tall and four boxes wide.

Punnett Square

	T	*t*
T	*TT*	*Tt*
t	*Tt*	*tt*

Probability

Punnett squares show all of the possible outcomes. Because chance is a factor in genetics, the actual results don't always match the Punnett square's probability. It is like flipping a coin. The probability is 50/50 that the coin will be heads or tails. However if you flip a coin 100 times, you cannot guarantee that 50 times it will be heads and 50 times it will be tails. It is the same in genetics. Even though a Punnett square may predict that one fourth of the offspring will have a particular genotype, in reality the amount could be higher or lower.

After You Read

Mini Glossary

allele (uh LEEL): alternative form of a gene for each trait of an organism

dominant: trait of an organism that is observed and that masks the recessive trait

fertilization: uniting of male and female gametes

gamete: male and female sex cells

genetics: study of heredity

genotype (JEE noh tipe): allele combination of an organism

heredity: passing on of characteristics from parents to offspring

heterozygous (heh tuh roh ZI gus): when there are two different alleles for a trait

homozygous (hoh moh ZI gus): when there are two identical alleles for a trait

hybrid: offspring of parents having different forms of a trait

law of independent assortment: Mendelian principle stating that genes for different traits are inherited independently of each other

law of segregation: Mendelian principle explaining that because every organism has two alleles for every gene, it can produce different types of gametes. During fertilization, these gametes randomly pair to produce four combinations of alleles.

phenotype (FEE nuh tipe): the way an organism looks and behaves

pollination: the transfer of pollen grains from male reproductive organs to female reproductive organs of plants

recessive: traits of an organism that can be masked by the dominant form of the trait

trait: characteristics that are passed from parents to offspring

zygote (ZI goht): cell formed when a male gamete unites with a female gamete

1. Review the terms and their definitions in the Mini Glossary above. Write a sentence using at least two of the terms.

____

____

2. Use the pyramid diagram to help you review what you have read. Arrange the steps Mendel used in his first experiment with pea plants. Place the letter of each step in the correct order in the pyramid.

a. Cross-pollinate P_1 generation to grow monohybrids
b. Observe results in F_2 generation
c. Find true breeders for a single trait
d. Allow F_1 generation to self-pollinate

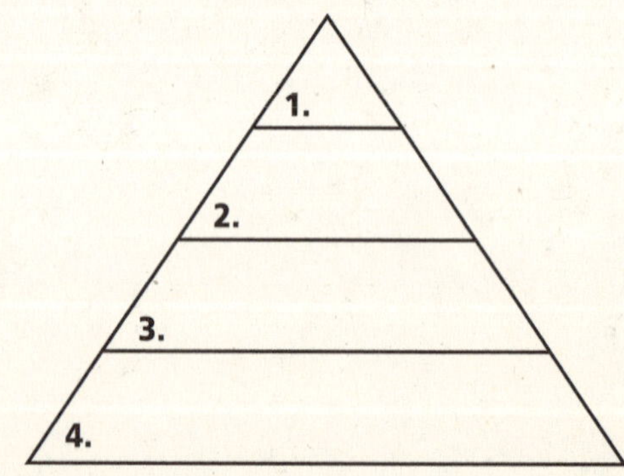

Section 10.1 Mendel's Laws of Heredity, *continued*

3. Use the cause and effect diagram below to explain Mendel's two laws. Write the effect of each law in the correct box.

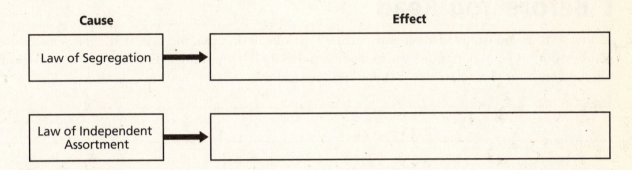

Visit the Glencoe Science Web site at **science.glencoe.com** to find your biology book and learn more about Mendel's laws of heredity.

Section 10.2 Meiosis

Before You Read

In this section you will read about a particular type of cell division, meiosis, that creates many gene combinations. Genes determine the individual traits of all living things. On the lines below, list three different traits or characteristics that you can see in a rose.

____

____

Read to Learn

Mark the Text **Identify Facts** Use one color to highlight each fact about diploid cells. Use another color to highlight each fact about haploid cells.

Reading Check

1. What do you call a cell with two of each kind of chromosome?

____

2. What do you call a cell with one of each kind of chromosome?

____

Genes, Chromosomes, and Numbers

All living things have thousands of genes. Genes determine individual traits. The genes do not just float around in a cell. They are lined up on chromosomes. A typical chromosome can have thousands of genes.

If you took a cell from one of Mendel's pea plants, you would see that it has 14 chromosomes, or seven pairs. In the body cells of most living things, chromosomes come in pairs. One of the chromosomes in each pair comes from the male parent. The other chromosome comes from the female parent. A cell that has two of each kind of chromosome is called a **diploid** cell. Using microscopes, scientists can now see the paired chromosomes and know that an allele for each trait is located on each paired chromosome. ✓

How are diploid cells different from haploid cells?

A diploid cell is a body cell. Organisms also produce a different kind of cell called a **haploid** cell. This cell contains only one of each kind of chromosome. A gamete, or sex cell, is a haploid cell. Remember, Mendel concluded that parents give one allele of each trait to their offspring. That happens because the gamete is a haploid cell containing half the number of chromosomes as a body cell. ✓

Every living thing has a set number of chromosomes. For example, every dog has 78 chromosomes, every human has 46 chromosomes, and every tomato plant has 24 chromosomes. As you can see, the number of chromosomes is not related to how complex an organism is. When the organism reproduces, it only passes on half the number of chromosomes. Earlier in this section you learned that genes are lined up on a chromosome. In a diploid cell the two chromosomes of each pair are called **homologous** (hoh MAH luh gus) **chromosomes.** Each chromosome of the pair has genes for the same traits in the same order.

Because there are different alleles for the same gene, it is possible that two chromosomes of a homologous pair will not be identical to each other. For example, garden peas have 14 chromosomes, or seven pairs. The pairs are numbered 1 through 7. Chromosome 4 has genes for three of the traits Mendel studied. Every pea plant has two copies of chromosome 4, one from each parent. Chromosome 4 contains the allele for height. As you know from the previous section, it is possible for one chromosome to contain *T* (tall), and the other chromosome to contain *t* (short). The two chromosomes in a homologous pair might not be identical.

How does a diploid cell become a haploid cell?

How can an organism pass on half its chromosomes? It does so in a process of cell division called meiosis. **Meiosis** (mi OH sus) produces gametes containing half the number of chromosomes as the parent's body cell. ✓

Meiosis is actually two separate divisions. The first division is meiosis I, and the second is meiosis II. The whole process begins with one diploid cell and ends with four haploid cells. The haploid cells are gametes, or sex cells. Male gametes are called **sperm.** Female gametes are called **eggs.** When a sperm unites with, or fertilizes an egg, the resulting zygote has a diploid number of chromosomes. Then, using mitosis, which you studied earlier, the zygote becomes a multicellular organism. This way of reproducing, which involves producing and then uniting haploid sex cells, is called **sexual reproduction.**

Reading Check

3. What does meiosis produce?

____

____

____

The Phases of Meiosis

Before meiosis begins, the chromosomes in a cell are replicated, or copied. Each chromosome consists of two sister chromatids connected by a centromere. A cell in prophase I of meiosis acts similarly to a cell in prophase of mitosis. The DNA of the

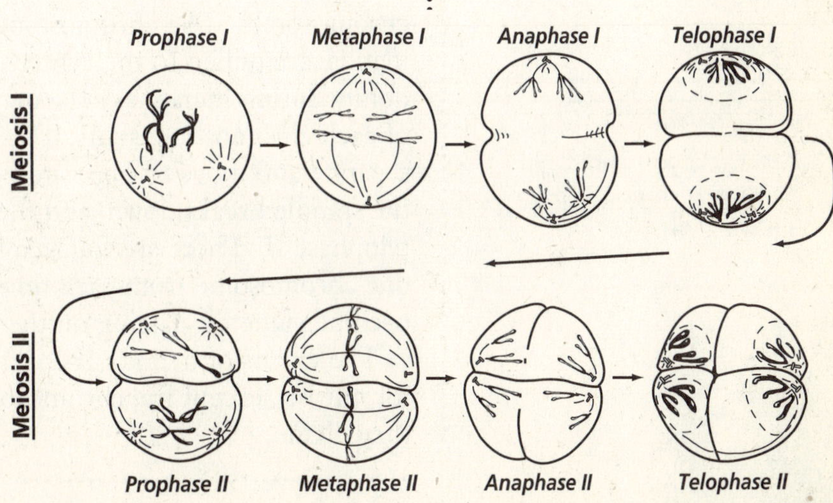

Reading Check

4. What happens during crossing over?

chromosomes coils up and a spindle forms. The chromosomes pair up with their homologous chromosomes near the middle of the cell. This pairing brings the two chromatids of each chromosome close together, making what is called a tetrad. The homologous chromosomes pair so tightly that sometimes a piece of a chromatid can break off. The piece changes places with a piece of the chromatid from the other homologous chromosome of the tetrad. This exchange is called **crossing over.** Crossing over can occur at several places at the same time.

During metaphase I of meiosis, the centromere of each chromosome becomes attached to the spindle fiber. The spindle fibers pull the chromosomes into the center of the cell. The spindle lines up the tetrads. If the cell were a globe, it would be as if one pair of sister chromatids were lying horizontally above the equator, and the other set of sister chromatids were lying horizontally below the equator. Let's stick with our globe illustration. During anaphase I the chromosomes begin to move apart from each other. It is as if the chromosomes above the equator move to the North Pole, and the chromosomes below the equator move to the South Pole. During telophase I of meiosis, the spindle is broken down and the chromosomes uncoil. The cytoplasm divides into two new cells.

How are haploid cells formed?

Each cell has half the genetic information that the original cell had because it has only one chromosome from each homologous pair. But remember that the chromosomes copied themselves at the beginning of the process. The new cells are going to have to divide in order to create haploid cells. This second division is called meiosis II. During prophase II, a spindle forms in each of the new cells. The spindle fibers attach to the centromeres of the chromosomes. The chromosomes, still made up of sister chromatids, are pulled to the center of the cell where they line up randomly during metaphase II. Anaphase II begins as the centromere of each chromosome splits. This allows the sister chromatids to separate and move to opposite poles of the cell. Nuclei re-form, the spindle breaks down, and the cytoplasm divides during telophase II. There are four haploid cells, each cell containing one chromosome from each homologous pair. These haploid cells become gametes, passing on the genes they contain to offspring.

The events of meiosis II are the same as those you studied for mitosis, except the chromosomes do not replicate before they divide.

Meiosis Provides for Genetic Variation

As you can see, the cell division that happens during meiosis creates many possible gene combinations. When crossing over occurs, even more variations are possible. The reassortment of genetic information that occurs during meiosis is called **genetic recombination.** It is a major source of variation among organisms. Meiosis also provides the physical basis for explaining Mendel's results. Mendel's laws and meiosis provide the foundation for heredity. ✓

Nondisjunction

Meiosis usually happens without any problems, but sometimes chromosomes do not separate correctly. When this happens, the gametes that form will either have too many chromosomes or not enough chromosomes. The failure of chromosomes to separate correctly is called **nondisjunction.** ✓

An organism with extra chromosomes may survive, but an organism that is missing one or more chromosomes does not usually develop. Surprisingly, in plants, extra chromosomes can actually be helpful. Often the flowers and fruits are larger, and the plant is healthier. Because of this, plant breeders have learned to cause nondisjunction by using chemicals.

Gene Linkage and Maps

Genes on the same chromosome are usually linked and inherited together instead of independently. It is the chromosomes that follow Mendel's law of independent assortment, not the genes. Linked genes can separate as a result of crossing over. Scientists have found that genes that are farther apart on a chromosome tend to cross over more often than genes that are close together. Using this information, scientists can make chromosome maps that show the sequence of genes on a chromosome.

✓ Reading Check

5. How does meiosis provide for genetic variation?

____

____

____

✓ Reading Check

6. What is nondisjunction?

____

____

____

After You Read

Mini Glossary

crossing over: pieces of homologous chromosomes change places during prophase I of meiosis; results in new allele combinations on a chromosome

diploid: cell with two of each kind of chromosome

egg: haploid female sex cell produced by meiosis

genetic recombination: reassortment of genetic information during meiosis that creates many possible gene combinations

haploid: cell with one of each kind of chromosome

homologous (hoh MAH luh gus) chromosome: each chromosome pair contains genes for the same traits arranged in the same order

meiosis (mi OH sus): type of cell division where one specialized body cell produces four gametes, each containing half the number of chromosomes as a parent's body cell

nondisjunction: failure of homologous chromosomes to separate correctly during meiosis; results in gametes with too many or too few chromosomes

sexual reproduction: pattern of reproduction that involves producing and subsequent uniting of haploid sex cells

sperm: haploid male sex cell produced by meiosis

1. Review the terms and their definitions in the Mini Glossary above. Then list the four terms that have to do with genetic variation on the lines below.

____

____

2. Fill in the boxes below explaining how each of the topics relates to meiosis.

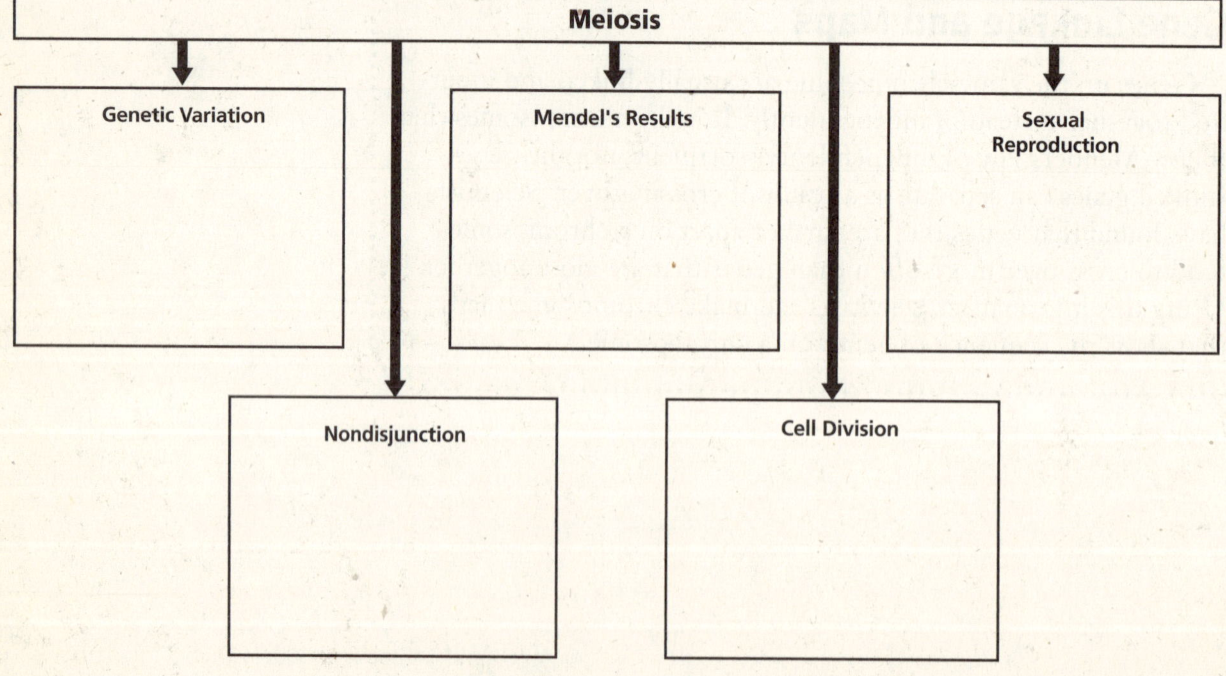

What does RNA look like?

RNA, like DNA, is a nucleic acid. But the structure of RNA is quite different. RNA is a single strand. It looks like one half of a zipper. DNA is a double strand. The sugar in RNA is different than the sugar in DNA. Finally, both RNA and DNA have four nitrogenous bases, but instead of thymine, RNA has uracil. Remember that in DNA, guanine binds with cytosine, and thymine binds with adenine. In RNA uracil (U) binds with adenine. The structure of RNA helps it do all the work of building proteins.

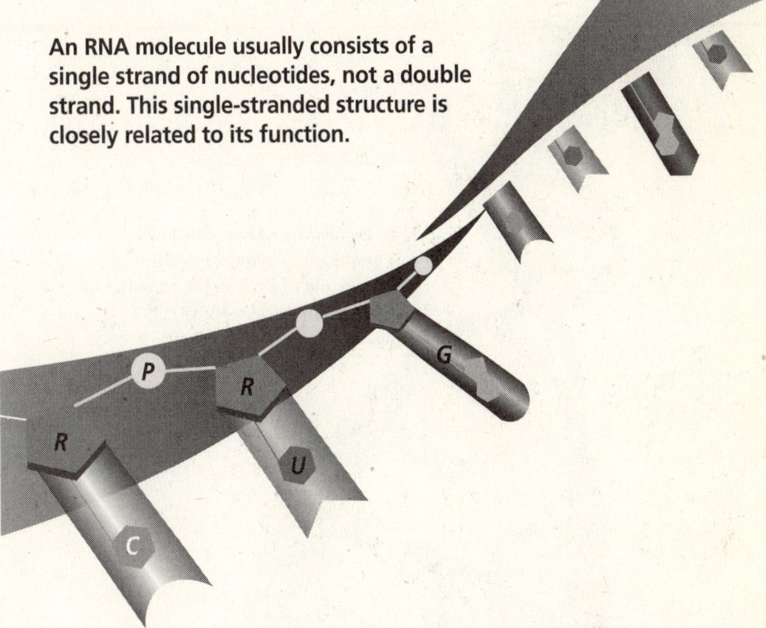

An RNA molecule usually consists of a single strand of nucleotides, not a double strand. This single-stranded structure is closely related to its function.

Transcription

In order to get the information to the cytoplasm, first messenger RNA has to be made. In this process, called **transcription** (trans KRIHP shun), RNA is made from part of a DNA strand. Use the illustration on page 120 to help you understand the process after you read the following description.

First, a portion of the DNA molecule unzips. Free RNA nucleotides pair with the nucleotides on the DNA strand. The mRNA strand is complete when the RNA nucleotides form a strand by bonding together. The mRNA strand breaks away and the DNA strands rejoin. The mRNA strand leaves the nucleus and enters the cytoplasm. You can see that transcription is similar to replication with one important difference—a single strand RNA molecule is created rather than a double strand DNA molecule. You can also see from the illustration that mRNA pairs guanine with cytosine, but pairs uracil with adenine.

Mark the Text **Identify Details** Circle the parts of the diagram on page 120 that illustrate each part of the transcription process.

RNA Processing

Not all of a DNA strand carries information to make proteins. There are long sequences of noncoding nucleotides on DNA strands. Enzymes cut out any noncoding sequences that may have been transcribed. In this way, the mRNA carries only information it needs to make protein.

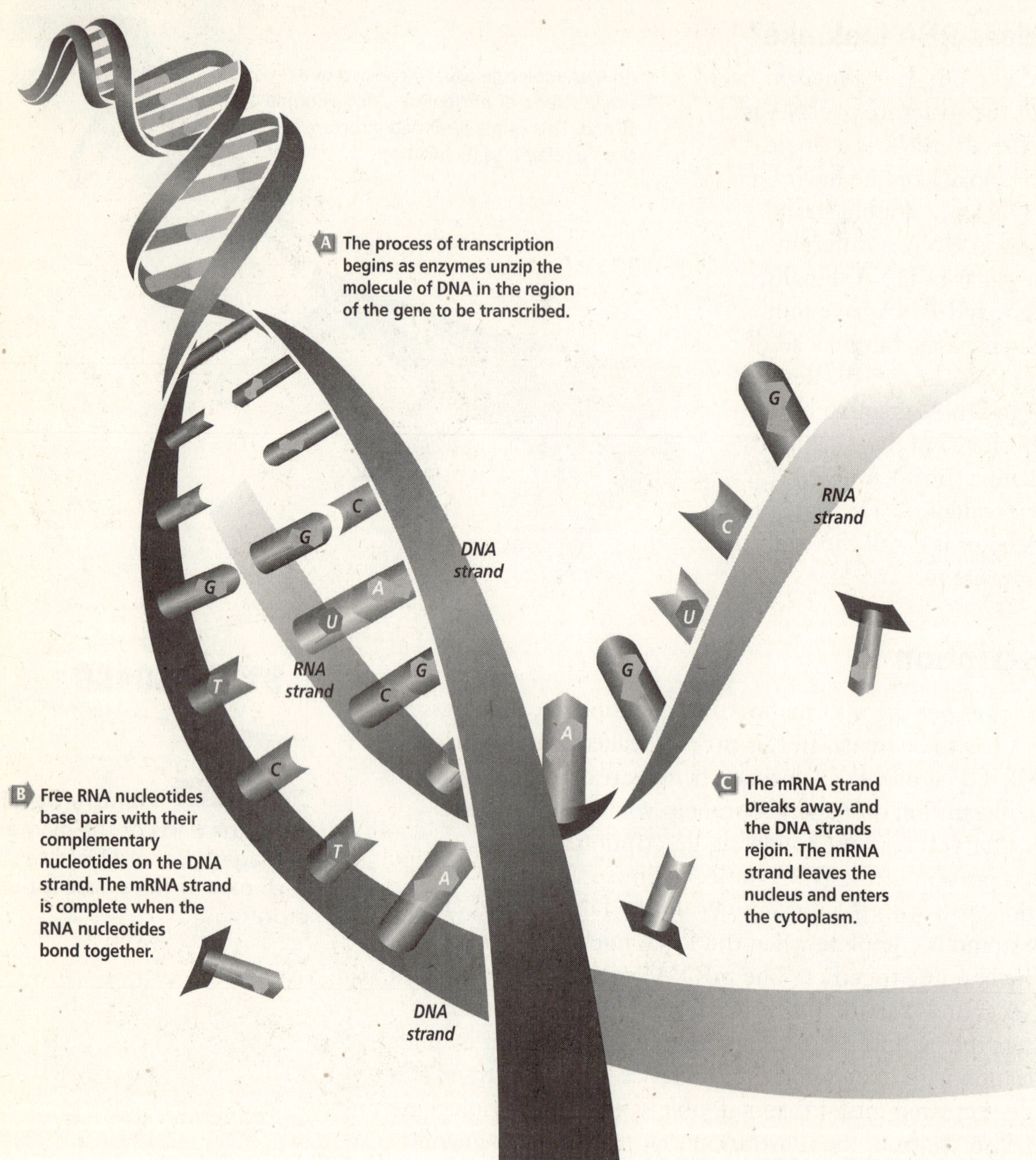
A The process of transcription begins as enzymes unzip the molecule of DNA in the region of the gene to be transcribed.
B Free RNA nucleotides base pairs with their complementary nucleotides on the DNA strand. The mRNA strand is complete when the RNA nucleotides bond together.
C The mRNA strand breaks away, and the DNA strands rejoin. The mRNA strand leaves the nucleus and enters the cytoplasm.
DNA strand
RNA strand
DNA strand
RNA strand
G
C
G
A
U
G
C
T
C
T
A
A
G
U
C
G

The Genetic Code

The nucleotide sequence transcribed from DNA to a strand of messenger RNA is a genetic message that has all the information needed to build a protein. The message is in a special language that uses nitrogenous bases as the alphabet. Remember that proteins are made up of amino acid chains. There are 20 different amino acids. These amino acids are made from only four nitrogenous bases. Scientists wondered how four nitrogenous bases could make a code for 20 amino acids.

Scientists were able to crack the genetic code when they discovered that it takes a group of three nitrogenous bases in mRNA to code for one amino acid. Each group of three nitrogenous bases is known as a **codon.** For example, the codon for the amino acid alanine is G-C-U. The codon for lysine is A-A-A. Every amino acid has a three-letter codon, each letter representing one of the four nitrogenous bases. That is how four nitrogenous bases can code for 20 amino acids. There is even a codon that tells the mRNA that this is the start of the amino acid chain and another codon that says this is the end. To simplify, those codons are called *start* and *stop.*

Reading Check

2. What is a codon?

____

____

____

Translation: From mRNA to Protein

Remember the factory example? Messenger RNA is the worker that brings the instructions for making protein to the cytoplasm. It takes two more kinds of RNA to actually make the protein. The process of changing the information in mRNA into an amino acid chain in protein is called **translation.**

Here is how it works.

1. The mRNA moves to the cytoplasm.
2. A ribosome (rRNA) attaches itself to the start codon, A-U-G, on the mRNA.
3. Transfer RNA (tRNA) molecules, carrying amino acids, approach the ribosome. The nitrogenous base sequence that is the complement to the mRNA sequence is the anticodon. If the mRNA codon is G-C-C, the tRNA anticodon is C-G-G. For every codon on mRNA there is an anticodon on tRNA.
4. The ribosome attaches the anticodon to the codon and the amino acids bond. The ribosome then slides to the next codon.
5. Again the ribosome attaches the anticodon to the codon, amino acids bond, and the ribosome slides over.
6. This translation process continues until the stop codon is reached. At this point the amino acids have formed a chain and when the stop codon is reached, the chain is released.

You can see from the illustration of the translation process on page 122 that the tRNA does not stay attached during the whole process. As soon as the amino acid bonds to the amino acid next to it, the tRNA that brought it moves away to bring another amino acid.

Think it Over

3. Analyze What is the difference between transcription and translation?

____

____

____

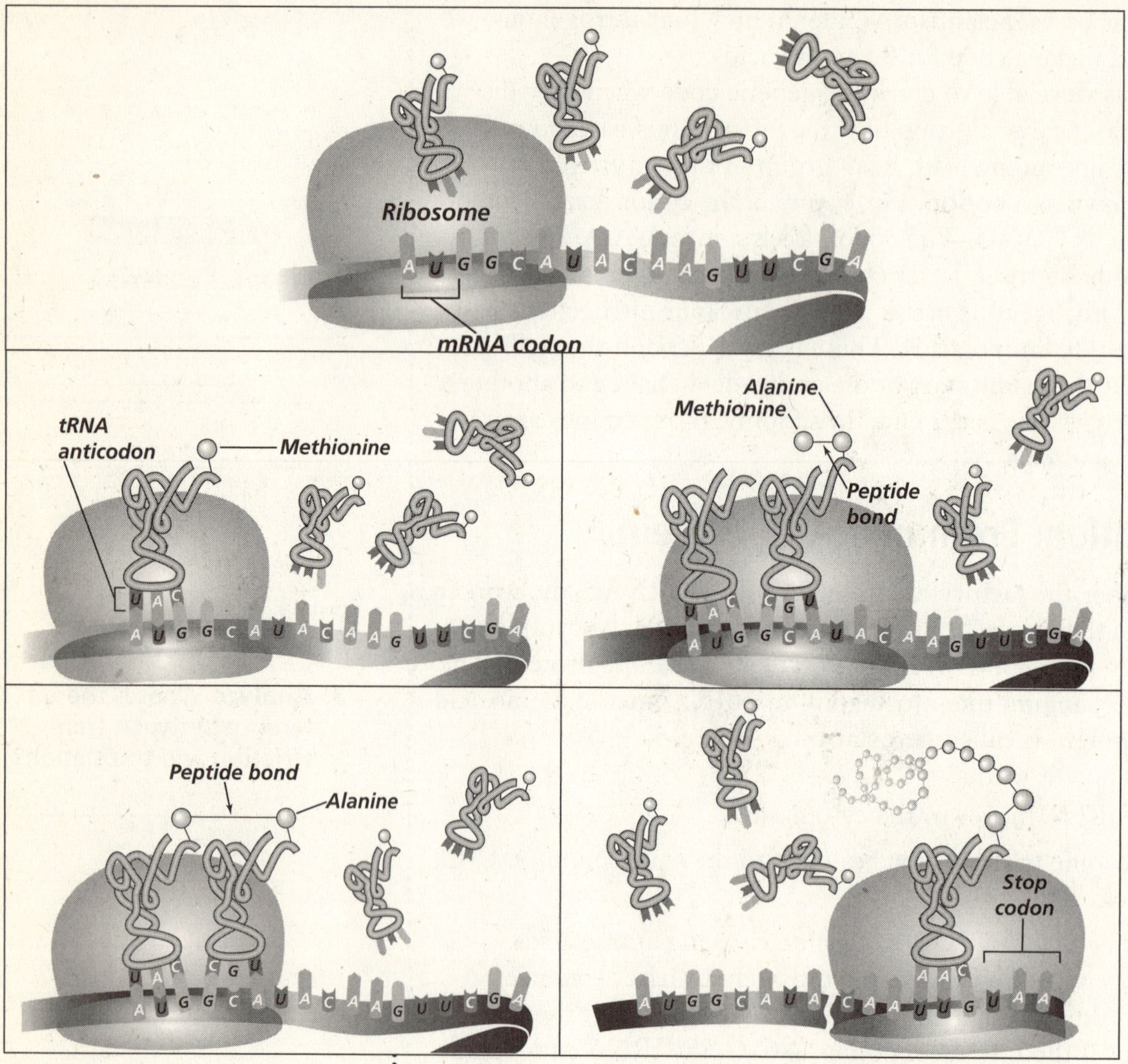

Section 11.2 From DNA to Protein, *continued*

Ribosome
A U G G C A U A C A A G U U C G A
mRNA codon

tRNA anticodon
Methionine

Alanine
Methionine
Peptide bond

Peptide bond
Alanine

Stop codon

Amino acid chains become proteins when they are freed from the ribosome. The amino acid chains twist and curl into complex three-dimensional shapes. Each protein chain forms the same shape every time it is produced. These proteins become enzymes and cell structures.

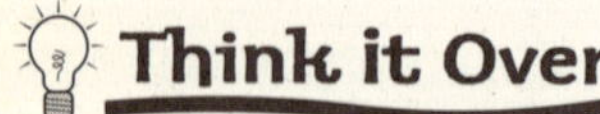

Think it Over

4. **Sequence** The pathway of information flows from (Circle your choice.)
   a. DNA to mRNA to protein.
   b. mRNA to DNA to protein.
   c. protein to mRNA to DNA.

What is the central dogma?

If you were to summarize the process of replication, transcription, translation, and protein formation you might say simply that the pathway of information flows from DNA to mRNA to protein. This process is called the central dogma of biology. This means that the same process occurs in every living thing, from the simplest bacteria to the most complex animal.

Section **11.2** **From DNA to Protein,** *continued*

After You Read

Mini Glossary

codon: group of three nitrogenous bases in mRNA that code for one amino acid

messenger RNA: RNA that carries information from DNA in the nucleus to the cell's cytoplasm

ribosomal RNA: RNA that makes up the ribosomes; binds to mRNA and uses its information to assemble amino acids in the right order

transcription (trans KRIHP shun): process in the cell nucleus where a copy of RNA is made from part of a DNA strand

transfer RNA: RNA that delivers amino acids to the ribosomes to be assembled into proteins

translation: process of changing the information in mRNA into an amino acid chain in a protein

1. Read the key terms and definitions in the Mini Glossary above. Then on the lines, write a definition of **transcription** and **translation** using your own words.

____

____

____

2. Under each type of RNA, write the words or phrases that tell something about it.

mRNA →	rRNA →	tRNA

codon
moves along mRNA
anticodon
transcription
translation
connects codon to anticodon
brings instructions
brings amino acid
uses instructions to assemble amino acids

Visit the Glencoe Science Web site at **science.glencoe.com** to find your biology book and learn more about DNA to protein.

Name Date Class

Section 11.3 Genetic Changes

Before You Read

If you have had an X ray taken, you may remember that before they took it they covered the part of your body not being X rayed with a heavy lead shield. Think about why the technician protects you from X rays and write the reason on the lines below. After you read this section, check your answer. You may want to add new information that you learned.

____

____

Read to Learn

Mark the Text **Identify Main Ideas** As you read this section, highlight the main idea in each paragraph. Stop after every paragraph and put what you just read into your own words.

Mutations

Every day millions of cells correctly transfer DNA information to proteins. Occasionally, however, there is a change in the DNA sequence. Any change in the DNA sequence is called a **mutation.** Mutations can be caused by errors in replication, transcription, or cell division. Forces outside of the cell can cause changes to DNA. You will learn about each of these mutations in this section.

Can a mutation be a good thing?

Imagine that the nucleotide sequence of a sperm or egg cell changes. If that sperm or egg cell results in fertilization, then the change in sequence would become part of the offspring. The mutation could result in a protein that does not work correctly and the offspring may not survive. Sometimes a mutation is helpful. The mutation may produce a new trait. Maybe the mutation results in the ability to see farther or run faster. The offspring may survive better in its environment. It can pass the new trait on to its own offspring. Later you will learn how mutations played a role in evolution.

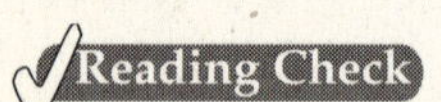

1. When is a mutation helpful?

____

____

____

Mutations in body cells are usually caused by outside forces. Radiation from the sun, X rays, or radioactive materials can change the DNA of skin, muscle, or bone cells. Since these cells are not sex cells, the changes in the DNA are not passed on to offspring. However, the organism may be harmed by the mutation. When mutated cells divide they pass on the mutation. For example, damage to a stomach cell may cause it to lose its ability to make the acid needed to digest food. When that cell divides, the new cells will have the same mutation.

Sometimes a mutation in a body cell affects the way the cell divides. This may cause the cells to grow and divide too quickly, producing cancer. Too much exposure to ultraviolet radiation in sunlight mutates skin cells, causing skin cancer.

2. Infer Why is a mutation in a body cell not passed on to offspring?

____

____

____

What is happening at the DNA level?

Recall how information flows from DNA to mRNA to protein. If the DNA is mutated, what will happen to the mRNA? It will take the changed information into the cytoplasm and tRNA will bring the wrong amino acids to the rRNA. The protein that is created will be based on the mutated information.

Mutations occur in two different ways at the DNA level. The first type of mutation is called a point mutation. A **point mutation** happens when there is a change in a single base pair in DNA. If the DNA sequence should be A-A-G U-U-U-G-G-C but is A-A-G-U-U-U-A-G-C then the amino acid chain is made with serine instead of glycine. A point mutation is a little like a letter change in a sentence.

THE DOG BIT THE CAT
THE DOG BIT THE CAR

As you can see, it changes the meaning of the whole sentence.

Identify Details Point to each type of mutation as you read about it.

A frameshift mutation involves more than a single codon. It happens when a nitrogenous base is deleted or added. The illustration on the right shows what happens when a base is deleted. It changes everything that follows it. If we use our sentences as an example, we can see what the result might be.

THE DOG BIT THE CAT
(correct)
THE DOB ITT HEC AT
(deleted base (G))
THE DOC GBI TTH ECA T
(added base (C))

As you can see, a **frameshift mutation** occurs when a single nitrogenous base is added or deleted from the DNA sequence. It shifts the reading of the codon by one base. A frameshift mutation is usually more harmful to an organism than a point mutation.

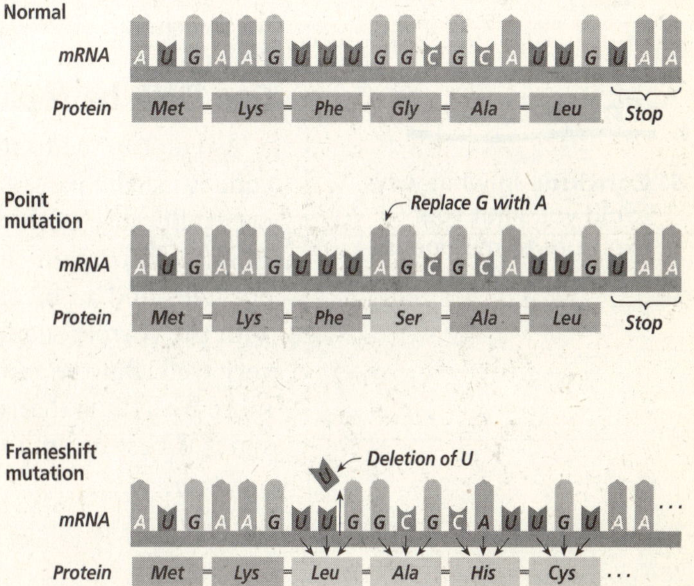

Chromosomal Alterations

Changes occur to chromosomes as well as the DNA sequence on the chromosomes. Sometimes parts of the chromosomes break off during mitosis or meiosis. The pieces may join to the wrong chromosomes, join backwards, or join in the wrong places. Occasionally the broken pieces get lost. These structure changes in chromosomes are called **chromosomal mutations.**

Chromosomal mutations can happen in any organism, but they are especially common in plants. As you remember from an earlier chapter, mutations affect the way genes are distributed during meiosis. Some of the gametes have too many chromosomes; some of the gametes don't have enough chromosomes. Few chromosomal mutations are passed on because the fertilized egg usually dies. If the organism does develop, it is often not able to reproduce, so the mutation is not passed on. ✓

Reading Check

3. Are chromosomal mutations more common in plants or animals?

____

____

____

Causes of Mutations

Some mutations seem to just happen. They are mistakes in base pairing during DNA replication. These mutations are said to be spontaneous. Many mutations are caused by environmental factors. Any outside agent that can cause a change in DNA is called a **mutagen** (MYEW tuh jun). Mutagens include radiation, chemicals, and high temperatures. ✓

Reading Check

4. What is a mutagen?

____

____

____

Some mutagens cause DNA to break apart. This can change the sequence of the bases. A base may disappear, or two bases may fuse together. Other mutagens cause one base to be substituted for another.

Can DNA be repaired?

As you can see from observing the world around you, the genetic code is usually passed on accurately. But mistakes or mutations can occur. Because of this, repair mechanisms are present in organisms. Cells contain enzymes that check the DNA sequence. If the enzymes find an incorrect sequence of nucleotides, they replace it with the correct sequence. The repair mechanisms usually work very well. But the more an organism is exposed to a mutagen, the more likely it is that a mistake will not be corrected. For this reason it is best to limit exposure to mutagens.

Think it Over

5. Conclude In what ways could you limit your exposure to mutagens?

____

____

____

Section 11.3 **Genetic Changes,** *continued*

After You Read

Mini Glossary

chromosomal mutation: mutation that occurs when parts of the chromosomes break off during mitosis or meiosis and join to the wrong chromosome, or join backwards or in the wrong place on the chromosome

frameshift mutation: mutation that occurs when a single nitrogenous base is added or deleted from the DNA sequence; causes a shift in the reading of codons by one base

mutagen (MYEW tuh jun): any outside agent that can cause a change in DNA; includes high temperatures, radiation, or chemicals

mutation: any change in a DNA sequence

point mutation: a change in a single base pair in DNA

1. Read the key terms and definitions in the Mini Glossary above. Why is a frameshift mutation usually more harmful than a point mutation? Write your answer on the lines below.

____

____

____

2. Use the partially completed outline below to help you review what you have read. Fill in the blanks where information is missing.

I. Mutations occur

- A. In reproductive cells
- B. In ____________________ cells

II. Types of Mutations

- A. ____________________ mutations
- B. ____________________ mutations
- C. ____________________ mutations

III. Causes of Mutations

- A. Just happens = Spontaneous
- B. Environmental factor = ____________________

Visit the Glencoe Science Web site at **science.glencoe.com** to find your biology book and learn more about genetic changes.

Section 12.1 Mendelian Inheritance of Human Traits

Before You Read

Certain genetic traits run in families. Skim the Read to Learn section and highlight two important facts about how diseases are passed from generation to generation. Then write the facts you highlighted on the lines below.

Read to Learn

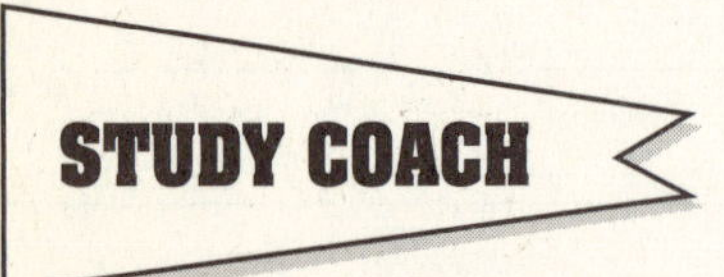

Create an Outline Using the headings, make an outline of the information in this section.

1. What is a pedigree?

Pedigree Symbols

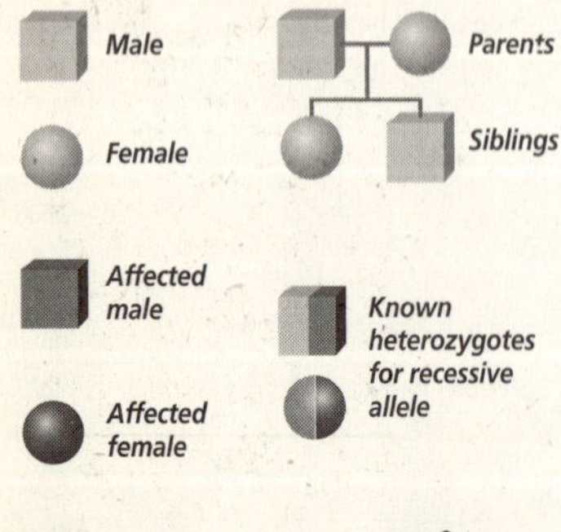

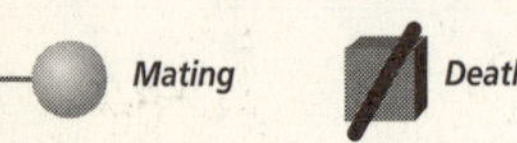

Making a Pedigree

Have you ever seen a family tree? What does a family tree show? It shows the relationships among family members. It shows how grandparents, parents, children, aunts, uncles, cousins, and siblings are related.

Geneticists are scientists who study how inherited or genetic traits pass from one generation to the next. Sometimes they use what is called a pedigree to study a family's genetic puzzle. A **pedigree** is a visual diagram of genetic inheritance used by geneticists to map genetic traits. A pedigree uses a set of symbols to identify males and females in a certain family that carry the genetic trait being studied. ✓

Why is a pedigree important?

Perhaps a family has members who suffer from a rare genetic illness. By studying the pedigree, the geneticist will be able to see if it is likely that the trait will be passed on to children or grandchildren. An individual can also study the pedigree to see if there is a risk that he or she will pass along a genetic disease. Studying a family pedigree gives a lot of genetic information about a family.

On a pedigree, circles represent females and squares represent males. Shaded circles and squares represent those who have the trait that is being studied. Unshaded circles and squares represent individuals who do not have the trait. A half-shaded circle or square represents a **carrier**—someone who has a recessive allele for a specific trait. As you have learned, an allele is one of the different forms of a gene that exists for a genetic trait. A circle and a square connected by a horizontal line represents parents. A vertical line connects parents with their offspring. ✓

Reading Check

2. What symbols represent males and females in a pedigree?

____

____

How is a pedigree analyzed?

Look at the illustration to the right. This pedigree shows how a rare, recessive allele is passed from generation to generation. Suppose individual III-1 on the pedigree wanted to know whether she would pass on this allele to her children. We can study the pedigree to answer this question.

We will begin at the top. Individuals I-1 and I-2 are both carriers of the recessive allele for this trait. We know this because they produced II-3, who shows the recessive phenotype. We also know that II-2 is a carrier because she has passed the allele to later generations (IV-2 and IV-4). Because III-1 has a parent (II-2) who is a carrier of the recessive allele, III-1 has a one-in-two chance of also being a carrier of the recessive allele.

The Punnett square to the right shows the mating of individuals I-1 and I-2. It also shows the probability of individuals II-2 through II-5 receiving the recessive allele.

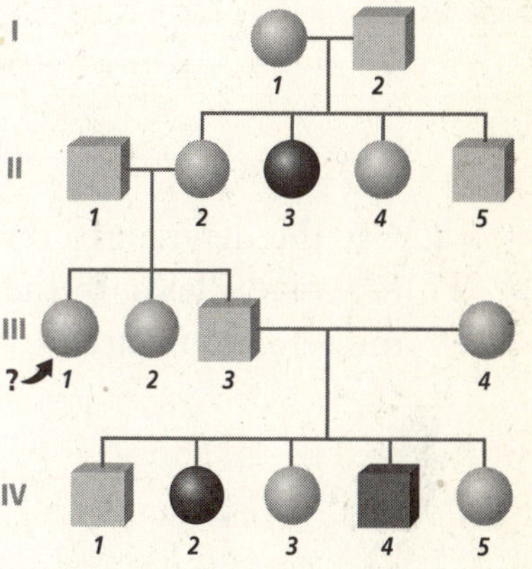

	R	*r*
R	***RR***	***Rr***
r	***Rr***	***rr***

Simple Recessive Heredity

Most genetic disorders are caused by recessive alleles. Diseases such as cystic fibrosis, Tay-Sachs (tay saks) disease, and phenylketonuria (fen ul kee tun YOO ree uh), also known as PKU, can be predicted by the use of a pedigree. For an offspring to inherit a recessive trait, both parents must have the recessive allele.

Simple Dominant Heredity

Remember that to inherit a dominant trait, only one parent needs to have the dominant allele for that trait.

A cleft chin, a widow's peak hairline, freely hanging earlobes, almond-shaped eyes, and thick lips are examples of dominant traits. Only one allele needs to be present for these traits to show up.

Huntington's disease is caused by a rare dominant allele. There is no effective treatment for Huntington's disease, which causes a breakdown in certain parts of the brain. Because Huntington's disease doesn't occur until a person is between the ages of 30 and 50, many people have already had children before they develop the disease. A pedigree could help people with Huntington's disease in their family better understand their own risks for the disease and for passing it on to future generations.

Think it Over

3. Analyze Which alleles cause genetic disorders such as PKU? (Circle your choice.)

a. recessive
b. dominant

After You Read

Mini Glossary

carrier: an individual who has a recessive allele for a specific trait

pedigree: visual representation of genetic inheritance used by geneticists to map genetic traits

1. Read the key terms and definitions in the Mini Glossary above. On the lines below, use the words **pedigree** and **carrier** in a sentence.

____

____

2. Use the diagram to review what you have learned about the genetic disorders caused by recessive alleles. In the box, fill in the name of the pattern of heredity illustrated. Then fill in the remaining circles with other genetic disorders.

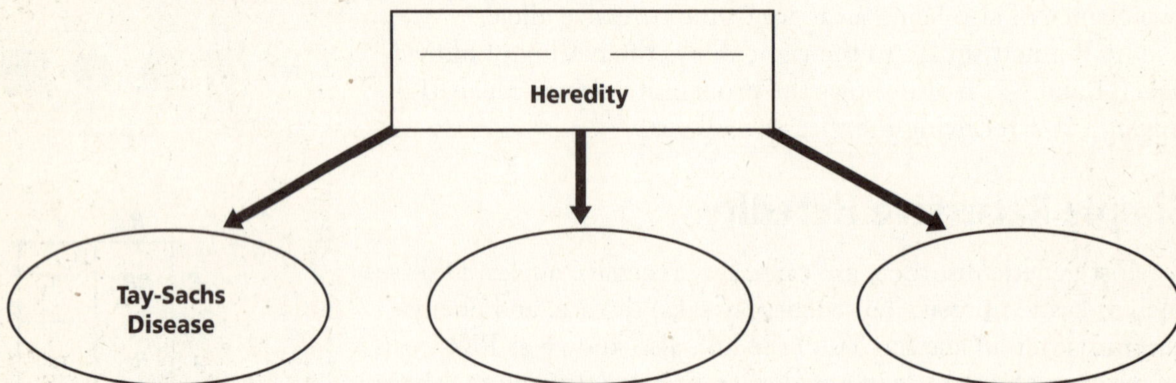

3. In the space below, draw each of the symbols used in a pedigree. Beside each symbol, write a definition of the symbol.

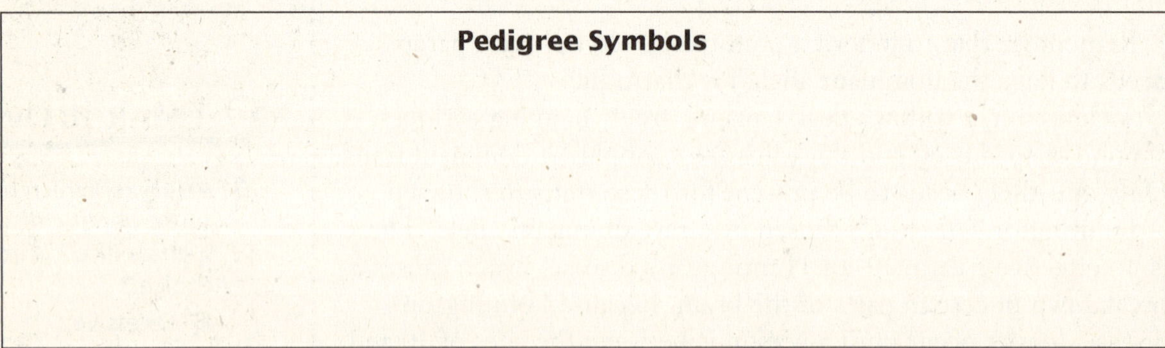

Visit the Glencoe Science Web site at **science.glencoe.com** to find your biology book and learn more about Mendelian inheritance of human traits.

Section 12.2 When Heredity Follows Different Rules

Before You Read

This section is about some of the ways in which inherited traits combine. Think about the plants in a garden or some of the animals you have seen in nature. How many different combinations of colors are there in plant and animal life? In the Read to Learn section, highlight examples of different combinations of inherited traits.

Read to Learn

Complex Patterns of Inheritance

Create a Quiz After you read this section, create a quiz based on what you have learned. Then be sure to answer the quiz questions.

Sometimes traits are not inherited through simple Mendelian genetics. Some traits are not simply dominant or recessive. When neither allele of the parents is completely dominant, the phenotype of the heterozygous offspring is a mix of the two parents. This pattern of inheritance is called **incomplete dominance.** For example, when a homozygous red snapdragon is crossed with a homozygous white snapdragon, the offspring's color will be a mix of the two. It will be pink.

Look at the Punnett squares below. The square on the left shows that the intermediate pink flower happens because neither allele of the pair is completely dominant. The square on the right shows the F_2 generation of snapdragons. Notice that when the pink flowered snapdragons are crossed with each other, the ratio of red to pink to white flowers in the F_2 generation is 1:2:1. That means that there will be one red, two pink, and one white snapdragon in the second generation. This follows Mendel's law of segregation.

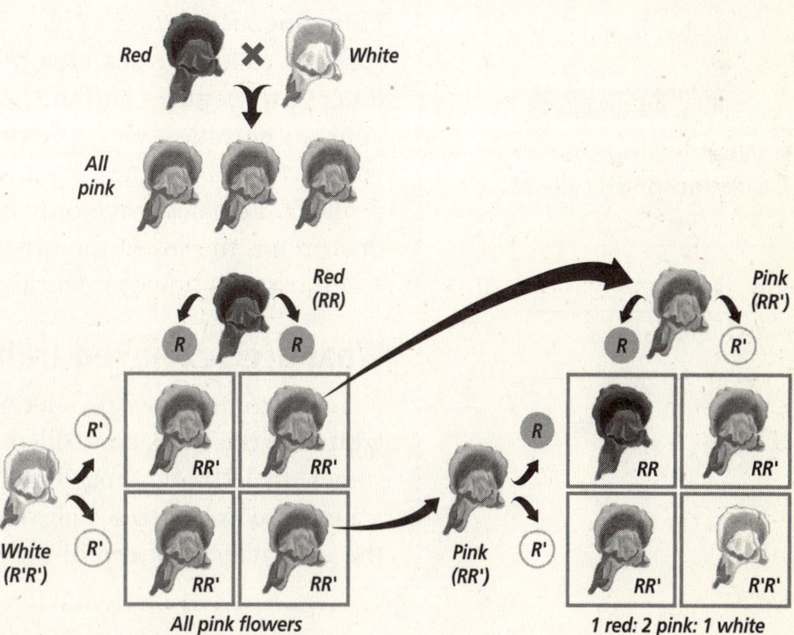

Reading Check

1. What is the term that describes when both alleles show up equally?

____

____

What is codominant inheritance?

In codominant inheritance, both alleles show up equally. **Codominant alleles** cause the phenotypes of both homozygote parents to be expressed equally in the heterozygote offspring. For example, when a certain variety of black chicken is crossed with a white chicken, all of the offspring are checkered. Some of the feathers are black and some of the feathers are white.

Think it Over

2. Conclude Which of the following is the best description of codominance? (Circle your choice.)

a. Ratio of the trait in the second generation is 1:2:1.

b. The recessive trait shows up equally with the dominant trait.

c. Phenotypes of both parents are expressed equally.

How do multiple alleles work?

In some populations, traits can be controlled by **multiple alleles.** This means there are more than two alleles for a genetic trait. We will use the pigeon population for our example. Each pigeon can only have two alleles for a genetic trait. There are multiple allele combinations for some genetic traits within the pigeon population. For example, many combinations of pigeon feather colors exist. The allele for ash-red colored feathers is dominant. The allele for wild-type blue feathers is recessive to the allele for ash-red feathers. The allele for chocolate-brown feathers is recessive to both the ash-red and the wild-type blue alleles. Sometimes there are as many as 100 alleles for a single trait!

Reading Check

3. What is the 23rd pair of chromosomes called?

____

____

____

What determines the sex of an organism?

Humans have 23 pairs of chromosomes. Twenty-two of these pairs of homologous chromosomes are called **autosomes.** Homologous autosomes look alike. The 23rd pair of chromosomes is called the **sex chromosomes** and is indicated by the letter X for females and the letter Y for males. If you are female, your sex chromosomes are homologous, XX. If you are male, your sex chromosomes are XY. Males make two kinds of gametes, X and Y. Females make only X gametes. The X or Y male gamete determines the sex of the offspring. The Punnett square on page 133 illustrates how this works.

What are sex-linked traits?

Sex chromosomes also determine sex-linked traits. **Sex-linked traits** are the traits controlled by genes located on sex chromosomes. In 1910, Thomas Hunt Morgan discovered that some traits were linked to sex chromosomes. Sex-linked traits follow the inheritance pattern of the sex chromosome on which they are found. Eye color in fruit flies is an example of an X-linked trait. This means eye color in fruit flies is determined by a gene on the X chromosome. X-linked traits are passed to both males and females. Y-linked traits are passed only to male offspring because the genes for these traits are on the Y chromosome.

Look at the Punnett square again. You will notice that any trait on a Y chromosome could only pass to a male offspring, since the male offspring are the only ones to receive a Y chromosome.

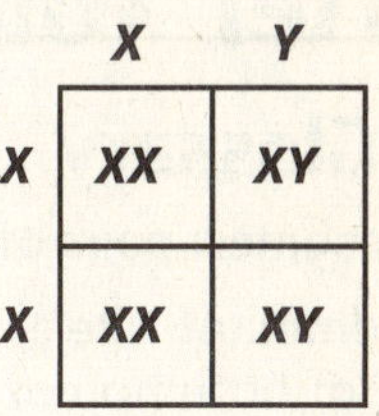

	X	Y
X	XX	XY
X	XX	XY

What is polygenic inheritance?

Some traits, such as skin color and height in humans, vary over a wide range. This is because the traits are controlled by many genes rather than by just one gene. **Polygenic inheritance** is the inheritance pattern of a trait that is controlled by two or more genes. The genes may be on the same or different chromosomes.

Think it Over

4. **Apply** Height in humans is an example of (Circle your choice.)
   a. polygenic inheritance.
   b. external environmental influences.
   c. internal environmental influences.

Environmental Influences

It is important to know that the genetic makeup of an organism at fertilization determines only the organism's potential to develop and function. Many factors can influence how the gene is expressed, or whether the gene is expressed at all. There are internal and external influences.

How does external environment affect organisms?

Nutrition, light, chemicals, infectious agents such as bacteria, fungi, parasites, and viruses, and other factors can all influence the ways genes are expressed. The arctic fox, for example, has gray-brown fur in warm temperatures. When temperatures fall, the fur becomes white. In this case, temperature is the external factor that affects the phenotype of fur color.

How does internal environment affect organisms?

The internal environments of males and females are different because of hormones and structural differences. Horn size in mountain sheep is expressed differently in males and females. In males, the horns are much heavier and more coiled than the horns of females.

The age of an organism can also affect the way genes function. The internal environment of an organism changes with age, but it is not clearly understood how these changes affect the function of genes.

After You Read

Mini Glossary

autosomes: pairs of homologous chromosomes

codominant alleles: pattern where phenotypes of both homozygous parent's alleles appear equally

incomplete dominance: neither allele of the parent is completely dominant but combine and display a new trait

multiple alleles: presence of more than two alleles for a genetic trait

polygenic inheritance: inheritance pattern of a trait controlled by two or more genes; genes may be on the same or different chromosomes

sex chromosomes: in humans, the 23rd pair of chromosomes, determine the sex of an individual and carry sex-linked traits

sex-linked traits: traits controlled by genes located on the sex chromosomes

1. Read the terms and definitions in the Mini Glossary above. Circle two terms that are related to each other. On the lines below, tell how these terms are related.

____

____

____

____

2. Fill in the partially completed outline to help you review what you have read.

I. Sex of an individual
   A. Determined by ______________________
   B. Females are represented by the letter ______________________
   C. Males are represented by the letter ______________________

II. Sex-linked traits
   A. First discovered by ______________________
   B. Genes are located on ______________________

Visit the Glencoe Science Web site at **science.glencoe.com** to find your biology book and learn more about when heredity follows different rules.

Section 12.3

Complex Inheritance of Human Traits

Before You Read

Before you read about family gene pools, think about your own family. Then, on the lines below, make a list of the ways the members of your family resemble each other.

Read to Learn

Codominance in Humans

You will remember that in codominance the phenotypes of both homozygote parents are expressed equally in the heterozygote offspring. One example of this in humans is a group of inherited red blood cell disorders called sickle-cell anemia.

Check for Understanding As you read this section, be sure to reread any parts you don't understand. Highlight each sentence as you reread it.

What is sickle-cell anemia?

Sickle-cell anemia is a major health problem in the United States and in Africa. It is most common in African Americans and in white Americans whose families came from countries around the Mediterranean Sea.

In a person who is homozygous for the sickle-cell allele, abnormal sickle-shaped red blood cells are produced. Normal red blood cells are disc shaped. Sickle-shaped cells occur in the body's narrow capillaries. They slow blood flow, block small vessels, and cause tissue damage and pain. Because of the short life span of the sickle cells, individuals with this disease have several related disorders.

Individuals who are heterozygous for the sickle-cell allele produce both normal and sickle-shaped red blood cells. This is an example of codominance. These individuals produce enough normal red blood cells that they do not have the serious health problems of those individuals who are homozygous for the allele. Individuals who are heterozygous for the allele can lead relatively normal lives.

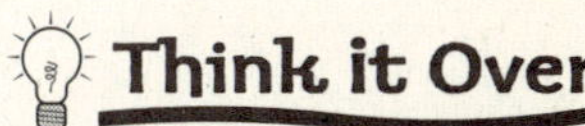

1. **Analyze** Which term describes when the phenotypes of both parents are expressed equally? (Circle your choice.)
   a. multiple alleles
   b. codominance

Possible Genotype Combinations	Phenotypes
A and A	A
A and B	AB
A and O	A
B and O	B
O and O	O

Multiple Alleles Govern Blood Type

The ABO blood group is a good example of a single gene that has multiple alleles in humans. Human blood types are determined by the presence or absence of certain molecules on the surface of red blood cells. Refer to the chart at the left to study the gene combination of blood types.

Think it Over

2. Conclude If a mother has type O blood and a father has type A blood, their child can have which of the following blood types? (Circle ALL that apply.)

a. A
b. O
c. AO

Why is blood typing important?

Blood typing determines the ABO blood group to which an individual belongs. It is necessary to know the blood type of a person before a blood transfusion can be given. If the wrong blood type is given, the red blood cells could clump together and cause death.

Blood typing can also be helpful in cases of disputed parentage. For example, if a child has type AB blood and his or her mother has type A blood, a man with type O blood could not be the father. Blood tests cannot prove that a man is the father, only that he could be. DNA tests are needed to determine actual parenthood.

Sex-Linked Traits in Humans

Genes that are carried on the sex chromosomes determine many human traits. Most of these genes are located on the X chromosome. You will remember that males pass an X chromosome to each daughter and a Y chromosome to each son. Females pass an X chromosome to both daughters and sons. If a recessive X-linked allele is passed to a son, the recessive phenotype will be expressed because there are no X-linked alleles on the Y chromosome received from the male.

Two traits that are determined by X-linked recessive inheritance in humans are red-green color blindness and the blood disorder, hemophilia. X-linked dominant and Y-linked human disorders are rare.

What is red-green color blindness?

People who have red-green color blindness are unable to tell the difference between these two colors. Color blindness is caused by the inheritance of a recessive allele on the X chromosome. One problem people with red-green color blindness have is the inability to identify red and green traffic lights by color.

What is hemophilia?

If you have ever cut yourself, you may have noticed that cuts usually stop bleeding quickly. The cut stops bleeding because the blood clots. Hemophilia is an X-linked disorder that keeps blood from clotting quickly. This means cuts do not stop bleeding very rapidly. It also means that a person could have internal bleeding from a bruise. ✓

Hemophilia can be treated with blood transfusions and injections of the blood-clotting enzyme that is absent in people with hemophilia. These treatments, however, are very expensive.

Reading Check

3. What is hemophilia?

____

____

____

Polygenic Inheritance in Humans

Think of the traits you have inherited from your parents. Many of these were inherited through simple Mendelian patterns or through multiple alleles, but others were determined by polygenic inheritance. This means two or more genes may control the inheritance pattern of a trait. Usually these traits vary over a wide range. Eye color and skin color are two examples of polygenic inheritance in humans. ✓

In the early 1900s, scientists studied data collected on skin color. They found that when light-skinned people mate with dark-skinned people, their children have intermediate skin colors. When the children have offspring, the skin colors in the offspring range from the light to the dark of the grandparents. Most of the offspring, however, have an intermediate skin color. This variation indicates that three to four genes are involved in determining skin color.

Reading Check

4. List two examples of polygenic inheritance.

____

____

Changes in Chromosome Numbers

Sometimes there are an abnormal number of cells in a set of chromosomes. To identify an abnormal number of chromosomes, a sample of cells is taken from an individual or a fetus (a developing mammal from nine weeks old to birth). The chromosomes are photographed and arranged in pairs by a computer. The pairs are arranged by length and location of the centromere. The chart showing the pairs is called a **karyotype.** It is very useful in identifying unusual chromosome numbers in cells. An example of a karyotype is at right.

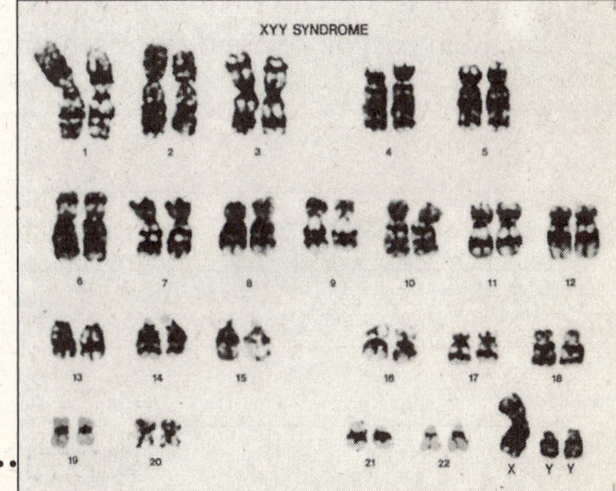

Reading Check

5. What is one example of an abnormal number of chromosomes in humans?

____

____

____

What is Down syndrome?

Most human abnormal chromosome numbers result in embryo death, usually even before a woman knows she is pregnant. Down syndrome is one example of an abnormal number of chromosomes in humans. Individuals with Down syndrome can survive to adulthood, unlike most others born with an abnormal number of chromosomes. Individuals with Down syndrome have at least some degree of mental retardation. The incidence of Down syndrome births is higher in older mothers, especially those over 40 years old.

What happens with abnormal numbers of sex chromosomes?

Sometimes an X chromosome will be missing (designated XO). There may also be an extra chromosome, such as in XXX or XXY. There may even be an extra Y chromosome (XYY). Any individual with a Y chromosome is a male. Any individual without a Y chromosome is a female. What happens in these cases? Most of these individuals lead normal lives but cannot have children. Some have varying levels of mental retardation.

After You Read

Mini Glossary

blood typing: determination of ABO blood group to which an individual belongs

karyotype (KAHR ee uh tipe): chart of chromosome pairs arranged according to length and location of the centromere; used to identify an abnormal number of chromosomes

1. Read the terms and their definitions in the Mini Glossary above. On the lines below, use the word **karyotype** correctly in a sentence.

____

____

____

____

____

Section
12.3 Complex Inheritance of Human Traits, *continued*

2. In Column 1 are some concepts you learned in this section. Column 2 gives a fact about each concept. Put the letter of the fact on the line next to the concept that matches it.

Column 1	Column 2
_____ 1. sickle-cell disease	a. human blood types
_____ 2. A, B, AB, O	b. an example of codominance
_____ 3. sex-linked traits	c. abnormal number of chromosomes
_____ 4. 47 chromosomes	d. mostly located on the X chromosome

Visit the Glencoe Science Web site at **science.glencoe.com** to find your biology book and learn more about the complex inheritance of human traits.

Name Date Class

Section 13.1 Applied Genetics

Before You Read

Skim the Read to Learn section below and highlight two important facts about applied genetics. Write those two facts in the space below.

Read to Learn

STUDY COACH

Create a Quiz After you have read this section, create a quiz based on what you have learned. After you have completed writing the quiz questions, be sure to answer them.

Selective Breeding

Humans were involved in applied genetics long before they knew what genes were. When our ancestors planted the seeds of the juiciest berries to grow new plants, they were using selective breeding techniques. When farmers breed the calves of the best milk-producing cow, they are using selective breeding.

Selective breeding takes time. Several generations of offspring need to be bred before the desired trait becomes common in the population. Selective breeding increases the frequency of desired traits, or alleles, in a population. An example of successful selective breeding is found in dairy cows. Today, the average dairy cow produces three times more milk than the average dairy cow did fifty years ago. This means that fewer than half the number of cows are needed to produce the same amount of milk.

Think it Over

1. **Conclude** How does selective breeding influence the frequency of desired traits in a population? (Circle your choice.)
   a. increases the frequency
   b. decreases the frequency

What is inbreeding?

To make sure that breeds always have the desired traits, and to eliminate undesired traits, breeders often use inbreeding. **Inbreeding** is mating between closely related individuals. Inbreeding creates individuals who are homozygous for most traits. Sometimes in inbreeding, harmful, recessive traits appear. That is because there is a greater chance that two closely related individuals both carry a recessive allele for a harmful trait.

Most people are familiar with pure breed dogs. A breed is a group of organisms within a species that has been bred for certain characteristics. For example, a German shepherd is a breed. All pure breed German shepherds have long hair that is black and tan, a black muzzle, and they resemble a wolf.

Another form of selective breeding is creating hybrids. A hybrid is the offspring of parents that have different forms of a trait. Plants are often hybrids. To get a fragrant, long-stemmed rose, breeders might cross a fragrant rose with a long-stemmed rose. The result of that cross is a hybrid. Often, plant hybrids are larger and healthier than the parent plants. ✓

Reading Check

2. What is a hybrid?

____

____

____

Determining Genotypes

Breeders must carefully select the plants or animals that have the greatest chance of passing on the desired traits. But you cannot always tell by looking at an organism if it is homozygous or heterozygous for a trait. How can breeders determine if an organism is heterozygous? They use the information that Mendel discovered in his experiments. Breeders perform a test cross to determine the combination of genes, or the genotype, of an organism. A **test cross** is a cross of an individual of unknown genotype with an individual of known genotype. ✓

Reading Check

3. What is a test cross?

____

____

____

How do you determine genotype?

Let's use Alaskan malamute dogs for our example of how to determine genotype. There is a recessive trait that causes dwarfism in Alaskan malamutes. For a malamute to be born a dwarf, the dog would have to have two recessive alleles (homozygous recessive) for that trait. Before breeding an Alaskan malamute, a breeder makes sure the dog does not carry the trait for dwarfism. Because the dog is normal size, the breeder knows the dog must have either two dominant alleles (homozygous dominant), or a dominant and a recessive allele (heterozygous). To perform a test cross, the breeder will mate a dwarf dog (homozygous recessive) with the unknown dog. What are the possible genotypes of the offspring?

A Punnett square will predict possible genotypes of the offspring (see the illustration on page 142). If the unknown dog has two dominant alleles for normal height (homozygous dominant), the offspring will all be heterozygous and normal height. If the unknown dog has one dominant and one recessive allele (heterozygous), half the offspring will be heterozygous and normal height. The other half will be dwarfs because they will have two recessive alleles (homozygous recessive).

A The unknown dog can be either homozygous dominant (DD) or heterozygous (Dd) for the trait.
?
×
dd

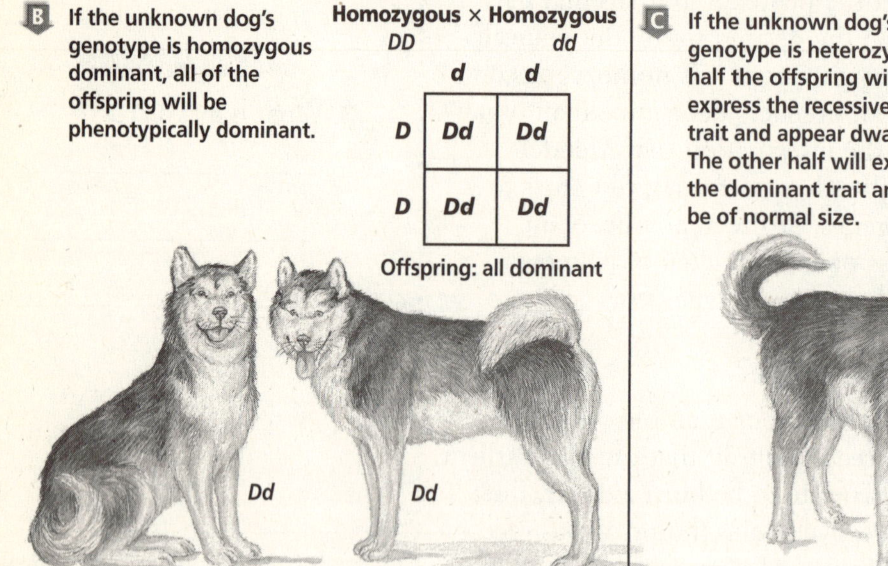
B If the unknown dog's genotype is homozygous dominant, all of the offspring will be phenotypically dominant.
Homozygous × Homozygous
DD dd
Offspring: all dominant
Dd
Dd

	d	d
D	Dd	Dd
D	Dd	Dd

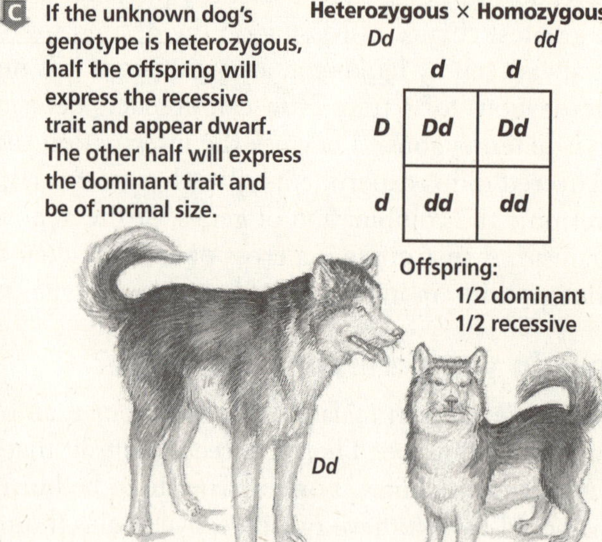
C If the unknown dog's genotype is heterozygous, half the offspring will express the recessive trait and appear dwarf. The other half will express the dominant trait and be of normal size.
Heterozygous × Homozygous
Dd dd
Offspring:
1/2 dominant
1/2 recessive
Dd
dd

	d	d
D	Dd	Dd
d	dd	dd

Remember from the beginning of this section that selective breeding increases the frequency of desired traits in a population. In our Alaskan malamute example, the breeder would only continue to breed the dog if the results of the test cross proved that the dog did not carry the recessive trait.

After You Read

Mini Glossary

inbreeding: mating closely related individuals; ensures that offspring are homozygous for most traits, but also brings out harmful, recessive traits

test cross: mating of one member of a species of unknown genotype with another member of the species of known genotype; can help determine the unknown genotype of the parent

Section 13.1 Applied Genetics, *continued*

1. Read the terms and their definitions from the Mini Glossary on page 142. Write a sentence for each term, using the term correctly.

2. Place the information found below the table under the correct heading in the table.

Selective Breeding	Determining Genotypes

a. Uses test crosses
b. Produces hybrids that are usually bigger and better
c. Uses inbreeding
d. Works best if known individual is homozygous recessive

3. Write a paragraph that explains how applied genetics has helped improve human life.

Visit the Glencoe Science Web site at **science.glencoe.com** to find your biology book and learn more about applied genetics.

Name Date Class

Section 13.2 Recombinant DNA Technology

Before You Read

This section discusses the technology that allows scientists to combine the DNA of one organism with the DNA of another organism. The result can be organisms with new characteristics. What would you think of a plant that glows like a firefly or a cotton crop that produces its own insecticide? On the lines below, list some examples of useful changes to an organism that you would like to see made through DNA technology.

____

____

____

Read to Learn

Mark the Text **Identify Details** Highlight each key term introduced in this section. Say the key term aloud. Then, highlight the sentence that explains the key term.

1. What advantages does genetic engineering have over selective breeding?

____

____

____

Genetic Engineering

In the previous section you learned that selective breeding increases the frequency of desired traits, or alleles, in a population. You also learned that selective breeding techniques such as inbreeding and creating hybrids take time. In many cases the offspring have to mature before the traits become obvious. Sometimes it takes several generations before the desired trait becomes common in the population.

There is a faster and more reliable way to increase the frequency of a desired allele in a population. It is called genetic engineering. In **genetic engineering,** very small pieces or fragments of DNA are cut from one organism and placed inside another organism. When DNA is made by connecting, or recombining, fragments of DNA from different sources, it is called **recombinant** (ree KAHM buh nunt) **DNA.** ✓

An organism uses the recombinant DNA as if it were its own. The DNA of two different species can even be combined. For example, inserting a specific part of the DNA of a firefly into the DNA of a plant will cause the plant to glow. When an organism contains recombinant DNA from a different species, it is called a **transgenic organism.** The glowing plant is an example of a transgenic organism.

What is the process for producing a transgenic organism?

Producing a transgenic organism is a three-step process. The first step is to cut the DNA fragment out of one organism. The second step is to connect the DNA fragment to a carrier. The third step is to insert the DNA fragment and its carrier into a new organism. Let's take a closer look at each step.

How is a DNA fragment cut from an organism?

Scientists have discovered that there are proteins called **restriction enzymes** that cut DNA. They are bacterial proteins that can cut both strands of the DNA molecule at a specific nucleotide sequence. There are hundreds of different restriction enzymes. Each one cuts DNA in a different place. In our example, Step 1 is picking a restriction enzyme that cuts the firefly DNA strand at the sequence that codes for making the enzyme that lights up the firefly.

How are DNA fragments connected to a carrier?

Organisms do not easily accept loose fragments of DNA from other organisms. For this reason, the DNA fragment needs a carrier to take it into the host cell. This is Step 2. In our example, the firefly DNA gets inserted into the carrier or vector DNA. A **vector** is the means by which DNA from another species can be carried into a host cell. ✓

Biological vectors include viruses and plasmids. A **plasmid** is a small ring of DNA found in a bacterial cell. The genes of a plasmid are different from those on the bacterial chromosome. In our example, the firefly DNA strand was inserted into a plasmid in a bacterial cell in Step 2. Now, in Step 3, the plasmid vector is inserted into a plant.

What are clones?

With the firefly DNA now a part of it, the plasmid reproduces within the bacterial cell, making up to 500 copies of itself. Every time the host cell divides, it copies the recombinant DNA as it copies its own DNA. Such genetically identical copies are called **clones.** Each identical recombinant DNA molecule is called a gene clone. Because the bacterial host cells in the plant will continue to copy the recombinant DNA, the plant will always have the firefly's DNA—and its light—within it.

✓ Reading Check

2. What is a vector?

____

____

____

Section 13.2 Recombinant DNA Technology, *continued*

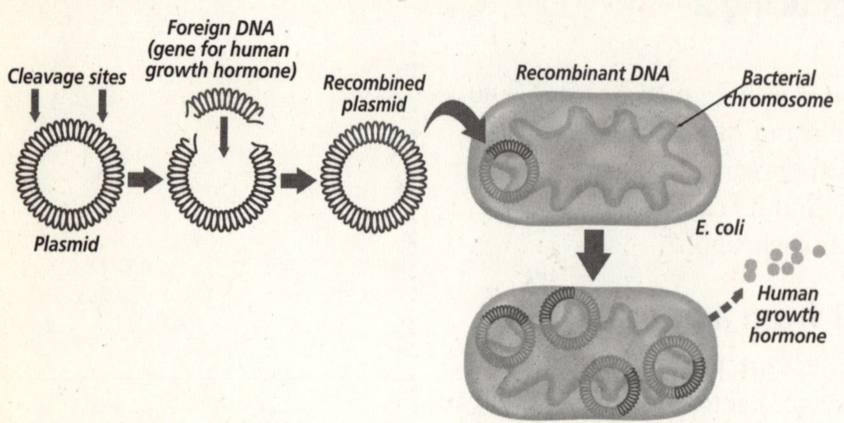

In some experiments, scientists insert particular types of recombinant DNA into host cells. This DNA has code within it to make a certain type of protein. Scientists then study what this protein does in cells that do not ordinarily produce it. At other times, scientists produce mutant forms of a protein. They then study how the mutation affects the function of the protein within a cell.

Technology has made gene cloning fairly simple. Scientists have built upon gene cloning to clone an entire animal. The most famous cloned animal was Dolly the sheep, cloned in 1997. The cloning process is not efficient, but scientists hope someday to use it so that ranchers and farmers can clone the most productive, healthy animals to increase and improve the food supply. ✓

Scientists developed a method of replicating DNA outside of living organisms, called polymerase chain reaction (PCR). This method uses heat to separate DNA strands from each other. An enzyme from a heat-loving bacterium is used to replicate the DNA when the correct nucleotides are added to a PCR machine. The PCR machine can make millions of copies of DNA in a day. Scientists analyze bacterial, plant, animal, and human DNA. Scientists use this type of DNA analysis in crime investigations and in the diagnosis of disease. Scientists also use PCR to provide pure DNA that is used to determine the correct sequence of DNA bases. This information helps scientists identify mutations.

✓Reading Check

3. Why do scientists want to clone animals?

____

____

____

Applications of DNA Technology

How can humans benefit from DNA technology? Three main areas seem to offer the greatest promise: industry, medicine, and agriculture. For example, scientists have changed the *E. coli* bacteria to produce the expensive blue dye used to color denim blue jeans. Scientists are also trying to develop corn that contains as much protein as beef. ✓

In medicine, recombinant DNA is used to produce insulin and the human growth hormone. The human gene responsible for clotting blood has been inserted into sheep chromosomes. The

✓Reading Check

4. Which three areas will most likely benefit from DNA technology?

____

____

____

sheep produce the clotting protein, which is then used for patients with hemophilia, a disease in which blood cannot clot quickly.

Researchers are discovering ways to increase the amount of vitamins in certain crops. That will help provide better nutrition. Some plants have already been developed that produce toxins to make them resistant to insects. That will limit the use of dangerous pesticides.

After You Read

Mini Glossary

clone: genetically identical copy of an organism or gene

genetic engineering: method of cutting DNA from one organism and inserting the DNA fragment into a host organism of the same or different species

plasmid: small ring of DNA found in a bacterial cell that is used as a biological vector

recombinant (ree KAHM buh nunt) DNA: DNA made by recombining fragments of DNA from different sources

restriction enzyme: DNA-cutting enzyme that can cut both strands of a DNA molecule at a specific nucleotide sequence

transgenic organism: organism that contains recombinant DNA from a different species

vector: means by which DNA from another species can be carried into the host cell; may be biological or mechanical

1. Read the terms and their definitions in the Mini Glossary above. Circle two terms that are related to one another. On the lines below, tell how these terms are related.

____

____

2. In the boxes below, list the steps for producing a transgenic organism.

Recombinant DNA Process for Producing a Transgenic Organism

Step 1	Step 2	Step 3

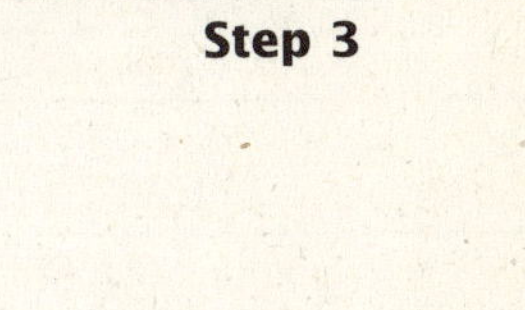

Visit the Glencoe Science Web site at **science.glencoe.com** to find your biology book and learn more about recombinant DNA technology.

Section 13.3 The Human Genome

Before You Read

This section discusses the effort to identify and map all the human genes. When the project is finished, it will be an incredible accomplishment. On the lines below, give an example of another important scientific achievement that you know about. Describe how the world might be different today if that achievement were never made.

Read to Learn

Mark the Text **Identify Main Ideas** As you read this section, stop after every few paragraphs and put what you have just read into your own words. Then highlight the main idea in each paragraph.

1. What is the Human Genome Project?

Mapping and Sequencing the Human Genome

The Human Genome Project is an international effort to completely map and sequence the human genome. The **human genome** is the approximately 35 000 to 40 000 genes on the 46 human chromosomes. Sequencing means putting each of these genes in the right order. The project began in 1990. Today, thousands of genes have been mapped to particular chromosomes. Half of the genome has been sequenced. But scientists still do not know the exact location of all the genes on the chromosomes. ✓

The genetic map that shows the relative locations of genes on a chromosome is called a **linkage map.** Imagine a map of your state that shows cities as being north, south, east, or west of each other, but does not show the exact location of each city. That is how a linkage map works. It shows relative location, not exact location.

Originally, the information used to assign genes to particular chromosomes came from linkage data of human pedigrees. Remember that in meiosis, genes sometimes cross over onto different chromosomes. Scientists know that genes that are farther apart cross over more often than genes that are close together. That information helps scientists create a linkage map. But mapping by linkage data is not efficient because scientists had to wait for individual humans to reproduce and mature in order to identify which genes were passed on.

Now, a faster more efficient way to map genes is available. Using polymerase chain reaction (PCR), millions of copies of DNA fragments are cloned in a day. Since scientists know the location of some genes and some segments of DNA, they are used as genetic markers. Because DNA segments that are near each other on a chromosome are often inherited together, markers are used to track the inheritance pattern of a gene that has not yet been identified.

Genes are sequenced by cutting DNA into fragments using restriction enzymes. The fragments are cloned, then put in the right order. The order is determined by overlapping matching sequences. Machines can perform this work, increasing the speed of map development.

Applications of the Human Genome Project

How will these chromosome maps be used? Doctors will be able to diagnose genetic disorders even in unborn babies. Gene therapies might be developed to correct genetic disorders. Law enforcement workers will be able to link suspects to evidence left at crime scenes.

Diagnosing genetic disorders has been an important benefit of the Human Genome Project. A diagnosis can be made before birth. Doctors take cells from the fluid surrounding the fetus and analyze the DNA. They can determine if the fetus will develop a genetic disorder. Now, thanks to DNA technology, doctors can use gene therapy to help individuals with genetic disorders. **Gene therapy** is the insertion of normal genes into human cells to correct genetic disorders. Doctors are conducting experiments involving gene therapy for cystic fibrosis, sickle-cell anemia, and hemophilia. Research is also going on to use gene therapy on cancer, heart disease, and AIDS. It is hoped that gene therapies will be developed to treat many different disorders. ✓

DNA technology is also helping law-enforcement workers solve crimes using DNA fingerprinting. All it takes is a small sample of hair, skin, blood, or other body tissue found at a crime scene. This sample's DNA is then compared with a DNA sample from a suspect. If the DNA samples match, the suspect most likely is guilty. DNA fingerprinting works because no two individuals (except identical twins) have the same DNA sequences, and because all cells of an individual (except gametes) have the same DNA. ✓

Geneticists are using polymerase chain reaction (PCR) to clone DNA from mummies to better understand ancient life. The DNA from fossils has been studied to compare extinct species with living species, or even two extinct species with each other. The uses of DNA technology are unlimited.

Reading Check

2. What is gene therapy?

____

____

____

Reading Check

3. Why does DNA fingerprinting work?

____

____

____

After You Read

Mini Glossary

gene therapy: insertion of normal genes into human cells to correct genetic disorders

human genome: map of the thousands of genes on 46 human chromosomes that when mapped and sequenced, may provide information on the treatment and cure of genetic disorders

linkage map: genetic map that shows the relative locations of genes on a chromosome

1. Read the key terms and definitions in the Mini Glossary above. Write a sentence using at least two of the terms correctly.

____

____

2. In the top row of boxes, write the ways that results from the Human Genome Project can be used. In the bottom row, give specific examples. One box from each row has been filled in for you.

Applications of the Human Genome Project

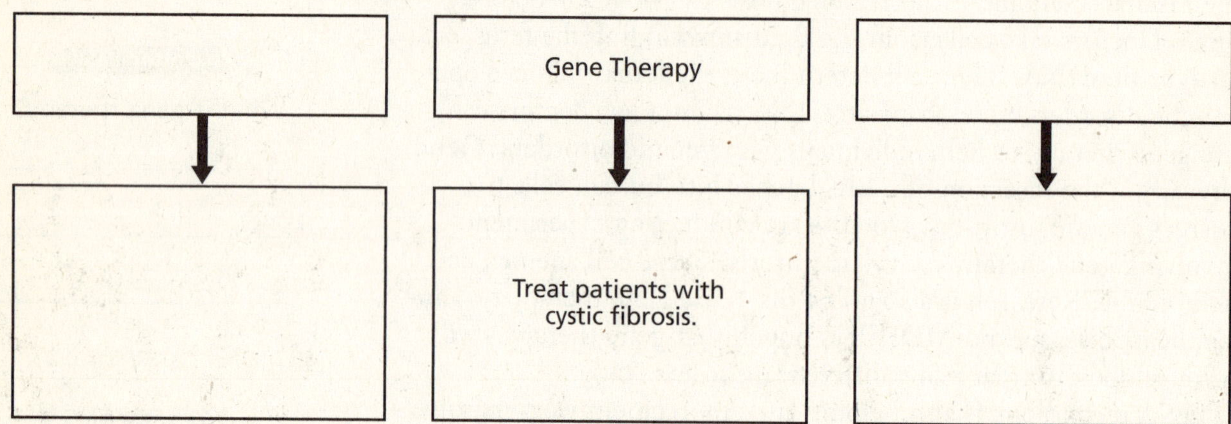

3. Circle T or F to indicate whether each of the following sentences is true or false.

T F 1. Humans have more than 75 000 genes located on 48 chromosomes.

T F 2. A linkage map shows the exact location of a gene on a chromosome.

T F 3. Using polymerase chain reaction (PCR), millions of copies of DNA fragments can be cloned in a day.

T F 4. Gene therapy is the insertion of abnormal genes into human cells to correct genetic disorders.

Visit the Glencoe Science Web site at **science.glencoe.com** to find your biology book and learn more about the human genome.

Section 14.1 The Record of Life

Before You Read

This section discusses how Earth itself contains a record of life. The evidence is found in rocks. In the space below, describe what evidence you think the rocks provide about life.

Read to Learn

Early History of Earth

What was early Earth like?

Some scientists think that early Earth was very hot. Earth's surface may have been hit hard and often by meteorites. These activities would have heated the surface. The inside of Earth was also under a lot of pressure. It became so hot that volcanoes violently gave off lava and hot gases that relieved the pressure. The gases helped to form the atmosphere, but it was not an atmosphere you could breathe. This early atmosphere did not have oxygen.

About 4.5 billion years ago, Earth cooled enough for the water vapor in the atmosphere to condense. This probably led to millions of years of thunderstorms. Enough water fell to create oceans. Some scientists think that some time after this, between 3.9 and 3.5 billion years ago, life formed in the oceans.

History in Rocks

Can scientists be sure how Earth formed? No, they cannot. Because rocks are constantly destroyed and new rocks formed, scientists cannot study any rocks that were created in Earth's earliest history. The oldest rocks found on Earth are about 3.9 billion years old. But rocks are still an important source of information about the many forms of life that existed on this planet. ✓

STUDY COACH

Make Flash Cards Making flash cards is a good way to learn. As you read the material, think of questions a teacher might ask on a test. Write each question on one side of a flash card. Then write the answer to the question on the other side. Use the flash cards to quiz yourself until you know the answers.

Reading Check

1. Why can scientists not be sure how Earth was formed?

Think it Over

2. **Infer** How might fossil teeth reveal an animal's diet?

What do fossils tell us?

Some rocks contain fossils. A **fossil** is evidence of an organism that lived long ago and that became preserved in Earth's rocks. From fossils, scientists have discovered that the millions of plants and animals living today are just a small part of all the species of plants and animals that have ever lived. In fact, scientists think that about 95 percent of all the species of plants and animals that ever lived on Earth are now extinct.

What do paleontologists study?

Paleontologists (pay lee ahn TAHL uh justs) are scientists who study ancient life. They are a lot like detectives. They use fossils as clues to things that happened long ago. A fossil can show the kinds of organisms that lived long ago. Fossils can even show something about the behavior of ancient animals. For example, fossil bones and teeth provide clues about the size of an animal, how it moved around, and what it ate.

Paleontologists also learn about ancient climate and geography. When they find a fossil of an ancient plant that looks like a plant we see today, they might conclude that the ancient plant needed the same type of climate as today's plant. If they find fossils of sea creatures in an area that is currently a desert, they might realize that the desert used to be an ocean.

How do fossils form?

Often, fossils formed when plants or animals died in mud, sand, or clay. They became covered with more mud, sand, or clay. Over time, the area where the living thing died became compressed or packed down. It hardened into a type of rock called sedimentary (sed uh MENT uh ree) rock. In a similar way, fossils are forming today at the bottoms of lakes, streams, and oceans. Fossils are also found frozen in ice or hardened into a tree sap called amber. Most fossils, though, are found in sedimentary rock.

How are fossils dated?

When scientists discover a fossil, they want to know how old it is. There are several ways to figure out the age of fossils. One is a method called relative dating. Sedimentary rocks form in layers, with new or younger layers added on top of older layers. The figure on page 153 shows an example of sedimentary layers. Fossils in lower layers are thought to be older than fossils in higher layers.

This does not tell scientists the exact age of the fossil. However, it does tell scientists the fossil's age in relation to other fossils above it or below it. ✓

What are radiometric dating techniques?

To find the specific age of a fossil or rock, scientists use radiometric dating techniques. These methods measure the rate of decay, or breakdown, of radioactive isotopes within the rocks. Sedimentary rocks cannot be dated this way. For that reason, scientists use other kinds of rocks that are near the sedimentary rocks.

In an earlier chapter you learned that a radioactive isotope is an atom with an unstable nucleus. The nucleus breaks down, or decays, over time, giving off radiation. When the nucleus is completely decayed, a new radioactive isotope has formed. Scientists use the decay rate like a clock. Scientists try to find out the age of a rock by comparing the amount of radioactive isotope it contains and the amount of new isotope it forms after decaying. Let's say there is a rock that contains a radioactive isotope that decays to half its original amount in one million years. That means that the half-life of that particular isotope is one million years. If the rock has equal amounts of the original isotope and the new isotope, it must be one million years old. The two most common isotopes that scientists use to date fossils and rocks are potassium-40 and carbon-14. Both are radioactive isotopes. ✓

Reading Check

3. Explain relative dating of fossils.

____

____

____

Reading Check

4. List the two most common radioactive isotopes used by scientists to date rocks and fossils.

____

____

A Trip Through Geologic Time

By looking at fossils and by figuring out the age of rocks, scientists have come up with something like a calendar of Earth's history. This calendar is called the geologic time scale. The scale is divided into four sections called *eras*. Each era represents a very long period of time. The four eras are:

1. Precambrian (pree KAM bree un)
2. Paleozoic (pay lee uh ZOH ihk)
3. Mesozoic (me zuh ZOH ihk)
4. Cenozoic (se nuh ZOH ihk).

Name Date Class

Section 14.1 The Record of Life, *continued*

Reading Check

5. What is a mass extinction?

___

___

___

Each era is subdivided into *periods.* Scientists selected the divisions based on information from fossils. Each division of the time scale is based on the kinds of organisms that lived during that time. The fossil record also shows several times when a mass extinction took place. A mass extinction is an event that happens when many organisms disappear from the fossil record at the same time. The figure below shows the four eras of Earth's geologic time scale.

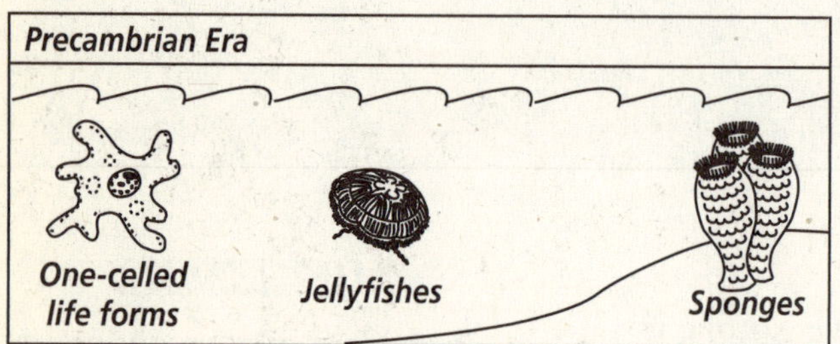

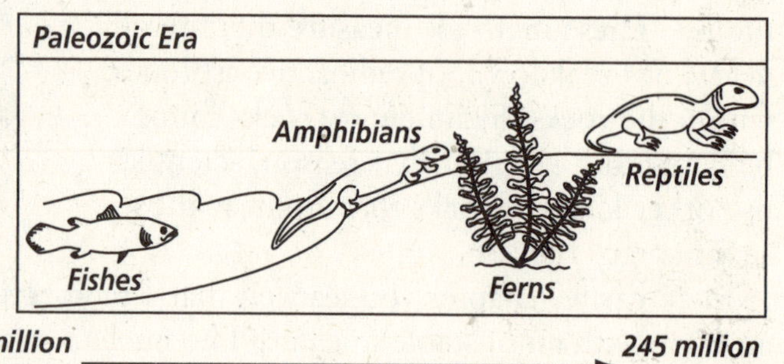

4.6 billion years ago → *544 million years ago* → *245 million years ago*

245 million years ago → *66 million years ago* → *present*

It is hard to understand the huge amounts of time that have passed since the early history of Earth. If we compare the geologic time scale to an imaginary calendar, it will be easier.

What life existed during the Precambrian era?

In your imaginary calendar, January 1 is the day Earth formed. The earliest fossils have been found in Precambrian rocks. Precambrian is toward the end of March on our calendar. Scientists found these fossils in the deserts of western Australia. They have found similar fossils on other continents. The fossils look like a modern species of photosynthetic cyanobacteria, which you will learn about later. These early fossils show that during the Precambrian era, living things were mostly one-celled organisms. By the end of the Precambrian era, however, multi-celled organisms such as sponges and jellyfishes were present. The Precambrian era makes up about 87 percent of Earth's history, which is the middle of October.

What life appeared during the Paleozoic era?

During the next era, the Paleozoic era, the number of different kinds of organisms increased greatly. In fact, there was such a variety of living things that the Paleozoic era has been divided into six different periods.

During one of the periods, the Cambrian, so many new types of living things appeared that paleontologists often refer to an "explosion." During this time, many different types of animals such as worms, sea stars, and arthropods lived in the oceans. Scientists also think that during the Paleozoic era fishes appeared. These were the first animals with backbones. ✓

Ferns, seed plants, and four-legged animals such as amphibians and reptiles also appeared during the Paleozoic era. However, the Paleozoic era ended with the largest mass extinction that shows up in the fossil record. At that time, about 90 percent of species that lived in the oceans and about 70 percent of land species disappeared.

What changes occurred during the Mesozoic era?

The third era, the Mesozoic era, would begin on December 10 on your calendar. Many changes happened during this time. The Mesozoic era is divided into three periods: the Triassic Period, the Jurassic Period, and the Cretaceous Period. Fossils show that mammals appeared on Earth during the Triassic Period. Many scientists think that birds evolved from dinosaurs during the Jurassic Period.

New types of mammals appeared during the Cretaceous Period. Also, flowering plants became plentiful on Earth. A mass extinction occurred at the end of the Mesozoic era. This was the end of the dinosaurs. Scientists estimate that at that time, two-thirds of all living species became extinct.

✓ Reading Check

6. What is the "explosion" paleontologists refer to?

____

____

____

Why did this extinction happen?

Some scientists propose that Earth was struck by a meteor at this time. The collision could have caused a huge, possibly poisonous dust cloud to fill the atmosphere. This may have changed the climate and so fewer species survived. There is a huge crater in the waters off eastern Mexico that scientists suggest was the impact site of this meteor.

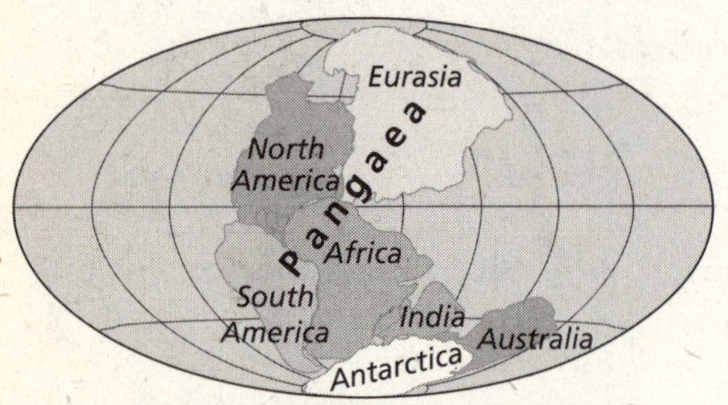

A About 245 million years ago, the continents were joined in a landmass known as Pangaea.

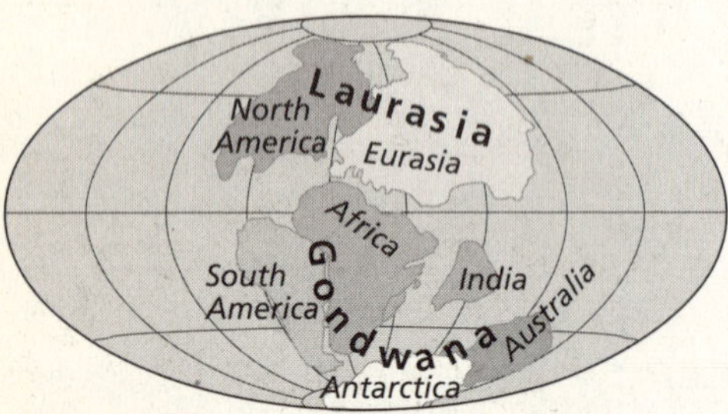

B By 135 million years ago, Pangaea broke apart resulting in two large landmasses.

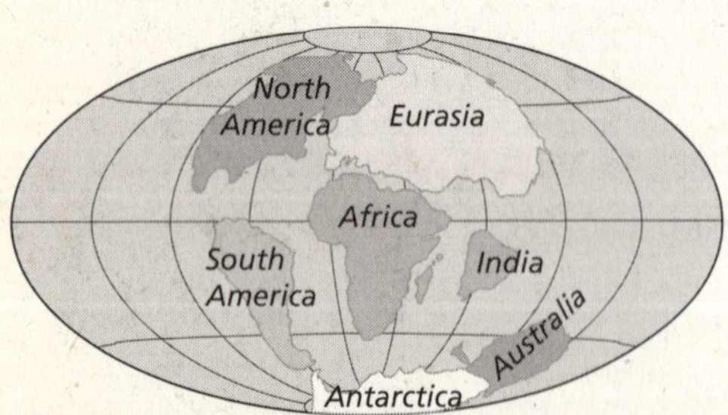

C By 65 million years ago, the end of the Mesozoic, most of the continents had taken on their modern shapes.

How did Earth change during the Mesozoic era?

It is thought that in the early part of the Mesozoic era, all the continents were joined together as one large landmass. Later, the landmass broke up and the continents began to drift apart. The figure to the left shows this idea. By the end of the Mesozoic era, the position of the continents was similar to what we know today.

What caused Earth to change? It may be that the continents are part of rigid plates on Earth's surface. The plates sit on top of a molten layer of rock, which allows the plates to move around. The theory of how continents move is called **plate tectonics** (tek TAH nihks). The plates are still moving. Today they move at a rate of about six centimeters per year. This is about the same rate at which hair grows.

What life appeared during the Cenozoic era?

The current era, the Cenozoic, began about 65 million years ago. That is December 26 on your calendar. We live in the Cenozoic era. During this time large mammals and humans first appeared. The modern human species, to which you belong, first appeared about 200 000 years ago. That is late in the evening of December 31 on the calendar.

After You Read

Mini Glossary

fossil: physical evidence of an organism that lived long ago that scientists use to study the past; evidence may appear in rocks, amber, or ice

plate tectonics (tek TAH nihks): geological explanation for the movement of continents over Earth's thick, liquid interior

1. Read the terms and their definitions in the Mini Glossary above. Select one of the terms and create a drawing in the space below that will help you remember its definition.

2. Place the statements below the table under the correct heading in the table.

The Record of Life

Fossils	Geological Time Scale

a. May be a bone, tooth, or imprint
b. Divided into eras
c. Provide information about the variety of life in the past
d. Eras are further divided into periods
e. Divisions based on kinds of organisms that lived then
f. Can provide information about ancient climates

3. Review the section, then read each sentence below and fill in the blanks. Use the following terms to fill in the blanks: **sedimentary, radiometric, plate tectonics, Cenozoic.**

a. Humans and large mammals appeared in the ______________________ Era.

b. The theory of ______________________ explains how continents move.

c. Most fossils are found in ______________________ rocks.

d. Scientists use ______________________ dating to determine the age of rocks and fossils.

Visit the Glencoe Science Web site at **science.glencoe.com** to find your biology book and learn more about the record of life.

Section 14.2 The Origin of Life

Before You Read

In this section, you will explore hypotheses that scientists have developed to explain how life on Earth began. Scientists develop experiments that help them test hypotheses. On the lines below, list something in biology that you would like to learn more about. Then write a sentence explaining an experiment that you could use to learn more about the topic.

Read to Learn

Mark the Text **Identify Key People** Underline the name of each scientist introduced in this section. Say the name aloud. Then highlight the sentence that explains the main contribution the person made to understanding the origin of life.

1. What did Francesco Redi's experiment prove?

Origins: The Early Ideas

Before the invention of microscopes and modern scientific equipment, ideas about the origin of life came from what people saw in the world they lived in. When people noticed maggots on meat, it was easy to assume that the maggots came from the meat. Likewise, when people found baby mice in sacks of grain, they thought that the grain produced the mice. These types of observations led people to believe in spontaneous generation. **Spontaneous generation** is the idea that something that is not alive can produce living things.

How was spontaneous generation disproved?

An experiment, performed in 1668, showed that maggots were not produced by decaying meat. An Italian doctor, Francesco Redi, put meat into different jars. He covered some of the jars, and left the other jars open. Flies buzzed around the tops of the covered jars but could not get to the meat. However, the meat in the open jars attracted many flies. Maggots were soon growing in those jars. The maggots developed into flies. No maggots or flies appeared in the covered jars. It was easy to conclude from this experiment that maggots do not come from spontaneous generation. Flies produce maggots, which become flies.

What did Pasteur's experiments prove?

With the development of the microscope, scientists could see that microorganisms live everywhere. They wondered if microorganisms came from the air. Was there some vital force in the air that caused microorganisms to arise spontaneously?

This was a more difficult idea to prove false. However, in the mid-1880s, Louis Pasteur designed an experiment that disproved the spontaneous generation of microorganisms.

Pasteur used a specially designed glass bottle that allowed air to enter the bottle, but kept microorganisms out. There was a broth with nutrients in the bottle. As long as Pasteur kept the microorganisms out, the broth remained clear.

Pasteur's experiment showed that microorganisms come from other microorganisms. This led to the concept of **biogenesis** (bi oh JEN uh sus), the idea that living things come only from other living things.

Think it Over

2. Describe How did Pasteur develop the concept of biogenesis?

____

____

____

Origins: The Modern Ideas

For more than 100 years biologists have accepted the idea of biogenesis. But that still does not answer the question: How did life begin on Earth? There are many theories, none of which have been proven scientifically. Scientists continue to look into theories and test hypotheses about what conditions existed on early Earth.

How did simple organic molecules form?

Scientists hypothesize that two things happened before life could appear on Earth. The first is that there had to be organic molecules—molecules containing carbon. The second is that the organic molecules must have formed into more complex molecules such as proteins, carbohydrates, and nucleic acids. These are the materials that are absolutely necessary for life. ✓

Reading Check

3. What two things had to happen before life could appear on Earth?

____

____

Scientists think that the early atmosphere of Earth did not contain any free oxygen. Instead, the atmosphere was made of water vapor, carbon dioxide, nitrogen, methane, and ammonia. How did this mixture combine and form the organic molecules that are found in all living things today?

In the 1930s, a Russian scientist named Alexander Oparin suggested that lightning striking Earth, heat from the sun, and Earth's own heat started chemical reactions using substances in Earth's atmosphere. The chemical reactions made small organic molecules that ended up in the ocean when it rained. This turned the ocean into an environment in which life could start. This hypothesis is illustrated in the figure at the top of page 160.

How was Oparin's hypothesis tested?

In 1953, two American scientists, Stanley Miller and Harold Urey, tested Oparin's hypothesis. They mixed the elements

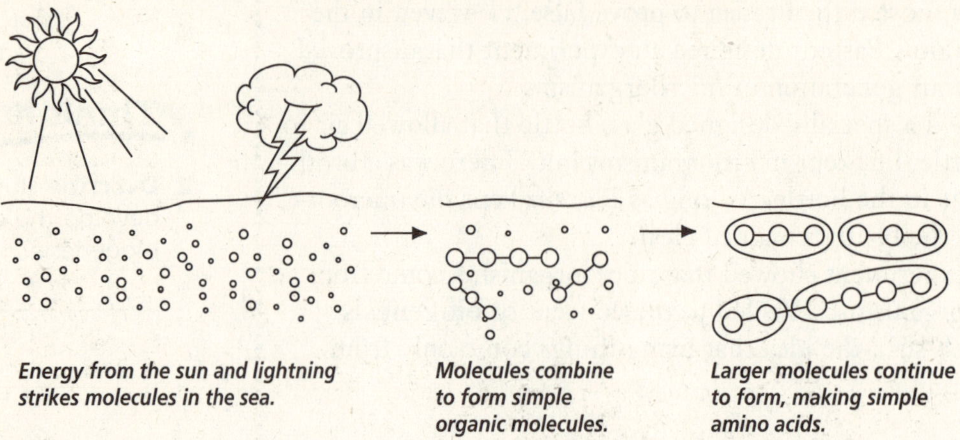

Energy from the sun and lightning strikes molecules in the sea.

Molecules combine to form simple organic molecules.

Larger molecules continue to form, making simple amino acids.

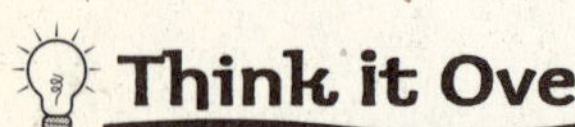

4. Analyze What happens when amino acids are heated? (Circle your choice.)

a. They become protocells.
b. They become proteins.
c. They melt.

thought to be in Earth's early atmosphere—water vapor, ammonia, methane, and hydrogen gases—and added electricity. It became hot. When the mixture cooled, it contained organic molecules such as amino acids and sugars. This result provides support for Oparin's hypothesis.

Scientists then tested whether it was possible for more complex molecules to form from the organic molecules. Many experiments were performed. Scientists discovered that when amino acids are heated, they turn into proteins. Many scientists were convinced that complex organic molecules had formed in small pools of warm water on Earth.

What is a protocell?

In 1992, Sidney Fox caused complex molecules to turn into protocells. A **protocell** is a large, organized structure, surrounded by a membrane. It is able to grow and divide. Growth and division are considered life activities.

The Evolution of Cells

In an earlier chapter you learned that prokaryotes are one-celled organisms that do not have internal membrane-bound structures. The first life-forms may have been prokaryotes that evolved from protocells. There was no oxygen available in the atmosphere at this time. So the first life-forms were able to survive without it. That means they were anaerobic. They probably were not able to make their own food but survived on the organic material in the ocean. Gradually they changed into organisms similar to present-day archaebacteria, which do make their own food. **Archaebacteria** (ar kee bac TEER ee uh) live in harsh environments such as deep-sea vents or hot springs. They make their food by chemosynthesis. In chemosynthesis, organisms turn compounds, such as sulfur compounds, into food.

What changes did photosynthesis cause?

The next organisms to develop may have been prokaryotes that could undergo photosynthesis. Photosynthesis releases oxygen. Oxygen began to enter the atmosphere. Organisms that use oxygen began to develop. The fossil record shows that 2.8 billion years ago there was an increase in the variety of prokaryotes.

Oxygen in the atmosphere had another effect. The sun's rays converted the oxygen to ozone molecules. This ozone layer shielded organisms from harmful ultraviolet radiation. That made it possible for even more complex organisms to evolve. ✓

Reading Check

5. What were two results of oxygen in the atmosphere?

____

____

What new organisms evolved?

The more complex organism that evolved was a eukaryote. A scientist named Lynn Margulis has a theory about how eukaryotic cells may have evolved. According to this theory, prokaryotes may have consumed bacteria. Some of these bacteria evolved into mitochondria, which released energy for the cell. Other types evolved into chloroplasts. Present day mitochondria and chloroplasts share similarities with bacteria. The DNA of chloroplasts and mitochondria resembles the DNA of prokaryotes instead of eukaryotes. The figure below illustrates this process.

A A prokaryote ingested some aerobic bacteria. The aerobes were protected and produced energy for the prokaryote.

B Over a long time, the aerobes become mitochondria, no longer able to live on their own.

C Some primitive prokaryotes also ingested cyanobacteria, which contain photosynthetic pigments.

D The cyanobacteria become chloroplasts, no longer able to live on their own.

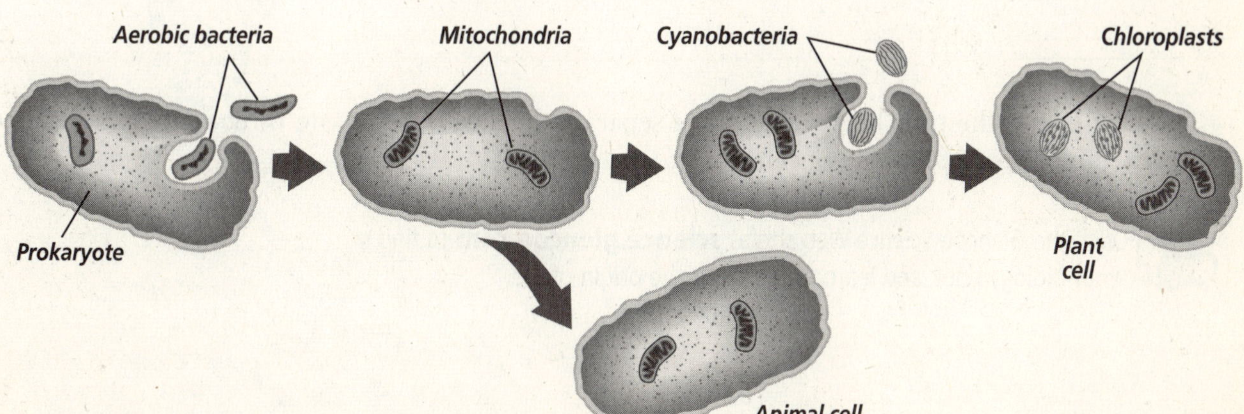

After You Read

Mini Glossary

archaebacteria (ar kee bac TEER ee uh): chemosynthetic prokaryotes that live in deep sea vents and hot springs

biogenesis (bi oh JEN uh sus): idea that living things come only from other living things

protocell: large, ordered structure, enclosed by a membrane that carries out some life activities, such as growth and division

spontaneous generation: mistaken idea that living things can arise from nonliving materials

1. Read the key terms and definitions in the Mini Glossary above. Highlight the two key terms that describe ideas about the origin of living things. On the lines below, explain which idea came first and why it was replaced by the other idea.

2. Place the following events in the order in which they occurred on the vertical time line. The first event and the last event have already been entered on the time line.

Ozone layer develops
Protocells develop
Organisms use photosynthesis
Complex organic molecules form
Prokaryotes appear
Organisms use chemosynthesis

Oceans contain organic molecules

Eukaryotes develop

3. Select one of the time line events. On a separate sheet of paper, write three sentences explaining the importance of this event.

Visit the Glencoe Science Web site at **science.glencoe.com** to find your biology book and learn more about the origin of life.

Section 15.1 Natural Selection and the Evidence for Evolution

Before You Read

In biology, evolution means that populations of a species change over time. In this section you will learn about Charles Darwin and his theory of evolution. Skim the Read to Learn section below and find three important facts about Darwin. Write those facts on the lines below. After you have written your facts, highlight the one you think is the most important.

Read to Learn

Charles Darwin and Natural Selection

Evolution describes the way populations change over time. The modern theory of evolution, in fact, is the main concept in biology. What you learn about evolution will make it easier for you to understand the subject of biology. A place to start is by learning about the ideas of Charles Darwin, an English naturalist who lived from 1809 to 1882. Darwin's ideas about evolution have been supported by fossil evidence.

Mark the Text Locate Information Underline every heading in the reading that asks a question. Then, use a different color to highlight the answers to those questions as you find them.

How did fossils shape ideas about evolution?

A fossil is evidence that an organism lived long ago. Scientists wondered how fossils formed. They wondered why many fossil species had died out or become extinct. They also wanted to know more about how extinct species and modern species might be related. There were many ideas about how species evolved. But the ideas of Charles Darwin became the basis of modern evolutionary theory. ✓

Reading Check

1. What forms the basis of modern evolutionary theory?

What did Darwin study?

Darwin spent five years on a research voyage around the world. He became interested in how species might be related to one another. While in the Galápagos (guh LAH puh gus) Islands off the west coast of South America, Darwin saw many species of plants and animals. He noted that these species looked similar to species he had seen in other places. He wondered if a species might be able to change over time. But at the time, he could not explain how such changes might happen.

After returning to England, Darwin spent twenty years doing research. He studied, experimented, read, and collected samples. He tried to figure out why some animals survive and others do not. Darwin bred pigeons and saw that there were small differences, or variations, in traits of individual pigeons. He also noticed that these traits could be inherited by offspring. Eventually, he conducted an experiment where he bred pigeons that had certain desirable traits. He observed that their offspring were born with the same desirable traits. Breeding organisms with a certain trait to produce offspring with identical traits is called **artificial selection.** Darwin decided that there must be a process in the natural world that works like artificial selection. Using evidence from his research, Darwin decided that that process in nature was natural selection.

Reading Check

2. What is the process in which organisms with favorable traits tend to survive and pass on these traits to their offspring?

____

____

____

What is natural selection?

In **natural selection,** organisms with favorable traits are able to reproduce and pass their traits on to their offspring, who then reproduce. Those without such favorable traits are more likely to die out before reproducing. For example, suppose fish that are slow get eaten before they can reproduce, while fish that are fast survive and reproduce. These offspring inherit the trait of speed from their parents. This way, they too are more likely to survive and pass on that trait to their offspring.

What have we learned since Darwin?

Much evidence supports Darwin's theories. However, it is hard to directly observe evolutionary processes that take place over millions of years. Despite this, much data has been gathered for many years from many sources. Most of today's biologists agree that evolution by natural selection best explains this data. The study of genetics adds even more to our understanding of evolution. We now know that traits are controlled by genes. All the genes that are available in a population are its gene pool. Changes in a population's gene pool over time play an important role in evolution.

Adaptations: Evidence for Evolution

An adaptation is anything that gives an organism a better chance of survival. The two main types of adaptations are structural adaptations and physiological (fih zee uh LAH jih kul) adaptations.

What are structural adaptations?

Structural adaptations take many different forms. Thorns, teeth, hair, beaks, and color are examples of structural adaptations that are inherited. Some adaptations take millions of years to become widespread in a population. Mole rats developed large teeth and claws. This structural adaptation helps them dig holes and protect themselves. Adaptations that keep predators from approaching an organism include a rose's thorns and a porcupine's quills.

Some animals develop coloring that helps them blend with their surroundings. This is an example of a subtle structural adaptation called **camouflage** (KA muh flahj). Camouflaged organisms survive and reproduce because they cannot be easily found by predators.

Mimicry (MIHM ih kree) is another type of structural adaptation. It occurs when one species looks like another species. In one form of mimicry, a harmless species takes on the look of a dangerous species. Predators that avoid the harmful species have a hard time telling the two species apart, and so they avoid both. In this way, the harmless species benefits. Another type of mimicry happens when two or more harmful species grow to resemble each other. For example, bees, wasps, and yellow jacket hornets all look alike and can sting. For this reason, some predators stay away from anything that has a bee-like appearance.

3. Analyze Which of the following is an example of mimicry? (Circle your choice.)

a. A harmless fly looks like a wasp.

b. A frog's color matches the tree it lives in.

c. A pesticide stops working on certain types of weeds.

What are physiological adaptations?

Some changes in gene pools can happen fairly quickly. A few medicines that have been developed within the last 50 years have begun to lose their effectiveness. The bacteria that the medicines used to treat have undergone physiological adaptations. These adaptations keep the bacteria from being killed off by various medications. Physiological adaptations are changes in an organism's metabolic processes. Some insects and weeds also have evolved to the point where they are not affected by chemical sprays.

Other Evidence for Evolution

Structural and physiological adaptations are considered direct evidence of evolution. But most of the evidence to support evolution is indirect. It comes from fossils and sciences such as anatomy, embryology (em bree AHL uh jee), and biochemistry. Scientists do not have fossils for all the changes that have taken place. However, fossils provide a big picture of how groups have changed.

Age	Paleocene 65 million years ago	Eocene 54 million years ago	Oligocene 33 million years ago	Miocene 23 million years ago	Present
Organism					
Skull and teeth				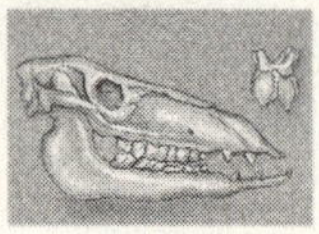	
Limb bones					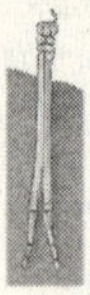

Fossils are important to the study of evolution because they provide a record of early life. When you compare an organism as it looks today with a fossil of that organism, you can see how it has changed over time. For example, scientists have learned from fossils that the ancestors of camels were as small as rabbits are today. This is illustrated on the table above.

Reading Check

4. What is the term for structural features that have a common evolutionary origin?

____

____

____

What can anatomy teach us about evolution?

Homologous Structures The anatomy of different organisms also shows evolutionary patterns. For example, some organisms have **homologous structures.** These are structural features with a common evolutionary origin. Such structures can be similar in arrangement, function, or both. ✓

The figure on page 167 shows how the forelimbs of three very different animals can be homologous. Biologists think that such similarities are evidence that these organisms evolved from a common ancestor.

Analogous Structures However, being structurally similar does not always mean that two species are closely related. For instance, birds and butterflies both have wings. But insects and birds evolved separately. When body parts of organisms do not have a common evolutionary origin but are similar in function, they are known as **analogous structures.**

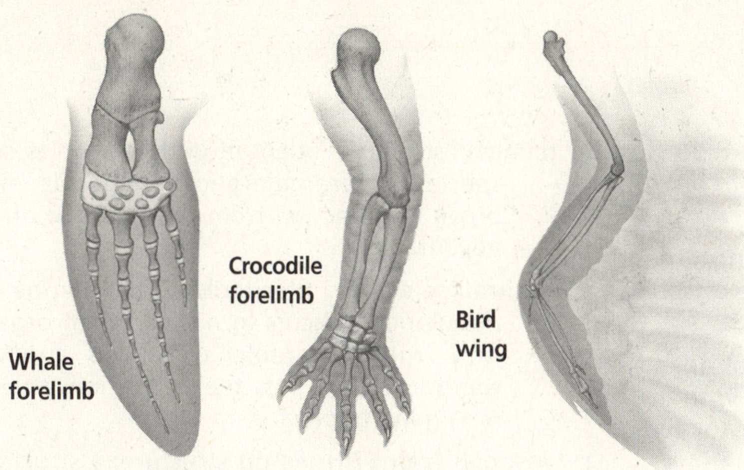

Although analogous structures do not have the same origin, they do provide evidence of evolution. For example, the ancestors of birds and insects both probably evolved wings separately while adapting to similar ways of life.

Vestigial Structures Another type of body feature that shows evolutionary relationship is a **vestigial** (veh STIH jee ul) **structure.** This is a body structure in a present-day organism that no longer serves its original purpose. The eyes of mole rats are an example. Mole rats still have eyes, but they are no longer used for sight. Vestigial structures are evidence of evolution because they show structural change over time.

Embryology An **embryo** is the earliest stage of growth and development of a plant or animal. Young embryos of fishes, birds, reptiles, and mammals have structures that suggest they all had a common ancestor.

What evidence does biochemistry provide for evolution?

Biochemistry also provides strong evidence for evolution. Nearly all organisms share DNA, ATP, and many enzymes in their chemical makeup. Groups that share more similarities in their biochemistry are considered to be more closely related. In the 1970s, biologists began to use RNA and DNA nucleotide sequences to construct evolutionary diagrams that show the levels of relationship among species. Today, scientists combine data from fossils and studies of anatomy, embryology, and biochemistry to interpret relationships among species.

Think it Over

5. **Compare** What is the difference between analogous and vestigial structures?

After You Read

Mini Glossary

analogous structures: structures that do not have a common evolutionary origin but are similar in function

artificial selection: process of breeding organisms with specific traits to produce offspring with the same traits

camouflage (KA muh flahj): structural adaptation that enables species to blend with their surroundings; allows a species to avoid detection by predators

embryo: the earliest stage of growth and development of a plant or an animal

homologous structures: structures with common evolutionary origin; can be similar in arrangement, function, or both

mimicry: structural adaptation that enables one species to resemble another species; may provide protection from predators or other advantages

natural selection: mechanism for change in populations; occurs in nature when organisms with favorable variations survive, reproduce, and pass their variations to the next generation

vestigial (veh STIH jee ul) structure: a structure in a present-day organism that no longer serves its original purpose, but was probably useful to its ancestor

1. Read the terms and their definitions in the Mini Glossary above. Then, choose a term that describes a type of structural adaptation. On the lines below, write a sentence using the term.

2. Use the table below to review what you have learned about adaptation. Write two types of adaptations you read about in the first column. Write a fact about the rate at which the adaptation occurs in the second column, and an example of each type of adaptation in the third.

Adaptation Table

Type of Adaptation	Rate at Which it Occurs	Example

Visit the Glencoe Science Web site at **science.glencoe.com** to find your biology book and learn more about natural selection and the evidence for evolution.

Name Date Class

Section 15.2 Mechanisms of Evolution

Before You Read

In this section you will learn about different ways that evolution occurs. For example, evolution can occur when a physical barrier divides a population into smaller groups that can no longer interact. What barriers might divide a population? Write examples on the lines below.

____

____

____

Read to Learn

Population Genetics and Evolution

In the previous section you learned about Darwin's theory of evolution by natural selection. Since Darwin's time, scientists have learned about genes and have modified Darwin's theories. Today, scientists look at the way genes act in plant and animal populations. This study is called population genetics. It is based on the thought that evolution happens when the genes in a population change over a long period of time.

Individuals in a population do not evolve—populations do. Individuals' genes do not change during their lifetime. But within a population, genes and their frequencies change over time. This is what causes evolution.

Mark the Text **Identify Main Ideas** Highlight the main idea of each paragraph.

What is genetic equilibrium?

How can a population's genes change over time? Picture all the alleles (the alternate forms of a gene) of a population's genes in a large group called a **gene pool.** The percentage of times any allele is in the gene pool is called **allelic frequency.** When this frequency stays the same over generations, **genetic equilibrium** exists. A population in genetic equilibrium is not evolving. Once a change happens, though, the population's genetic equilibrium is disrupted and evolution takes place.

Think it Over

1. **Analyze** Which of the following is an example of genetic equilibrium? (Circle your choice.)
   a. Generation after generation of a population of roses are red.
   b. A mutation in a population of red roses results in some yellow offspring.

What can change genetic equilibrium?

One way genetic equilibrium is disturbed is by mutation. A mutation is any change or random error in a DNA sequence. Some mutations simply occur by chance. Radiation and chemicals can also cause mutation.

Think it Over

2. Analyze On what population would genetic drift most likely have the greatest impact? (Circle your choice.)

a. a population of twelve turtles on a small, isolated island
b. the population of humans in the United States
c. a population of 5000 woodpeckers in western Canada

Genetic drift is another way that a population's genetic equilibrium can be disrupted. **Genetic drift** is the change of allelic frequencies by chance events. This change can greatly affect small populations made up of descendants of a small number of organisms.

For example, in Pennsylvania, there is a small Amish population of about 12 000 people. The Amish marry only other members of their community. Of the original 30 settlers in this community, at least one carried a recessive allele that resulted in offspring with short arms and legs and extra fingers and toes. Today, the frequency of this allele in this population is high—1 in 14. But, in the rest of the United States, the frequency is lower, only 1 in 1000.

Gene flow also can upset genetic equilibrium. Gene flow occurs when an individual leaves or enters a population. This individual's genes either leave or enter the gene pool as a result.

Mutation, genetic drift, and gene flow primarily affect small and isolated gene pools. The impact is much smaller in larger, less isolated gene pools.

The Evolution of the Species

How do changes in a gene pool bring about evolution of a new species? Remember that a species is a group of living things that look alike and can mate with each other to produce fertile offspring. The evolution of a new species is called **speciation** (spee shee AY shun). Speciation occurs when members of similar populations no longer mate with each other to produce fertile offspring. A new species could develop when part of a population has been geographically cut off from the rest of its population. The figure on page 171 illustrates this idea.

An example would be when a river forms a physical barrier and divides a population in two. This is called **geographic isolation.** The separated parts of the population can no longer mate. Over time, the gene pools of the now separate populations become very different. In this way, natural selection results in new species.

Reading Check

3. What is one way that reproductive isolation can occur?

____

____

____

What happens in reproductive isolation?

As populations become increasingly different from each other, **reproductive isolation** occurs. This happens when organisms that at one time mated with each other and produced fertile offspring are no longer able to do so. This can be because the genetic material of the populations becomes so different that fertilization cannot occur. Reproductive isolation also occurs if the mating seasons of similar populations are at different times of year. ✓

A Tree frogs are a single population.

B The formation of a river may divide the frogs into two populations. A new form may appear in one population.

C Over time, the divided populations may become two species that may no longer interbreed, even if reunited.

What role do chromosomes play in the development of a new species?

Chromosomes can also be important in the development of new species. Many new types of plants and some types of animals evolve as a result of what is called polyploidy (PAH lih ploy dee). Any individual or species with a multiple (an extra set) of the normal set of chromosomes is known as a **polyploid.** Mistakes during mitosis or meiosis result in polyploid individuals. Some polyploids cannot produce offspring capable of reproducing. But still others develop into adults that can interbreed and a new species results. Many flowering plants and some important crops—such as wheat, cotton, and apples—originated by polyploidy.

How much time does it take to develop a new species?

Although the developing of new species by polyploidy takes only one generation, most other types of speciation take much longer. **Gradualism** is the idea that species originate through a gradual change of adaptations. For example, fossil evidence shows that sea lilies evolved slowly and steadily over time.

Punctuated equilibrium is a theory that speciation occurs quickly, in rapid bursts. There are long periods of genetic equilibrium between the bursts. In this theory, environmental changes like higher temperatures or a competitive species moving into a population's habitat lead to fast changes in the population's gene pool. Fossil evidence shows several elephant species may have evolved by punctuated equilibrium.

Patterns of Evolution

Biologists have observed that different patterns of evolution occur in different environments. These patterns support the idea that natural selection is important for evolution. An example of this occurs in the Hawaiian honeycreepers.

Hawaiian honeycreepers are all similar in body size and shape, but they differ in color and beak shape. They also live in different habitats. Despite their differences, scientists hypothesize that these birds evolved from a single species that lived on the

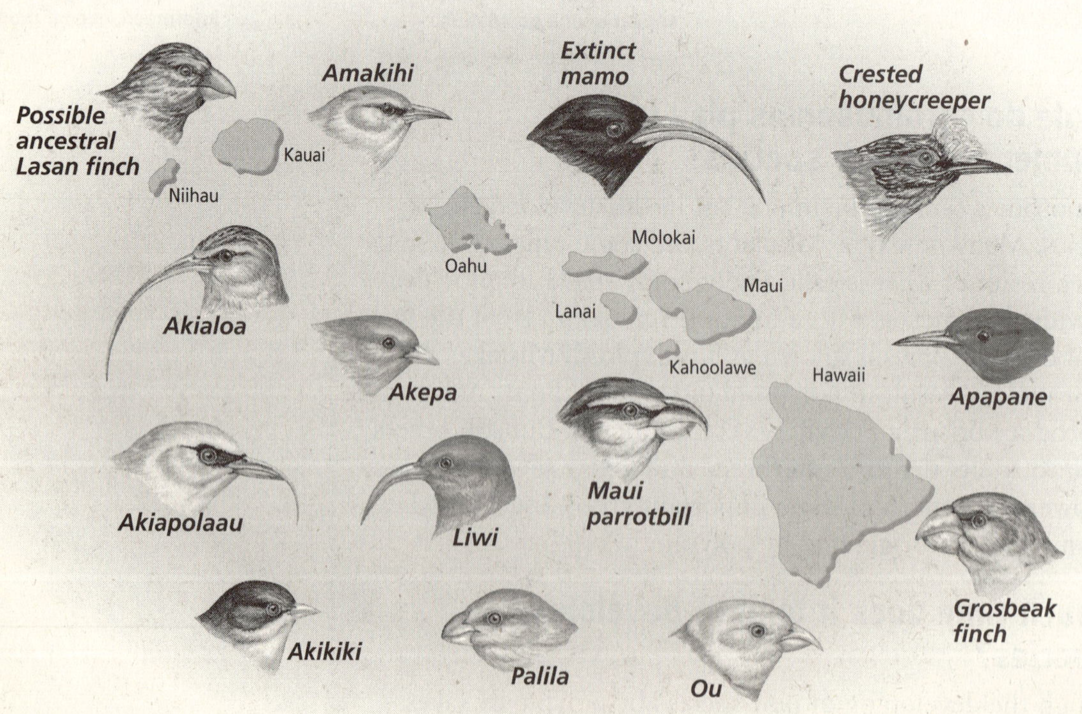

Hawaiian Islands long ago. When a single ancestral species evolves into many different species that fit a number of different habitats, the result is called adaptive radiation. **Adaptive radiation** is a type of divergent evolution. **Divergent evolution** occurs as populations that were once similar to an ancestral species change and adapt to different living conditions. These populations eventually become new species.

Another pattern that can occur is convergent evolution. **Convergent evolution** occurs when unrelated species live in similar environments in different parts of the world. Because they have similar environmental pressures, they share similar pressures of natural selection. As a result, they have similarities. For example, there is an organ pipe cactus that grows in the deserts of North and South America and a plant that looks similar and lives in African deserts. These plants are not related, but their environments are similar. Both plants have fleshy bodies and no leaves. Convergent evolution has apparently occurred. ✓

Reading Check

4. What features do unrelated species that develop similar traits in different parts of the world demonstrate?

____

____

____

After You Read

Mini Glossary

adaptive radiation: divergent evolution in which ancestral species evolve into a variety of species that fit diverse habitats

allelic frequency: percentage of any specific allele found in a population's gene pool

convergent evolution: evolution in which unrelated organisms evolve similar traits; occurs when unrelated species occupy similar environments

divergent evolution: evolution in which species that once were similar to an ancestral species diverge; occurs when populations change as they adapt to different environmental conditions, eventually resulting in new species

gene pool: all of the alleles available in a population

genetic drift: alteration of allelic frequencies in a population by chance events; disrupts a population's genetic equilibrium

genetic equilibrium: condition in which the frequency of alleles in a population remains the same over generations; no evolution occurs

geographic isolation: occurs whenever a physical barrier such as a river divides a population; results in individuals of the population no longer being able to mate; can lead to the formation of new species

gradualism: idea that species originate through a gradual change of adaptations

polyploid: any species with multiple sets of the normal set of chromosomes; results from errors during mitosis or meiosis

punctuated equilibrium: idea that periods of speciation occur relatively quickly with long periods of genetic equilibrium between

reproductive isolation: occurs when formerly interbreeding organisms can no longer produce fertile offspring due to an incompatibility of their genetic material or by differences in mating behavior

speciation (spee shee AY shun): process of evolution of new species that occurs when members of similar populations no longer interbreed to produce fertile offspring within their natural environment

1. Read the terms and their definitions from the Mini Glossary on page 173. Circle the two terms that refer to different ideas about the rate in which speciation occurs. Then, choose one of these terms and use it correctly in a sentence.

____

____

2. Use the statements under the diagram to fill in the results of each type of isolation.

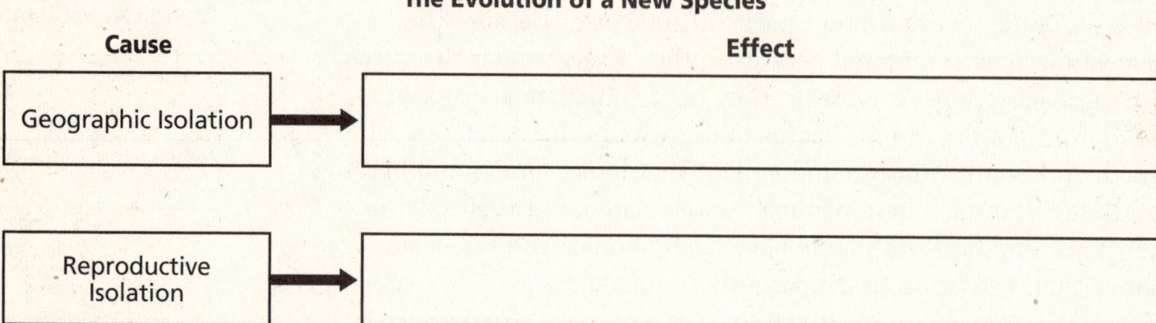

- Members of the same species no longer mate because they cannot reach each other.
- The genetic material of the populations becomes so different that fertilization cannot occur.
- A barrier divides a population.
- Populations have different mating seasons.

3. Review the section, then fill in the blanks below using the following terms: **punctuated equilibrium, speciation, genetic equilibrium, polyploid, mutation.**

1. When the frequency of alleles for a specific trait remains the same for generations, it is called ____________________.
2. One of the factors that can interrupt genetic equilibrium is ____________________.
3. ____________________ occurs when members of similar populations can no longer interbreed to produce fertile offspring within their natural environment.
4. The idea that a sudden environmental change can cause rapid changes in a population's gene pool is called ____________________.
5. A ____________________ is an individual with a multiple of the normal set of chromosomes.

Visit the Glencoe Science Web site at **science.glencoe.com** to find your biology book and learn more about the mechanisms of evolution.

Section 16.1 Primate Adaptation and Evolution

Before You Read

You can see primates (monkeys, apes, gorillas, and lemurs) on display at most zoos. At the zoo you will also see primates walking around and learning about the primates on display. Who are these primates that are not confined to exhibits? They are humans. Think about behaviors you have observed in non-human primates and then compare those to behaviors you have observed in humans. Write your thoughts on the lines below.

____

____

Read to Learn

What is a primate?

Primates are a group of mammals that includes lemurs, monkeys, apes, and humans. Primates come in many shapes and sizes but share common traits.

Unlike most other mammals, most primates have rounded heads with flattened faces. Primate brains are large compared to their body size. The fact that primates are social and have a variety of behaviors shows that primate brains are complex.

Mark the Text **Locate Information** Highlight in one color all the features that distinguish primates.

What adaptations help primates live in trees?

Most primates live in trees and have several adaptations that help them live there. Even primates that do not live in trees, such as you, still have these adaptations. For example, all primates have flexible shoulder and hip joints. These joints help some primates climb trees and swing through branches.

Primate hands and feet are unique among mammals. The fingers and toes have nails instead of claws. Most primates have an **opposable thumb**—the thumb can cross the palm of the hand and touch the fingertips. This allows primates to grab and cling to objects, such as tree branches. Primates can also hold and manipulate tools.

Primate eyes face forward so they see an object through both eyes at the same time. This helps primates see depth and judge distances, which is helpful

when jumping from tree to tree or driving a car. Primates also see in color. Seeing in color helps with depth perception, helps detect predators, and helps the primate find ripe fruit.

Primate Origins

All primates have many things in common, therefore scientists believe they share an evolutionary history. Scientists use fossil evidence and compare modern primates to come up with ideas about how primates are related and how they evolved. Primates are classified, or divided, into two major groups: strepsirrhines and haplorhines.

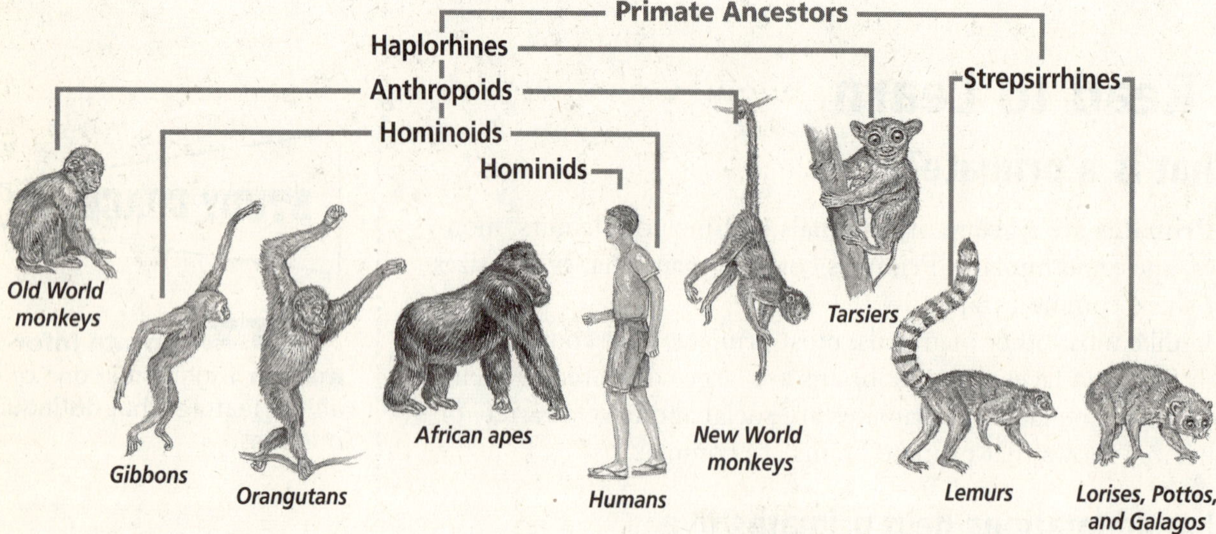

The chart above shows the ways that primates are related to one another. Present-day strepsirrhines are small primates. Most have large eyes, sleep during the day, and are active at night. They live in the tropical forests of Africa and Southeast Asia. ✓

The other living primates are part of the haplorhine group. A small portion of the haplorhine group is the tarsiers, tiny primates living in Southeast Asia. The rest of the haplorhine group is made up of the **anthropoids** (AN thruh poydz), the humanlike primates.

Reading Check

1. Where do strepsirrhines live?

____

____

Who are the anthropoids?

As you can see from the chart, anthropoids include hominoids, Old World monkeys, and New World monkeys. The hominoids are divided into apes and humans. Anthropoids have many features in common. For example, their brains are more complex than strepsirrhines. Their skeletons are larger and more upright. ✓

Reading Check

2. List the three divisions of the anthropoid group.

____

____

____

Section
16.1 Primary Adaptation and Evolution, *continued*

Monkeys are divided into two groups, New World monkeys and Old World monkeys. New World monkeys live in the trees of rain forests in South and Central America. Many New World monkeys have a long, muscular tail. This is called a **prehensile** (pree HEN sul) **tail.** They use this tail much like an arm or leg, grasping and wrapping it around branches as they move from tree to tree. Only New World monkeys have a prehensile tail.

Old World monkeys are generally larger than New World monkeys. Some, such as the colobus monkey, live in trees. Others, such as baboons, remain primarily on the ground. Macaques live in the trees and on the ground. Old World monkeys have adapted to many environments and climates. They range from the hot, dry savannas of Africa to the cold mountain forests of Japan.

The hominoid group consists of apes and humans. Orangutans, chimpanzees, gibbons, and gorillas are classified as apes. Apes do not have tails. They have adapted to life in trees using long, muscled arms for climbing in trees, swinging from branches, or walking on two legs with support from their hands. Although many apes live in trees, most also spend time on the ground. The social interactions among apes indicate a large brain capacity.

Humans have an even larger brain capacity. We also walk upright. There will be more to learn about humans in the next section.

Where did anthropoids evolve?

Scientists suggest that monkeys, apes, and humans share a common ancestor based on their structural and social similarities. Using the fossil record, scientists track the evolution of the anthropoid group. The oldest anthropoid fossils are 37 to 40 million years old. They were found in Africa and Asia. New World and Old World monkeys evolved separately. The oldest New World monkey fossils are 30 to 35 million years old. The oldest Old World monkey fossils are 20 to 22 million years old.

According to the fossil record, a global cooling occurred at the same time hominoids evolved in Africa and Asia. The apes adapted and diversified. Using DNA, scientists have come up with a probable order in which the different apes and humans evolved. Gibbons were probably the first apes that evolved, followed by orangutans, finally the African apes, gorillas and chimpanzees. Structural and DNA data reveal that chimpanzees share the closest common ancestry with humans.

Reading Check

3. What unusual feature do many New World monkeys have?

______

______

______

Think it Over

4. Analyze On which of the following continents would you not expect to see an Old World monkey? (Circle your choice.)

a. Africa
b. Asia
c. North America

Reading Check

5. Which two hominoids were the last to evolve?

______

______

After You Read

Mini Glossary

anthropoids (AN thruh poydz): humanlike primates that include New World monkeys, Old World monkeys, and hominoids

opposable thumb: primate characteristic of having a thumb that can cross the palm and meet the other fingertips; enables animal to grasp and cling to objects

prehensile (pre HEN sul) tail: long muscular tail used as a fifth limb for grasping and wrapping around objects; characteristic of many New World monkeys

primate: group of mammals including lemurs, monkeys, apes, and humans that evolved from a common ancestor; shared characteristics include a rounded head, a flattened face, fingernails, flexible shoulder and hip joints, opposable thumbs or big toes, and a large, complex brain

1. Read the terms and their definitions in the Mini Glossary above. Highlight the key terms that apply to humans. Then choose one of the highlighted words and write a sentence using the word.

____

____

2. In each box write one important fact about the primate group named in that box.

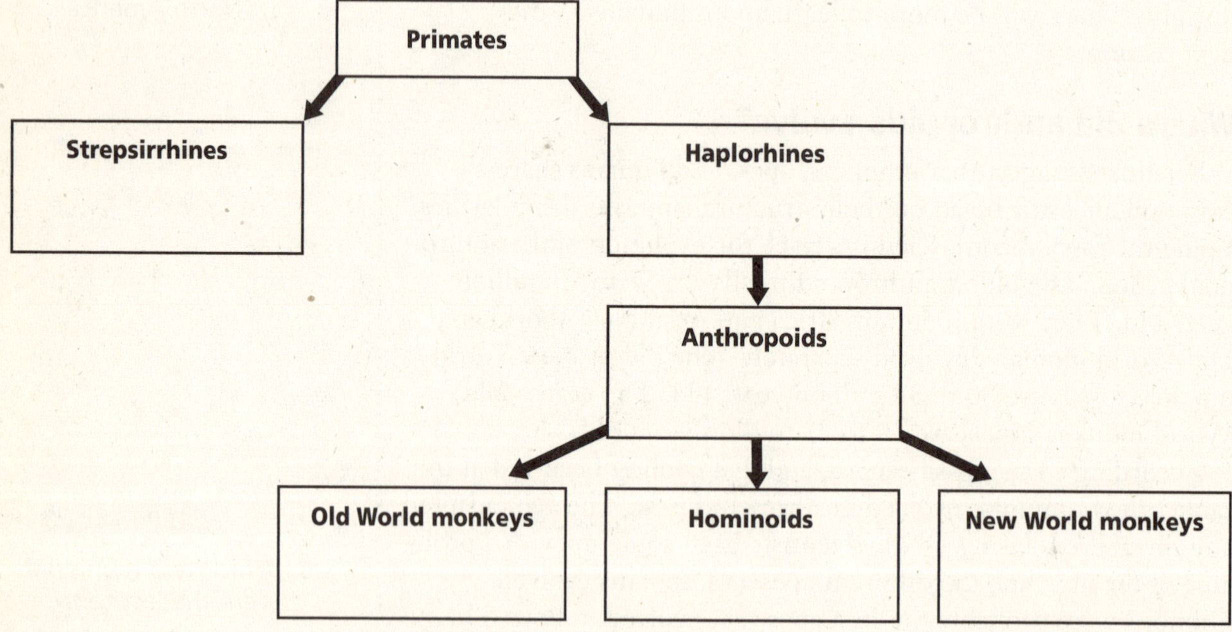

Visit the Glencoe Science Web site at **science.glencoe.com** to find your biology book and learn more about primate adaptation and evolution.

Section 16.2 Human Ancestry

Before You Read

In the previous section you learned that chimpanzees are closely related to humans. Yet there are many things that humans can do that chimpanzees cannot do. On the lines below, list some of the things that humans can do that other animals cannot do.

____

____

____

____

Read to Learn

Hominids

You learned in the previous section that humans are part of a primate group called the hominoids. **Hominoids** (HAH mih noydz) are primates that can walk upright on two legs.This group includes gorillas, chimpanzees, bonobos, and humans.

Some scientists suggest that the separation in the hominoid population is a result of environmental changes. These changes caused some hominoid ancestors to leave the trees and move onto the ground to find food. It may have been helpful to be **bipedal,** or able to walk upright on two legs. Walking upright on two legs increases speed. It also leaves the arms and hands free for feeding, protecting young, and using tools. This bipedal ability created a division of the hominoid group. **Hominids** (HAH mih nudz) are bipedal primates that include modern humans and their direct ancestors.

We learn about our hominid ancestors from the fossil record. While the record is incomplete, more and more hominid fossils are found every year.

Mark the Text **Identify Details** Skim the section and highlight the name of human ancestors (they appear in italics). Then say each name aloud.

What discovery did Raymond Dart make?

In 1924, Raymond Dart discovered a skull with an unusual feature. The skull had mostly apelike features but the opening in the skull for the spinal cord was in the same place as a human skull. The location of the opening, on the bottom of the skull, suggested that this hominoid walked upright. Dart classified the organism as a new species of primate. He called it *Australopithecus africanus* (aw stray loh PIH thuh kus • a frih KAH nus), which means "southern ape from Africa."

Reading Check

1. What was different about the skull Raymond Dart discovered?

Reading Check

2. Who is "Lucy"?

Think it Over

3. Analyze What is the significance of tool making?

The skull has been dated to about 2.5 to 2.8 million years old. Since then, many more australopithecine fossils have been found. An **australopithecine** (ah stra loh PIH thuh sine) is an early hominid from Africa with both humanlike and apelike features.

Who is "Lucy"?

There are several different species of australopithecines. "Lucy" is the most famous fossil of the *Australopithecus afarensis* species. Donald Johanson discovered Lucy's skeleton in 1974. Lucy has been dated to about 3.2 million years ago. *Australopithecus afarensis* had shoulders and arms that were apelike. The pelvis indicates they walked on two legs, or were bipedal, like humans. They had a small brain, similar to the apes. Because of the combination of apelike and humanlike features, it may be that Lucy and hominids like her ate and slept as family groups in trees, but walked upright on the ground to travel.

Australopithecines disappeared between 2 and 2.5 million years ago. Some scientists suggest that they are the ancestors to modern humans.

The Emergence of Modern Humans

Modern humans use tools and have a larger brain than australopithecines. When did hominids first begin to use tools? When did the larger brain develop?

What is *Homo habilis?*

In 1964, Louis and Mary Leakey discovered skull pieces in Tanzania, Africa. This skull was more humanlike than those of australopithecines. The braincase was larger, and the teeth and jaws were smaller, more like those of modern humans. The Leakeys classified this new species with modern humans, in the genus *Homo.* Simple stone tools were found near the fossil skull. Because of the tool-making ability they named the species *Homo habilis*, which means "handy human." The fossils have been dated between 1.5 and 2.5 million years ago. *Homo habilis* is the earliest known hominid to make and use stone tools. These tools suggest that *Homo habilis* might have been a scavenger who used the stone tools to cut meat from dead bodies of animals that had been killed by other animals.

Some scientists propose that *Homo habilis* or another early hominid population gave rise to *Homo erectus*, a new hominid species that appeared 1.5 to 1.8 million years ago. *Homo erectus* means "upright human." The face of *Homo erectus* was more humanlike than that of *Homo habilis*. Evidence suggests that *Homo erectus* used fire, made stone axes, and lived in caves.

The fossil pattern shows that *Homo erectus* migrated from Africa, spreading into Asia and possibly Europe, before becoming extinct between 130 000 and 300 000 years ago.

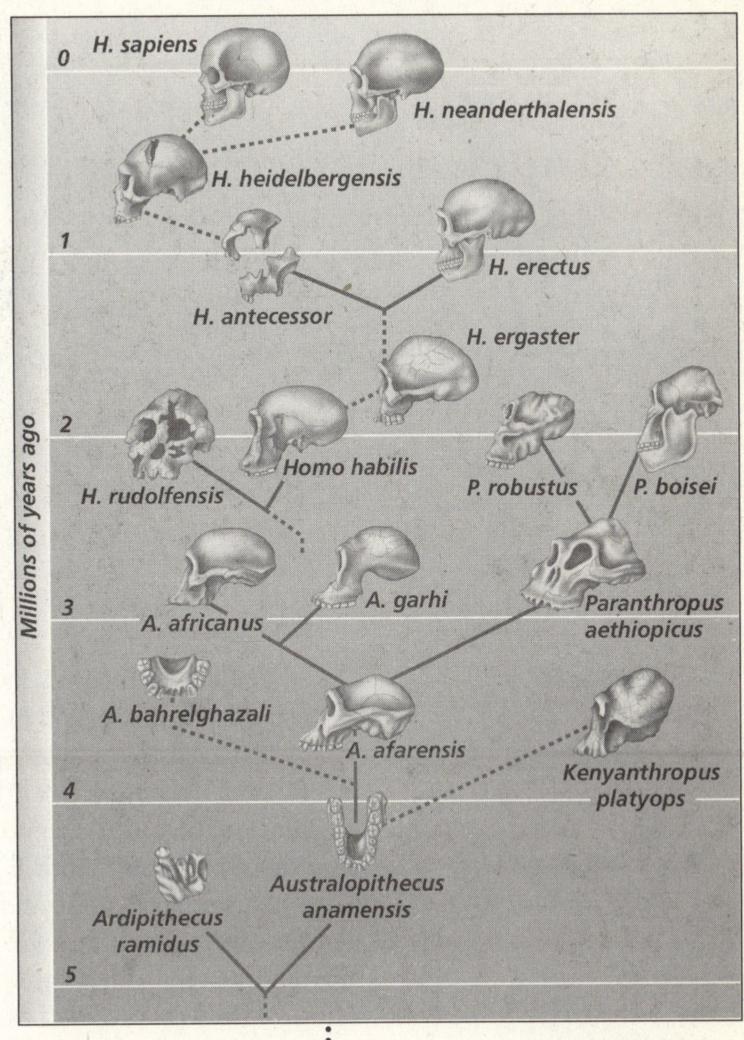

Where did *Homo sapiens* come from?

Several hypotheses have been suggested to explain how modern humans, *Homo sapiens*, may have emerged. The hypotheses were formed after studying fossil bones and teeth, and the results of DNA studies. The figure at the right illustrates a possible pathway for the evolution of *Homo sapiens*. A description of the most popular hypothesis follows.

The fossil record shows that *Homo sapiens* appeared in Europe, Africa, the Middle East, and Asia about 100 000 to 500 000 years ago. Two species are thought to have appeared before *Homo sapiens*. One is called *Homo antecessor*, the other is called *Homo heidelbergensis* (hi duhl berg EN sus). Scientists do not know which is the direct ancestor to *Homo sapiens*. More studies will have to be completed. Both the earlier species had smaller teeth and larger brains than *Homo erectus*. ✓

Reading Check

4. What two species appeared before *Homo Sapiens?*

____

____

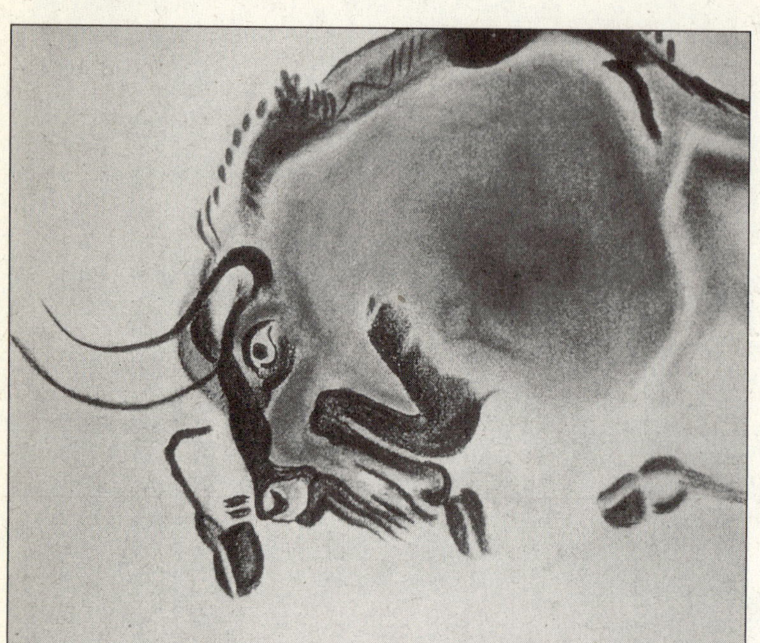

Who were the Neandertals?

You may have heard of the **Neandertals** (nee AN dur tawlz). They were another early *Homo* species. Neandertals lived from 35 000 to 100 000 years ago in Europe, Asia, and the Middle East. The brains of Neandertals were at least as large as the brains of modern humans. Their bones were thicker than modern humans and they had larger faces. Neandertals lived in caves during the ice ages of their time. Scientists have found tools, carved figures, flowers, pollen, and other evidence that suggests that Neandertals may have communicated through a spoken language and had a religion.

Could Neandertals have evolved into modern humans? Most scientists think not. The DNA and fossil evidence suggests that Neandertals are a sister species of *Homo sapiens* called **Cro-Magnon** (kroh MAG nun). Both DNA evidence and fossil measurements indicate that Cro-Magnon was more likely our direct ancestor. Cro-Magnons were identical to modern humans in height, skull structure, tooth structure, and brain size. They appeared about the same time as Neandertals. Cro-Magnons were toolmakers and artists. There are also indications that they used language. The illustration shown here is an example of a Cro-Magnon cave painting.

After You Read

Mini Glossary

australopithecine (ah stra loh PIH thuh sine): early African hominid, genus *Australopithecus,* that had both apelike and humanlike characteristics

bipedal: ability to walk upright on two legs; leaves arms and hands free for other activities such as hunting, protecting young, and using tools

Cro-Magnon: species of *Homo sapiens* that spread throughout Europe between 35 000 to 40 000 years ago; were identical to modern humans in height, skull and tooth structure, and brain size

hominid (HAH mih nud): a group of bipedal primates that includes modern humans and their direct ancestors

hominoid (HAH mih noyd): a group of primates that can walk upright on two legs; includes gorillas, chimpanzees, bonobos, and humans

Neandertal (nee AN dur tawl): *Homo* species that lived from 35 000 to 100 000 years ago in Europe, Asia, and the Middle East; had thick bones and large faces with prominent noses and brains at least as large as those of modern humans

1. Read the terms and their definitions in the Mini Glossary above. Circle the three key terms that identify particular early hominid species.

2. Use the table below to help you review what you have read. In the second column list the characteristics of the hominid species described in the first column.

Australopithecines 3 to 4 million years ago	**Characteristics:**
Homo habilis, Homo erectus 1.5 to 2 million years ago	**Characteristics:**
Neandertals 35 000 to 100 000 years ago	**Characteristics:**
Cro-Magnons 35 000 to 40 000 years ago	**Characteristics:**

Visit the Glencoe Science Web site at **science.glencoe.com** to find your biology book and learn more about human ancestry.

Name Date Class

Section 17.1 Classification

Before You Read

This section explains how scientists have organized the study of living things. This organization makes it easy to tell which organisms share characteristics and which are related to each other. How organized are you? Do you organize your clothes or CD collection? On the lines below, describe the advantages of being organized.

____

____

____

Read to Learn

Create a Quiz After you have read this section, create a five-question quiz based on what you have learned. After you have written the questions, be sure to answer them.

1. What is taxonomy?

____

____

____

How Classification Began

Organizing items not only makes them easier to find, it can also make them easier to understand. One tool biologists use to organize and understand living organisms is classification. **Classification** is the grouping of objects or information based on similarities. The branch of biology that groups and names organisms based on their characteristics is called **taxonomy** (tak SAH nuh mee). ✓

Hundreds of years ago, the Greek philosopher Aristotle developed the first widely accepted biological classification systems. He started with two groups, plants and animals. Then he divided plants into three groups, herbs, shrubs, and trees depending on size and structure. He divided animals into groups according to characteristics such as physical differences and habitat or where they lived. Aristotle's system was useful for a while. But he had grouped together some organisms that really had very little in common. For example, he grouped birds, bats, and flying insects together because they could fly.

In the early 1700s, Carolus Linnaeus developed a system of classifying organisms that is still used today. Linnaeus based his system on physical and structural similarities of organisms. For example, he classified plants according to their flower structures. This method of classifying showed the relationships of organisms.

How are species identified?

Linnaeus developed a two-word naming system called **binomial nomenclature.** This system identifies specific species. The first word identifies the genus. A **genus** (JEE nus; plural, genera) is a group of similar species. The second word is called the **specific epithet,** which often describes a characteristic of the organism. The species name is made up of both the genus name and the specific epithet. For example, the species name for modern humans is *Homo sapiens. Homo* is the name of the genus humans are in. *Sapiens* is a Latin word for "wise." ✓

Reading Check

2. In the species name *Homo sapiens,* what is the genus? What is the specific epithet?

____

____

____

Modern Classification

Scientists who study taxonomy are called taxonomists. They try to identify and classify organisms based on a number of things. These include the external and internal structures of the organism as well as where the organism lives. Taxonomists also consider the genetic makeup of organisms to reveal their evolutionary relationships to other organisms.

What are some uses of taxonomy?

Classification gives biologists a framework that allows them to study the relationships between living and extinct organisms. For example, this framework allows biologists to study the relationship between birds and dinosaurs.

Biologists have found that the bones of some dinosaurs have large internal spaces. So do the bones of birds. Dinosaur and bird skeletons have other similarities. Because of these findings, some biologists believe that dinosaurs are more closely related to birds than to reptiles.

Taxonomy can be a useful tool for scientists who work in such areas as agriculture, forestry, and medicine. For example, a child might eat berries from a plant in his or her backyard. The parents would rush the child and a sample of the plant to the hospital. A scientist in the poison control center could identify the plant and the doctors would then determine what treatment might be necessary.

Taxonomy can also help the economy. Taxonomists can discover new sources of lumber, medicines, and energy. For example, a taxonomist might know that a certain species of pine tree contains chemicals that make good disinfectants. It is possible that a closely related pine species could have the same useful substances. So instead of having one source of chemicals, there may be two sources.

Think it Over

3. Apply For which of the following could taxonomy be used? (Circle your choice.)

a. to determine whether a plant is safe to be planted in a school yard
b. to find a new source for a medicine that comes from plants
c. both a. and b.
d. neither a. or b.

How Living Things Are Classified

A classification system makes it easier to find things. For example, one of the ways the books in a library are classified is by fiction and nonfiction. The fiction books might then be divided into adult, youth, and children's books to make it easy for readers to find just the right book.

Living organisms are divided by groups called taxa (singular, taxon). The taxa range from having very broad characteristics to much more specific characteristics. The smallest taxon is species. At the species level, organisms look alike and are able to breed with one another. The next largest taxon is genus. At the genus level, there is a group of similar species that are closely related.

Reading Check

4. Which taxonomic group is a group of similar families?

____

____

____

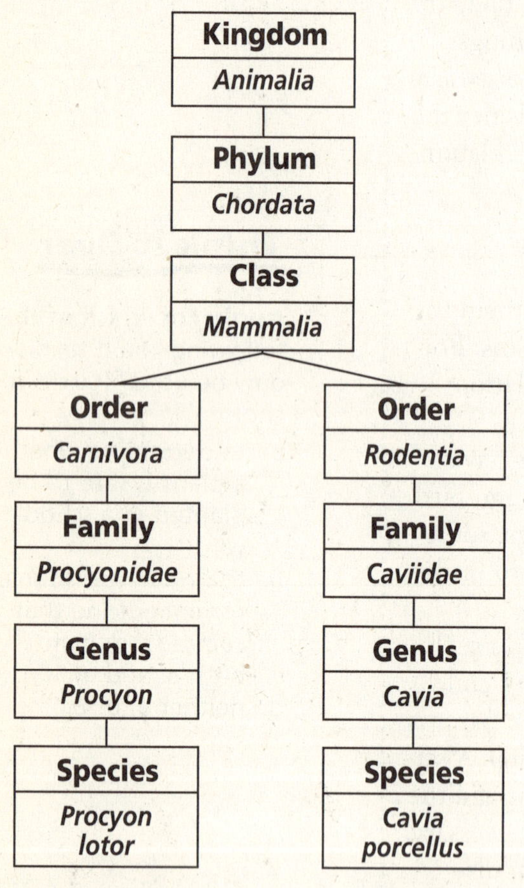

What are the different taxonomic groups?

As you can see from the figure to the left, a species is the most specific group. A genus is a group of similar species. A **family** is a taxon of similar genera. An **order** is a taxon of similar families. A **class** is a taxon of similar orders. A **phylum** is a taxon of similar classes. (Plant taxonomists use the taxon **division** instead of phylum.) A **kingdom** is a taxon of similar phyla (plural for phylum). Try this memory tool to help you remember the order of the groups from kingdom to species: *k*eep *p*utting *c*hocolate *o*ut *f*or goodness *s*ake. The first letter of each word stands for one of the taxa. ✓

Look at the figure to the left again. The raccoon and guinea pig share only the groups with the broadest characteristics. They are both part of the kingdom Animalia, the phylum Chordata, and the class Mammalia. But the similarities between the raccoon and guinea pig end at that point because the raccoon and guinea pig are really not very similar.

Are there other ways to classify plants and animals?

We will use the six-kingdom classification system. However, there are several other classification systems. For example, some taxonomists use the six-kingdom system but add a taxon called domains. The domain includes all six kingdoms.

After You Read

Mini Glossary

binomial nomenclature: two-word system developed by Carolus Linnaeus to name species; first word identifies the genus of the organism, the second word is the specific epithet and often describes a characteristic of the organism

class: taxonomic grouping of similar orders

classification: grouping of objects or information based on similarities

division: taxonomic grouping of similar classes; term used instead of phyla by plant taxonomists

family: taxonomic grouping of similar genera

genus: first word of a two-part scientific name used to identify a group of similar species

kingdom: taxonomic grouping of similar phyla

order: taxonomic grouping of similar families

phylum: taxonomic grouping of similar classes

specific epithet: the second word of a species name, which often describes a characteristic of an organism

taxonomy: branch of biology that groups and names organisms based on studies of their shared characteristics

1. Read the terms and their definitions in the Mini Glossary above. Find two terms in the Mini Glossary that make up the two-word naming system called binomial nomenclature. Write the terms on the lines provided.

____

____

2. Fill in the lines of the triangle to show the taxonomic classification system.

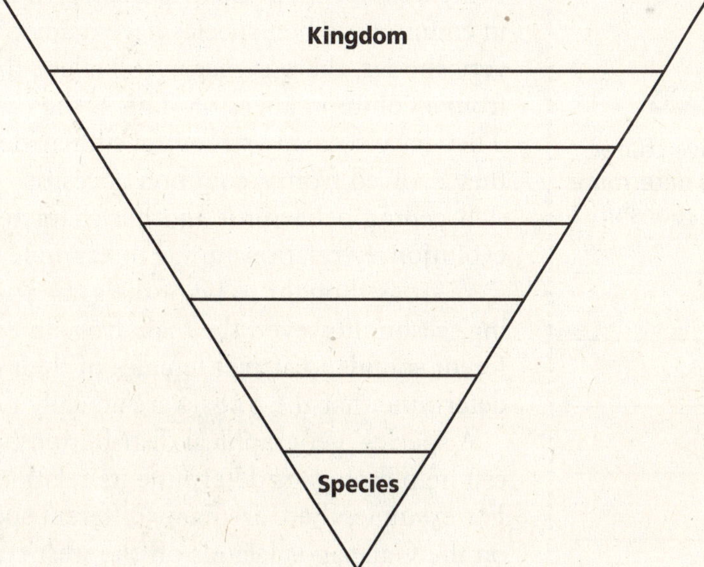

Visit the Glencoe Science Web site at **science.glencoe.com** to find your biology book and learn more about classification.

Name Date Class

Section 17.2 The Six Kingdoms

Before You Read

This section is about the six kingdoms into which all living things are classified. Things are classified according to characteristics. Take a look at a family. What tells you these people are related? Is their hair the same color? Do they speak the same way? On the lines below, make a list of characteristics you would use to identify members of the same family.

____

____

____

Read to Learn

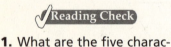

Mark the Text **Identify Main Ideas** Highlight the main idea in each paragraph.

1. What are the five characteristics used to determine evolutionary relationships?

____

____

____

____

____

How are evolutionary relationships determined?

Evolutionary relationships are determined based on similarities in structure, breeding behavior, geographical distribution, chromosomes, and biochemistry. These characteristics give clues about how species evolved. They also show probable evolutionary relationships. ✓

Structural similarities are the physical features that species have in common. When species have many physical features that are very similar, there is reason to believe that they may have evolved from a common ancestor. This is the case for bobcats and lynxes. They have so many structural similarities that taxonomists suggest they evolved from a common ancestor.

Breeding behavior is another characteristic used to determine evolutionary relationships. For example, in one pond area there were frogs thought to be of the same species. During the breeding season, however, the male frogs in each group made very different sounds to attract females of their own group. This helped determine that the frogs were actually two different species.

A species' geographical distribution (where a species lives) can help biologists determine its relationship with other species. For example, there are many different species of birds called finches on the Galápagos Islands off the west coast of South America. Biologists suggest that in the past there was only one species of finch on the islands. The finches probably moved to different parts of the islands and evolved into distinct species over time.

Another characteristic scientists use to determine evolutionary relationships is chromosome similarity. Many species do not look alike, but they do have similar chromosomes. When this is true, scientists believe the species are related. For example, even though the plants broccoli, kale, cabbage, and cauliflower do not look alike, they do have chromosomes that are almost identical in structure.

Biochemistry also is used to determine evolutionary relationships. Scientists study the DNA of species and find that closely related species have similar DNA sequences and, therefore, similar proteins. Sometimes these biochemical studies reveal that species once thought to be closely related are not. This is the case with the red panda and the giant panda. The red panda is actually more closely related to raccoons. The giant panda is more closely related to bears. Biochemical studies of the pandas show them not to be closely related to each other.

2. Analyze Finding that two species have similar DNA is an example of studying which characteristic to determine evolutionary relationships? (Circle your choice.)

a. biochemistry
b. geographical distribution
c. structural similarities

Phylogenetic Classification: Models

The evolutionary history of a species is called its **phylogeny** (fy LAH juh nee). There are different models that show the phylogeny of a species.

One system of classification based on phylogeny is **cladistics** (kla DIHS tiks). Scientists who use cladistics assume that as groups of organisms evolve from a common ancestor, they keep some unique inherited characteristics. Taxonomists call these unique inherited characteristics derived traits. Biologists use these derived traits to make a branching diagram called a **cladogram** (KLA deh gram).

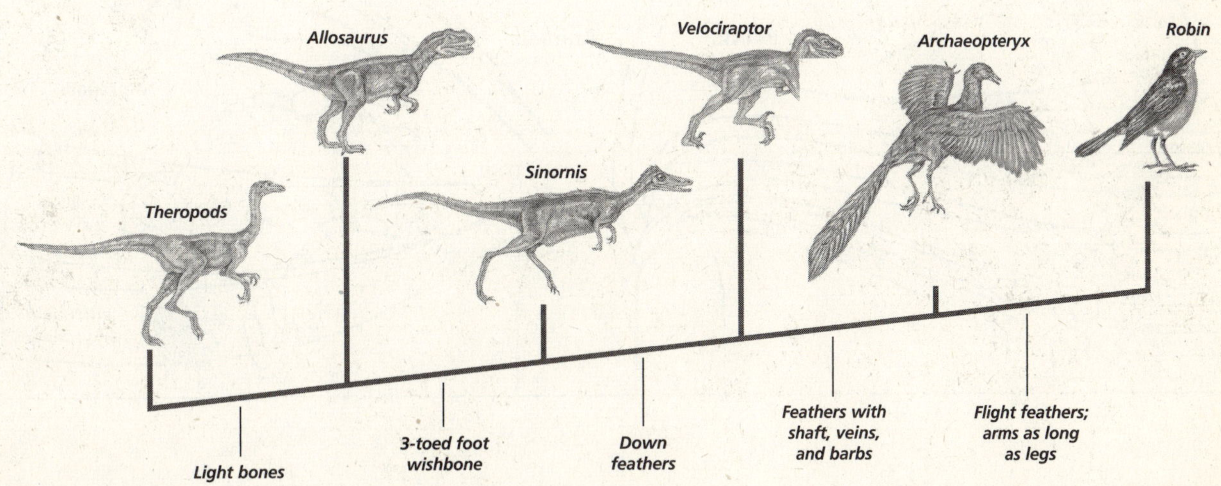

An example of a cladogram appears on page 189. It shows the possible ancestry of modern birds, such as the robin. Notice the derived traits listed across the bottom of the diagram. The more derived traits organisms share, the more closely related they are thought to be. The closer two organisms are on a cladogram, the more likely it is that they share a recent common ancestor. So the *Archaeopteryx* and the robin are thought to be more closely related than the *Velociraptor* and the robin. Remember, though, that cladograms show only probable evolutionary relationships; they are not definite relationships.

Another model scientists use to show the phylogeny of a species is fan shaped. Unlike the cladogram, the fanlike diagram may show the time organisms became extinct. It may also show the number of species in a group. The fanlike diagram uses information from fossils and structural, biochemical, and cladistic studies. An example is shown below. It shows the six-kingdom classification system. In this model, species that are closer together are probably more closely related.

Think it Over

3. Conclude Using the figure below, decide which species are most closely related. (Circle your choice.)

a. mammals and sea stars
b. ferns and conifers
c. sponges and arthropods
d. flowering plants and birds

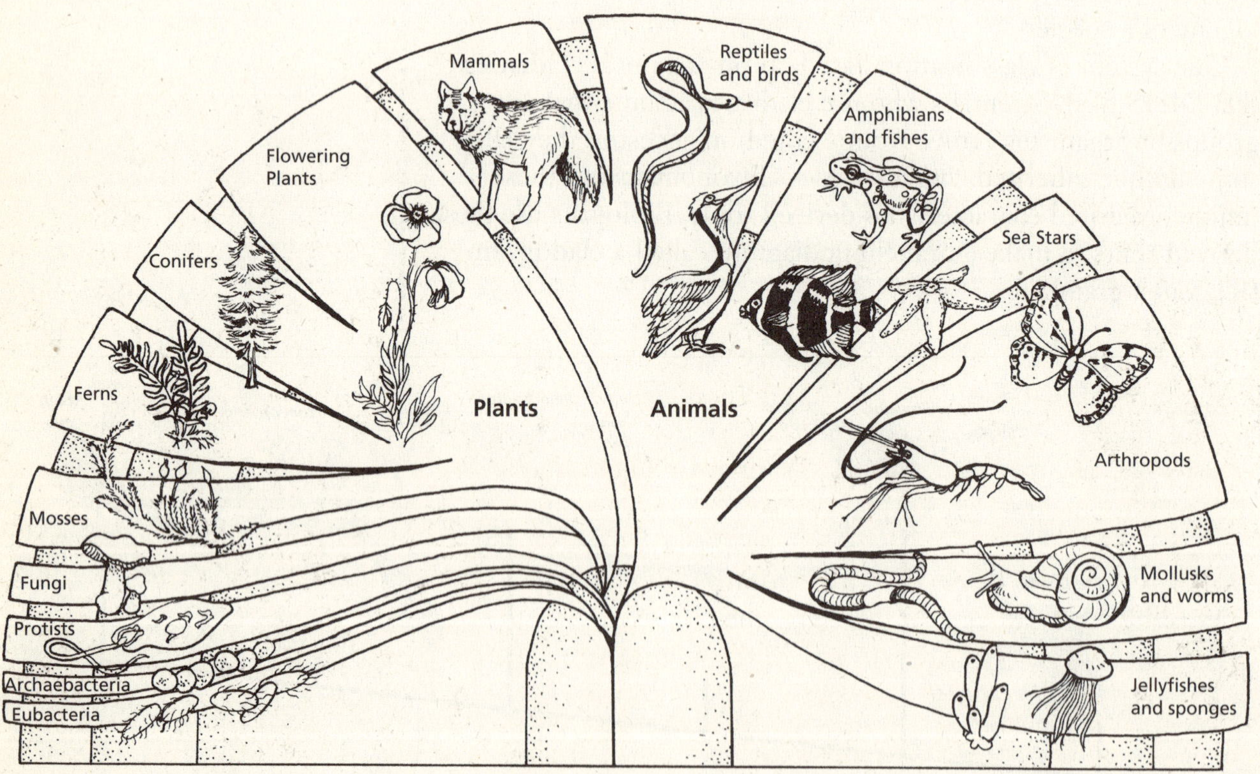

The Six Kingdoms of Organisms

As you can see in the figure on page 190, the six kingdoms of organisms are archaebacteria, eubacteria, protists, fungi, plants, and animals. There are two main characteristics that separate the six kingdoms. Those characteristics are cell structure and the way in which the organisms get energy.

Eubacteria (yew bak TEER ee uh) have strong cell walls. Some eubacteria eat other organisms for energy; they are heterotrophs. Other eubacteria make their own energy; they are autotrophs. Some eubacteria make their own energy the way plants do; they are photosynthetic. Others make energy by chemical reactions; they are chemosynthetic. The eubacteria live in most habitats, except the most extreme. Some eubacteria cause diseases, like strep throat and pneumonia. Most eubacteria, however, are not harmful and many are helpful.

Archaebacteria (ar kee bac TEER ee uh) have very different cell walls than eubacteria. But like eubacteria, archaebacteria make their own energy. They are chemosynthetic and photosynthetic. Archaebacteria live in extreme environments. They live in such places as swamps, deep-ocean hot-water vents and seawater evaporating ponds. The environments in which the archaebacteria live often have no oxygen.

Protists have simple organ systems. They live in moist environments. Protists can be unicellular, and have one cell or be multicellular, and have many cells. Some protists are autotrophs, similar to plants. Others are heterotrophs, similar to animals. Still others are funguslike heterotrophs that produce reproductive structures like fungi.

Fungi (singular, fungus) can be either unicellular or multicellular. Fungi are stationary; they do not move from one place to another. Fungi are heterotrophs; they do not make their own energy. Instead, they absorb nutrients from organic materials in their environment.

Plants are multicellular. Their cells have walls. They do not move from place to place; they are stationary. Plants are autotrophs; they make their own energy by the photosynthesis process. Photosynthesis also provides oxygen in the atmosphere. There are over 250 000 species of plants. These include flowering plants, mosses, ferns, and evergreens. ✓

✓Reading Check

4. List two ways fungi and plants are different.

____

____

Animals are also multicellular. But their cells do not have walls. Animals' cells are organized into tissues, tissues are organized into organs, and organs are organized into organ systems. Most animals are able to move from place to place. Animals are heterotrophs; they eat other organisms for energy.

After You Read

Mini Glossary

cladistics: biological classification system based on phylogeny; assumes that as groups of organisms diverge and evolve from a common ancestral group, they retain derived traits

cladogram: branching diagram that models the phylogeny of a species based on the derived traits of a group of organisms

eubacteria (yew bak TEER ee uh): group of prokaryotes with strong cell walls and a variety of structures; may be autotrophs (chemosynthetic or photosynthetic) or heterotrophs

fungus: group of unicellular or multicellular heterotrophic eukaryotes that do not move from place to place; absorb nutrients from organic materials in the environment

phylogeny (fy LAH juh nee): evolutionary history of a species based on comparative relationships of structures and comparisons of modern life forms with fossils

protist: diverse group of multicellular or unicellular eukaryotes that lack complex organ systems and live in moist environments; may be autotrophic or heterotrophic

1. Read the key terms and definitions in the Mini Glossary above. On the lines below, explain how the terms **cladistics, cladogram,** and **phylogeny** are related.

____

____

2. On the lines below, list the six kingdoms of organisms. Next to each kingdom, write one fact about that kingdom.

____

____

____

____

____

____

Visit the Glencoe Science Web site at **science.glencoe.com** to find your biology book and learn more about the six kingdoms.

Name Date Class

Section 18.1 Viruses

Before You Read

Has your doctor ever told you that your sore throat was caused by a virus? Have you received a shot to help your body defend itself against the flu? What are some of the things you think about when you hear the word *virus?* On the lines below, write some examples of health problems that are caused by viruses.

Sometime you have flu and you get dizzy and sometimes I recieved a shot

Read to Learn

What is a virus?

Viruses are particles that are not alive. They cause diseases and infections. Viruses are made up of nucleic acids, either DNA or RNA, surrounded by a protein coat. They are smaller than the tiniest bacterium. Most biologists agree that viruses are not alive because they don't grow, develop, or carry out respiration.

All viruses replicate, or make copies of themselves. However, viruses need the help of living cells to copy themselves. In order to copy itself, a virus must enter a living cell. The cell in which a virus replicates is called the **host cell.** ✓

What is the structure of a virus?

Viruses differ in size and shape. Some viruses contain as few as four genes. Other viruses have hundreds of genes. Every virus has an inner core of nucleic acid. This core contains the virus's genetic material. Some viruses contain RNA, some contain DNA. The nucleic acid contains the instructions for making copies of the virus.

The outer protein coat surrounding the virus is called a **capsid.** The arrangement of proteins in the capsid gives the virus its shape. The protein arrangement also determines what cell can be infected and how the virus infects the cell. Some large viruses contain another layer called an envelope.

Mark the Text **Identify Ideas** Circle each part of the virus in the diagram below as you read about it.

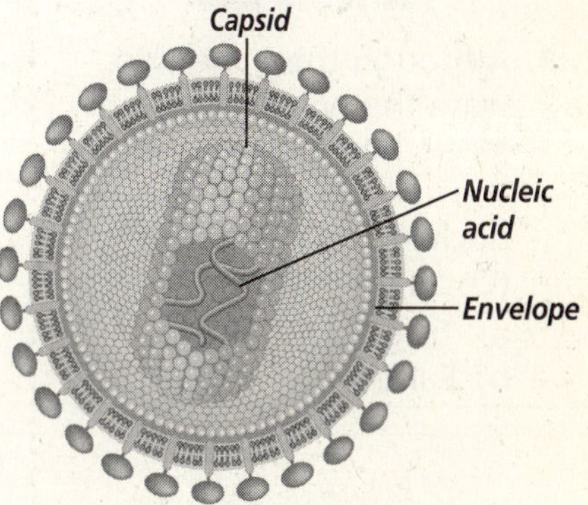

✓ Reading Check

1. What is the cell in which a virus reproduces called?

host cell

Think it Over

2. **Analyze** Another layer around some large viruses is called the (Circle your choice.)
   a. capsid.
   b. envelope.

How does a virus attach to a host cell?

Before a virus can replicate, it has to enter a host cell. In order to do this, it must first attach itself to the host cell. Every virus has a specially shaped device called an attachment protein. Because of the specific shape, each virus can only attach to a few kinds of cells. The attachment process is like two pieces of a jigsaw puzzle fitting together.

A virus that infects a bacterium is called a **bacteriophage** (bak TIHR ee uh fayj), or phage for short. A protein located in the tail fibers of the bacteriophage recognizes and attaches to its bacterial host cell.

Do viruses infect more than one species?

Many viruses are species specific. They infect only certain species. For example, smallpox infects only humans. Other viruses are not species specific. The virus that causes the flu infects both humans and animals. Viruses that are not species specific are more difficult for scientists to eliminate.

Viral Replication Cycles

Once a virus attaches itself to the host cell, the virus enters the cell and takes over its metabolism. Only then can the virus replicate itself. There are two ways that viruses get into the cells. First, the virus can inject its nucleic acid into the host cell. When this takes place, the capsid of the virus stays attached to the outside of the host cell. An enveloped virus enters a host cell in another way. After attachment, the plasma membrane of the host cell surrounds the virus. This produces a virus-filled vacuole inside the host cell's cytoplasm. Recall that a vacuole is a membrane-bound compartment used for temporary storage of materials within a cell. Once the vacuole is in the cytoplasm, the virus bursts out of the vacuole and releases its nucleic acid into the cell. ✓

Reading Check

3. Although there are two ways that viruses get into a cell, both ways result in the virus releasing its nucleic acid into the cell. (Circle your choice.)
   a. True
   b. False

What is a lytic cycle?

Once the virus is inside the host cell, the virus's genes are expressed. The substances that are produced take over the host cell's genetic material.

When viruses take over the cell, the cell stops producing the materials it needs to live. It uses its own enzymes and energy to make new viruses. The new viruses then burst from the host cell and the host cell dies. The new viruses can then infect and kill other host cells. This process is called a **lytic** (LIH tik) **cycle.** A lytic cycle is illustrated at right.

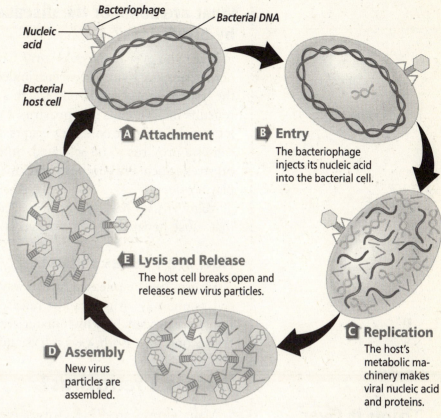

The host's metabolic machinery makes viral nucleic acid and proteins.
D Assembly
New virus particles are assembled.

What happens during a lysogenic cycle?

Not all viruses kill the cells they infect. Some viruses go through a lysogenic cycle. A **lysogenic cycle** is a replication cycle in which the virus's nucleic acid is integrated into the chromosome of the host cell. ✓

The cycle begins in the same way as a lytic cycle. The virus attaches to the host cell and its nucleic acid enters the cell. Instead of taking over the genetic material of the host, the viral DNA is included in the host cell's chromosome. Once that happens, the viral DNA is called a **provirus.** The cell continues to carry out its own metabolic activity, but every time the host cell reproduces, the provirus is replicated as well. This means that every cell that comes from an infected host cell has a copy of the provirus. ✓

The lysogenic cycle can last for many years. At any time, though, the provirus can be activated and enter a lytic cycle. When that happens, the virus will replicate and kill the host cell.

Reading Check

4. What is the cycle called during which the virus does not destroy the host cell?

lysogenic cycle

Reading Check

5. What is the name for viral DNA that is integrated into the host cell's chromosome?

provirus

What are some of the diseases caused by proviruses?

Have you ever had a cold sore? Cold sores are caused by the herpes simplex I virus. Reoccurring cold sores are an example of the lysogenic process. Even though a cold sore heals, the virus remains in your cells as a provirus. When the provirus enters a lytic cycle, another cold sore erupts. No one knows for sure what causes a provirus to be activated. Some scientists think that physical stress, such as a sunburn, and emotional stress, such as anxiety, play a role.

Many viruses have lysogenic cycles. For example, the viruses that cause hepatitis B and chicken pox are lysogenic viruses.

How are viruses released?

Viruses can be released in two ways. Lysis, the bursting of a cell, is one way viruses can be released. Exocytosis, the active transport process by which materials are expelled from a cell, is the second way. Both ways release new viruses from the host cell.

What is a retrovirus?

6. Analyze Why is the presence of the enzyme reverse transcriptase a sign of infection by a retrovirus?

is only
one nuclei
acid

Many viruses are RNA viruses, meaning that RNA is their only nucleic acid. The human immunodeficiency virus (HIV) that causes AIDS is an example of an RNA virus. The RNA virus with the most complex replication cycle is the **retrovirus** (reh tro VY rus). Retroviruses have an enzyme that changes their RNA into DNA.

How does that happen? Once inside a host cell, the retrovirus makes DNA from its RNA. To do this, it uses **reverse transcriptase** (trans KRIHP tayz), an enzyme it carries inside its capsid. This enzyme helps produce DNA from viral RNA. The viral DNA is integrated into the host cell's chromosome and becomes a provirus. If reverse transcriptase is found in a person, it is evidence for infection by a retrovirus.

What is HIV?

Once it gets into a human host, HIV infects white blood cells. Because the viral genetic material is a provirus, new viruses are produced slowly.

An HIV-infected person might experience no AIDS symptoms for a long time. Most people with an HIV infection eventually get AIDS because more and more white blood cells are infected and produce new viruses over time. Infected host cells function until the proviruses enter a lytic cycle and kill their host cells. White blood cells are part of the body's disease-fighting system. When they are lost, a body cannot protect itself from organisms that cause disease. This is a symptom of AIDS.

Cancer and Viruses

Some viruses have been linked to cancer in humans and animals. The viruses disrupt the normal growth and division of cells in a host. This causes abnormal growth and can create tumors. Examples of viruses that have been linked to cancer include the hepatitis B virus, which has been shown to play a role in causing liver cancer.

What are prions and viroids?

Researchers have recently found particles that behave somewhat like viruses and cause infectious diseases. **Prions** are made up of proteins but have no genetic material. They are thought to cause other proteins to malfunction. Prions are responsible for many animal diseases, such as mad cow disease and its human equivalent.

Viroids are a single circular strand of RNA with no protein coat. They have been found to cause infectious diseases in several plants.

Think it Over

7. **Identify** Which of the following is an example of a viroid? (Circle your choice.)
   a. the virus that causes mad cow disease
   b. a plant disease
   c. tobacco mosaic virus

What is a plant virus?

More than 400 viruses infect a variety of plants. These viruses cause diseases that stunt growth and reduce production in their host plants. Plant viruses enter and infect a host plant through wounds or insect bites. However, not all viral plant diseases are harmful or fatal. Some mosaic viruses cause beautiful patterns of color in the flowers of plants.

Origin of Viruses

Because they are relatively simple in structure, you might think that viruses represent an early form of life. This is probably not true. Since viruses need host cells to replicate, scientists suggest that viruses might have originated from their host cells. Some scientists suggest that viruses may be nucleic acids that break free from their host cells while still being able to replicate parasitically within the host cells.

After You Read

Mini Glossary

bacteriophage (bak TIHR ee uh fayj): also called phages, viruses that infect bacteria

capsid: outer coat of protein that surrounds a virus's inner core of nucleic acid; arrangement of capsid proteins gives a virus its shape

host cell: living cell in which a virus replicates

lysogenic cycle: viral replication cycle in which a virus's nucleic acid is integrated into a host cell's chromosome; a provirus is formed and replicated each time the host cell reproduces; the host cell is not killed until the lytic cycle is activated

lytic (LIH tic) cycle: viral replication cycle in which a virus takes over a host cell's genetic material and uses the host cell's structures and energy to replicate until the host cell bursts, killing it

prion: a virus-like infectious agent composed of only protein, with no genetic material

provirus: viral DNA that is integrated into a host cell's chromosome and replicated each time the host cell replicates

retrovirus (reh tro VY rus): the RNA virus with the most complex replication style

reverse transcriptase (trans KRIHP tayz): enzyme carried in the capsid of a retrovirus that helps produce viral DNA from viral RNA

viroid: a virus-like infectious agent that is composed of only a single, circular strand of RNA

virus: a disease-causing, nonliving particle made up of nucleic acids enclosed in a protein coat; inside living cells called host cells

1. Read the key terms and their definitions in the Mini Glossary above. Then use the space below to write a sentence for each of the following terms: **provirus** and **virus.**

The provirus is a viral DNA that is in the cells chromosome. A virus is a disease.

2. Choose one of the question headings in the Read to Learn section. Write the question in the space below. Then write your answer to that question on the lines that follow.

Question: What is virus?

Answer: Virus is a particle that are not alive that causes diseases and inflections.

Section 18.1 Viruses, *continued*

3. Use the pyramid diagram below to help you review what you have read. Arrange the steps in the replication process of the retrovirus in the correct order by placing the letters below next to the numbers on the pyramid.

a. DNA is made from viral RNA

b. Virus enters cell

c. Host cell and provirus reproduce; new virus forms

d. Viral DNA (provirus) integrates into host cell's chromosomes

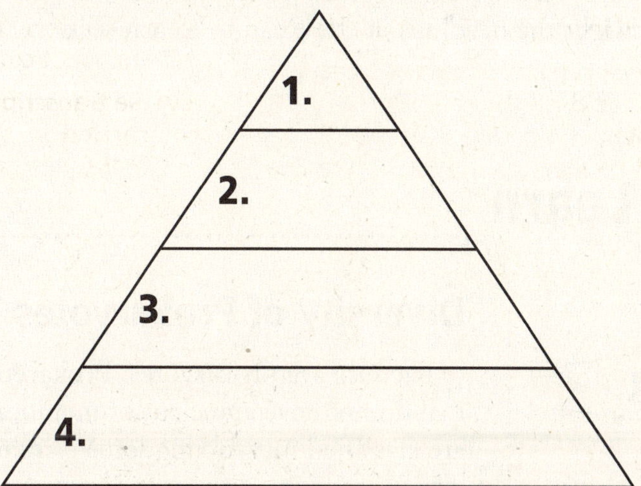

Visit the Glencoe Science Web site at **science.glencoe.com** to find your biology book and learn more about viruses.

Section 18.2 Archaebacteria and Eubacteria

Before You Read

What do you think about when you hear the word *bacteria?* You probably think about germs or something that is bad for you. Did you know that some bacteria are actually helpful? For example, some bacteria help with human digestion, while other bacteria help produce cheese, yogurt, and sourdough bread. In this section, you will learn about types of bacteria and what they do. Before you read this section, highlight the headings in the Read to Learn section. These are the topics you will be learning about.

Read to Learn

Mark the Text **Identify Main Ideas** Skim the reading and highlight the main idea of each paragraph.

Diversity of Prokaryotes

Bacteria are prokaryotes. Prokaryotes are unicellular organisms that do not have a nucleus or membrane-bound organelles. Bacteria are classified in two kingdoms—archaebacteria and eubacteria. There are many differences between archaebacteria and eubacteria. For example, their cell walls and the lipids in their plasma membranes differ. In addition, the structure and function of the genes of archaebacteria are more similar to those of eukaryotes than those of eubacteria.

While they are so different, many scientists propose that archebacteria and eubacteria came from a common ancestor several billion years ago.

What are archaebacteria?

Archaebacteria live mainly in extreme habitats where there is usually no free oxygen supply. There are three different types of archaebacteria. The first type lives in oxygen-free environments and produces methane gas. They live in marshes, lake bottoms, and the digestive tracts of some mammals, such as cows. They also are found at sewage disposal plants, where they help with the breakdown of sewage.

A second type of archaebacterium lives only in water with high concentrations of salt, such as in Utah's Great Salt Lake and the Middle East's Dead Sea. A third type lives in the hot, acidic waters of sulphur springs and thrives near cracks deep in the ocean floor. ✓

1. What are the three environments where archaebacteria live?

Utah great salt lake Middle east

What are eubacteria?

Eubacteria, the other kingdom of bacteria, live in less severe places than do archaebacteria. When people speak of bacteria, they are usually referring to the more common eubacteria. One type of eubacteria is heterotrophic. Remember that a heterotrophic organism cannot make its own food. Heterotrophic eubacteria live almost everywhere and use organic molecules as their food source. Some bacterial heterotrophs are parasites, obtaining their nutrients from living organisms. Others are saprophytes—organisms that feed on dead organisms or organic waste.

A second type of eubacterium is the photosynthetic autotroph. These eubacteria use the sun's energy to make food for themselves. Cyanobacteria are photosynthetic autotrophs. They can be found in ponds, streams, and moist areas of land.

A third type of eubacterium is the chemosynthetic autotroph. These bacteria also make organic molecules that are their food, but they do not need sunlight to do so. Instead, they break down and release the energy of inorganic compounds containing sulfur and nitrogen. This process is called **chemosynthesis** (kee moh SIHN thuh sus).

Reading Check

2. Identify the process that bacteria use to make food from the breakdown of inorganic compounds.

chemosynthesis

What is a bacterium?

Although tiny, a bacterial cell has all the structures necessary to carry out its life functions. The illustration on page 202 shows many of the important parts of a cell. One of the most important parts might be the cell wall. The cell wall protects the bacterium by preventing it from bursting.

Most bacteria live in an environment where there are more water molecules outside the cell than inside the cell. That means water is always trying to get inside the bacterial cell. If there is an opening in the cell wall, water enters the cell and it bursts. Scientists have used that important fact to develop medicines that fight bacterial infections.

If you have a bacterial infection, a doctor will often give you an antibiotic. Some antibiotics work by preventing bacteria from making cell walls. Water enters the cells, and they burst and die. The first antibiotic discovered was penicillin.

Section 18.2 Archaebacteria and Eubacteria, *continued*

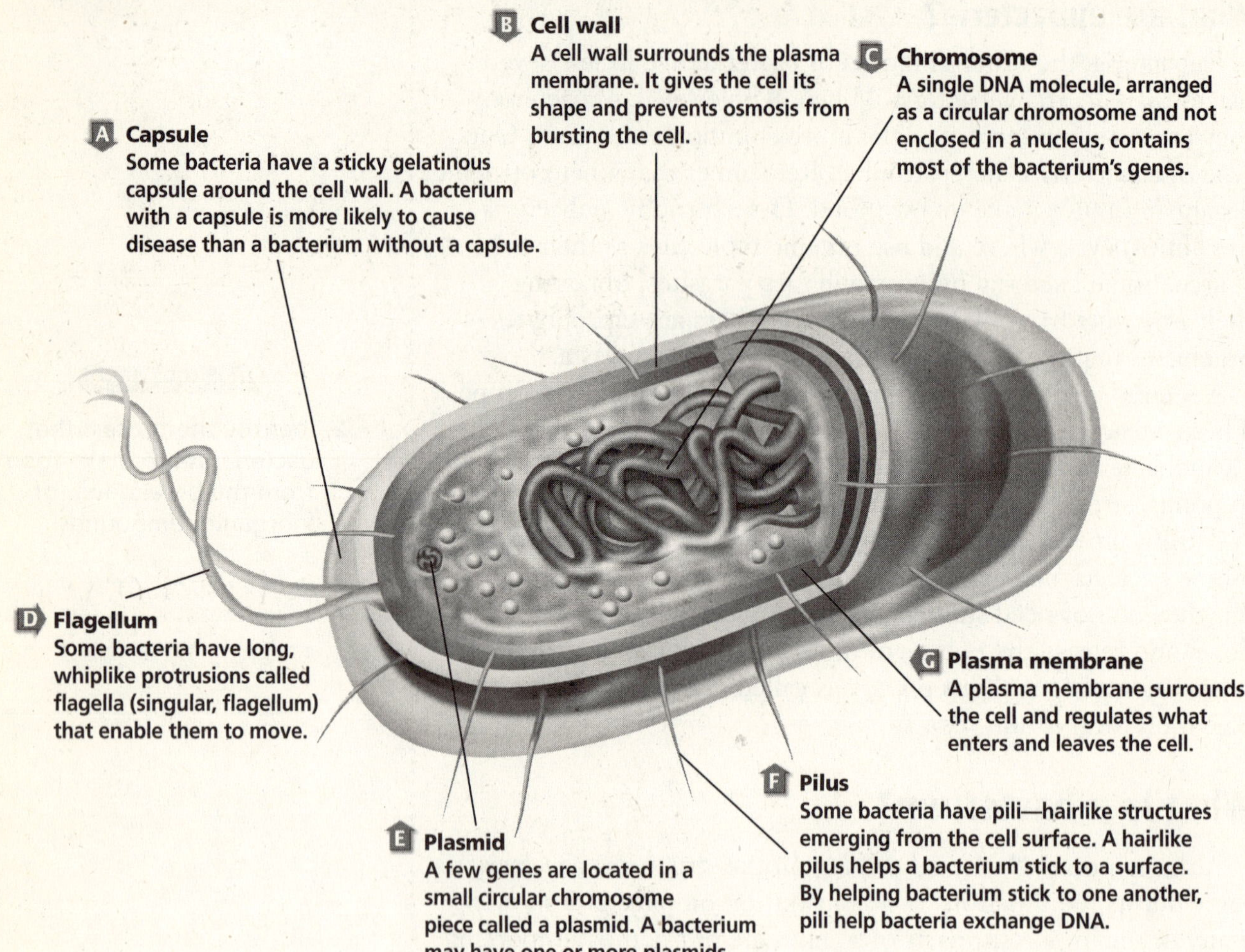

B Cell wall
A cell wall surrounds the plasma membrane. It gives the cell its shape and prevents osmosis from bursting the cell.
C Chromosome
A single DNA molecule, arranged as a circular chromosome and not enclosed in a nucleus, contains most of the bacterium's genes.
A Capsule
Some bacteria have a sticky gelatinous capsule around the cell wall. A bacterium with a capsule is more likely to cause disease than a bacterium without a capsule.
D Flagellum
Some bacteria have long, whiplike protrusions called flagella (singular, flagellum) that enable them to move.
G Plasma membrane
A plasma membrane surrounds the cell and regulates what enters and leaves the cell.
F Pilus
Some bacteria have pili—hairlike structures emerging from the cell surface. A hairlike pilus helps a bacterium stick to a surface. By helping bacterium stick to one another, pili help bacteria exchange DNA.
E Plasmid
A few genes are located in a small circular chromosome piece called a plasmid. A bacterium may have one or more plasmids.

How can different bacteria be identified?

Scientists use different ways to identify bacteria. From a medical point of view, it is important to identify bacteria because some antibiotics can destroy only particular bacteria. Gram staining is a common method of identifying bacteria by checking for differences in the structure of their cell walls. When bacteria are Gram stained, some will turn purple and others will turn pink. Bacterial cells that turn purple are called Gram-positive. Gram-negative cells turn pink. Based on the results, doctors know which antibiotic to use.

Shape is another way to identify bacteria. The three most common shapes are spheres, rods, and spirals. The scientific term for a sphere-shaped bacterium is coccus; for a rod-shaped bacterium, bacillus; and a spiral-shaped bacterium, spirillum.

Scientists also use the growth pattern of bacterial cells to identify them. They use the prefix *diplo–* to refer to cells that grow in pairs.

They use the prefix *staphylo–* to refer to cells that resemble grape clusters, and they use the prefix *strepto–* to refer to cells that grow in long chains. Using this information you can probably picture *Staphylococcus* bacteria.

How do bacteria reproduce?

All the cells you have studied to this point reproduced by mitosis or meiosis. However, bacteria do not have a nucleus or a pair of chromosomes. Bacteria have only one circular chromosome and some smaller circular pieces of DNA called plasmids.

Bacteria reproduce asexually in a process called **binary fission.** First a bacterium makes a copy of its chromosome. Now there are two chromosomes in the cell. The cell begins to grow larger and, eventually, the chromosomes move to opposite ends of the cell. A wall begins to form in the middle. This separates the cell into two new cells. The two new cells are genetically identical.

Bacteria also reproduce sexually using **conjugation** (kahn juh GAY shun). This method produces bacteria with a whole new genetic combination. Use the illustration of the bacteria at the right to help you understand the process. The pilus (plural, pili) is the bridge-like structure that one bacterium uses to connect to another bacterium. The first bacterium transfers all or part of its chromosome to the second bacterium through the pilus. Now the second bacterium has a mix of genetic material. This bacterium reproduces by binary fission and now there are two cells with the new genetic material.

Conjugation

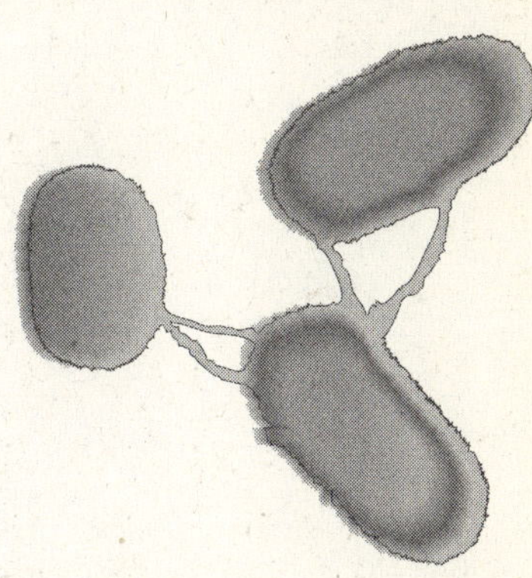

Think it Over

3. Identify On the lines below list the two ways that bacteria can reproduce. Draw a circle around the method that produces genetically identical cells.

____

____

Adaptations in Bacteria

Bacteria have adapted in some amazing ways. Based on fossil evidence, some scientists propose that bacteria were probably among the first organisms to undergo photosynthesis. Recall that photosynthesis produces both food and oxygen. As oxygen levels in Earth's atmosphere increased, some bacteria probably adapted to use it for respiration. Bacteria that must have oxygen to survive are called **obligate aerobes.** Not all bacteria adapted to use oxygen, however. Bacteria that die in the presence of oxygen are called **obligate anaerobes.** There is a third type of bacteria that can live with or without oxygen. ✓

Bacteria have been around for millions of years. As you remember from previous sections, the environment of Earth has changed a lot in that time. Bacteria have a survival mechanism when faced with unfavorable environmental conditions. They produce endospores. An **endospore** is a tiny structure that contains a

Reading Check

4. What type of bacteria must have oxygen in order to survive?

____

____

____

bacterium's DNA and a small amount of cytoplasm. The endospore has a tough outer covering that resists drying out, temperature extremes, and harsh chemicals.

What happens to the endospore?

The endospore rests until environmental conditions improve. Then, similar to a seed, the endospore germinates, or produces a cell that begins to grow and reproduce. Endospores can stay in the resting state for thousands of years before they germinate.

Endospores can create problems for people. One endospore is produced by the bacterium that causes botulism, a type of food poisoning. Another is produced by the bacterium that causes anthrax. Botulism can appear in food that has not been properly processed during canning. If the endospore for botulism is present, it grows in the closed environment of the can and produces a **toxin,** which is a powerful and deadly poison.

Anthrax is caused by a bacterium that lives in soil. Cattle or sheep occasionally contract anthrax. If people handle the animals, they may get a harmless anthrax infection of the skin. Sometimes the anthrax endospores become airborne. If many endospores are inhaled into the warm moist environment of the lungs, they can germinate, causing an often-fatal illness. Because anthrax can spread easily through the air, it has been used as a biological weapon to intentionally harm people.

The Importance of Bacteria

Most people tend to think of bacteria in terms of illness and disease. However, there are actually only a few disease-causing bacteria compared to the number of harmless and beneficial bacteria. In fact, we could not survive without bacteria.

What do bacteria do for us?

All living things need nitrogen. It is part of the proteins, DNA, RNA, and ATP of organisms. Nitrogen makes up about 80 percent of the atmosphere. Yet most living things, including plants, cannot directly use nitrogen from the air. This is where the bacteria come in. Some bacteria convert nitrogen from the air into ammonia. The process is called **nitrogen fixation.** Other bacteria then convert ammonia into forms of nitrogen that plants can use. Bacteria are the only organisms that can perform these chemical changes. ✓

Another important process that bacteria carry out is returning nutrients, both organic and inorganic, to the environment.

✓ Reading Check

5. How do bacteria help plants?

___

___

___

Without bacteria to return nutrients to the environment, there would not be any nutrients for the organisms at the bottom of the food chain. Recall that these autotrophs use the nutrients in the food they make. Without autotrophs producing oxygen as a byproduct of making food, there would not be oxygen in the atmosphere. So you can see that all of life depends on bacteria.

In what other ways are bacteria helpful?

Bacteria produce products that have distinctive flavors and smells. Swiss cheese, yogurt, pickles, and sauerkraut are made with bacteria. Bacteria live in your intestines and produce vitamins and enzymes to help digest food.

Earlier in this section you learned that antibiotics can kill bacteria that cause illnesses. Some bacteria are used to produce the antibiotics that kill other bacteria. Antibiotics have been an important factor in helping people live longer.

How does bacteria cause disease?

Bacteria cause diseases in plants, animals, and humans. Disease-causing bacteria enter the human body through openings such as the mouth. Bacteria are carried in air, food, and water. Sometimes bacteria enter the body through skin wounds.

There are two ways bacterial diseases harm people. First, the growth of bacteria can interfere with the normal function of body tissues. Second, the bacteria can release a toxin that directly attacks the host.

In the past, bacterial diseases had a greater effect on human populations than they do now. Antibiotics and improved health-care systems have reduced the number of deaths caused by bacteria in the United States.

Think it Over

6. **Defend** If someone tells you that bacteria are bad, how would you respond?

____

____

____

After You Read

Mini Glossary

binary fission: reproductive process in which one cell divides into two separate, genetically identical cells

chemosynthesis (kee moh SIHN thuh sus): autotrophic process in which organisms receive energy from the breakdown of inorganic compounds containing sulfur and nitrogen

conjugation (kahn juh GAY shun): form of reproduction in some bacteria where one bacterium transfers all or part of its genetic material to another through a bridge-like structure called a pilus

endospore: structure formed by bacteria during life-threatening conditions that contains DNA and a small amount of cytoplasm; has a tough outer covering

Name Date Class

nitrogen fixation: process by which bacteria convert nitrogen in the air into ammonia

obligate aerobe: bacteria that must have oxygen to survive

obligate anaerobe: bacteria that die in the presence of oxygen

toxin: poison produced by a bacterium

1. Review the terms and their definitions in the Mini Glossary. Circle the two words that describe how bacteria reproduce. Then underline the word that describes the type of reproduction that results in bacteria with new genetic material. On the lines below describe how that might be useful to bacteria.

____

____

2. Write the statements under the heading they best relate to.

The Importance of Bacteria

Helpful	Harmful

a. Endospores germinate in human lungs.
b. Provide nitrogen in a usable form for plants
c. Oxygen is a byproduct of making food.
d. Cause infection in humans
e. Flavor food
f. Create toxins

3. Bacteria come in three different shapes: spheres, rods, and spirals. Write their scientific names below each figure.

Spheres

1. ______________

Rods

2. ______________

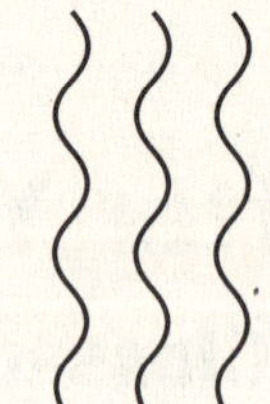

Spirals

3. ______________

Visit the Glencoe Science Web site at **science.glencoe.com** to find your biology book and learn more about archaebacteria and eubacteria.

Name Date Class

Section 19.1

The World of Protists

Before You Read

You have learned all life is organized into six kingdoms. Without using your notes, name all six on the lines below. Some of the organisms you will learn about in this section are plantlike, and some are animal-like. Still others have characteristics like fungi. Some of them were placed in different kingdoms before they were finally classified as Kingdom Protista. As you read this section keep in mind how much variety there is in the world of protists.

____

____

____

____

Read to Learn

What is a protist?

There is no such thing as a typical protist. In fact, Kingdom Protista contains the most diverse organisms of all the kingdoms. There are single-celled (unicellular) protists as well as many-celled (multicellular) protists. Some are microscopic, others are very large. Some can make their own food, some cannot. Protists have only one thing in common—they are all eukaryotes. That means most of their metabolic processes (chemical reactions) take place inside their membrane-bound organelles. Other than that, organisms classified as protists are quite different from each other.

Some protists, called protozoans, seem to be like animals except that they only have one cell. Others, called **algae,** seem to be like plants except they do not have roots, stems, or leaves. Algae are photosynthetic and autotrophic. Unicellular algae are the basis of aquatic food chains and produce much of the oxygen in Earth's atmosphere. Still other protists seem to be like fungi except that they do not have the same kind of cell walls that fungi have. You can see why protists have been difficult to classify.

Mark the Text **Identify Main Ideas** As you skim the section, draw a circle around the name of each of the four groups that make up the protozoans.

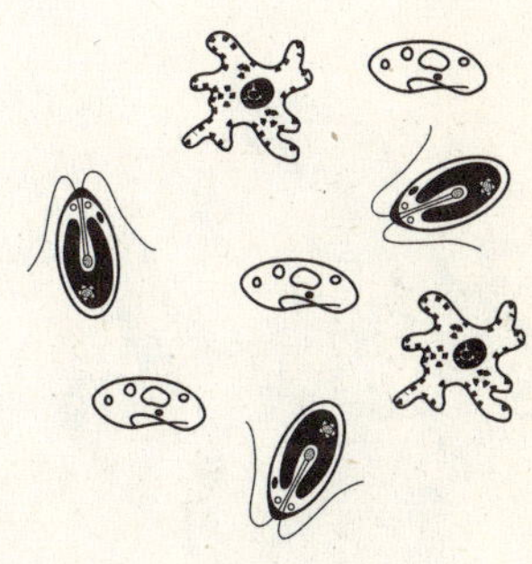

Name Date Class

Reading Check

1. What are protozoans?

____

____

____

Reading Check

2. How are protozoans grouped?

____

____

What is a protozoan?

A **protozoan** is a unicellular animal-like protist. Protozoans are found in moist environments. If you were to pick up wet decaying leaves from the edge of a pond and place them under a microscope, you would discover the small world inhabited by protozoans. Protozoans are a diverse group, but all feed on other organisms or dead organic matter. This means they are all heterotrophs. ✓

Diversity of Protozoans

Protozoans have been separated into four groups: amoebas, flagellates, ciliates, and sporozoans. Three of the groups—amoebas, flagellates, and ciliates—are grouped according to the way they move. The fourth group, sporozoans, are grouped together because they are parasites. ✓

Can amoebas change their shape?

Amoebas send out extensions of their plasma membranes to move and feed. These extensions are called **pseudopodia** (sew duh POH dee uh). Amoebas can do this because they do not have a cell wall. The shape of the amoeba changes as pseudopodia form. The illustration below shows how an amoeba feeds on small organisms. The pseudopodia surround the food.

Because amoebas live in moist places, nutrients dissolved in the water can pass directly through the cell membrane and into the organism. Freshwater amoebas (and other protozoans) sometimes absorb too much water. They have pumps in the cytoplasm that collect and pump out excess water. These pumps are called contractile vacuoles.

Remember that protozoans are found in moist environments. Amoebas that live in the sea are part of plankton. Plankton is an assortment of organisms, many of which are microscopic,

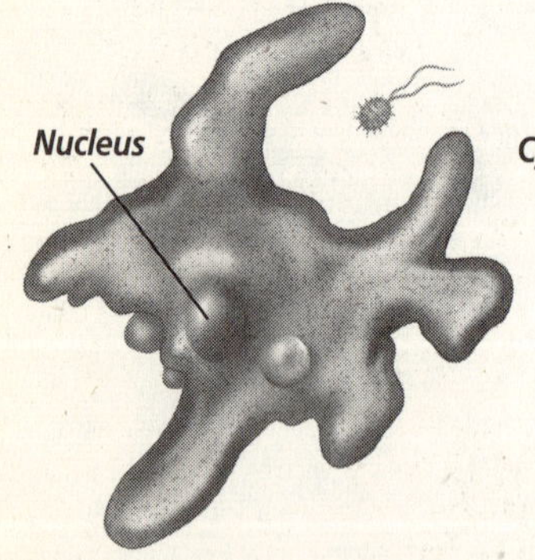

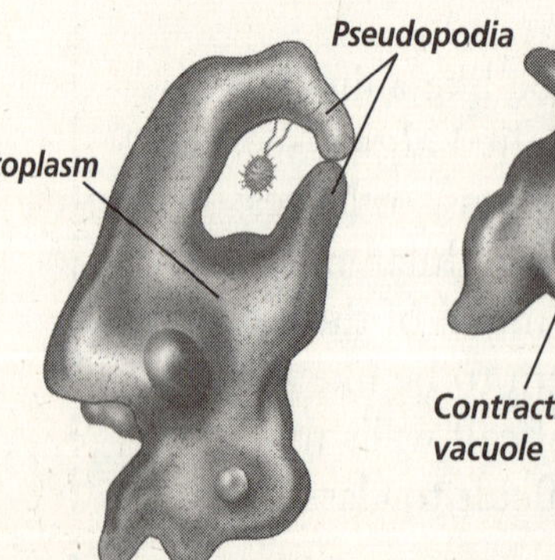

Contractile vacuole

Food vacuole

A As an amoeba approaches food, pseudopodia form and eventually surround the food.

B The food becomes enclosed in a food vacuole.

C Digestive enzymes break down the food, and the nutrients diffuse into the cytoplasm.

that float in the ocean and form the base of the ocean's food chain. One group of amoebas, called radiolarians, are an important part of marine plankton.

Most amoebas reproduce by asexual reproduction. In **asexual reproduction,** the amoeba divides into two cells, producing identical offspring.

How did flagellates get their name?

Flagellates (FLAJ uh luts) got their name because they are protists that have one or more flagella. Flagella are whip-like organelles that move from side to side and enable the protozoans to move about. Some flagellates are parasites. One type causes sleeping sickness in humans. Another flagellate lives inside the guts of termites. This protozoan produces an enzyme that digests wood, allowing the termite and the flagellate to use nutrients from the wood.

What are the ciliates?

Ciliates got their name from cilia, the hairlike projections that cover their bodies. They beat the cilia to move through the watery places in which they live.

Paramecia are one of the largest unicellular organisms. The illustration on page 210 shows the parts of a paramecium. You can see that it has many organelles and structures. Many structures may work together to perform one important life function. For example, a paramecium uses its cilia, oral groove, gullet, and food vacuoles for digestion.

Paramecia reproduce asexually by dividing into two cells. They also are able to reproduce through conjugation, which is a form of sexual reproduction. In an earlier chapter you learned that conjugation happens when an organism places its genetic material in another organism. In this process two paramecia join and exchange genetic material. Then they separate and each one divides through asexual reproduction, passing on the new genetic mixture.

What are the sporozoans?

Most of the organisms in the group of protozoans called **sporozoans** produce structures called spores. A **spore** is a reproductive cell with a hard outer coat that produces a new organism without fertilization.

Think it Over

3. Summarize Name the two ways that paramecia can reproduce.

____

____

Section 19.1 The World of Protists, *continued*

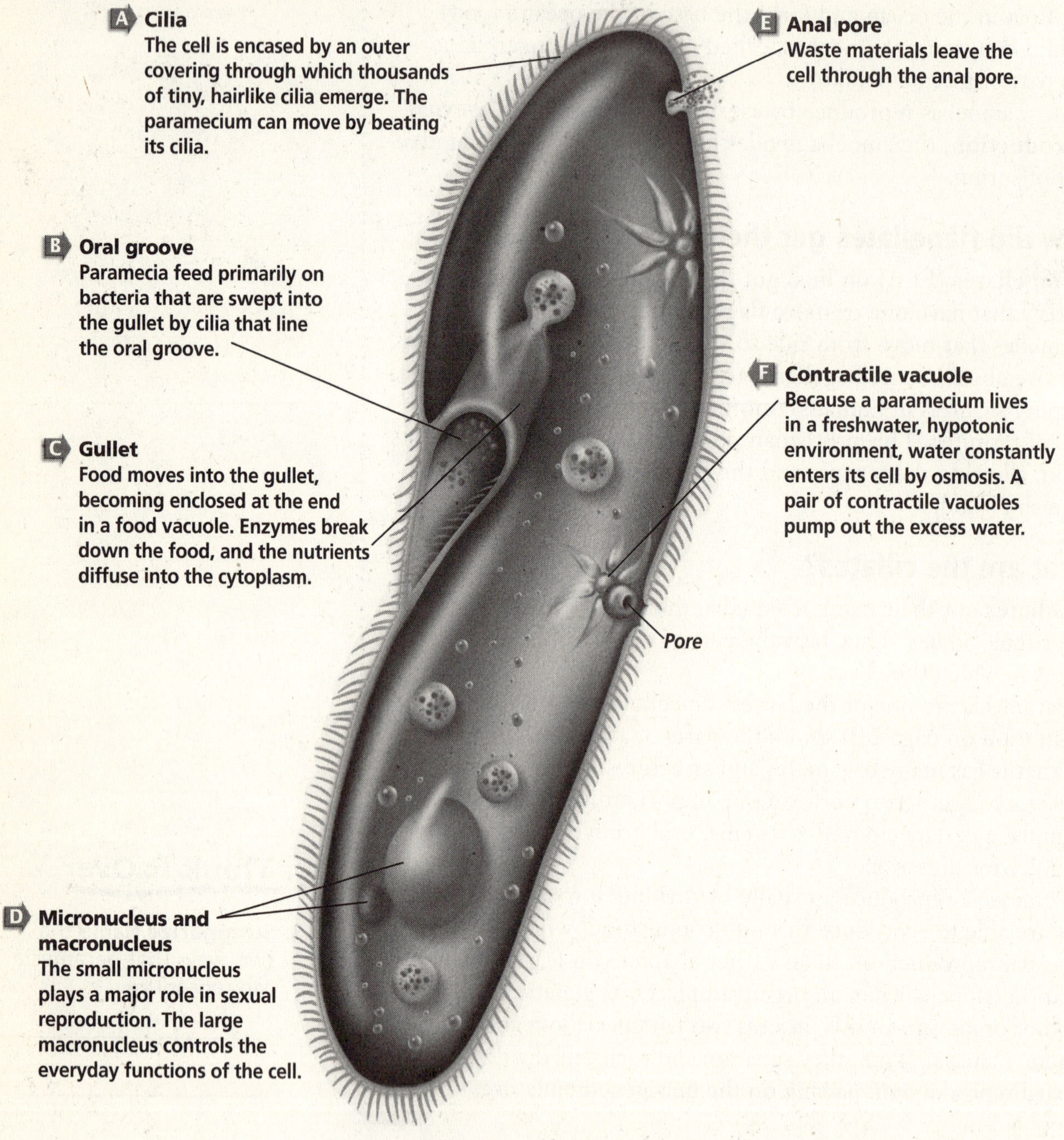
A Cilia
The cell is encased by an outer covering through which thousands of tiny, hairlike cilia emerge. The paramecium can move by beating its cilia.
B Oral groove
Paramecia feed primarily on bacteria that are swept into the gullet by cilia that line the oral groove.
C Gullet
Food moves into the gullet, becoming enclosed at the end in a food vacuole. Enzymes break down the food, and the nutrients diffuse into the cytoplasm.
D Micronucleus and macronucleus
The small micronucleus plays a major role in sexual reproduction. The large macronucleus controls the everyday functions of the cell.
E Anal pore
Waste materials leave the cell through the anal pore.
F Contractile vacuole
Because a paramecium lives in a freshwater, hypotonic environment, water constantly enters its cell by osmosis. A pair of contractile vacuoles pump out the excess water.
Pore

All sporozoans are parasites. They usually live inside a host organism at the site of a steady food supply, such as in the animal's blood or intestines. Malaria is a disease caused by a sporozoan. Usually occurring in tropical climates, malaria is spread by certain types of mosquitoes. When an infected mosquito bites someone, the sporozoan that causes malaria is passed from the mosquito to the person it bites.

After You Read

Mini Glossary

algae: photosynthetic, plantlike, autotrophic protists

asexual reproduction: type of reproduction where one parent produces one or more identical offspring by dividing into two cells

ciliates: group of protozoans with short hairlike cilia that aid in movement

flagellates: protists that have one or more flagella

protozoans (proh tuh ZOH uhnz): unicellular, heterotrophic, animal-like protists

pseudopodia (sew duh POH dee uh): in protozoans, extensions of the plasma membrane; aid in movement and feeding

spore: type of cell with a hard outer coat that forms a new organism without fertilization

sporozoans: group of parasitic protozoans that reproduce by spore production

1. Read the key terms and their definitions in the Mini Glossary above. Circle the key terms that identify three of the four groups that protozoans are divided into. Write the name of the missing group on the line below.

____

2. Fill in the numbered boxes to complete the chart.

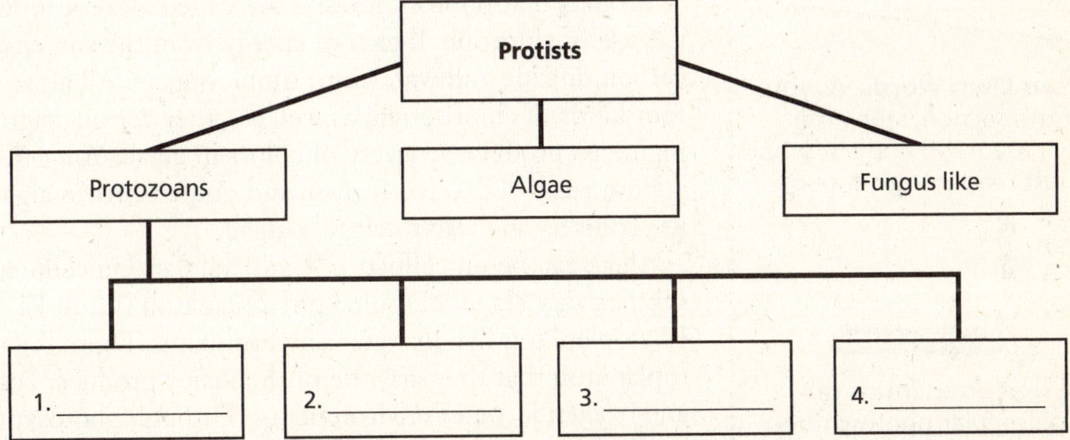

3. Write a fact for each numbered box on the lines below.

____

____

____

____

Visit the Glencoe Science Web site at **science.glencoe.com** to find your biology book and learn more about the world of protists.

Name Date Class

Section 19.2 Algae: Plantlike Protists

Before You Read

You probably come into contact with algae every day. Diatoms are a type of algae whose remains become a powdery, porous rock called diatomite. Diatomite is highly absorbent. It is used in pet litter and to clean up chemical spills. It also is used as an abrasive in household cleaners. It is even added to paint to add sparkle. Now that you know that diatomite is absorbent, sparkling, and abrasive, see if you can imagine some additional uses for it. Write your ideas on the lines below.

____

____

____

Read to Learn

In Your Own Words As you read this section, stop after every paragraph and put what you just read into your own words.

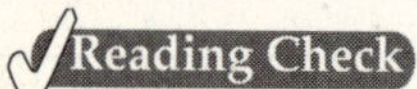

1. What do scientists call unicellular photosynthesizing algae?

____

____

____

What are algae?

Protists that photosynthesize are called algae. Photosynthesis is the use of chlorophyll to trap energy from the sun and convert carbon dioxide and water into simple sugars. All algae have up to four kinds of chlorophyll, as well as other various pigments. The pigments produce a variety of colors in algae. You can find purple, yellow, rusty-red, olive-brown, and golden-brown algae. Scientists use color as one way to classify algae.

Algae can be unicellular or multicellular. Unicellular photosynthesizing algae are called phytoplankton (fi toh PLANK tun). Phytoplankton live in water environments. There is so much phytoplankton that they are one of the major producers of nutrients and oxygen in water environments. Through photosynthesis, algae produce much of the oxygen used on Earth. ✓

Seaweed is a multicellular algae. It looks like a plant, but it is not. Algae do not have roots, stems, or leaves.

Diversity of Algae

Algae are classified into six different phyla. Three of the phyla—the euglenoids, diatoms, and dinoflagellates—include only the unicellular species. In the other three phyla—red, green, and brown algae—most of the species are multicellular.

What are the euglenoids?

Euglenoids (yoo GLEE noydz) are unusual algae because they have both plant and animal characteristics. They are plantlike because most contain chlorophyll and photosynthesize. However, they do not have a cell wall made of cellulose. They also differ from plants because, when light is not available, they can ingest food in ways that are similar to protozoans. They also have one or more flagella to help them move around. They use the flagella to move toward light or food.

What type of algae has shells?

Diatoms (DI uh tahmz) are members of another phylum. They are unicellular photosynthesizing organisms that make up a large part of the phytoplankton in both freshwater and sea water ecosystems. The fact that they have shells made of silica sets diatoms apart from other algae.

The delicate shells of diatoms actually come in two parts, similar to a box with a lid. Each species of diatom has its own unique design and shape.

Diatoms contain chlorophyll and another pigment called carotenoid. This gives the diatoms a golden-yellow color. The food that diatoms make is stored as oil in the organism. The oil in the diatom helps the organism to float near the surface where more sunlight is available. Fishes who feed on diatoms have an oily taste. ✓

Reading Check

2. How does storing food as oil help diatoms?

____

____

____

How do diatoms reproduce?

Diatoms have two different ways of reproducing, asexually and sexually. In asexual reproduction, the two halves of the shells (the box and the lid) separate. Each half then produces a new half to fit inside itself. This means that one of the new diatoms will be smaller than the original. As they continue reproducing this way, half of the diatoms will get smaller and smaller. When diatoms are one-quarter of their original size they begin to reproduce sexually. That means they will produce gametes that join to become a fertilized cell known as a zygote. The process of asexual and sexual reproduction is pictured on page 214.

When diatoms die, their shells sink to the ocean floor. The shells are then dredged up and used to make the many products mentioned in the Before You Read activity.

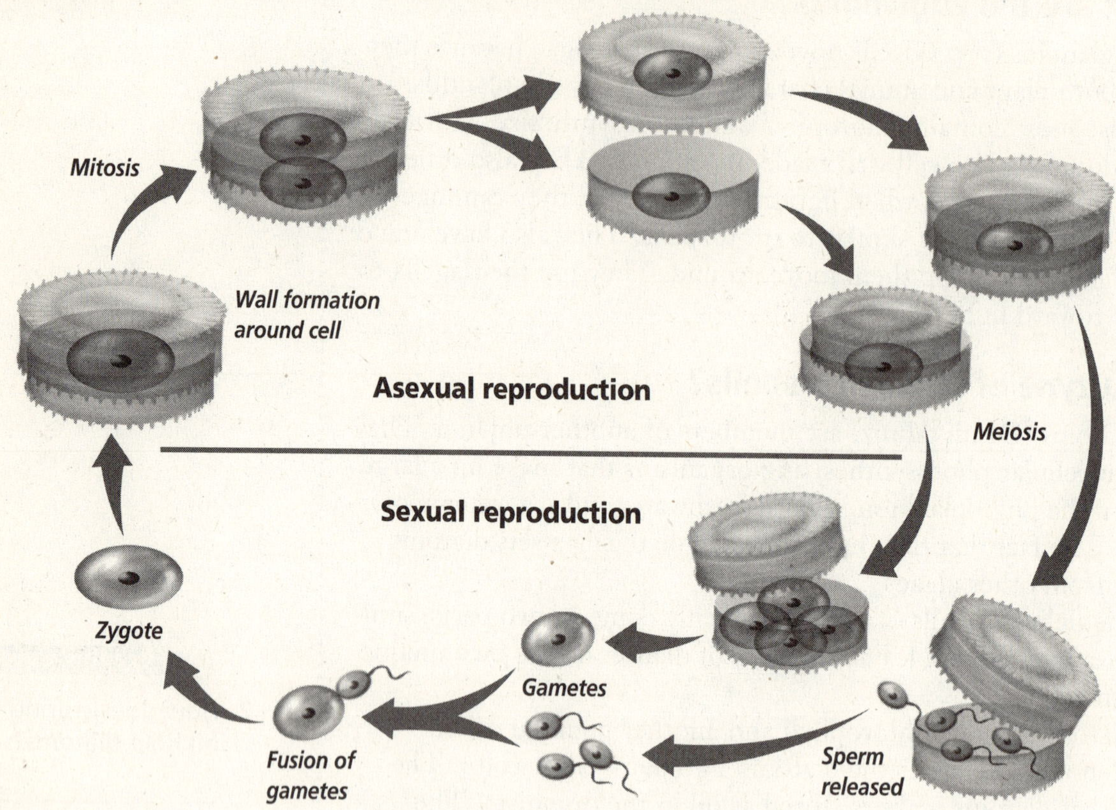

Think it Over

3. Identify Describe the food chain that leads from dinoflagellates to humans.

____

____

____

What are the dinoflagellates?

Dinoflagellates get their name from two flagella located in grooves. The cell spins slowly as the flagella beat.

Most of the dinoflagellates live in salt water as phytoplankton. Some species of dinoflagellates produce toxins, or poisons. Occasionally they produce enough toxins to kill large numbers of fishes. One of the toxic species is responsible for an event that is called red tide. In a red tide, the organisms become so numerous that the ocean turns a reddish color. During a red tide, humans run the risk of being poisoned if they eat shellfish, such as clams or mussels, which have fed on the toxic algae. Because of this, harvesting shellfish is usually banned during a red tide.

How can red algae survive in deep water?

The next three phyla, red algae, brown algae, and green algae, are multicellular organisms. Red algae are seaweeds. The body of the seaweed is called a **thallus.** Seaweeds lack roots, stems, or leaves. Red algae use structures called holdfasts to attach to rocks. Red algae grow in tropical waters or along rocky coasts in cold water.

Red algae can survive in deep water where most other seaweeds cannot live. This is because red algae have some photosynthetic pigments that absorb green, violet, and blue light. That is the only part of the light spectrum that can penetrate water below 100 m. ✓

Reading Check

4. What do red algae have that allows them to survive in deep water?

Can algae create a forest?

The second of the multicellular phyla are the brown algae. They are usually found in cool ocean water. Many types of brown algae have air bladders that keep their bodies floating near the surface where light is available. The largest of the brown algae is kelp. It can grow to be 60 m long. In some places kelp forms huge, dense, underwater forests. Kelp forests are ecosystems that provide homes for many saltwater organisms.

What are the characteristics of green algae?

Most green algae live in freshwater, so you may have seen them at the edges of ponds or lakes. Surprisingly, green algae can live in many different environments. Some live in salt water, some in snow, and some on tree trunks. There is even one species of green algae that lives in the fur of an animal.

Green algae are the most diverse of all the algae. They can be unicellular or multicellular. They may live in colonies. A **colony** is a group of cells that lives together in close association.

There is a species of green algae called *Volvox.* It is composed of hundreds of cells. Each cell has flagellates. Together the cells form a hollow, ball-shaped structure. The cells are connected by cytoplasm. The flagella of the individual cells face the outside of the ball-shaped structure. To move the colony, the flagella beat together, spinning the ball through the water. Small balls of daughter colonies form inside the larger ball. Eventually the larger ball breaks open and releases the daughter colonies.

How do green algae reproduce?

Green algae can reproduce sexually or asexually. One method of asexual reproduction in green algae is called **fragmentation.** During fragmentation an individual organism breaks into pieces and each piece forms a new organism. ✓

Reading Check

5. What is one method of asexual reproduction in green algae?

Reading Check

6. What is the name of the life cycle of green algae?

____

____

____

Alternation of Generations

Green algae have a complex life cycle. This life cycle alternates between individuals that produce spores and individuals that produce gametes. This pattern is called **alternation of generations.** It means the generations alternate between haploid and diploid. Remember that haploid forms of cells are sex cells or gametes. They contain half the number of chromosomes as the parent organism. Diploid forms are the body cells. They contain the same amount of chromosomes as the parent.

How does alternation of generations work?

Use the illustration below to help you understand alternation of generations. The haploid form of the algae is called the **gametophyte** because it produces gametes. The gametes join to form a zygote, which is a fertilized cell. From the zygote, the diploid form of the algae will develop. The diploid form is called the **sporophyte.** Certain cells in the sporophyte undergo meiosis. These spores are haploid spores that develop into new gametophytes.

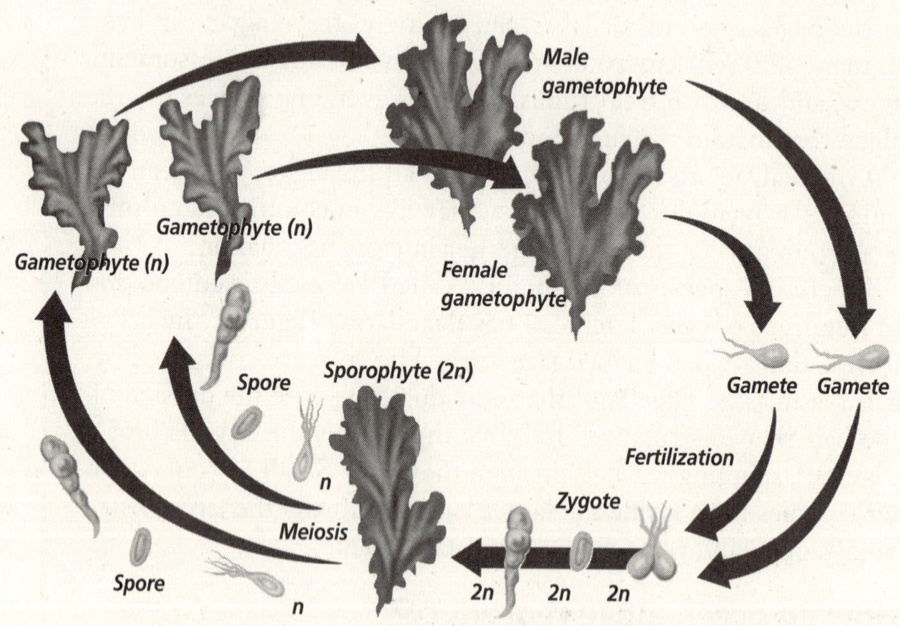

After You Read

Mini Glossary

alternation of generations: type of life cycle found in green algae and all plants where an organism alternates between a haploid gametophyte generation and a diploid sporophyte generation

colony: group of cells that live together in close association

fragmentation: type of asexual reproduction in algae where an individual breaks into pieces and each piece grows into a new individual

gametophyte: in algae, the haploid form of an organism in alternation of generations that produces gametes

sporophyte: in algae, the diploid form of an organism in alternation of generations that produces spores

thallus: body structure produced by plants and algae that lacks roots, stems, and leaves

1. Review the terms and their definitions in the Mini Glossary above. Circle the three words that relate to the life cycle of green algae. Then describe the process on the lines below using your own words.

____

____

2. Place the six phyla of algae in the appropriate column in the idea map below. One of the algae will be used twice.

red algae	euglenoids	diatoms
green algae	brown algae	dinoflagellates

Plantlike Protists: Algae

- Major producer of oxygen
- Important part of food chain

One-celled	Multicellular
______________	______________
______________	______________
______________	______________
______________	

Visit the Glencoe Science Web site at **science.glencoe.com** to find your biology book and learn more about algae, the plantlike protists.

Name Date Class

Section 19.3 Slime Molds, Water Molds, and Downy Mildews

Before You Read

You may have seen commercials on television about household cleaners that can rid a home of mold and mildew. From what you may have observed, where do you think molds and mildew most often are found? Write on the lines below what conditions might play a part in making a good environment for molds and mildew.

Read to Learn

Create a Quiz After you have read this section, create a quiz based on what you have learned. After you have written the quiz questions, be sure to answer them.

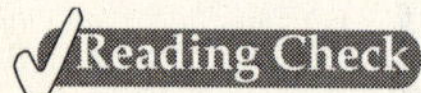

1. Name the two different types of slime mold.

What are funguslike protists?

Certain groups of protists have features like those of a fungus. Funguslike protists consist of the slime molds, the water molds, and the downy mildews. Like fungi, the funguslike protists decompose organic material for food.

There are three phyla of funguslike protists. Two of the phyla consist of slime molds. The third phylum is made up of the water molds and downy mildews.

Slime Molds

Many slime molds are beautifully colored, ranging from brilliant yellow or orange to rich blue, violet, and jet black. They live in cool moist places where they grow on damp, organic matter, such as rotting leaves or decaying tree stumps and logs.

There are two major types of slime molds. One type, the plasmodial slime molds, belong to one phylum. The other type of slime molds, the cellular slime molds, belong to a different phylum.

Slime molds are animal-like during much of their life cycle. They move about and surround food in a way similar to amoebas. However, like fungi, slime molds make spores to reproduce.

What is plasmodial slime?

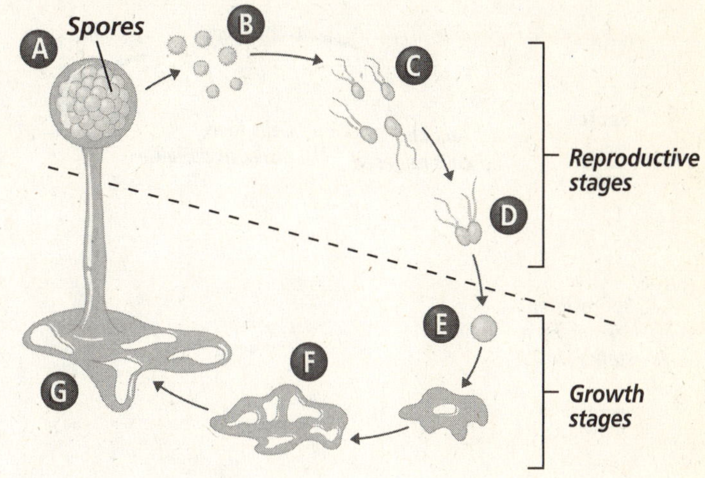

Plasmodial slime molds get their name because they form a plasmodium (plaz MOH dee um). A **plasmodium** is a mass of cytoplasm that contains many diploid nuclei but no cell walls or membranes. When a plasmodial slime mold looks like a slimy mass, it is in the feeding stage. The plasmodium eats by moving along decaying logs or leaves and surrounding its food. Faster plasmodiums travel at 2.5 centimeters per hour. At that rate, a plasmodium would cross this page in about eight hours.

A plasmodium can be more than a meter in diameter. If food and moisture become scarce, a plasmodium will change by transforming itself into many separate stalks. Each stalk is a spore-producing structure. That means that meiosis takes place in the stalk and produces haploid spores. When the wind blows, the spores are carried away. The illustration above shows the life cycle of plasmodial slime.

How do cellular slime molds differ?

Cellular slime molds have a different life cycle than plasmodial slime molds, so they have been placed in a different phyla. Cellular slime molds spend part of their life cycle as single amoeba-like cells. They feed, grow, and divide by cell division. ✓

When food becomes scarce, these independent cells join with hundreds or thousands of others to reproduce. When they get together, they look like a plasmodium. However, this mass of cells is multicellular. That means it is made up of many individual cells, each with a distinct cell membrane. This multicellular mass forms a single stalk that produces spores. You can see the process illustrated on page 220.

✓Reading Check

2. How do cellular slime molds differ from plasmodial slime molds?

____

____

____

Water Molds and Downy Mildews

The third phylum of the funguslike protists contains the water molds and downy mildews. These organisms live in water or moist places. Some feed on dead organisms and others are plant parasites.

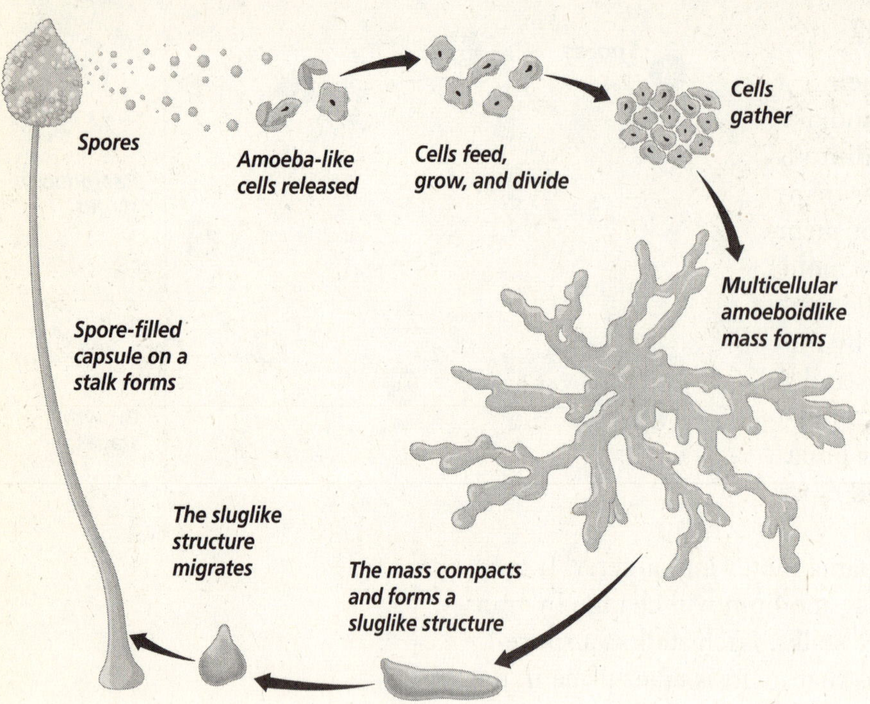
Spores
Amoeba-like cells released
Cells feed, grow, and divide
Cells gather
Multicellular amoeboidlike mass forms
Spore-filled capsule on a stalk forms
The sluglike structure migrates
The mass compacts and forms a sluglike structure

Most water molds look like fuzzy, white growth on decaying matter. They resemble some fungi. However, at some point in their life cycle, water molds produce reproductive cells with flagellates—something that fungi never do.

Downy mildew led to a serious famine that struck Ireland in the mid 1800s. At the time, the major food crop for the people of Ireland was the potato. A downy mildew destroyed most of the potatoes. During the famine that followed, many Irish citizens moved to other countries, including America.

Think it Over

3. **Analyze** How are water molds and downy mildews different from slime molds?

____

____

Reading Check

4. What evidence suggests that ancient green algae was the ancestor of modern plants?

____

____

Origin of Protists

Now that you have studied the various protists, you may be wondering how protists are related to the animals, plants, and fungi they resemble.

Currently there is not conclusive evidence that suggests that ancient protists were the evolutionary ancestors of fungi, plants, or animals, nor is there enough evidence to say that protists emerged as a separate evolutionary line. However, after analyzing RNA, many biologists agree that ancient green algae were probably the ancestors to modern plants. ✓

After You Read

Mini Glossary

plasmodium: in plasmodial slime molds, the mass of cytoplasm that contains many diploid nuclei but no cell walls or membranes

1. Review the term and its definition in the Mini Glossary above. Write a sentence using the term **plasmodium** on the lines below.

2. Using the idea map below and information from this section, answer the following questions.

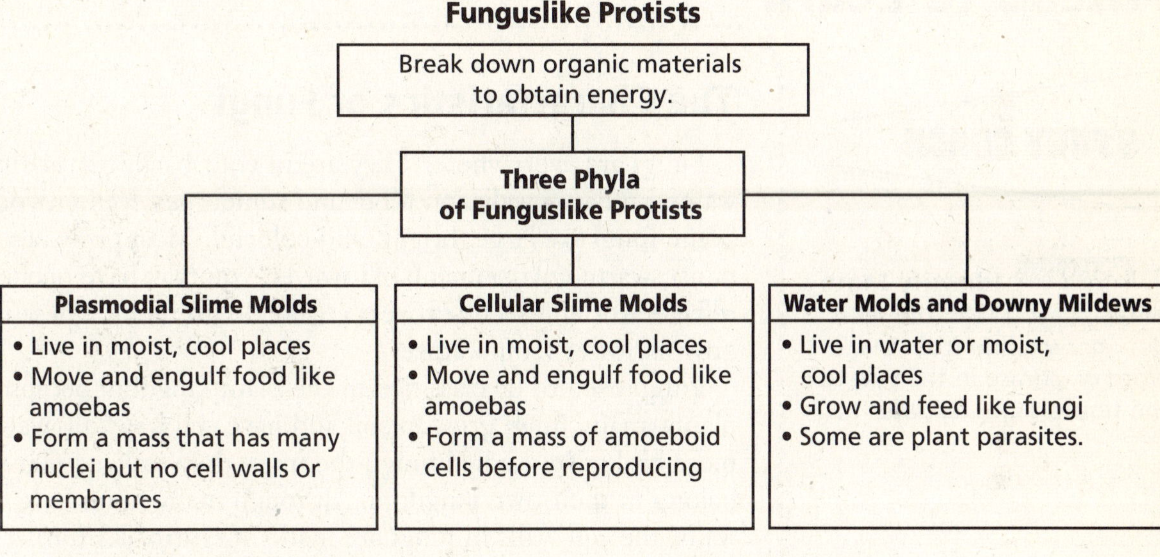

a. How are plasmodial slime molds and cellular slime molds different?

b. How are some funguslike protists harmful to plants?

Visit the Glencoe Science Web site at **science.glencoe.com** to find your biology book and learn more about slime molds, water molds, and downy mildews.

Name Date Class

Section 20.1 What is a fungus?

Before You Read

You have studied three of the six kingdoms so far, archaebacteria, eubacteria, and protists. In the next two sections, you will learn about the kingdom Fungi. Fungus is the singular of fungi. Mushrooms are types of fungi. Think about places you have seen mushrooms growing. What do those places have in common? Were they hot, dry, cool, or damp? Did the mushrooms appear suddenly or grow slowly over time? Write your thoughts on the lines below.

____

____

Read to Learn

Mark the Text **Identify Main Ideas** As you read the section, draw a line from the word or phrase in the text to the figure that illustrates it.

The Characteristics of Fungi

Fungi are everywhere. They are in your backyard, in air and water, on damp walls, on food, and sometimes even on your body. Some fungi are large, bright, and colorful. Most grow best in moist, warm environments. However, you may have opened your refrigerator and seen leftovers covered with mold. These molds grow at lower temperatures.

Fungi used to be classified in the plant kingdom because, like plants, many fungi grow in soil and have cell walls. However, the more biologists studied fungi, the more they realized that fungi belong in their own kingdom. Although many fungi have cell walls, the cell walls in fungi are made of **chitin** (KI tun), a complex carbohydrate. Plants have cell walls made of cellulose.

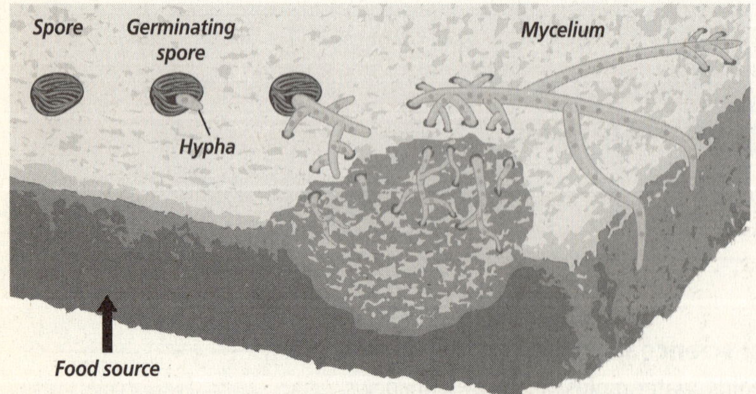

A germinating fungal spore produces hyphae that branch to form mycelium.

What is the structure of fungi?

There are a few unicellular fungi such as yeasts. Most fungi, however, are multicellular. The basic structural units of multicellular fungi are long, threadlike filaments called **hyphae** (HI fee) (singular, hypha). The hyphae grow from the spore of a fungus. As they grow, they branch to form a network of filaments called a **mycelium** (mi SEE lee um). The illustration to the left shows the spore, hyphae, and mycelium.

What is it like inside the hyphae?

There are different types of hyphae in the mycelium. The differences are based on function. Some hyphae anchor the fungus, some enter into the food source, and others are formed only for the purpose of reproduction.

Hyphae are made of cytoplasm containing many nuclei. If you looked at a hypha under a microscope, you would see hundreds of nuclei streaming along in a continuous flow of cytoplasm. The flow of cytoplasm sends nutrients throughout the fungus. Some hyphae have walls called septa that divide the hypha into cells with one or more nuclei. The septa are usually porous, which means cytoplasm, organelles, and nutrients can pass through them.

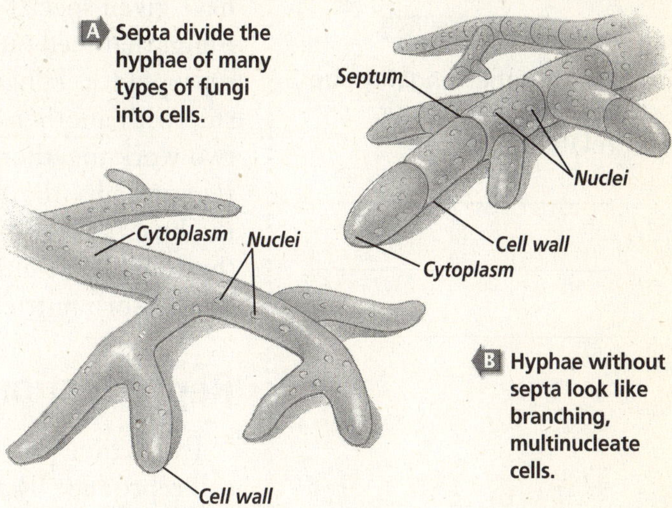

A Septa divide the hyphae of many types of fungi into cells.

B Hyphae without septa look like branching, multinucleate cells.

Adaptations in Fungi

Some fungi cause food to spoil, others cause diseases, and some are even poisonous. However, fungi are important and beneficial. Without fungi, the world would be overrun with huge amounts of waste, dead organisms, and dead plants. Thanks to many fungi, some bacteria, and protists, the organic material is broken down and recycled into the raw materials that other living organisms need.

How do fungi get food?

Fungi cannot make their own food. They are heterotrophs. Fungi use a process called extracellular digestion to obtain nutrients. This means food is digested outside a fungus's cells and then the digested food is absorbed. For example, some hyphae of a fungus will grow into an orange. They release digestive enzymes into the orange that break down the large organic molecules into smaller molecules. These small molecules are absorbed into the hyphae and move into the flowing cytoplasm.

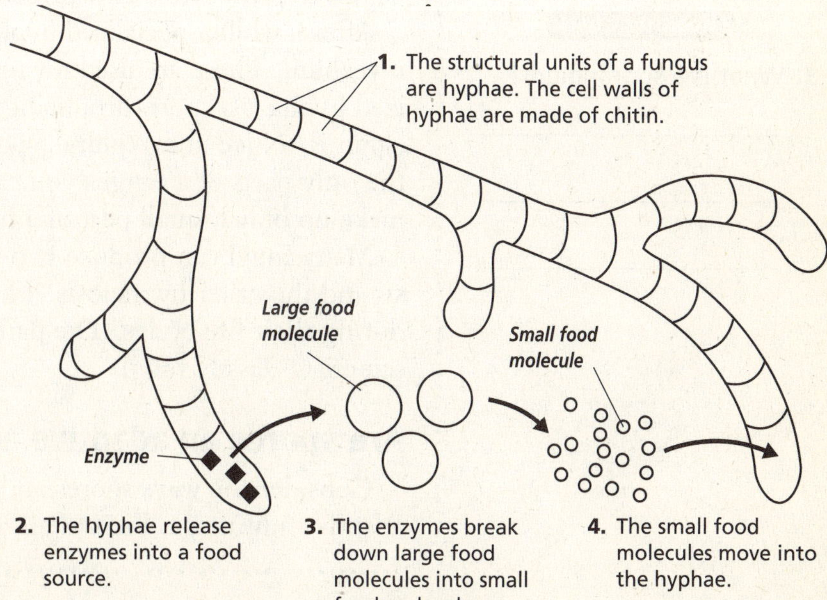

1. The structural units of a fungus are hyphae. The cell walls of hyphae are made of chitin.
Large food molecule
Small food molecule
Enzyme
2. The hyphae release enzymes into a food source.
3. The enzymes break down large food molecules into small food molecules.
4. The small food molecules move into the hyphae.

Section
20.1 What is a fungus?, *continued*

Reading Check

1. Identify the fungal group that has a symbiotic relationship.

Fungi have different types of food sources. In fact, scientists have given special names to fungi depending on their food source. Fungi that feed on waste and dead organic material are called saprophytes. Fungi that have a symbiotic (cooperative) relationship with another organism, such as algae, are mutualists. The two work together to get food. There also are parasitic fungi. Parasites absorb nutrients from the cells of their hosts, often killing the host. Parasitic fungi have special hyphae called **haustoria** (huh STOR ee uh) that grow directly into the host's cells where they absorb nutrients and minerals from the host.

Reproduction in Fungi

Reading Check

2. In what type of fungi does budding occur?

Every time you dig in the garden, you are helping fungi in the soil reproduce by fragmentation. In fragmentation, pieces of hyphae break off a mycelium and grow into new mycelia. Every time you dig in the soil, your shovel slices through mycelia, breaking, or fragmenting them.

Fragmentation is one form of asexual reproduction in fungi. Another form is budding. Unicellular fungi called yeasts often reproduce by budding. During **budding** mitosis occurs and a new individual pinches off from the parent.

How do fungi reproduce by spores?

Reading Check

3. What is a sporangium?

Most fungi produce spores. A spore is a reproductive cell that can develop into a new organism. When a fungal spore arrives in a place with favorable growing conditions, a hypha comes out and begins to grow into a new mycelium.

After a while, specialized hyphae grow away from the rest of the mycelium. These are used for reproduction. At the tip of a specialized hypha is a spore-producing sac or case called a **sporangium** (spuh RAN jee uhm) (plural, sporangia). The sporangia are usually the only parts of a fungus you can see. However, the sporangia make up only a small part of a fungus.

Many fungi can produce two types of spores, one type by mitosis and the other by meiosis. This will happen at different times during their life cycles. The patterns of reproduction are one way scientists classify fungi.

Are spores an adaptive advantage?

Consider the ways spores and their production give fungi an adaptive advantage. First of all, the sporangia protect the spores and keep them from drying out until they are ready to be released. Second, most fungi produce a large number of spores at one time. For example, a puffball produces about 1 trillion spores. Producing so many spores increases the survival chances for the species. Finally, spores are small, lightweight, and can be easily carried by wind, water, and animals. Spores carried on the wind can travel hundreds of kilometers.

After You Read

Mini Glossary

budding: type of asexual reproduction in which a cell or group of cells pinch off from the parent to form a new individual

chitin: complex carbohydrate that makes up the cell walls of fungi

haustoria: in parasitic fungi, hyphae that grow into host cells and absorb nutrients and minerals from the host

hyphae: threadlike filaments that are the basic structural units of multicellular fungi

mycelium: in fungi, a complex network of branching hyphae

sporangium: in fungi, a sac or case of hyphae in which spores are produced

1. Read the terms and the definitions in the Mini Glossary above. Circle one key term, then in the space provided, write the definition of the term in your own words.

2. Write facts you have learned about fungi to complete the web diagram below.

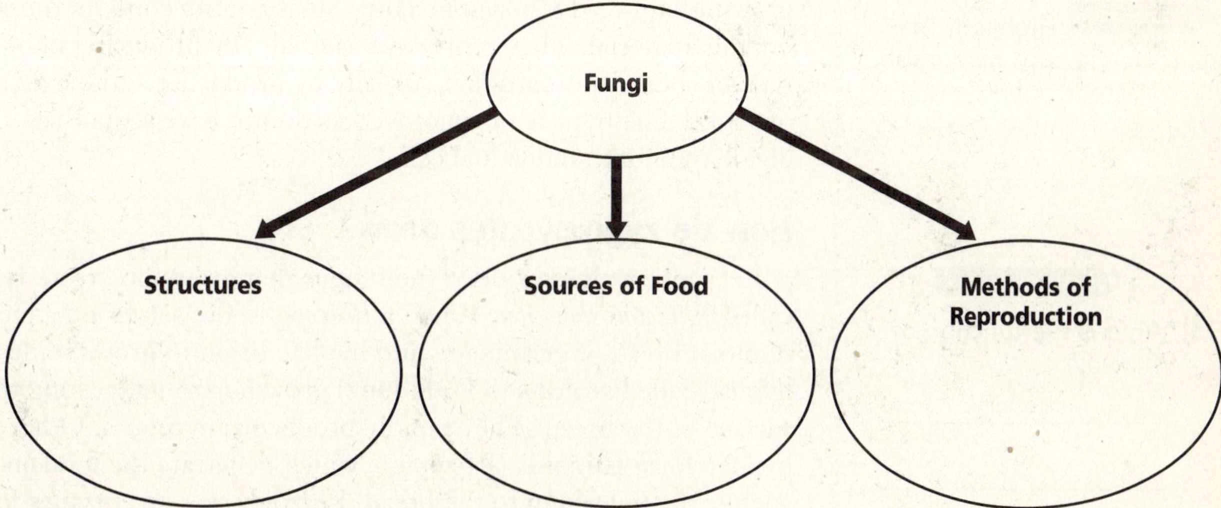

Visit the Glencoe Science Web site at **science.glencoe.com** to find your biology book and learn more about fungi.

Section 20.2 The Diversity of Fungi

Before You Read

Fungi face some interesting survival challenges. Like plants, they grow in one place and depend on the dispersal of their reproductive cells to grow in new locations. Plants are usually taller than fungi and some depend on the wind to carry seeds long distances. Fungi are usually so low to the ground that the air around them is still. As a result, they have evolved some unusual features to get their spores into the air. Some fungi, such as puffballs, shoot spores into the air. Truffles are underground fungi. They have a scent that animals love, so the animals dig into the ground to find the truffles, spreading the spores. A fungus called stinkhorn smells like rotting meat. It attracts flies that then fly away covered with spores. As you read this section, think of the methods by which fungi spread their spores.

Read to Learn

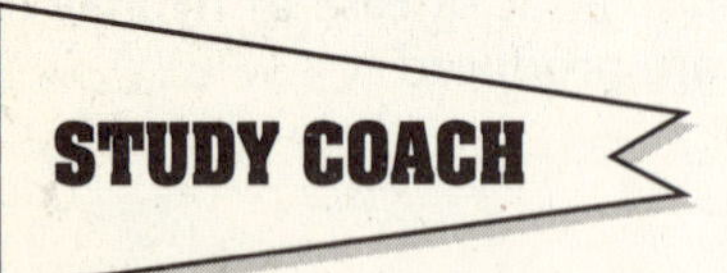

Identify Main Ideas As you read this section, stop after every few paragraphs and put what you have just read into your own words.

Mark the Text Highlight the main idea in each paragraph.

Zygomycotes

You may recall from the previous section that scientists classify fungi according to how they reproduce. Some fungi reproduce both asexually and sexually during their life cycles. This section covers several divisions in fungi.

The first group of fungi you will learn about are the members of the phylum Zygomycota (zy goh mi KOH tuh). There are about 1000 species of zygomycotes. The mold that you find growing on bread is a zygomycote. Most zygomycotes decompose organic material. They reproduce asexually by producing one type of spore and reproduce sexually by producing another type of spore. The hyphae of zygomycotes do not have septa that divide them into individual cells.

How do zygomycotes grow?

Let's take a closer look at the fungus that grows on bread. It is called *Rhizopus stolonifer.* When a *Rhizopus* spore settles on a piece of moist bread, it germinates and hyphae begin to grow. Some hyphae called **stolons** (STOH lunz) grow horizontally along the surface of the bread. They rapidly produce a mycelium. Other hyphae form **rhizoids** (RI zoydz), which penetrate the food and anchor the mycelium to the bread. Rhizoids secrete enzymes for extracellular digestion. Extracellular digestion occurs when enzymes break down large molecules of the food into small molecules that are absorbed by the hyphae. ✓

1. What is the function of rhizoids?

______

______

______

How do zygomycotes reproduce asexually?

Asexual reproduction begins as hyphae grow upward and develop sporangia at the tips. Spores develop in the sporangia and then are released. The spores that land on a moist food supply germinate, form hyphae, and reproduce asexually again.

How do zygomycotes reproduce sexually?

What if the bread on which *Rhizopus* was growing began to dry out? That unfavorable condition might cause the fungus to reproduce sexually. When zygomycotes reproduce sexually, they produce zygospores. **Zygospores** (ZI go sporz) are spores with thick walls that can withstand unfavorable conditions.

Sexual reproduction in *Rhizopus* happens when haploid hypha from two compatible mycelia, called plus and minus, fuse. Each of them forms a gametangium. A **gametangium** (ga muh TAN ghee uhm) is a structure containing a haploid nucleus. When the haploid nuclei of the two gametangia fuse, a diploid zygote forms. The zygote has a thick wall and becomes a dormant zygospore. The illustration below shows the difference between asexual and sexual reproduction.

2. Infer What adaptive advantages are evident in a zygospore?

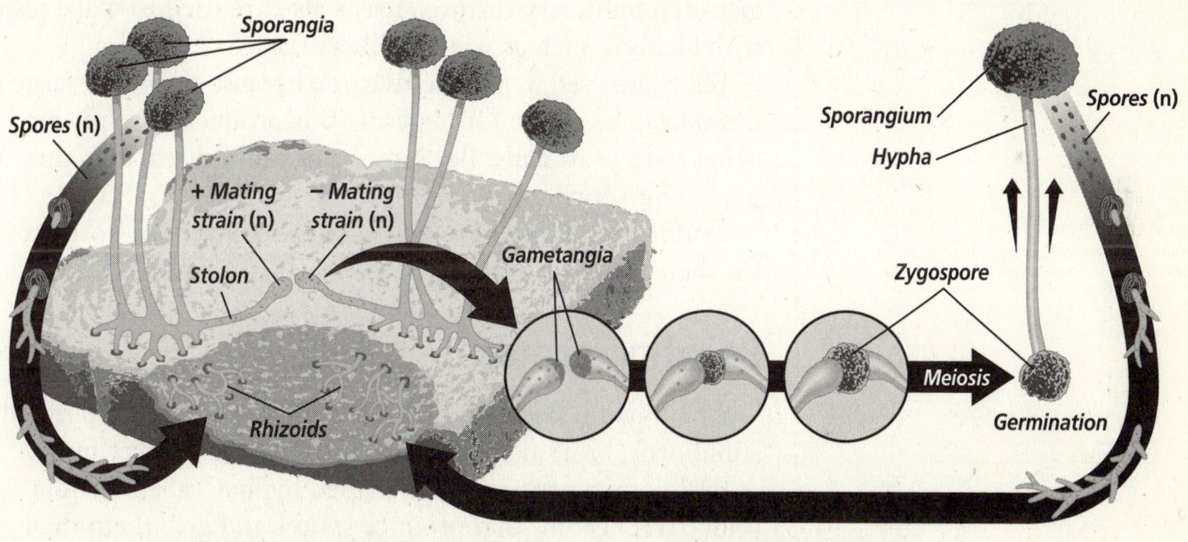

How long can a zygospore remain dormant?

A zygospore may remain dormant for many months. It can survive without water and through periods of cold and heat. When conditions become favorable again, the zygospore absorbs water, undergoes meiosis, and produces a hypha with a sporangium. Each haploid spore in the sporangium can grow into a new mycelium.

Ascomycotes

The Ascomycota is the largest phylum of fungi. They are sometimes called sac fungi. They have tiny saclike structures, each called an **ascus,** in which the sexual spores of the fungi develop. The sexual spores are called **ascospores** because they are formed inside an ascus.

A different kind of spore is produced during asexual reproduction. Hyphae grow up from the mycelium, becoming longer to form **conidiophores** (kuh NIH dee uh forz). Then chains or clusters of asexual spores called **conidia** develop from the tips of the conidiophores. Wind, water, and animals spread the spores.

Reading Check

3. What ascomycote might you encounter every day?

___

___

___

How are sac fungi important?

You encounter sac fungi nearly every day. Yeasts, a type of sac fungi, are anaerobes and ferment sugars to produce carbon dioxide and ethyl alcohol. They are used in bread making because they produce carbon dioxide, a gas that causes bread to rise and gives it a light, airy texture. Yeasts also are used to make products with alcohol, such as wine and beer. ✓

Yeasts are used in genetic research because they have large chromosomes. A vaccine for hepatitis B is produced by splicing human genes with yeast cells. Because yeasts multiply rapidly, they are an important source of the vaccine.

Truffles and morels are two edible ascomycotes. Both are prized for their flavor.

Basidiomycotes

The phylum Basidiomycota contains mushrooms, puffballs, stinkhorns, bird's nest fungi, rust, smut, and bracket fungi.

Basidiomycotes have club-shaped hyphae called **basidia** (buh SIHD ee uh) that produce spores and give them their common name—club fungi. Basidia usually develop on short-lived reproductive structures that can easily be seen. The spores are called **basidiospores.**

A basidiomycote, like the mushroom in the illustration on page 229, has a complicated reproductive cycle. Study the illustration to understand this complicated cycle.

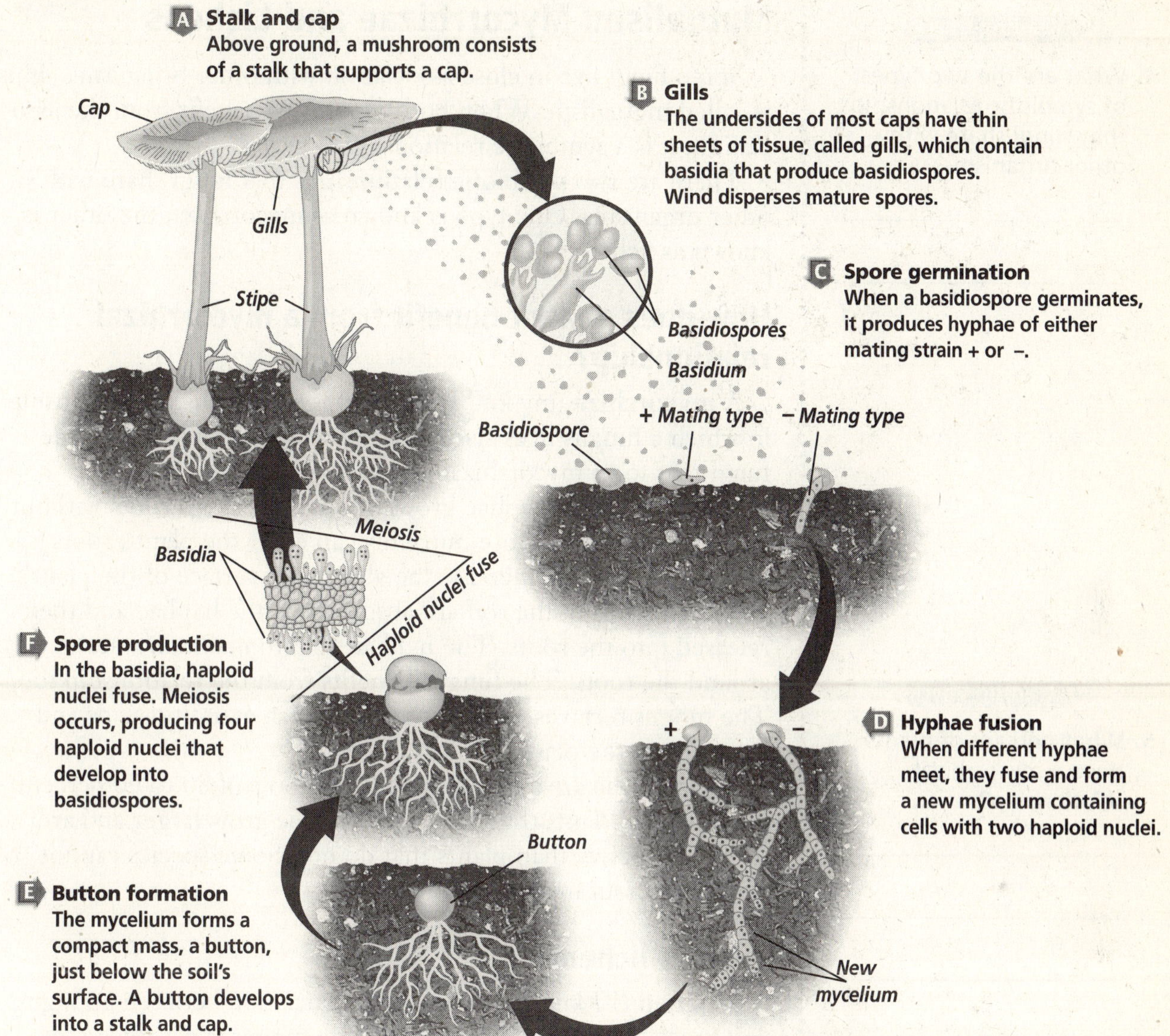

A Stalk and cap
Above ground, a mushroom consists of a stalk that supports a cap.
Cap
Gills
Stipe
B Gills
The undersides of most caps have thin sheets of tissue, called gills, which contain basidia that produce basidiospores. Wind disperses mature spores.
Basidiospores
Basidium
C Spore germination
When a basidiospore germinates, it produces hyphae of either mating strain + or –.
Basidiospore
+ Mating type
– Mating type
Meiosis
Basidia
Haploid nuclei fuse
F Spore production
In the basidia, haploid nuclei fuse. Meiosis occurs, producing four haploid nuclei that develop into basidiospores.
+
–
D Hyphae fusion
When different hyphae meet, they fuse and form a new mycelium containing cells with two haploid nuclei.
Button
E Button formation
The mycelium forms a compact mass, a button, just below the soil's surface. A button develops into a stalk and cap.
New mycelium

Deuteromycotes

Unlike the zygomycotes, ascomycotes, and basidiomycotes, the deuteromycotes only reproduce asexually. It may be possible that they have a sexual reproduction phase, but scientists have not observed it.

Deuteromycotes have many commercial uses. They are used to make foods, such as soy sauce and bleu cheese. They also are used to produce citric acid. Citric acid gives jams, jellies, fruit-flavored candies, and soft drinks a tart taste. Penicillin, an antibiotic prescribed by doctors to treat bacterial infections like strep throat, is produced from a deuteromycote.

Reading Check

4. What are the two types of symbiotic relationships that fungi share with other organisms?

____

____

Mutualism: Mycorrhizae and Lichens

Some fungi live in close association with other organisms. This is called mutualism. When both organisms benefit from the association, it is a symbiotic relationship.

There are two symbiotic relationships that fungi share with other organisms. One type is known as mycorrhiza, the other is known as lichen. ✓

How does a plant benefit from a mycorrhizal relationship?

A **mycorrhiza** (my kuh RHY zuh) is a mutualistic relationship in which a fungus lives symbiotically with a plant. Most of the fungi that form mycorrhizae are basidiomycotes.

Fine, threadlike hyphae grow around the plant's roots without harming the plant. More nutrients can enter the plant's roots because the hyphae increase the absorptive surface of the plant's roots. Minerals in the soil are absorbed by the hyphae and then released into the roots. The hyphae help to maintain water around the roots. The fungus benefits from the relationship too. The fungus receives organic nutrients such as amino acids and sugars from the plant.

Mycorrhizae are associated with the roots of 80 to 90 percent of all plants. Plants that have mycorrhizae grow larger and are more productive than plants that do not. Some species cannot survive without mycorrhizae. ✓

Reading Check

5. What percent of plants have a mycorrhizal relationship?

____

____

____

What is lichen?

A **lichen** (LI kun) is a symbiotic association between a fungus and green algae or cyanobacteria. The fungus is usually an ascomycote. The algae or cyanobacteria are photosynthetic.

The fungus portion of the lichen forms a dense web of hyphae. The algae or cyanobacteria grow inside the web. Together, the fungus and its photosynthetic partner form a structure that looks like a single organism.

Lichens need only light, air, and minerals to grow. Here is how the relationship works. The photosynthetic partner provides food for both organisms. The fungus provides its partner with water and minerals that it absorbs from rain and the air. The fungus protects the partner from changes in environmental conditions.

Lichens are found all over the world. They are often the first to colonize an area. They can live in dry deserts or on bare rocks exposed to cold wind. On the arctic tundra, lichens are the most common form of vegetation. Caribou graze on lichen just as cattle graze on grass.

Lichens indicate pollution levels in the air. The fungus easily absorbs material from the air. If pollutants are present, the fungus dies. Without the fungal part of the lichen, the photosynthetic partner dies.

Origins of Fungi

Scientists hypothesize that zygomycotes evolved first. Basidiomycotes and ascomycotes evolved later from a common ancestor. Scientists use fossils to provide clues as to how and when organisms evolved. However, fossils of fungi are rare because the fungi are made of soft materials. Scientists have discovered fungi fossils that are over 400 million years old. ✓

✓Reading Check

6. Why are fossils of fungi so rare?

____

____

____

After You Read

Mini Glossary

ascospores: sexual spore of ascomycote fungi that develop within an ascus

ascus: tiny, saclike structure in ascomycotes in which ascospores develop

basidia: club-shaped hyphae of basidiomycote fungi that produce spores

basidiospores: spores produced in the basidia of basidiomycotes during sexual reproduction

conidia: chains or clusters of asexual ascomycote spores that develop on the tips of conidiophores

conidiophores: in ascomycotes, elongated, upright hyphae that produce conidia at their tips

gametangium: structure that contains a haploid nucleus; formed by the fusion of haploid hyphae

lichen: organism formed from a symbiotic association between a fungus, usually an ascomycote, and a photosynthetic green algae or cyanobacteria

mycorrhiza: mutualistic relationship in which a fungus lives symbiotically with a plant

rhizoids: fungal hyphae that penetrate food and anchor a mycelium; secrete enzymes for extracellular digestion

stolons: fungal hyphae that grow horizontally along a surface and rapidly produce a mycelium

zygospores: thick-walled spores of zygometes that can withstand unfavorable conditions

Section
20.2 The Diversity of Fungi, *continued*

1. Review the terms and their definitions in the Mini Glossary on page 231. Circle the two terms that have to do with symbiotic relationships and mutualism. Then on the lines below use those terms in a sentence that explains the interaction between organisms.

2. Fill in the numbered boxes to complete the chart below. Use the following terms: **sporangia, rhizoids, hyphae, stolons, mycelium.**

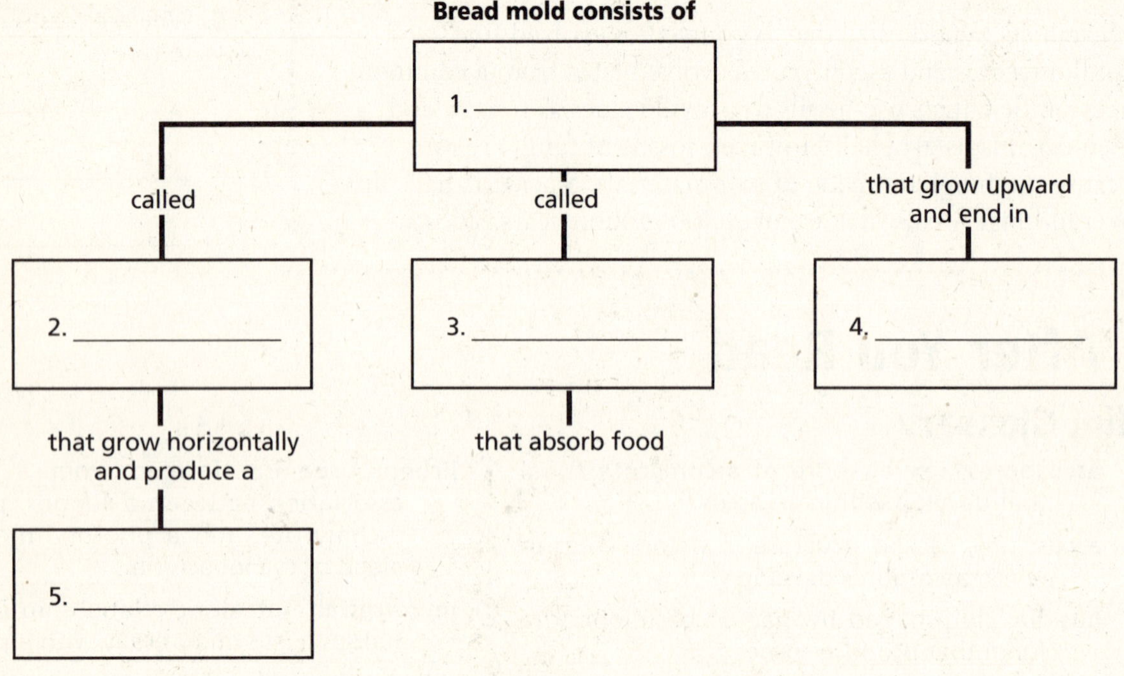

3. When you transplant flowers, shrubs, or trees, why should you not disturb the soil around a plant's roots? Write your response on the lines below.

Visit the Glencoe Science Web site at **science.glencoe.com** to find your biology book and learn more about the diversity of fungi.

Name Date Class

Section
21.1 Adapting to Life on Land

Before You Read

Think of all the things that plants provide for us. They are an important source of food. They also provide oxygen through photosynthesis. Some plants are valuable sources of medicine. What are some other things that plants provide? Write your ideas on the lines below.

____

____

____

Read to Learn

Origins of Plants

In earlier sections you learned that some organisms that are not plants have plantlike characteristics. In this section you will learn exactly what a plant is and more about the plant kingdom.

Plants have many characteristics. A plant is a multicellular eukaryote. That means it is made of cells that have a true nucleus and membrane-bound organelles. Plants have thick cell walls made of a carbohydrate called cellulose. The stems and leaves of plants have a waxy waterproof coating called a **cuticle** (KYEWT ih kul). The cuticle helps in reducing water loss. Most plants are able to produce their own food through the process of photosynthesis. ✓

Make Flash Cards For each paragraph, think of a question your teacher might ask on a test. Write the question on one side of a flash card. Then write the answer on the other side. Quiz yourself until you know the answers.

✓ Reading Check

1. What are the main characteristics of plants?

____

____

____

Where did plants come from?

A billion years ago there were no plants. In fact, the land was quite bare. However, the shallow waters of the oceans contained bacteria, algae, and simple animals such as corals, sponges, jellyfish, and worms. Scientists hypothesize that the algae at the edges of the inland seas and oceans adapted to life on land.

How might that have happened? It is possible that plants evolved from threadlike green algae that lived in ancient oceans. Some of the evidence to support this hypothesis can be found in modern members of the algae and plant groups. Modern green algae and plants have cell walls that contain cellulose. Both groups use the same type of chlorophyll in photosynthesis and store food in the form of starch.

Plants first began to appear in the fossil record over 440 million years ago. The plants were simple and did not have leaves.

Adaptations in Plants

Life on land has challenges. All living things need water to survive. Threadlike green algae floating in a pond absorb water and nutrients directly into cells. How will land-based plants get water and nutrients? For most land plants, the only available water and nutrients are in the soil. Only the part of the plant that is in the soil can absorb the nutrients. Land plants had to adapt to absorb water and nutrients.

2. What is the purpose of the cuticle?

You already have learned about one adaptation. The waxy cuticle creates a barrier that helps prevent water in the plant's tissues from evaporating. You can feel the smooth, somewhat slippery, waxy cuticle if you run your fingers over the leaves or stems of some houseplants.

Reproduction on land was another challenge. Algae release gametes, or male and female sex cells, into the water where fertilization takes place. Once on land, plants had to adapt to protect the gametes from drying out. Another problem was exposure to wind and weather. Land plants also had to develop structures to help them grow against the force of gravity. Over the past 440 million years, plants have developed many adaptations.

Mark the Text **Identify Main Parts** Highlight the appropriate part of the illustration after you read about it.

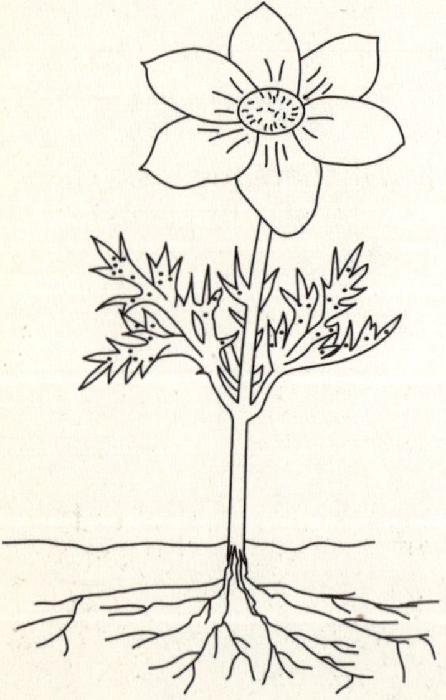

What are the basic plant organs?

A **leaf** is a plant organ that grows from a stem. Photosynthesis usually occurs in leaves. Leaves come in many different sizes and shapes. They can even vary on the same plant. Each plant species has unique leaves or leaflike structures.

A **root** is a plant organ that absorbs water and nutrients directly from the soil. Roots contain tissues that transport those nutrients to the stem. Roots anchor a plant to the ground. Some roots function as food storage. Radishes and sweet potatoes are two examples of food storage structures in roots.

A **stem** is a plant organ that provides support for growth. The stem moves water from the roots to the leaves, and moves sugars produced in the leaves to the roots. The stem contains tissues for transporting food, water, and other materials from one part of the plant to another. Sometimes stems serve as organs for food storage. In green stems, chlorophyll-containing cells can carry out photosynthesis.

Section 21.1 **Adapting to Life on Land,** *continued*

Most of the plants you are familiar with, such as roses, ferns, ivy, pine trees, and maple trees, are vascular plants. The stem contains vascular tissue. **Vascular** (VAS kyuh lur) **tissue** is made of tubelike, long cells through which water and nutrients travel. Plants that have vascular tissues are called **vascular plants.** ✓

There also are nonvascular plants such as mosses, hornworts, and liverworts. **Nonvascular plants** do not have vascular tissues. The body of a nonvascular plant is often only a few cells thick. Water and nutrients travel from cell to cell by the processes of osmosis and diffusion. (For a review of these processes, see Section 6.2 for a discussion of diffusion and Section 8.1 for a discussion of osmosis.)

Vascular tissues were an important structural adaptation. Vascular plants can live farther away from water than nonvascular plants. Vascular tissues with thickened cells called fibers also allow vascular plants to grow taller than nonvascular plants.

✓ Reading Check

3. What is vascular tissue?

What is a seed?

An important adaptation in some land plants was the development of seeds. A **seed** is a plant organ that contains an embryo and a food supply covered in a hard protective coat. The seed protects the embryo from drying out. Land plants reproduce by either spores or seeds. Remember that a spore is a reproductive cell that forms without fertilization and produces a new organism. Spores have a hard outer wall and are haploid. Haploid cells contain only one of each kind of chromosome. Seeds are diploid since they contain two of each kind of chromosome.

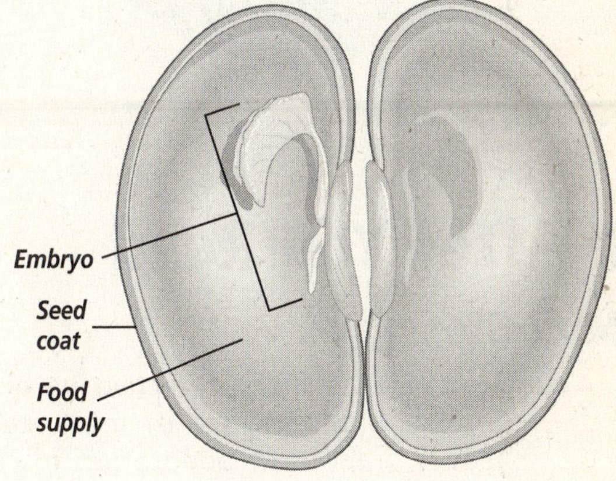

In non-seed plants such as mosses and ferns, the sperm require a film of water on the plant in order to reach the egg. In seed plants, which include all conifers and flowering plants, sperm reach the egg without using a film of water. This is one reason non-seed plants require wetter habitats than most seed plants.

What is alternation of generations in plants?

The lives of all plants include two stages, or alternating generations. One generation is the gametophyte generation. The other generation is the sporophyte generation. All cells of the gametophyte generation are haploid cells, including the gametes produced

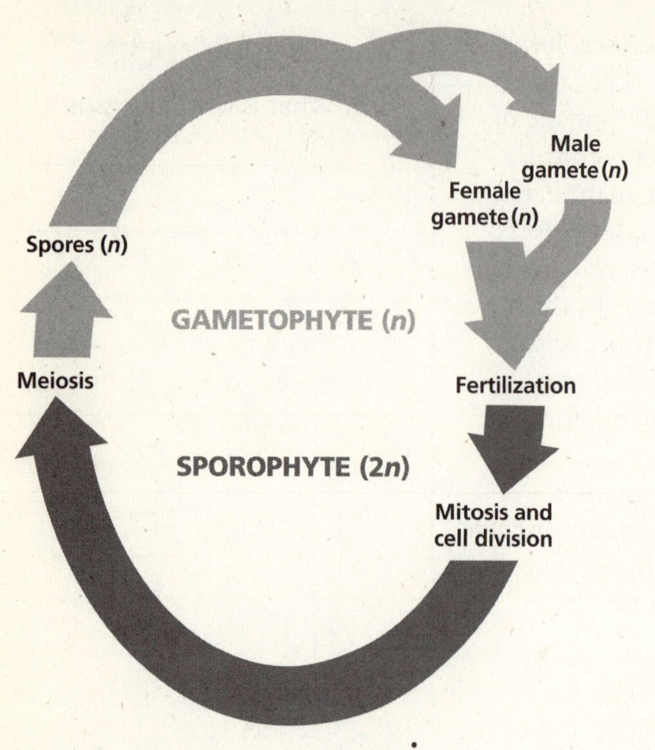

during this generation. All cells of a sporophyte generation are diploid. The spores produced in the sporophyte generation, however, are haploid.

In non-seed vascular plants such as ferns, spores are released into the environment. They grow into haploid gametophyte plants. These plants produce male and female gametes. After fertilization, a sporophyte plant develops and grows from the gametophyte plant.

In seed plants such as sunflowers, spores develop inside the sporophyte and become the gametophyte generation. The gametophytes are made up of only a few cells. Male and female gametes are produced by these gametophytes. This takes place within the sporophyte plant. After fertilization, a new sporophyte develops within a seed. The seed is released and the new sporophyte plant grows when conditions are favorable.

After You Read

Mini Glossary

cuticle (KYEWT ih kul): protective, waxy coating on the outer surface of most stems and leaves; important adaptation in reducing water loss

leaf: the plant organ that grows from a stem in which photosynthesis usually occurs

nonvascular plant: plants that do not have vascular tissues

root: plant organ that absorbs water and nutrients usually from soil; contains vascular tissues; anchors plant; can be a storage organ

seed: a plant organ of seed plants consisting of an embryo, a food supply, and a protective coat; protects the embryo from drying out

stem: plant organ that provides support and growth; contains tissues that transport food, water, and other materials; organ from which leaves grow. Can serve as food storage organ; green stems can carry out photosynthesis

vascular plant: plants that have vascular tissues; enables taller growth and survival on land

vascular (VAS kyuh lur) tissue: tissues found in vascular plants composed of tubelike, elongated cells through which food, water, and other materials are transported throughout the plant

Section 21.1 Adapting to Life on Land, *continued*

1. Review the terms and their definitions in the Mini Glossary on page 236. Then on the lines below, explain how the development of the cuticle and seeds were adaptations to life on land.

2. Choose one of the question headings in the Read to Learn section. Write the question in the space below. Then write your answer to that question on the lines that follow.

Question:

Answer:

Visit the Glencoe Science Web site at **science.glencoe.com** to find your biology book and learn more about adapting to life on land.

Section 21.2 Survey of the Plant Kingdom

Before You Read

Name or picture in your mind at least ten different plants. You might think of roses, cacti, ferns, oak trees, and grasses to name a few. They are all plants, yet they look different from each other. As in the other kingdoms you have studied, the plant kingdom has divisions based on shared characteristics. If you were to place plants in divisions, what characteristics would you use? Write your ideas for classifying plants on the lines below.

Read to Learn

STUDY COACH

Mark the Text **Identify Facts** As you read this section, say the name of each plant division aloud. Underline a fact about each plant division.

Reading Check

1. What are the highlights of plant evolution?

Phylogeny of Plants

Phylogeny is the evolutionary history of a species. Once plants adapted to life on land, they were not finished adapting. Many geological and climate changes have taken place since the first plants evolved. Landmasses have moved from place to place over Earth's surface. Climates have changed. Bodies of water have formed and then disappeared. Hundreds of thousands of plant species have evolved and many of them became extinct as conditions changed. As plant species evolved in this changing landscape, they kept many of their old characteristics and developed new ones.

Botanists (scientists who study plants) use plant characteristics to classify plants into divisions. Highlights of plant evolution include:

1. origins of plants from green algae,
2. the production of a waxy cuticle,
3. the development of vascular tissue and roots, and
4. the production of seeds.

Producing seeds is one way to separate plants into two groups—non-seed plants and seed plants. ✓

Non-seed Plants

Non-seed plants produce reproductive cells called spores. Some non-seed plants are vascular, and some are non-vascular. There are seven divisions within the non-seed plant group.

What are hepaticophytes?

Hepaticophytes (heh PAH tih koh fites) make up one division of non-seed plants. Hepaticophytes include plants commonly known as liverworts. They are small plants with flattened bodies that remind some people of the shape of an animal's liver. Liverworts are nonvascular plants. They grow only in moist environments. Because they are nonvascular, they move water and nutrients throughout the plant by osmosis and diffusion. Some studies suggest that liverworts may be the ancestors of all plants. ✓

Reading Check

2. What do studies suggest about liverworts?

____

____

____

What are anthocerophytes?

Anthocerophytes (an THOH ser oh fites) also are small, non-vascular plants that grow in damp, shady habitats. They rely on osmosis and diffusion to transport nutrients. The sporophytes of these plants resemble the horns of an animal, giving the plants the common name of hornworts.

What are bryophytes?

Bryophytes (BRI uh fites) are commonly called mosses. They are nonvascular plants. However, some mosses have cells that transport water and sugars. Moss plants are usually less than 5 cm tall. They have leaflike structures that are only one or two cells thick.

What are psilophytes?

Psilophytes, known as whisk ferns, are vascular plants. They are unique among vascular plants because they do not have roots or leaves. They consist of thin, green stems. Psilophytes are found in tropical and subtropical environments. Only one genus of psilophytes can be found in the southern United States.

What are lycophytes?

Once lycophytes (LI koh fites) were 30 m tall and formed a large part of the forests that existed during the Paleozoic Era. Lycophytes, the club mosses, are usually less than 25 cm high. Lycophytes are vascular plants that have stems, roots, and leaves.

What are arthrophytes?

Arthrophytes (AR throh fites), often called horsetails, are vascular plants. They have hollow, jointed stems. Circling the stem are leaves that resemble scales. All modern horsetails are small, but their fossil relatives were the size of trees.

What are pterophytes?

Pterophytes (TER oh fites) are the most well-known and diverse group of non-seed vascular plants. You know them as ferns. Ferns were abundant in forests of the Paleozoic and Mesozoic Eras. They have leaves called **fronds** that vary in length from 1 cm to 500 cm. The pterophytes are the only group in the non-seed vascular plants that have complex leaves.

Reading Check

3. What do all seed plants have in common?

____

____

____

Seed Plants

Seed plants produce seeds. A seed consists of a plant embryo and a food supply covered by a hard protective coat. In dry conditions, seeds are better adapted for reproduction than spores. All seed plants have vascular tissues. ✓

What are cycads?

Cycads (SI kuds) were abundant during the Mesozoic Era. Today there are about 100 species of cycads. They look like palm trees, have scaly trunks, and come in various heights. Cycads produce male and female cones on separate trees. **Cones** are scaly structures that support male or female reproductive structures. Seeds are produced in female cones. Male cones produce clouds of pollen.

What are gnetophytes?

Gnetophytes (NEE toh fites) are divided into three groups. One group contains tropical trees and climbing vines. The second group includes shrubs that grow in dry, desert regions. The third is a single species that lives in the deserts of southwest Africa.

Section 21.2 **Survey of the Plant Kingdom,** *continued*

What are ginkgophytes?

There is only one living species in this division, the *Ginkgo biloba.* Ginkgoes (GING kohs) are small trees with fan-shaped leaves. Like cycads, ginkgoes have male and female reproductive structures on separate trees. Because the seeds produced on the female tree have an unpleasant smell, male ginkgoes are usually planted. ✓

Reading Check

4. How many different species are in the ginkgophyta division?

What are conifers?

Conifers (KAH nuh furz) are cone-bearing trees such as pine, fir, cypress, and redwood. Conifers produce seeds in cones. Many conifers have needlelike leaves. Bristlecone pines, the oldest living trees in the world, are conifers. The Pacific yew, another type of conifer, is a source of cancer-fighting drugs.

What are anthophytes?

Anthophytes (AN thoh fites) are the flowering plants. There are approximately 250 000 species of anthophytes. Anthophytes produce flowers from which fruits develop. A fruit usually contains one or more seeds. Anthophyta is broken into two classes. You will learn about those in a later chapter.

After You Read

Mini Glossary

cones: scaly structures produced by some seed plants that support male or female reproductive structures and are the sites of seed production

frond: fern leaves; may vary in length from 1 cm to 500 cm

1. Review the terms and their definitions in the Mini Glossary above. Then on the blank lines below, use each term in a sentence.

Section 21.2 Survey of the Plant Kingdom, *continued*

2. Read each sentence below the diagram. If the statement describes non-seed plants, write its letter in the Non-seed box. If the statement describes seed plants, write its letter in the Seed box.

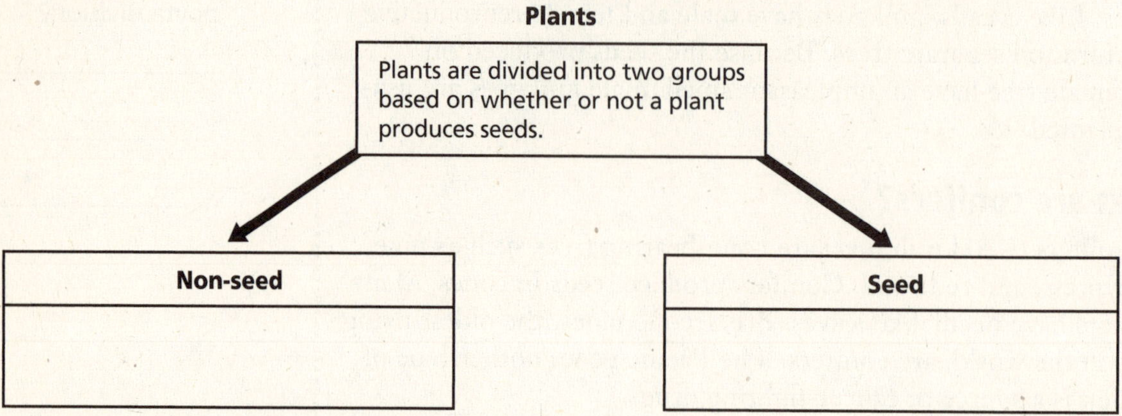

Plants
Plants are divided into two groups based on whether or not a plant produces seeds.
Non-seed
Seed

a. There are seven divisions of these plants.
b. These plants release seeds.
c. These plants release reproductive cells called spores.
d. There are five divisions of these plants.
e. These plants may be either vascular or nonvascular.
f. These plants contain a plant organ made up of an embryonic plant and a food supply covered by a hard protective coat.
g. All of these plants are vascular.

Visit the Glencoe Science Web site at **science.glencoe.com** to find your biology book and learn more about surveying the plant kingdom.

Name Date Class

Section 22.1 Nonvascular Plants

Before You Read

You will learn more about mosses in this section. Mosses are among the first plants to grow in an area after a fire or where the ground has been disturbed in some way. For this reason, moss is often called a pioneer species. How might a habitat benefit by pioneer species such as moss? Write your thoughts on the lines below.

Read to Learn

What is a nonvascular plant?

If you live in a moist environment, you are likely to see nonvascular plants. That is because nonvascular plants need to be near water for life functions such as reproduction and photosynthesis. A steady supply of water is not available everywhere, therefore nonvascular plants are limited to moist habitats near streams and rivers, or in temperate and tropical rain forests. Nonvascular plants do not have roots.

Mark the Text **Highlight Main Ideas** Underline the name of each nonvascular plant division in this section. Say the name aloud. Then highlight the words or phrases that describe that plant division.

How is alternation of generations unique in nonvascular plants?

Remember that the life cycle of all plants includes an alternation of generations between a diploid sporophyte and a haploid gametophyte. In nonvascular plants, the gametophyte generation is dominant. Sporophytes depend on gametophytes for water and other substances. Sporophytes grow attached to the gametophytes. The illustration at right shows sporophytes growing attached to the gametophytes.

Gametophytes of nonvascular plants produce two kinds of sexual reproductive structures. The male reproductive structure is called the **antheridium** (an thuh RIH dee um). The sperm are produced in this structure. The **archegonium** (ar kih GOH nee um) is the female reproductive structure. Eggs are produced in this structure. Fertilization happens in the archegonium and begins the sporophyte generation.

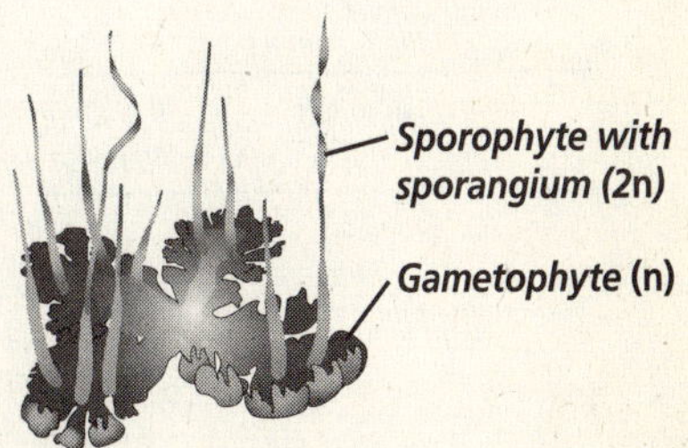

Adaptations in Bryophyta

There are several divisions of nonvascular plants. You will look at the mosses, or bryophytes, first. Of all the nonvascular plants, people are most familiar with bryophytes.

Mosses are small plants with leafy stems. The leaves are usually one cell thick. Instead of roots, mosses have rhizoids, colorless multicellular structures, to help anchor the plant to the soil. Mosses usually grow in a thick ground covering of hundreds of plants. Large areas of growing moss help prevent erosion.

Mosses grow in a variety of habitats. They even grow in the arctic during the brief growing season when enough moisture is present.

Sphagnum, a well-known moss also known as peat moss, is used as fuel. Dried peat moss absorbs large amounts of water, so florists and gardeners use it to increase the water-holding ability of soil.

Adaptations in Hepaticophyta

Another division of nonvascular plants is the liverworts, or hepaticophytes. Like mosses, liverworts grow as clumps or masses in moist environments. The division hepaticophyta derives its name from *hepar*, which refers to the liver. The shape of the liverwort gametophyte is thought to look like an animal's liver. Liverworts are found all over the world.

There are two classifications of liverworts. One is called thallose; the other is called leafy. The body of a thallose liverwort is called a thallus. It is broad and resembles a fleshy, lobed leaf. Thallose liverworts are usually found growing on damp soil. ✓

Leafy liverworts are common in tropical jungles and places that are often blanketed by fog. They look less like a liver than the thallose variety. The stems have three rows of flat, thin leaves. Liverworts have rhizoids made up of one long cell to help anchor the plant to the soil.

Reading Check

1. What are the two classifications of liverworts?

____

____

Adaptations in Anthocerophyta

Anthocerophytes are known as hornworts. They are similar to liverworts in several ways. Like some liverworts, hornworts have a thallose body. The sporophyte of a hornwort resembles the horn of an animal, which is why they are called "hornworts." ✓

Hornworts have one to several chloroplasts in each cell of the sporophyte, depending on the species. Unlike other nonvascular plants, the sporophyte, not the gametophyte, produces most of the food used by both generations.

Reading Check

2. What kind of body do hornworts have?

____

____

____

Origins of Nonvascular Plants

Fossil and genetic evidence suggests that liverworts were the first land plants. Fossils from the Paleozoic Era, more than 440 million years ago, have been identified as nonvascular plants. However, scientists hypothesize that nonvascular plants were growing even earlier than that. Both nonvascular and vascular plants may share a common ancestor. That ancestor probably had alternating sporophyte and gametophyte generations, cellulose in the cell walls, and chlorophyll for photosynthesis. ✓

Reading Check

3. What do scientists hypothesize were the first land plants?

____

____

____

After You Read

Mini Glossary

antheridium (an thuh RIH dee um): male reproductive structure in which sperm develop

archegonium (ar kih GOH nee um): female reproductive structure in which eggs develop

1. Read the terms and definitions in the Mini Glossary above. On the lines below, tell how the two key terms are related.

____

____

____

2. Use the table to give an example of each division of nonvascular plants. Then list the shared characteristics of these plants.

Nonvascular Plants		
Division	**Example**	**Shared Characteristics**
Bryophyta		
Hepaticophyta		
Anthocerophyta		

Visit the Glencoe Science Web site at **science.glencoe.com** to find your biology book and learn more about nonvascular plants.

Name Date Class

Section
22.2 Non-Seed Vascular Plants

Before You Read

Have you ever gone camping or visited another part of the world where plumbing was not available? What did you miss most about not having plumbing? Write your thoughts on the lines below. The main purposes of plumbing are to bring water and carry away wastes. Some plants have plumbing. This plant plumbing—called vascular tissues—distributes water and other dissolved substances throughout a plant.

Read to Learn

Create a Quiz After you have read this section, create a quiz based on what you have learned. After you have completed writing the quiz questions, be sure to answer them.

What is a non-seed vascular plant?

In this section you will learn about three divisions of non-seed vascular plants: Lycophyta, Arthrophyta, and Pterophyta. The main difference between vascular and nonvascular plants is that vascular plants have vascular tissue. Vascular tissue is made of tubelike, long cells through which water and nutrients travel. This makes it possible for vascular plants to live in a variety of habitats.

How do vascular plants exhibit alternation of generations?

Like all plants, vascular plants have an alternation of generations. Unlike nonvascular plants, however, the dominant generation in vascular plants is the sporophyte. The sporophyte is larger in size than the gametophyte. The sporophyte does not depend on the gametophyte for water or nutrients.

An important advance in non-seed vascular plants was the adaptation of leaves to protect the developing reproductive cells. In some of the plants, leaves that bear spores form a compact cluster called a **strobilus** (stroh BIH lus). The spores are released from the strobilus and can grow to form gametophytes.

The illustration on page 247 shows a fern gametophyte called a **prothallus** (proh THA lus). Gametophytes are relatively small. They live in or on the soil.

Section 22.2 **Non-Seed Vascular Plants,** *continued*

Reproductive structures—the archegonium and the antheridium—develop on the gametophyte. Sperm are released from the antheridia. Sperm need a continuous film of water to reach and fertilize the eggs in the archegonia. The zygote can grow into a large, dominant sporophyte.

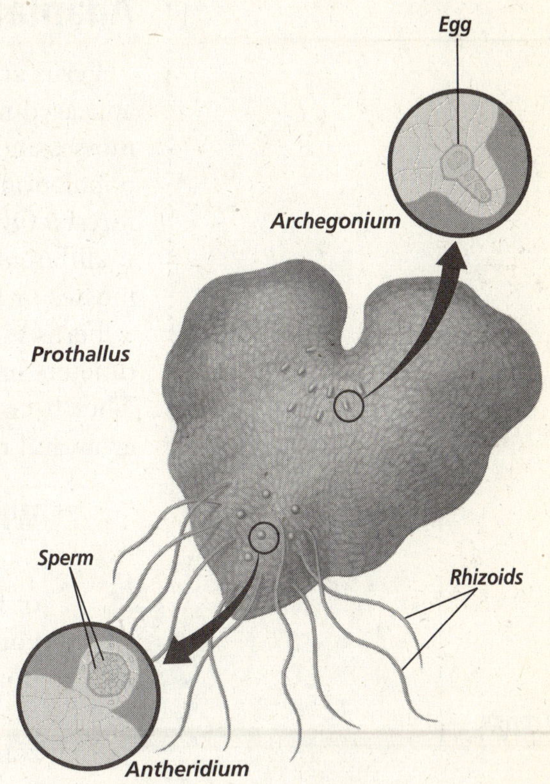

Adaptations in Lycophyta

Fossil evidence shows that ancient lycophytes were the size of trees and part of the early forest community. Modern lycophytes are much smaller and are called club mosses or spike mosses. The leafy stems look like moss gametophytes, and their reproductive structures are club or spike shaped. However, unlike mosses, the sporophyte generation of lycophytes is dominant. The sporophyte has roots, stems, and small leaflike structures. A single vein of vascular tissue runs through each leaflike structure. The stems of lycophytes may be sticking up or along the ground and have roots growing from the base of the stem.

Adaptations in Arthrophyta

Like the lycophytes, arthrophytes (horsetails) were once tree-sized members of the early forest community. Today they grow to about 1 m tall. The name horsetail refers to the bushy appearance of some species. There are only about 15 species of arthrophytes in existence. Sometimes they are called scouring rushes because they contain silica, an abrasive material. In the past, people used horsetails to scrub pots and pans.

Most horsetails grow in damp soil in shallow ponds, in marshes, and on stream banks. The ribbed and hollow stem is not like other vascular plants. It appears jointed, and at each joint there is a circle of tiny scalelike leaves. ✓

Arthrophyte spores are produced in strobili. After the spores are released, they grow into gametophytes with antheridia and archegonia.

Think it Over

1. **Infer** What might a drought do to the long-term survival of non-seed vascular plants?

____

____

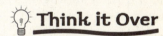

2. Where do most horsetails grow?

____

____

Adaptations in Pterophyta

Ferns are pterophytes. According to the fossil record, ferns appeared nearly 375 million years ago during the time when club mosses and horsetails were the main members of Earth's plant population. Ancient ferns grew tall and treelike and formed forests. Over time, ferns evolved into many species. They adapted to different environments. Today there are more ferns than club mosses or horsetails.

Ferns range in size from a few meters tall to only a few centimeters in diameter. Some ferns are able to live in dry areas. They become dormant, or inactive, when water is limited, then grow and reproduce when water is available again.

Fronds

Root

Rhizome

What is the structure of ferns?

As with most vascular plants, the sporophyte generation of ferns has roots, stems, and leaves. When you see a plant you know as a fern, you are seeing the sporophyte generation. In most ferns, the main stem is underground. The thick, underground stem is called a **rhizome.** The rhizome contains many starch-filled cells for food storage. Leaves of the fern are called fronds. Fronds grow upward from the rhizome. Ferns are the first of the vascular plants to have evolved leaves with branching veins of vascular tissue. The branched veins in ferns transport water and food to and from all the cells in the plant.

If you turn the frond of a fern over, you may find tiny circles of brown or rust-colored dust. These are clusters of spore-producing sporangia. Each cluster forms a **sorus** (plural, sori).

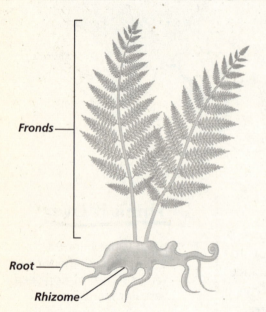

Think it Over

3. Analyze Why are non-seed vascular plants smaller and less widespread that their prehistoric ancestors?

____

____

Origins of Non-Seed Vascular Plants

The earliest evidence of non-seed vascular plants comes from the Devonian period around 375 million years ago. During the Carboniferous Period, large, tree-sized lycophytes, arthrophytes, and pterophytes were abundant in warm, moist forests. Then about 280 million years ago, Earth's climate became cooler and drier. Many species of non-seed vascular plants died out. Today non-seed vascular plants are much smaller and less widespread than their prehistoric ancestors. The evolution of vascular tissue helped these plants to live on land and to have larger body sizes in comparison to the nonvascular plants.

After You Read

Mini Glossary

prothallus (proh THA lus): fern gametophyte

rhizome: thick, underground stem of a fern and other vascular plants; often functions as an organ for food storage

sorus: clusters of sporangia usually found on the underside of fern fronds

strobilus (stroh BIH lus): compact cluster of spore-bearing leaves produced by some non-seed vascular plants

1. Read the terms and definitions in the Mini Glossary above. Write a sentence using at least two of the terms.

2. Fill in the missing information in the table.

Non-seed Vascular Plants	
Name of dominant generation	
Name of reproductive cell that grows into gametophyte	
Name of thick underground root	
First plant to have leaves with vascular tissue	

3. Explain why most non-seed vascular plants live in moist habitats.

Visit the Glencoe Science Web site at **science.glencoe.com** to find your biology book and learn more about non-seed vascular plants.

Name Date Class

Section 22.3 Seed Plants

Before You Read

Seed plants are an important part of our economy. Farmers produce corn, wheat, and oats. Apples, oranges, olives, and figs grow in orchards. In some states there are fields of strawberries, or acres planted with grapes. Look on your kitchen shelves. See how many products originate from seeds. List those products here.

____

____

Read to Learn

Mark the Text **Identify Main Ideas** Highlight every sentence in this section that has the word *seed* in it. Say the sentence aloud after you highlight it.

What is a seed plant?

A seed plant is a vascular plant that produces seeds. The seed is actually a small sporophyte plant inside a protective coat. The seed may be surrounded by fruit or carried on the scales of a cone.

In seed plants, the sporophyte generation is dominant. However, alternation of generations still occurs. The sporophyte produces spores. The spores develop into male and female gametophytes. In seed plants, the male gametophyte is inside a structure called the **pollen grain.** The pollen grain contains sperm cells, nutrients, and a protective outer coating. The female gametophyte, which produces the egg cell, is inside a sporophyte structure called an **ovule.** The ovule forms the seed after fertilization.

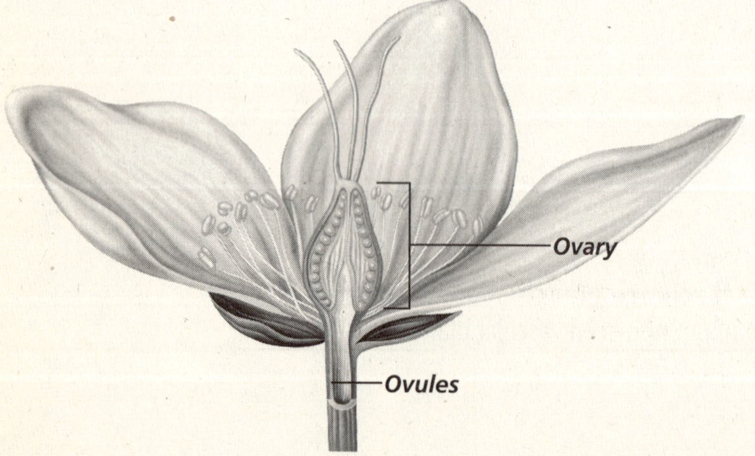

Where does the sporophyte come from?

The union of the sperm and egg is called fertilization. Fertilization forms the sporophyte zygote. In most seed plants this does not require a continuous film of water. Remember, in nonvascular and non-seed vascular plants the sperm have to swim to the egg on a continuous film of water.

Section 22.3 Seed Plants, *continued*

Since seed plants do not need this film of water for fertilization, they can grow and reproduce in a wide variety of habitats, even those with limited water.

After fertilization, the zygote develops into an embryo. An **embryo** is an early stage of development in an organism. An embryo is the young diploid sporophyte of a plant. Embryos of seed plants include one or more cotyledons. **Cotyledons** (kah tuh LEE dunz) usually store or absorb food for the developing embryo. In conifers, or cone-bearing trees, and many flowering plants, cotyledons are the leaflike structures on the plant's stem when the plant comes through the soil.

What are the advantages of seeds?

A seed is an embryo and its food supply inside a tough, protective coat. Seed plants have important advantages over non-seed plants. The seed contains food for the young plant until the leaves develop enough to carry out photosynthesis. In conifers and some flowering plants, the food supply is stored in the cotyledons. The embryo is protected during harsh conditions by the tough seed coat. Some seeds have adaptations that help them move away from the parent plants. The feathery tufts on dandelion seeds or milkweed seeds are two examples of such seed adaptations. When the seeds are spread out, away from the parent plants, the young plants do not have to compete with the parent plants for sunlight, water, living space, and soil nutrients.

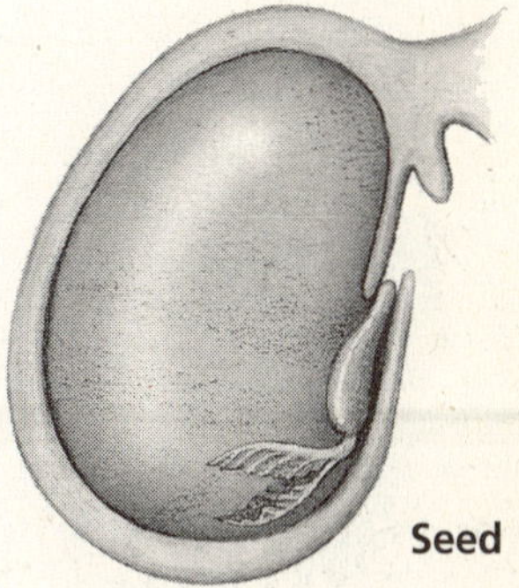

Seed

Is there diversity in seed plants?

Seed plants are divided into two groups: those whose seeds are protected by fruit, and those whose seeds are not protected by fruit. Botanists, or biologists who study plants, call plants whose seeds are not protected by fruit, gymnosperms. The term gymnosperm means "naked seed." An example of seeds not protected by fruits are the seeds released by pinecones. The gymnosperm plant divisions are Cycadophyta, Ginkgophyta, Gnetophyta, and Coniferophyta.

Plants that produce seeds protected by fruit are called angiosperms. A **fruit** includes the ripened ovary of a flower. The fruit provides protection for seeds and helps in seed dispersal. The Anthophyta division contains all fruit producing plants. ✓

Reading Check

1. What do you call plants that produce seeds protected by fruit?

______

______

______

Think it Over

2. **Identify** What two divisions of plants have male and female reproductive structures on separate plants?

Adaptations in Cycadophyta

There are about 100 species of cycads today. They live in the tropics and subtropics. Cycads have male and female reproductive systems on separate plants. On the male plants, cones produce pollen grains that produce motile (capable of movement) sperm. Cycads are one of the few seed plants that produce motile sperm. The female plant has cones that produce ovules. The trunks and leaves of many cycads look like palm trees. However, they are not closely related because palms are anthophytes.

Adaptations in Ginkgophyta

Today this division is represented by only one living species, *Ginkgo biloba.* Like cycads, ginkgo male and female reproductive systems are on separate plants. The male ginkgo produces pollen grains from cones that grow from the bases of leaf clusters. Like cycads, ginkgo pollen grains produce motile sperm. The female ginkgo produces ovules which, when fertilized, grow fleshy, apricot-colored seed coats. These soft seed coats give off a bad smell if they are broken or crushed. Male ginkgoes are often planted in cities because they tolerate smog and pollution.

Adaptations in Gnetophyta

Most gnetophytes are found in the deserts or mountains of Asia, Africa, North America, and Central or South America. There are three genera within Gnetophyta. One is composed of tropical climbing plants. The second contains shrublike plants. This is the only gnetophyte genus found in the United States. The third genus is a strange-looking plant found only in South Africa. The plant may live 1000 years. It grows close to the ground and has a large root. It only has two leaves that continue to lengthen and curl as the plant grows older.

Adaptations in Coniferophyta

Conifers are trees and shrubs with needlelike or scalelike leaves. They are a part of forests throughout the world and include pine, fir, spruce, juniper, cedar, redwood, yew, and larch.

Section 22.3 **Seed Plants,** *continued*

The reproductive structures of most conifers are produced in cones. Most conifers have male and female cones on different branches of the same tree. Male cones produce pollen. They are small and easy to overlook. Female cones are much larger. They stay on the tree until the seeds have matured.

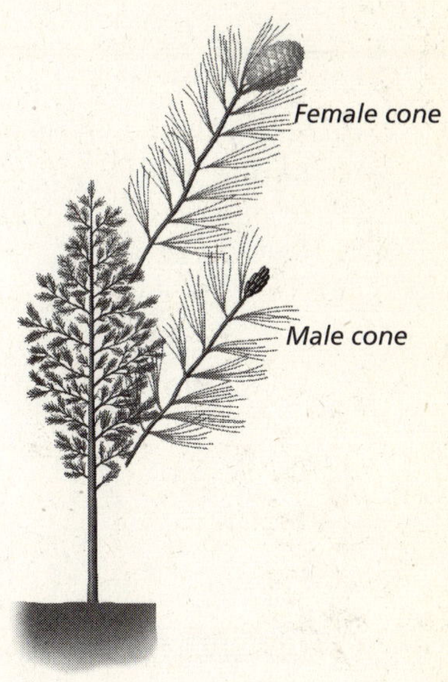

Why do conifers keep their leaves year-round?

Most conifers keep their leaves for more than one year. As leaves age or are damaged they drop to the ground. The conifer, however, never loses all of its leaves at one time. For this reason conifers are sometimes called evergreens.

Plants that keep their leaves year-round can photosynthesize whenever favorable conditions exist. This is an advantage where the growing season is short. Also, plants that keep their leaves year-round do not use food reserves each spring to produce a new set of leaves.

Evergreen leaves usually have a heavy coating of cutin, a waxy, waterproof material that helps reduce water loss. The shape of conifer leaves, needlelike or scalelike, also helps reduce water loss.

Are all conifers evergreen?

A few conifers, including larches and bald cypress trees, are deciduous. **Deciduous plants** drop all their leaves each fall or when water is scarce. Plants lose most of their water through the leaves. Very little water is lost through the bark or roots. It is an advantage to drop leaves in times of scarce water. However, without leaves, the plant cannot photosynthesize, and must stay dormant. ✓

Reading Check

3. How does dropping leaves help keep a plant from losing water?

____

____

____

Adaptations in Anthophyta

All flowering plants are in the division Anthophyta. They are the most well-known plants on Earth, with more than 250 000 identified species. Like other seed plants, anthophytes have roots, stems, and leaves. Unlike other seed plants, they produce flowers and enclose seeds in a fruit.

A fruit develops from the female reproductive structure in a flower. Sometimes other parts of the flower become part of the fruit. In pineapples, more than one flower is needed to develop the fruit. An advantage of fruit-enclosed seeds is the added protection the fruit provides for the embryo.

Fruit aids in the spread of seeds. Animals eat the fruits or carry them away to store for food. Seeds of some species that pass through the animal's digestive track unharmed are spread as the animal wanders. In fact, some seeds have to pass through a digestive track before they can begin to grow a new plant. Brightly colored fruit or fragrant fruit attracts birds and small animals. The large coconut floats, which allows it to spread far.

What are monocots and dicots?

Reading Check

4. How can you identify if a seed plant is a monocot or a dicot?

____

____

Earlier you learned that seeds contain cotyledons, which store or absorb food for the developing embryo. Now you will learn that Anthophyta is divided into two classes. They are named for the number of cotyledons contained in the seeds. **Monocotyledons** (mah nuh kah tuh LEE dunz) have one seed leaf. **Dicotyledons** (di kah tuh LEE dunz) have two seed leaves. They are often called monocots and dicots.

	Seed Leaves	Vascular Bundles in Leaves
Monocots	One cotyledon	Usually parallel
Dicots	Two cotyledons	Usually netlike

There is a simple way to tell the difference between monocot and dicot plants: by examining their leaves. Most monocot leaves have veins that run parallel to each other, like the veins in a blade of grass or a leaf from a corn plant or a palm tree. Dicots, on the other hand, have veins that branch to form a network or look like a net. Almost all trees, cacti, herbs, vegetables, wildflowers, and garden flowers are dicots.

What is the life span of a plant?

Why do some plants live longer than people, and others live only a few weeks? The life span of a plant is genetically determined.

Annual plants live for only one year or less. They sprout from seeds, grow, reproduce, and then die. Most annuals have green stems and do not have any woody tissue. Examples of annuals are corn, peas, beans, as well as many of the weeds found in gardens. The seeds of annuals can survive drought and winter conditions.

Biennial plants have two-year life spans. During the first year, biennials grow many leaves and develop a strong root system. Over winter, the part of the plant that is above ground dies back, but the roots remain alive. During the second spring, food stored in the roots is used to produce new shoots, flowers, and seeds. Examples of biennial plants are carrots, beets, and turnips.

Perennials live for several years. They produce flowers and seeds periodically, usually once a year. Some survive harsh conditions by dropping their leaves or dying back to soil level. Their woody stems and roots remain dormant. Examples of perennials are columbine, strawberries, asparagus, and brambles. ✓

Reading Check

5. What is the name for plants that have a life span of more than two years?

____

____

Origins of Seed Plants

Seed plants first appeared about 360 million years ago during the Paleozoic Era. Ancient relatives of ginkgoes and cycads shared the forest with dinosaurs during the Mesozoic Era. About 65 million years ago, most members of Ginkgophyta died out along with other organisms during a mass extinction.

Fossil evidence suggests that conifers emerged around 250 million years ago. During the Jurassic Period, conifers became the most common forest plant, as they still are today. Anthophytes first appeared about 140 million years ago, late in the Jurassic Period of the Mesozoic Era.

6. Sequence Which of the following best describes the order in which plants likely appeared on earth? (Circle your choice.)

a. anthophytes then conifers

b. conifers then anthophytes

After You Read

Mini Glossary

annual: anthophyte that lives for one year or less

biennial: anthophyte that has a life span of two years

cotyledons (kah tuh LEE dunz): structure of seed plant embryo that stores or absorbs food for the developing plant

deciduous plant: a plant that drops all of its leaves each fall or when water is scarce or unavailable; an adaptation for reducing water loss when water is unavailable

dicotyledon (di kah tuh LEE dun): class of anthophytes that have two seed leaves

embryo: the young diploid sporophyte of a plant

fruit: seed-containing ripened ovary of an anthophyte

monocotyledon (mah nuh kah tuh LEE dun): class of anthophytes that have one seed leaf

Section 22.3 Seed Plants, *continued*

ovule: in seed plants, the sporophyte structure surrounding the developing female gametophyte; forms the seed after fertilization

perennial: anthophyte that lives for several years

pollen grain: in seed plants, structure in which the male gametophyte develops; consists of sperm cells, nutrients, and a protective outer coating

1. Read the terms and definitions in the Mini Glossary above. Circle the three terms that categorize plants by their life spans. Write the terms on the lines below and provide two examples of plants for each category.

____

____

____

____

2. Complete the diagram by writing in seven adaptations of seed plants. Note: This section discussed more than seven adaptations.

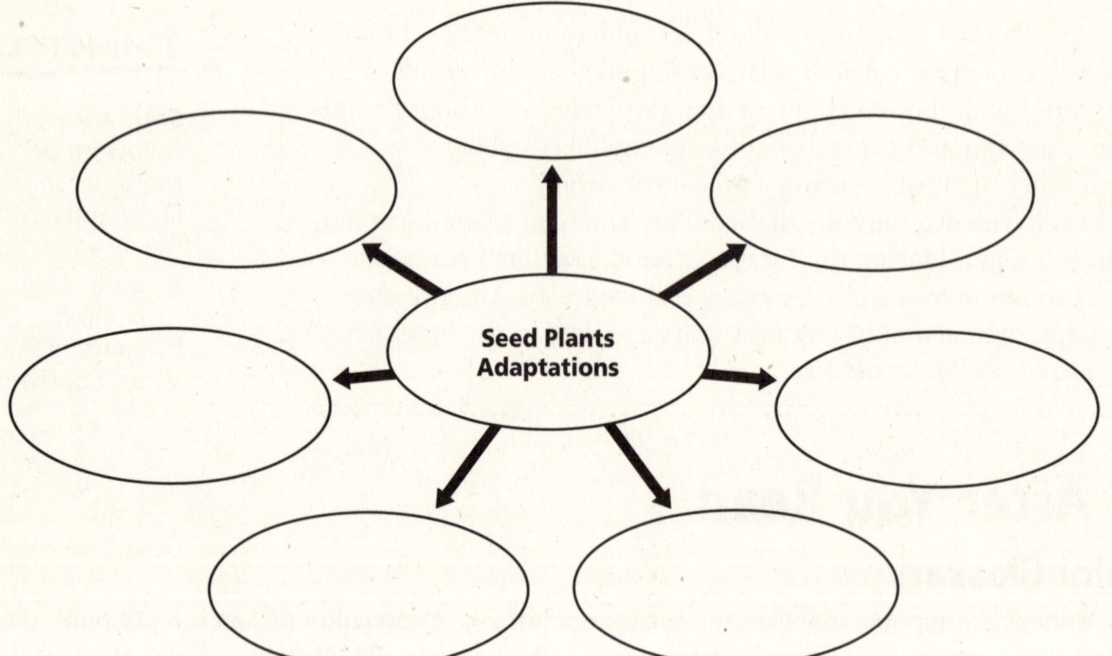

Visit the Glencoe Science Web site at **science.glencoe.com** to find your biology book and learn more about seed plants.

Name Date Class

Section 23.1 Plant Cells and Tissues

Before You Read

In health class you may have learned about the circulatory system. Most arteries in your body carry oxygenated blood throughout your body. Veins return blood to the heart, then the blood moves to the lungs where it receives more oxygen. Similarly, vascular tissues in plants have specific functions. You can see vascular tissues at work if you take fresh celery and slice off the ends of several stalks. Place them in a glass of water. Add a few drops of food coloring to the water. Check back in a couple of hours. Can you predict what might happen? Write your prediction on the lines below.

____

____

Read to Learn

Types of Plant Cells

In this section you will learn about different types of plant cells and tissues. Plant cells are different from animal cells. Plant cells have a cell wall, a central vacuole, and can contain chloroplasts. A vacuole is a membrane-bound, fluid-filled space in the cytoplasm. Chloroplasts are organelles that capture light energy and convert it to chemical energy.

Mark the Text **Locate Information** Draw a line from the words or phrases in the text to the corresponding part of the illustration.

What is the function of parenchyma cells?

The plant cells that are the most numerous are the **parenchyma** (puh RENG kuh muh) cells. They are found throughout the tissues of a plant. These sphere-shaped cells have thin, flexible walls. Most parenchyma cells have a large central vacuole, which sometimes is filled with a fluid called sap.

Parenchyma cells have two main functions: storage and food production. The large vacuole can be filled with water, starch grains, or oils. The part we eat of many fruits and vegetables is made mostly of parenchyma cells. The many chloroplasts sometimes found in parenchyma cells produce glucose during photosynthesis.

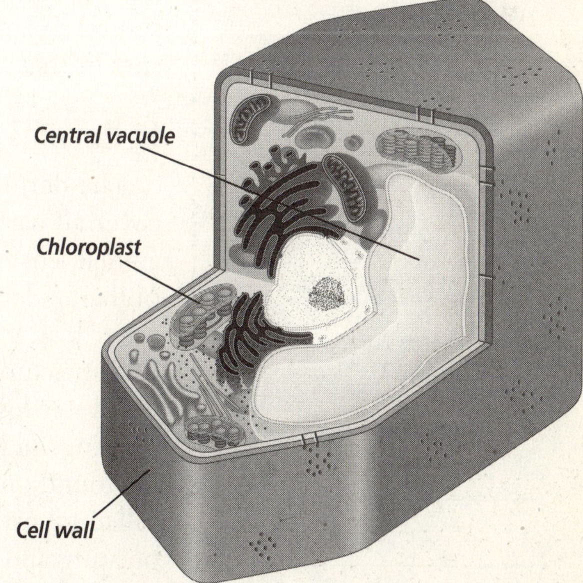

Section 23.1 **Plant Cells and Tissues,** *continued*

What are collenchyma cells?

Collenchyma (coh LENG kuh muh) cells are long cells with unevenly thickened cell walls. The walls of collenchyma cells can stretch as the cells grow. Collenchyma cells are arranged in tube-like strands, or cylinders, that provide strength and support for the surrounding tissue. The long, tough strands found in celery are made of collenchyma cells.

What do sclerenchyma cells do?

The walls of **sclerenchyma** (skle RENG kuh muh) cells are thick and rigid. When sclerenchyma cells mature, they often die. However, they still have a function in the plant. The cytoplasm breaks into smaller and smaller pieces until it disappears, but the strong, thick cell walls remain. They provide support for the plant. There are two types of sclerenchyma cells found in plants, fibers and sclerids (SKLER idz).

Fibers are long, thin cells that form strands. They provide support and strength for the plant. They are the source of fibers used for making rope and linen. Sclerids are irregularly shaped and often found in groups. The gritty texture you find in pears is caused by sclerids. They also are a major part of the pits found in peaches and other fruits. ✓

Reading Check

1. How are fibers useful to people?

______

______

______

Plant Tissues

A tissue is a group of cells that function together to perform an activity. Tissues can be referred to as plant subsystems. There are several tissue types in plants. You will learn about four plant tissues in this section.

Is dermal tissue like animal skin?

The dermal tissue, or **epidermis,** is made of flattened cells that cover all parts of the plant. It functions much like the skin of an animal, covering and protecting the body of the plant. The cells of the epidermis produce the waxy cuticle that helps prevent water loss.

A structure called a stoma in the epidermal layer also helps control water loss. **Stomata** (STOH mah tuh) (singular, stoma) are openings in leaf tissue that control the exchange of gases. Stomata are found on green stems and on the surfaces of leaves. Two cells called **guard cells** surround each stoma. Guard cells control the opening and closing of stomata. The opening and closing of stomata controls the flow of water vapor from leaf tissues. ✓

Reading Check

2. What is the function of the stomata?

______

______

______

The dermal tissue that covers roots may have root hairs. Root hairs are parts of individual epidermal cells that help the root absorb water and dissolved minerals. If you see hairlike projections on stems and leaves, they are called trichomes. **Trichomes** (TRI kohmz) give a stem or leaf a fuzzy look. They help reduce the evaporation of water from the plant. In some plants, trichomes release poisonous substances that help protect the plant from plant-eating animals and insects.

What do the vascular tissues transport?

Food, dissolved minerals, and water are moved throughout the plant by vascular tissue. Xylem and phloem are the two types of vascular tissue. **Xylem** is made of tubelike cells that transport water and dissolved minerals from the roots to the rest of the plant. In seed plants, xylem is made of parenchyma, fibers, tracheids, and vessel elements. ✓

Reading Check

3. What is transported by xylem?

____

____

____

Tracheids (TRA kee uhdz) are tubelike cells tapered at each end. The cell walls between tracheids have small openings through which water and dissolved minerals flow.

Vessel elements also are tubelike cells in the xylem that transport water and dissolved minerals from the roots to the stem. They are wider and shorter than tracheids and have openings in their end walls. In some plants, mature vessel elements lose their end walls completely, making it easier for water and dissolved minerals to flow from cell to cell.

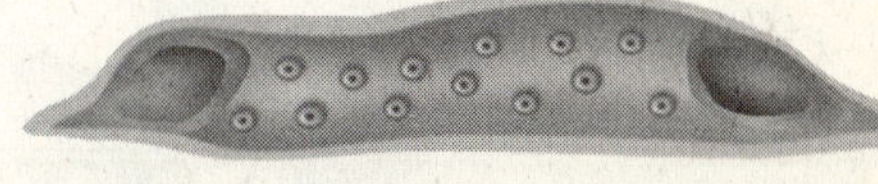

Vessel element

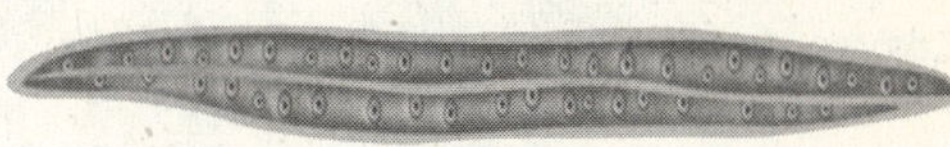

Tracheid

Almost all vascular plants have tracheids. Vessel elements are most common in anthophytes. This may be one reason anthophytes, which are true flowering plants, are the most successful plants on Earth. The vessel elements in anthophytes can transport water more efficiently than tracheids.

The other type of vascular tissue is **phloem.** Phloem transports sugars and other carbon compounds. Like xylem, phloem has long tubelike cells that are arranged end to end. These cells are called **sieve tube members.** Sieve tube members are alive at maturity. They are unusual because they contain cytoplasm but do not have a nucleus or ribosomes. Next to each sieve tube member is a companion cell.

Companion cells are cells with nuclei that help move materials through the sieve tubes. In anthophytes, the end walls between two sieve tube members are called sieve plates. The sieve plates have large pores or openings that allow sugars and organic compounds to move from sieve tube member to sieve tube member. Phloem tissue also contains fibers.

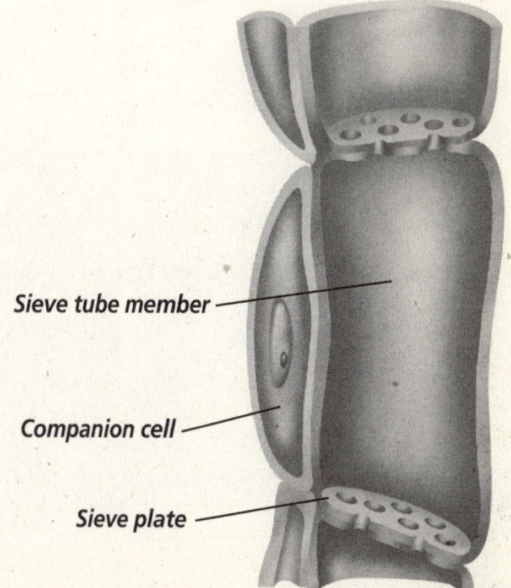

The fibers do not transport materials, but are important because they provide support for the plant.

What is ground tissue?

Reading Check

4. What are the functions of ground tissue?

____

____

____

Despite the name, ground tissue is found throughout a plant. Ground tissue is made mostly of parenchyma cells, but can include collenchyma and sclerenchyma cells. The functions of ground tissue include photosynthesis, storage, and support. The cells of ground tissue in leaves and some stems contain many chloroplasts that carry on photosynthesis. Ground tissue cells in some stems and roots contain large vacuoles that store starch grains and water.

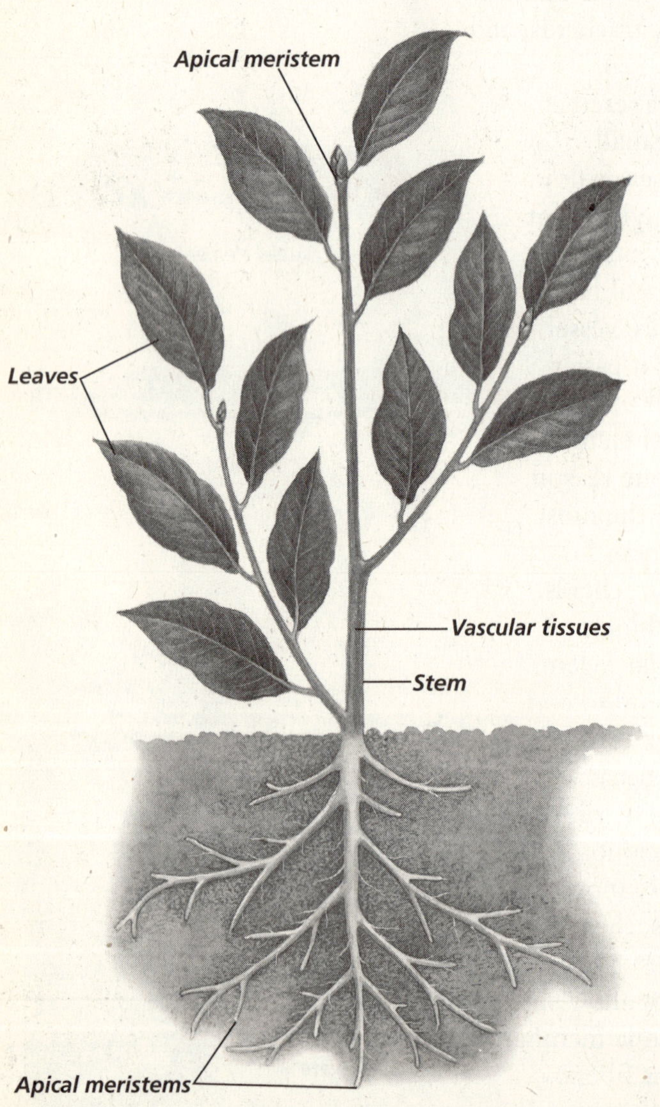

What are the meristematic tissues?

A growing plant produces new cells in areas called meristems. **Meristems** are areas of actively dividing cells. Meristematic cells are parenchyma cells with a different shape. They also have large nuclei.

There are different types of meristems. **Apical meristems** are found at the tips of roots and stems. They produce cells that allow the roots and stems to increase in length. **Lateral meristems** are cylinders of dividing cells located in roots and stems. The production of cells by the lateral meristems increases the diameter, or thickness, of roots and stems. Woody plants have two kinds of lateral meristems—vascular cambium and cork cambium. The **vascular cambium** produces new xylem and phloem cells in the stems and roots. The **cork cambium** produces cells with tough cell walls. These cells cover the surface of stems and roots. The outer bark of a tree is produced by the cork cambium.

A third type of lateral meristem is found in grasses, corn, and other monocots. This meristem adds cells that lengthen the part of the stem between the leaves. These plants do not have a vascular or cork cambium.

After You Read

Mini Glossary

apical meristem: regions of actively dividing cells near the tips of roots and stems; allow roots and stems to increase in length

collenchyma (coh LENG kuh muh): long plant cells with unevenly thickened cell walls

companion cells: cells with nuclei that help transport sugars and other organic compounds through the sieve tube members of the phloem

cork cambium: lateral meristem that produces a tough protective covering for the surface of stems and roots

epidermis: in plants, the outermost layer of flattened cells that covers and protects all parts of the plant; produces the waxy cuticle

guard cells: cells that control the opening and closing of the stomata; regulate the flow of water vapor from leaf tissue

lateral meristem: cylinders of dividing cells located in roots and stems; allows roots and stems to increase in diameter

meristems: regions of actively dividing cells in plants

parenchyma (puh RENG kuh muh): most abundant type of plant cell; spherical cells with thin, flexible walls and a large central vacuole, important for storage and food production

phloem: vascular plant tissue made up of tubular cells joined end to end; transports sugars and other organic compounds to all parts of the plant

sclerenchyma (skle RENG kuh muh): plant cells with rigid, thick cell walls; die when they are mature, but still provide support for the plant

sieve tube members: tubular cells in phloem; each cell lacks a nucleus and ribosomes

stomata (STOH mah tuh): openings in leaf tissues that control gas exchange and help control water loss

tracheids (TRA kee udz): tubular cells in the xylem that have tapered ends; have small openings for transport of water and dissolved minerals

trichomes (TRI kohmz): hairlike projections that extend from a plant's epidermis; help reduce water evaporation and may provide protection from predators

vascular cambium: lateral meristem that produces new xylem and phloem cells in the stems and roots

vessel elements: hollow, tubular cells in the xylem; transport water and dissolved minerals from the roots to the stem; have open ends through which water passes freely from cell to cell

xylem: vascular plant tissue composed of tubular cells that transport water and dissolved minerals from the roots to the rest of the plant

1. Read the key terms and definitions in the Mini Glossary above. On the lines below, tell how the key terms **stomata** and **guard cells** are related.

Section 23.1 Plant Cells and Tissues, *continued*

2. Use the Venn diagram to compare and contrast the cells that make up xylem and phloem. Arrange the characteristics you list according to whether they are true for xylem only, true for phloem only, or true for both.

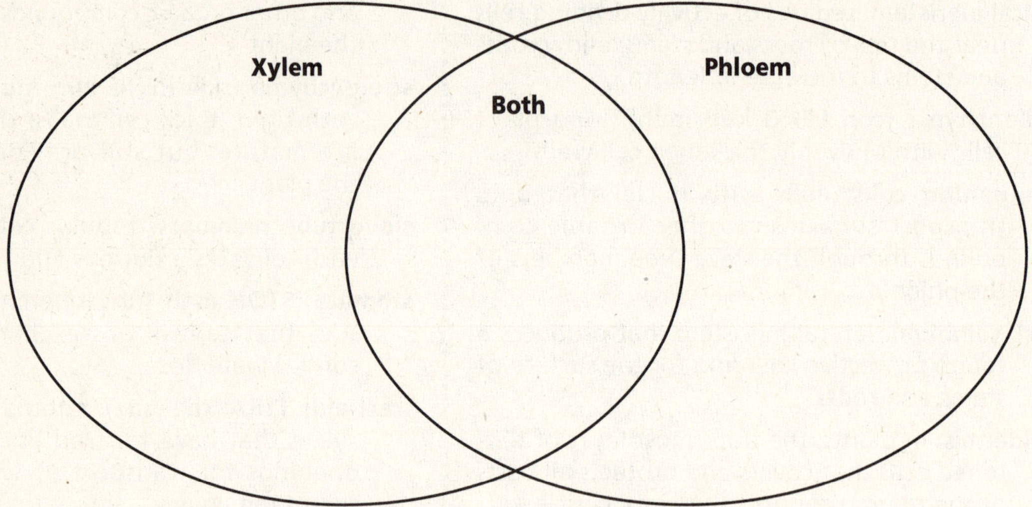

Visit the Glencoe Science Web site at **science.glencoe.com** to find your biology book and learn more about plant cells and tissues.

Section 23.2 Roots, Stems, and Leaves

Before You Read

Salad ingredients include lettuce, tomatoes, radishes, cucumbers, onions, bean sprouts, carrots, and celery. Botanists might look at a fresh salad and see plant organs. Think of salad ingredients you are familiar with. Decide whether they are roots, stems, or leaves and list them on the appropriate line below.

Roots: ______________________________

Stems: ______________________________

Leaves: ______________________________

Read to Learn

Roots

A root is a plant organ. Roots anchor a plant, usually absorb water and dissolved minerals, and contain vascular tissues that move minerals to and from the stem. All roots are not alike. They can be short or long, thick or thin, massive or threadlike. The surface area of a plant's roots can be as much as 50 times greater than the surface area of its leaves. Most roots grow in the soil, but some do not.

There are two main types of root systems—taproots and fibrous roots. Carrots and beets are taproots. Taproots are single, thick structures with smaller branching roots. Taproots accumulate and store food. Fibrous roots have many small branching roots that grow from a central point.

Mark the Text **Identify Details** As you skim the section, highlight the structures of roots, stems, and leaves in one color. Highlight the functions of roots, stems, and leaves in another color.

Taproots

Fibrous roots

Prop roots

Other types of roots include prop roots, aerial roots, and pneumatophores. Corn has prop roots. They begin above ground and help support the plant. Some climbing plants have aerial roots. The aerial roots do not touch the ground. Instead, they cling to walls to provide support for the climbing stems. Bald cypress trees grow mostly in swampy soils. They produce modified roots called pneumatophores, or "knees." The knees grow upward from the mud, and out of the water. They help supply oxygen to the roots.

What is the structure of roots?

Locate the root hair on the diagram below, which shows a cross-section of a root. You can see that a root hair is an extension of an epidermal cell. The epidermal cells are in the epidermis, the outermost layer of the root. Root hairs increase the surface area of a root that touches the soil. Root hairs absorb water, oxygen, and dissolved minerals.

Moving inward from the epidermis, the next layer of cells in the root is the **cortex.** The cortex is ground tissue. It helps move water and dissolved minerals into the vascular tissue. The cortex is made up of parenchyma cells that sometimes store food and water.

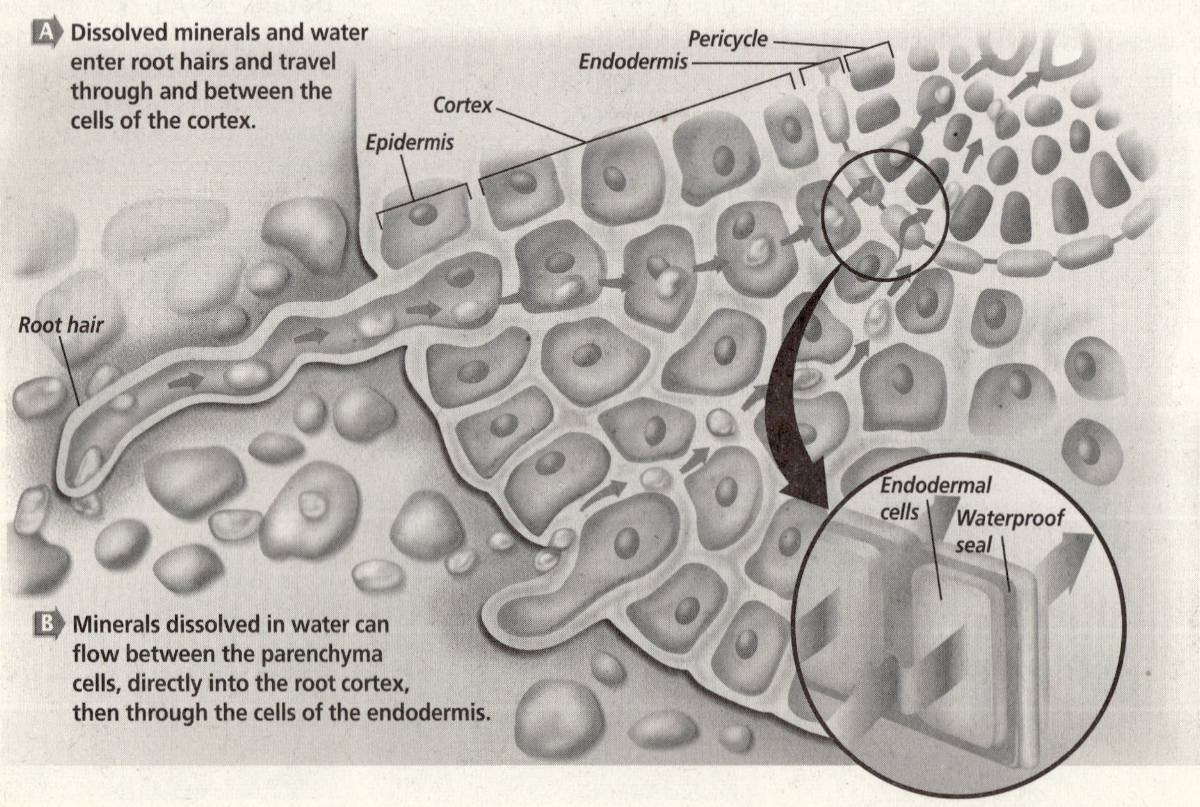
A Dissolved minerals and water enter root hairs and travel through and between the cells of the cortex.
Pericycle
Endodermis
Cortex
Epidermis
Root hair
Endodermal cells
Waterproof seal
B Minerals dissolved in water can flow between the parenchyma cells, directly into the root cortex, then through the cells of the endodermis.

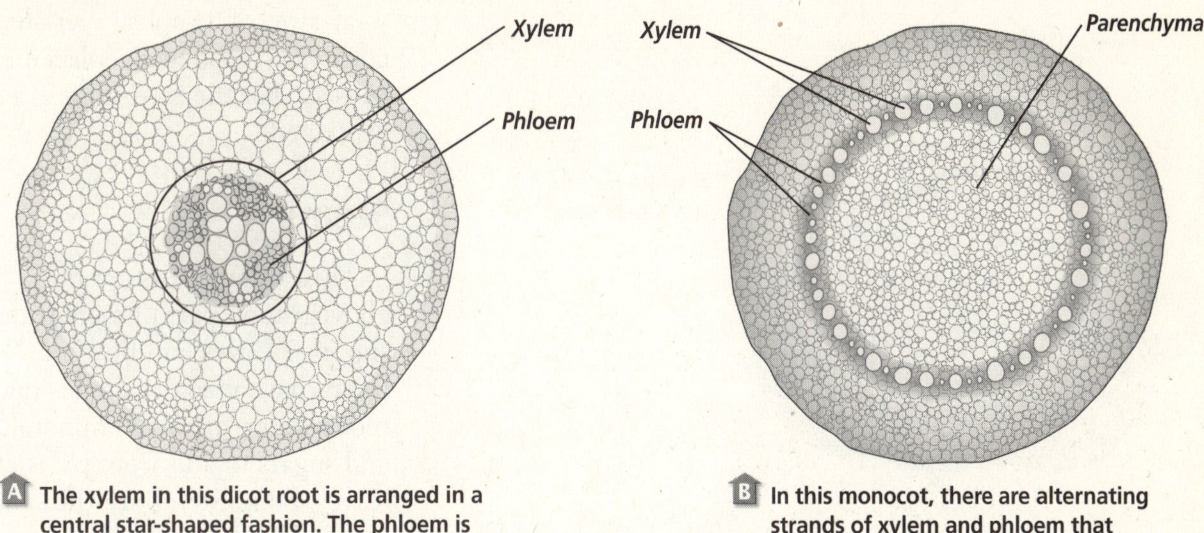

A The xylem in this dicot root is arranged in a central star-shaped fashion. The phloem is found between the points of the star.

B In this monocot, there are alternating strands of xylem and phloem that surround a core of parenchyma cells.

The next layer inward from the cortex is the **endodermis.** This is a layer of waterproof cells that form a seal around the root's vascular tissues. The waterproof seal of the endodermis forces the water and dissolved minerals to pass through the cells of the endodermis. The endodermis controls the flow of water and dissolved minerals into the root. Next to the endodermis is the **pericycle.** Lateral roots develop from the pericycle in older roots.

Xylem and phloem are located in the center of the root. One of the major differences between monocots and dicots is the arrangement of the xylem and phloem tissue. In dicot plants, the xylem forms a central star with phloem cells between the rays of the star. In monocot plants, strands of xylem alternate with strands of phloem. The alternating strands form a ring. Sometimes the center of the ring contains parenchyma cells called a pith.

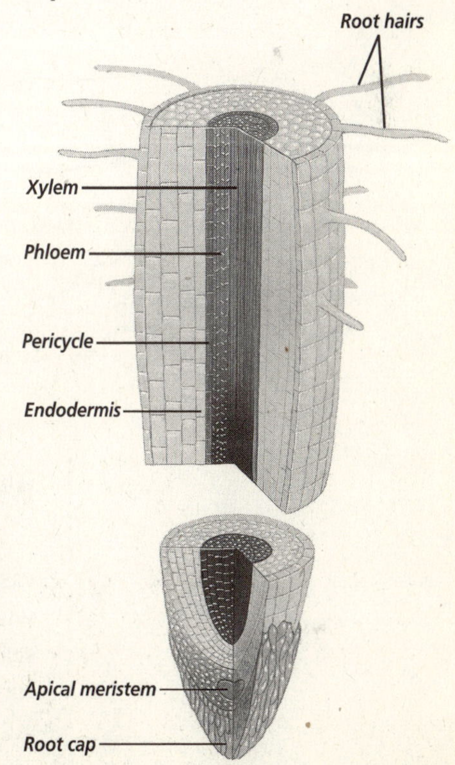

How do roots grow?

There are two areas of cell growth in roots. The root apical meristem produces cells that cause a root to increase in length. As these cells begin to mature, they develop into different types of cells. In dicots, the area between the xylem and the phloem becomes vascular cambium. The additional cells in the vascular cambium add to the root's diameter.

Each layer of new cells produced by the root apical meristem is left farther behind as new cells are added. The new cells cause the root to grow forward through the soil. A protective layer of parenchyma cells called the **root cap** covers the tip of each root. As the root grows through the soil, the cells of the root cap

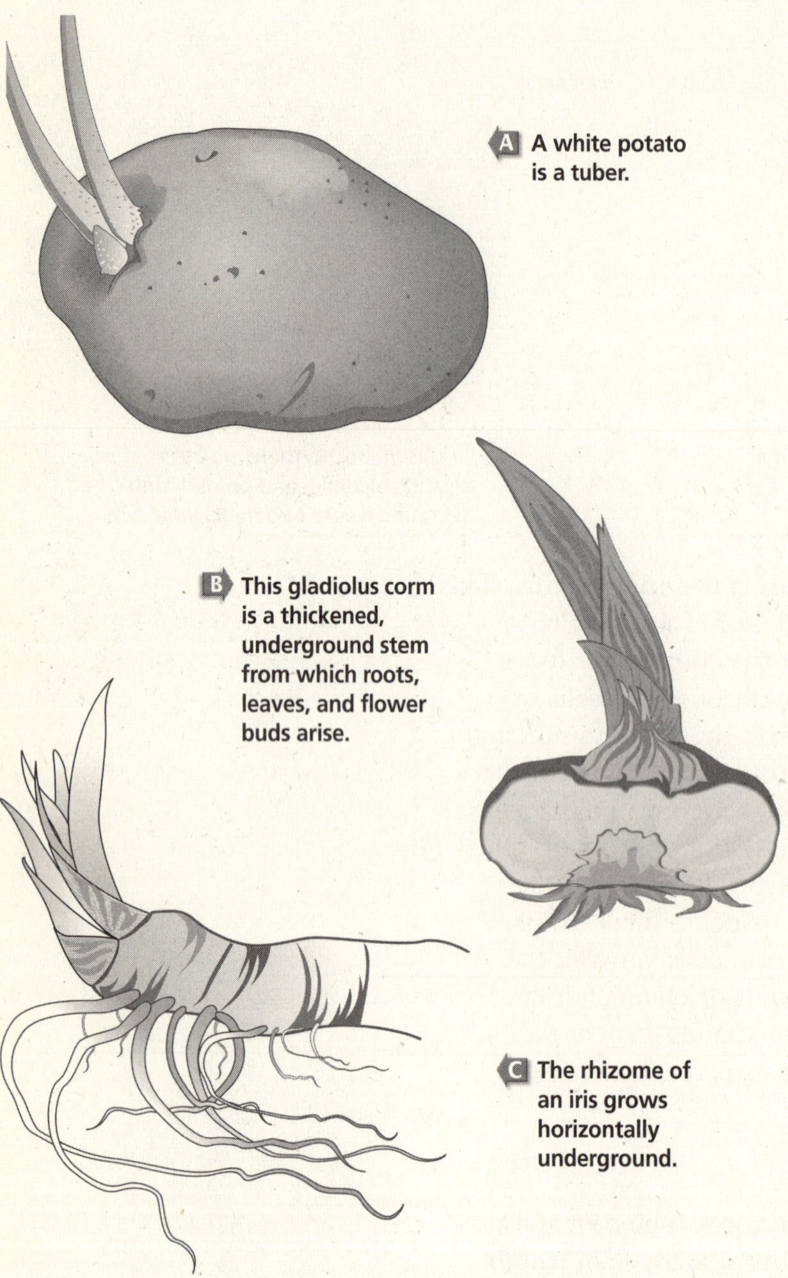

A A white potato is a tuber.

B This gladiolus corm is a thickened, underground stem from which roots, leaves, and flower buds arise.

C The rhizome of an iris grows horizontally underground.

wear away. The apical meristem constantly produces replacement cells so that the root tip is always protected.

Stems

Most often stems are the aboveground part of a plant that supports leaves and flowers. Stems have vascular tissues that move water, dissolved minerals, and sugars to and from roots and leaves. Some stems are thin and green, like those of pansies, basil, and carnations. Green stems are soft and flexible. They often carry out photosynthesis. Trees and shrubs have thick, woody stems. The stems of trees can be quite large. Woody stems are hard and rigid. They have cork and vascular cambiums.

Some stems store food. This helps the plant to survive drought or cold. Stems that act as food-storage organs are corms, tubers, and rhizomes. All three are underground stems. A corm is a short thickened stem surrounded by leaf scales. A tuber is a swollen stem that has buds from which new plants grow. Rhizomes also store food. The figure at left shows examples of stems that act as food-storage organs.

What is inside a stem?

The vascular tissue inside stems is arranged differently than the vascular tissue inside roots. Stems have bundles of vascular tissue within parenchyma tissue. In monocots, the vascular bundles are scattered throughout the stem. In dicots, the vascular bundles form a ring in the cortex.

Woody Stems Many plants, such as conifers, produce thick, sturdy stems that last several years or decades. As the stems of woody plants grow taller, they also grow in thickness. This added thickness, called secondary growth, is the result of cell divisions in the vascular cambium of the stem. The xylem tissue produced by secondary growth also is called wood. The annual growth rings in tree trunks are the layers of vascular tissue produced each year by secondary growth.

As secondary growth continues, the outer portion of the woody stem develops bark. Bark is made of phloem cells and the cork cambium. Bark is a tough, corky tissue that protects the stem from damage by insects or plant-eating animals. ✓

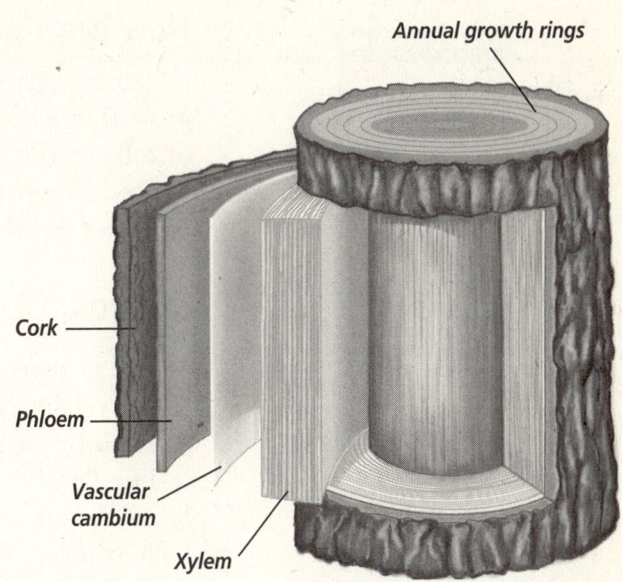

What does the stem do?

The stem transports water, sugars, and other compounds. Xylem moves water and dissolved minerals from the roots to the leaves. Water that is lost as water vapor through the stomata is continually replaced by water moving upward in the xylem. Water forms an unbroken column within the xylem. As water moves up through the xylem, it carries dissolved minerals to all living plant cells.

The phloem carries dissolved sugars. The sugars come from the photosynthetic tissues that are usually in the leaves. Any part of the plant that stores these sugars is called a **sink,** such as the parenchyma cells that make up the cortex in the root. The movement of sugars in the phloem is called **translocation** (trans lo KAY shun). The illustration at right shows the movement of materials in the vascular tissues of a carrot.

Reading Check

1. How does bark protect a plant?

____

____

____

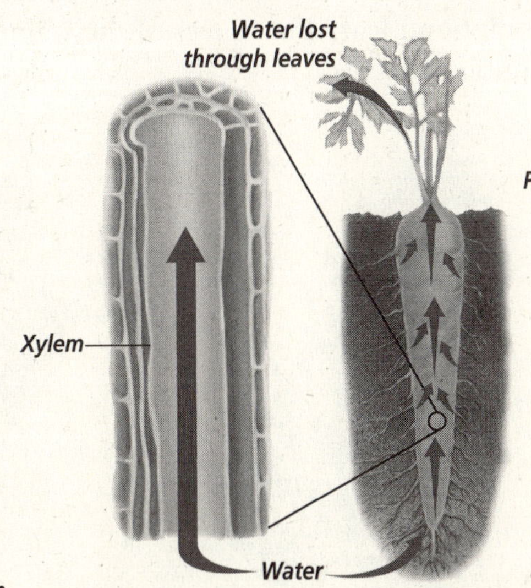

A The open ends of the xylem vessel cells form complete pipelike tubes.

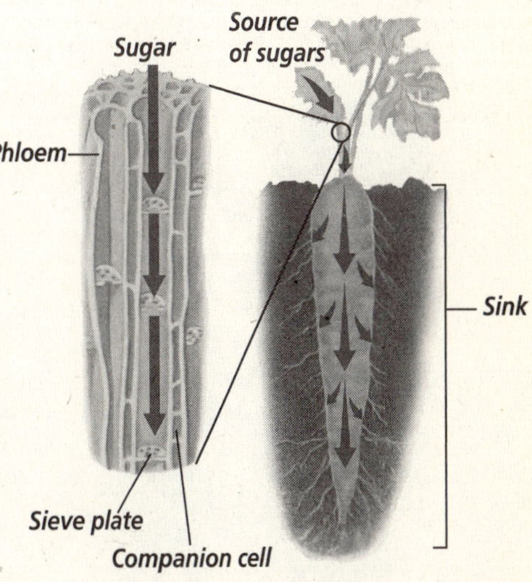

B Sugars in the phloem of this carrot plant are moving to sinks.

Reading Check

2. Does primary growth lengthen or thicken a stem?

How does a stem grow?

Primary growth lengthens a stem. It is similar to primary growth in a root. The increase in length is due to the production of cells by the apical meristem at the tip of the stem. Meristems located along the stem, called nodes, are where leaves and branches develop.

Leaves

The primary function of the leaves is photosynthesis. Most leaves have a relatively large surface area that receives sunlight. Sunlight passes into the photosynthetic tissues just beneath the leaf surface.

The flat, broad, green surface of a leaf is called the leaf blade. Sizes, shapes, and types of leaves vary a great deal. Some leaves, such as those of the Victoria water lily, can be more than two meters in diameter. Other leaves, such as duckweed, are measured in millimeters. Some plants produce different forms of leaves on the same plant.

A blade of grass is joined directly to the stem, but in other leaves, a stalk joins the leaf blade to the stem. The stalk is called a **petiole** (PE tee ohl) and is actually part of the leaf. The petiole contains vascular tissues that stretch out from the stem into the leaf and form veins. You can often see the veins if you look closely at a leaf. They look like lines or ridges.

Leaves vary in shape. A simple leaf is one with a blade that is not divided. Maple leaves and tulip poplar leaves are examples of simple leaves. When a leaf blade is divided into leaflets it is called a compound leaf. The leaves of the walnut are an example of compound leaves.

A The leaves of the walnut are compound with many leaflets.

B The needlelike leaves of the evergreen yew can receive sunlight year round.

C The tulip poplar is a deciduous tree with broad, distinctive, simple leaves.

Not all leaves grow on a stem in the same way. When two leaves grow directly opposite each other on a stem, the arrangement is called opposite. If the leaves are on opposite sides of the stem, but one is higher than the other, the arrangement is called alternate. When three or more leaves grow around a stem at the same position, the arrangement is called whorled.

What is the structure of a leaf?

Use the illustration at right as you study the internal structure of a typical leaf. Notice the cuticle and the upper and lower epidermis. Between the epidermal layers are two layers of mesophyll. **Mesophyll** (MEH zuh fihl) is the photosynthetic tissue of a leaf. It is usually made up of two types of parenchyma cells—palisade mesophyll and spongy mesophyll. Most photosynthesis takes place in the palisade mesophyll. These cells receive maximum exposure to sunlight and have many chloroplasts. The spongy mesophyll cells are surrounded by many air spaces that allow carbon dioxide, oxygen, and water vapor to flow freely. Gases also can move in and out of the stomata.

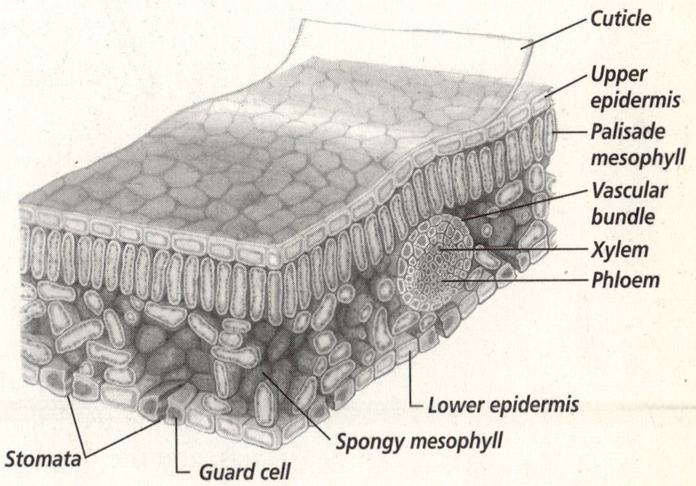

What is transpiration?

The waxy cuticle and stomata help reduce water loss. Guard cells control the size of a stoma. The loss of water through the stomata is called **transpiration.** The opening and closing of guard cells regulates transpiration. The proper functioning of guard cells is important because plants lose up to 90 percent of all the water they transport from the roots by transpiration. ✓

Reading Check

3. What regulates transpiration?

____

____

____

What are some modifications in leaves?

The pattern of veins in leaves is one way to distinguish different groups of plants. Veins are vascular tissues that run through the mesophyll of a leaf. The patterns can be parallel or netlike as shown on page 270. Parallel veins are found in many monocots. Many dicots have netlike veins. The leaves of ginkgoes are dichotomously veined.

Section
23.2 Roots, Stems, and Leaves, *continued*

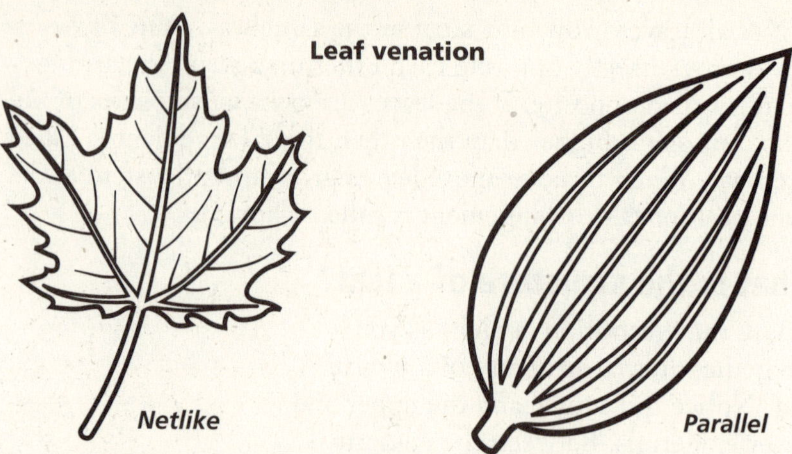

Many plants have leaves that have functions besides photosynthesis. Some leaves, when crushed or broken, release substances that cause pain or swelling. Animals and humans learn to avoid such leaves. Cactus spines are modified leaves that help reduce water loss and provide protection from plant-eaters.

Carnivorous plants have leaves with adaptations that can trap insects or other small animals. Other leaf modifications include the curling tendrils in sweet peas, the colorful bracts on poinsettias, and the overlapping scales that enclose and protect buds.

Leaves often act as water or food storage sites. The leaves of *Aloe vera*, for instance, store water. A bulb, on the other hand, provides an example of leaves storing food. A bulb consists of a shortened stem, a flower bud, and thick immature leaves. Food is stored in the base of the immature leaves. Onions, tulips, and narcissus grow from bulbs.

After You Read

Mini Glossary

cortex: layer of ground tissue in the root that is involved in the transport of water and dissolved minerals into the vascular tissue of the root

endodermis: single layer of cells that forms a waterproof seal around a root's vascular tissue; controls the flow of water and dissolved minerals into the root

mesophyll (MEH zuh fihl): photosynthetic tissue of a leaf

pericycle: in plants, the layer of cells just within the endodermis that gives rise to lateral roots

petiole (PE tee ohl): in plants, the stalk that joins the leaf blade to the stem

root cap: tough, protective layer of parenchyma cells that covers the tip of a root

sink: any part of a plant that stores sugars produced during photosynthesis

translocation (trans loh KAY shun): movement of sugars in the phloem of a plant

transpiration: in plants, the loss of water through stomata

Section 23.2 Roots, Stems, and Leaves, *continued*

1. Read the key terms and definitions in the Mini Glossary on page 270. Circle any three of the terms that describe the structure or function of leaves. In the space provided explain in your own words how those three terms relate to leaves.

2. For each plant part listed in the first column, write the name of one related structure in the second column and one function in the third column.

	Structure	**Function**
Roots		
Stems		
Leaves		

Visit the Glencoe Science Web site at **science.glencoe.com** to find your biology book and learn more about roots, stems, and leaves.

Name Date Class

Section 23.3 Plant Responses

Before You Read

Plants respond to light and touch. Have you ever noticed a houseplant growing toward the light? Have you ever pulled weeds for a long time and noticed that your hands felt itchy or that you developed a mild rash? Use the lines below to write how you think plants respond to light and touch and why this happens.

____

____

Read to Learn

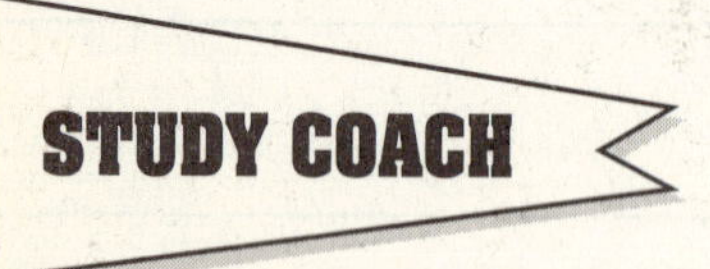

Mark the Text **Identify Details** As you read this section, highlight the hormone or group of hormones discussed in one color. Underline the action of the hormone in a different color.

Plant Hormones

Plants, like animals, have hormones that regulate growth and development. A **hormone** is a chemical that is produced in one part of an organism and transported to another part, where it causes a change. Only a small amount of the hormone is needed to make this change. In this section, you will learn about different groups of hormones and some individual hormones that affect plants.

Are auxins responsible for stem growth?

The group of plant hormones called **auxins** (AWK sunz) cause cells to lengthen or elongate. Indoleacetic (in doh luh SEE tihk) acid (IAA), an auxin, is produced in apical meristems of plant stems. IAA weakens the cell wall, allowing a cell to stretch and grow longer. Stem growth is the result of new cells from the apical meristem and their increasing cell lengths.

Auxins have other effects on plant growth and development. Auxin produced in the apical meristem keeps side branches from growing. Removing the stem tip reduces the amount of auxin and allows branches to develop.

Auxin also delays fruit formation and keeps fruit from dropping off the plant. When the amount of auxin in the plant decreases, ripe fruit drops to the ground and trees begin to shed their leaves.

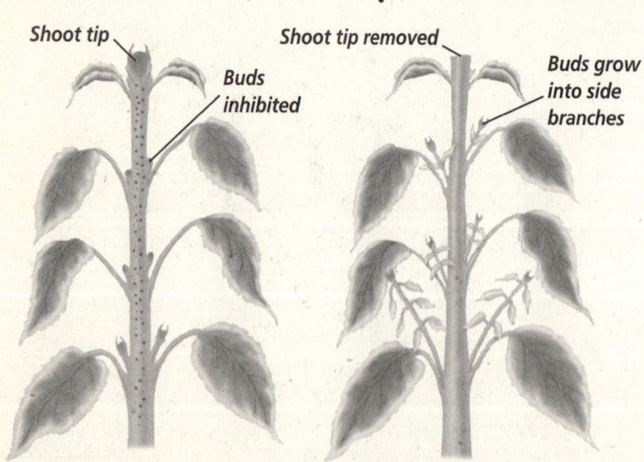

A Auxin produced in the tip of the main shoot inhibits the growth of the side branches.

B Once the tip is removed, the side branches start to grow.

How do gibberellins affect plants?

Another group of plant growth hormones called **gibberellins** (jih buh REH lunz) help cells elongate. Often dwarf plants are short because the plant does not produce gibberellins. If gibberellins are applied to the tips of these dwarf plants, they will grow taller. Gibberellins increase the rate at which seeds begin to grow and buds develop. Farmers use gibberellins to increase the formation of fruit. Florists use gibberellins to cause flower buds to open.

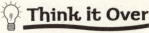

Think it Over

1. **Infer** Why might a florist need to use gibberellins?

____

____

____

What other hormones affect plant growth?

The hormones called **cytokinins** (si tuh KI nihnz) stimulate the production of proteins needed for mitosis. Without cytokinins, plant cells would grow larger but never divide. The effects of cytokinins are often enhanced by the presence of other hormones.

The plant hormone **ethylene** (EH thuh leen) is a gas made of carbon and hydrogen. Ethylene is produced and released by fruit while it ripens. It causes cell walls to weaken and soften. Ethylene speeds the ripening of fruits and causes the breakdown of complex carbohydrates to simple sugars. This gives ripe fruit its sweet taste. Many farmers often pick unripe fruit and then use ethylene to ripen the fruit later.

Think it Over

2. **Infer** How might placing a piece of ripe fruit in a brown paper bag with unripe fruit cause the unripe fruit to ripen?

____

____

____

Plant Responses

Tropism is a change in a plant's growth because of an external stimulus. The tropism is called positive if the plant grows toward the stimulus. The tropism is called negative if the plant grows away from the stimulus.

The growth of a plant toward light is called phototropism. It is caused by the hormone group auxins. There is more auxin on the side of the stem that is away from the light. This causes the cells on that side to lengthen. The difference in size between the lengthened cells and the cells on the other side causes the stem to bend toward the light.

Another tropism causes roots to grow downward and stems to grow upward. Gravitropism is plant growth in response to gravity. Positive gravitropism causes roots to grow into the soil and anchor the plant. Stems exhibit negative gravitropism as they grow away from the ground.

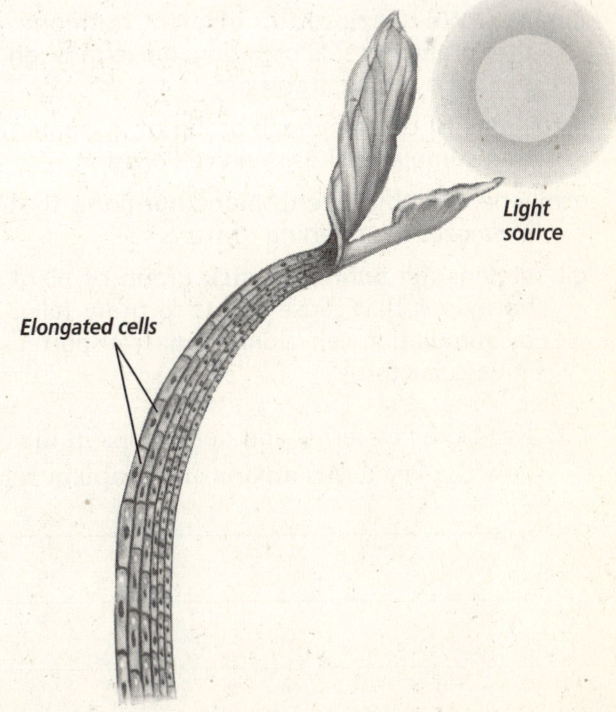

Reading Check

3. Why are tropisms not reversible?

Growth response to touch is called thigmotropism. During early growth, if the tendrils of sweet peas and passion vines touch a fence or trellis, they coil around it.

Tropisms involve growth and are not reversible. The position of a stem that has grown in a particular direction cannot be changed. However, if the direction of the stimulus is changed, the stem will begin growing in another direction.

A **nastic movement** is a responsive movement of a plant that is not dependent on the direction of the stimulus. An example of a nastic movement is the response of *Mimosa pudica* leaflets when they are touched. Touching causes a drop in water pressure in the cells at the base of each leaflet. The cells become limp and the leaflets change direction. A nastic movement is reversible because it does not involve plant growth.

Another example of nastic movement is the sudden closing of the hinged leaf of a Venus's-flytrap. If an insect triggers sensitive hairs on the inside of the leaf, the leaf snaps shut. In both examples, the leaves return to their original positions once the stimulus is removed.

After You Read

Mini Glossary

auxins (AWK sunz): group of plant hormones that promote cell elongation; move through a plant by active transport

cytokinins (si tuh KI nihnz): group of hormones that stimulate mitosis and cell division

ethylene (EH thuh leen): plant hormone that promotes the ripening of fruits

gibberellins (jih buh REH lunz): group of plant hormones that cause plants to grow taller by stimulating cell elongation; transported in vascular tissue

hormone: chemical produced in one part of an organism and transported to another part, where it causes a physiological change

nastic movement: responsive movement of a plant not dependent on the direction of the stimulus

tropism: change in a plant's growth because of an external stimulus

1. Read the key terms and definitions in the Mini Glossary above. On the lines below explain how the key terms **auxins** and **tropism** relate to each other.

2. In Column 1 are some new concepts from this section. Column 2 gives an example of each concept. Write the letter of the example on the line next to the concept that matches it.

Column 1	Column 2
_____1. plant hormones	a. growth responses to external stimuli
_____2. tropisms	b. sometimes caused by changes in cell pressure
_____3. nastic movement	c. affect plant growth and functions
_____4. phototropism	d. growth response to touch
_____5. thigmotropism	e. growth of plant toward light

Visit the Glencoe Science Web site at **science.glencoe.com** to find your biology book and learn more about plant responses.

Section 24.1 Life Cycles of Mosses, Ferns, and Conifers

Before You Read

Previous sections have briefly covered alternation of generations. Write any questions you have about alternation of generations on the lines below. If, after you read this section you still have questions, be sure to ask your teacher.

Read to Learn

Create a Quiz After you have read this section, create a quiz based on what you have learned. After you have completed writing the quiz questions, be sure to answer them.

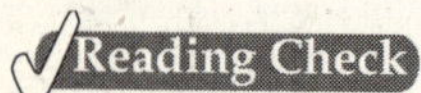

1. Is the gametophyte or the sporophyte the dominant generation in most plant species?

Alternation of Generations

Most plant cycles include an alternation of generations. As you learned earlier, alternation of generations consists of a sporophyte (spore-producing) generation, and a gametophyte (gamete producing) generation.

All cells of a sporophyte are diploid, containing two of each kind of chromosome. To create haploid spores, containing one of each kind of chromosome, certain cells of a sporophyte undergo meiosis. The haploid spores undergo cell divisions and form a multicellular, haploid gametophyte. The gametophyte produces haploid gametes, sex cells. The female gamete is an egg and the male gamete is a sperm. When a sperm fertilizes an egg, a diploid zygote forms. This is sexual reproduction. The zygote can undergo cell divisions and form an embryo sporophyte. If the embryo develops to maturity, the cycle can begin again.

This basic life cycle pattern is the same for most plants. However, the pattern may vary. For example, recall that in mosses, the gametophyte is the familiar form. In flowering plants, the sporophyte is the familiar form. Most people have never seen the female gametophyte of a flowering plant. Botanists refer to the bigger, visible plant as the dominant generation. The dominant generation lives longer and can stay alive without the other generation. In most plant species, the sporophyte is the dominant generation. ✓

How are new plants reproduced asexually?

In a process called **vegetative reproduction,** new plants are produced from existing plant organs or parts of organs. The new plants have the same genetic makeup as the original plant. For example, some thallose liverwort gametophytes can produce cuplike structures. Inside, tiny pieces of tissue called gemmae (JEH mee) develop. If gemmae fall from the cup to the ground, they can grow into other liverwort gametophytes. They are genetically identical to the liverwort that produced the gemmae.

Life Cycle of Mosses

The gametophyte generation is the dominant generation in mosses. The diagram below shows the life cycle of mosses.

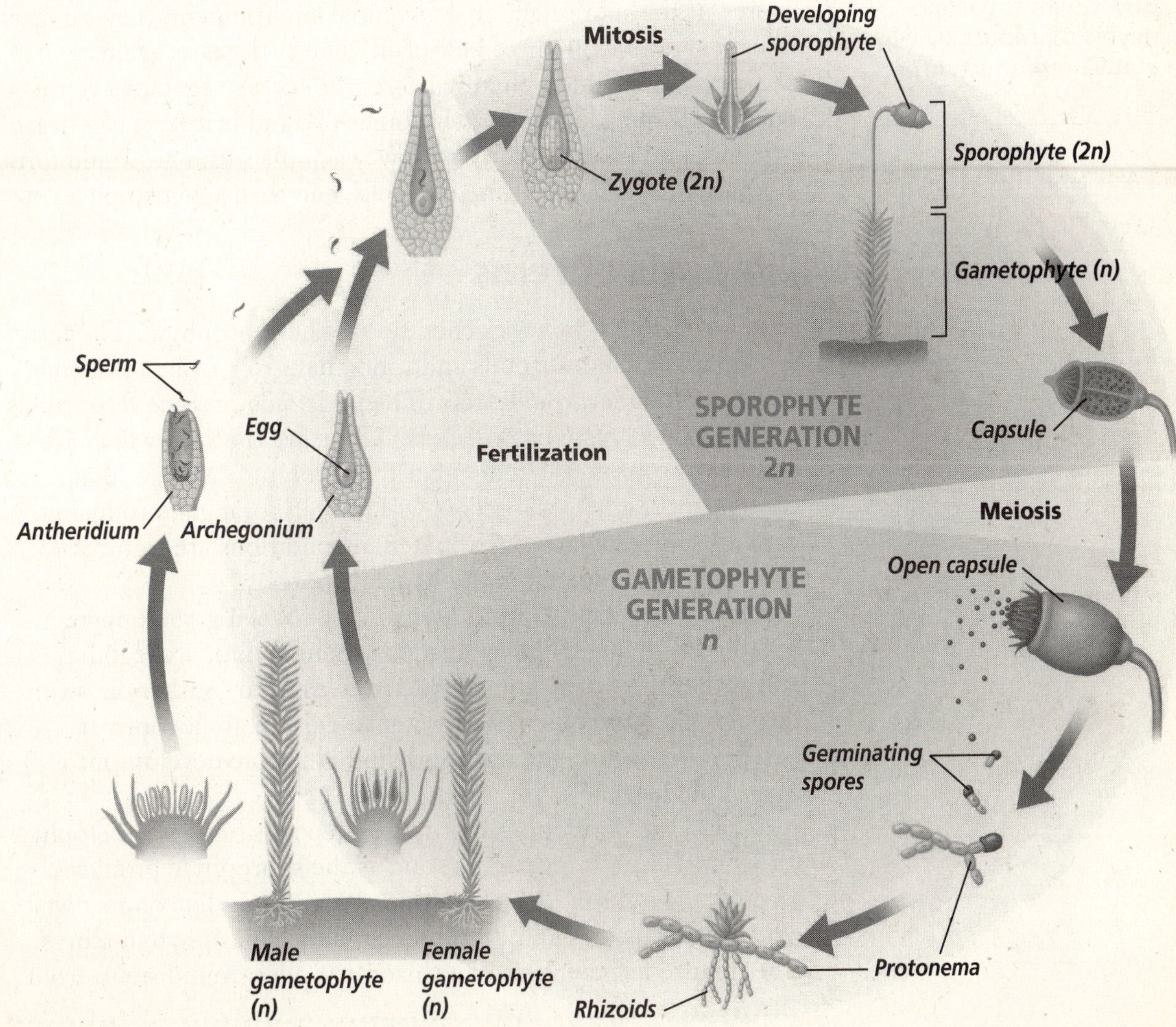

✓Reading Check

2. How do moss gametophytes reproduce by vegetative reproduction?

A haploid spore can germinate and grow to form a protonema. A **protonema** (proh tuh NEE muh) is a small green filament, or strand, of cells that can develop into the gametophyte. In some mosses, male and female reproductive structures form on separate gametophytes. In others, male and female reproductive structures are on the same gametophyte. The archegonium is the egg-producing female reproductive structure. The antheridium is the sperm-producing male reproductive structure.

Sperm from an antheridium swim in a continuous film of water to an egg in an archegonium. If fertilization occurs, a diploid zygote forms. The zygote undergoes cell divisions to become the sporophyte. The sporophyte is a stalk with a capsule on top. The sporophyte is attached to and depends on the gametophyte. Cells in the sporophyte capsule undergo meiosis, producing haploid spores. When the capsule is mature, it bursts open and releases spores. If the spores land in a favorable environment, they can germinate and develop into a new plant. The cycle starts again.

Some moss gametophytes also reproduce by vegetative reproduction. If the gametophyte becomes dry and brittle, it can break into pieces. When moisture returns, each piece can grow and form a protonema, a green filament of cells, and then a gametophyte. ✓

Life Cycle of Ferns

In ferns, the dominant generation is the sporophyte. The fern sporophyte includes fronds and roots that grow from a rhizome, which is an underground stem. The underside of some fern fronds are covered in rusty brown spots. These are the sori, which are clusters of sporangia. Remember that sporangia are sacs that contain spores. Meiosis occurs within the sporangia, producing haploid spores. When environmental conditions are right, the sporangia open, releasing the haploid spores.

A spore develops to form a tiny heart-shaped gametophyte called a prothallus. The prothallus produces both archegonia and antheridia on its surface. Sperm released by antheridia swim through a continuous film of water to eggs in archegonia. If fertilization occurs, the resulting diploid zygote develops into a sporophyte.

At first, the developing sporophyte depends on the gametophyte for its nutrition. However, as soon as the sporophyte produces green fronds, it can carry on photosynthesis and live on its own. The prothallus dies and decomposes as the sporophyte matures. The mature fern consists of a rhizome (underground stem) from

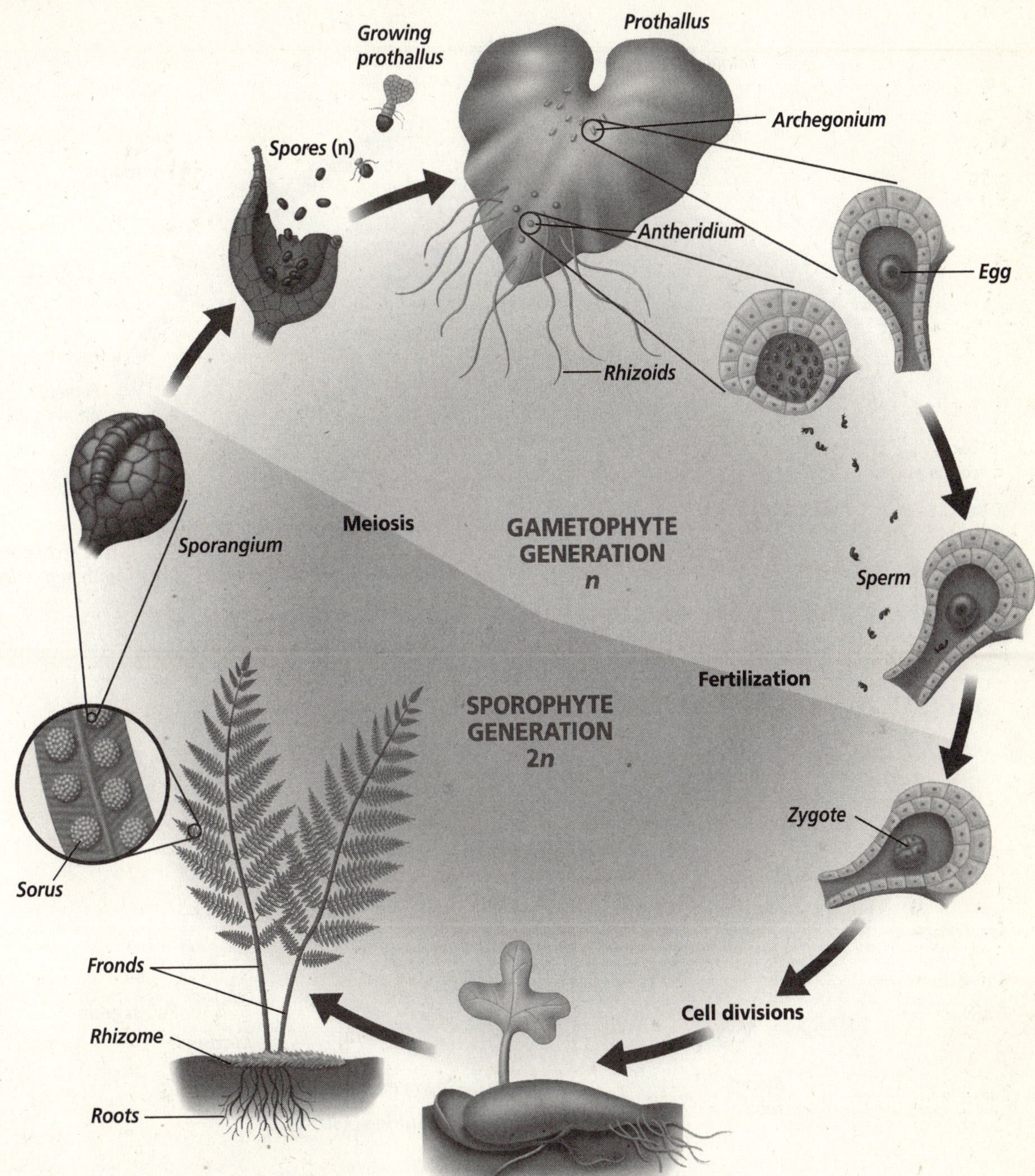

which roots and fronds grow. If pieces of the rhizome break off, new fern plants can develop from the pieces by vegetative reproduction. The cycle continues when sporangia develop on the fronds and spores are released. The figure above shows the life cycle of the fern.

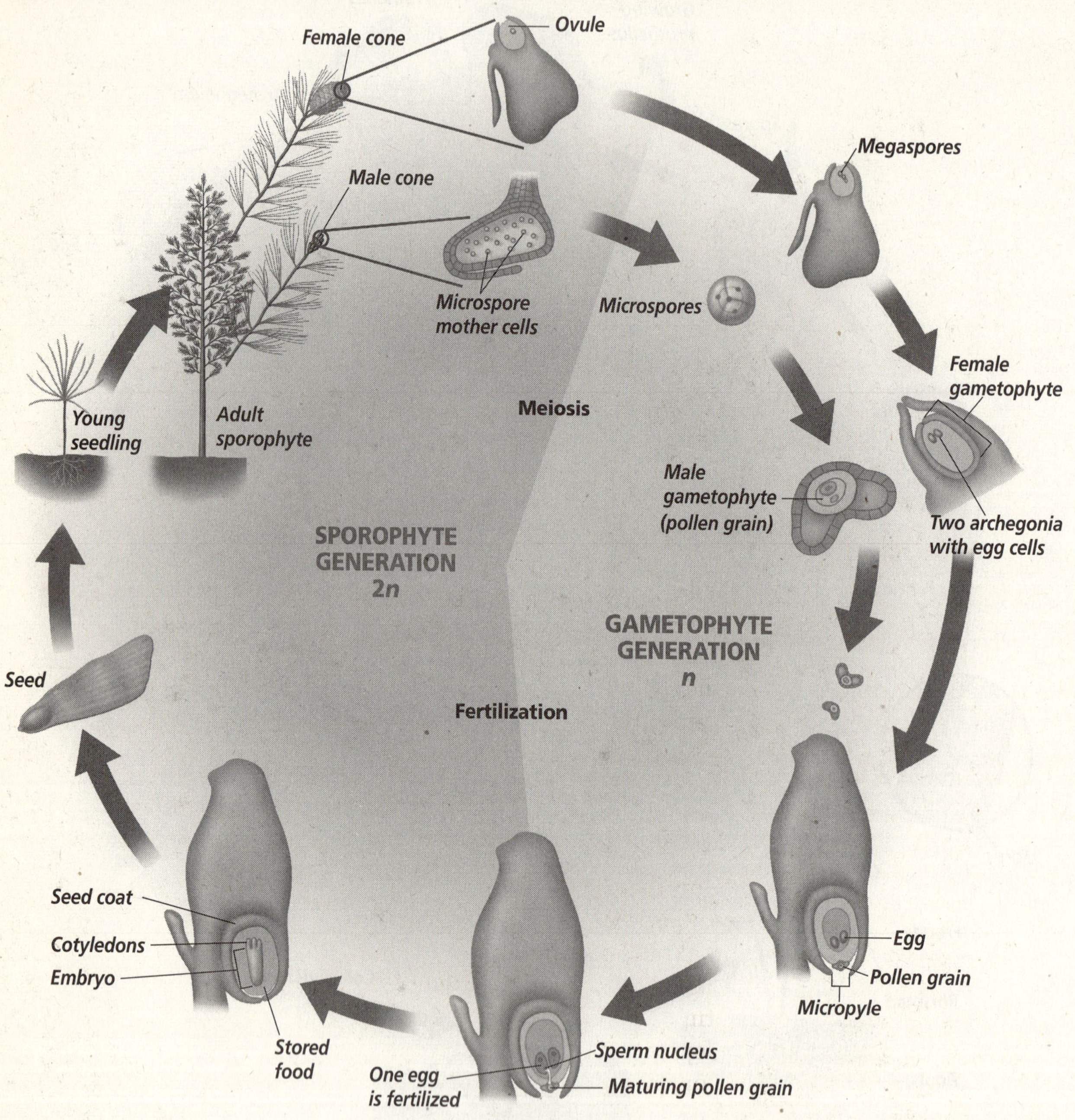

The Life Cycle of Conifers

The dominant generation in conifers is the sporophyte. Unlike ferns, conifers produce seeds. The stages of a conifer's life cycle are shown in the figure above. Adult conifer plants produce male and female cones on separate branches of the same plant. Cones contain sporangia on their scales.

Section 24.1 Life Cycles of Mosses, Ferns, and Conifers, *continued*

Female cones are larger than male cones. On the upper surface of each scale of a female cone, two ovules develop. Each ovule contains a sporangium with a diploid cell. The diploid cell undergoes meiosis and produces four megaspores. A **megaspore** is a female spore that eventually becomes the female gametophyte. One of the four megaspores lives and grows by cell division into the female gametophyte.

The female gametophyte depends on the sporophyte for protection and nutrition. Inside the female gametophyte are two or more archegonia, each containing an egg.

Male cones have sporangia that undergo meiosis to produce male spores called **microspores.** Each microspore can develop into a male gametophyte, or pollen grain. Each pollen grain has a hard, water-resistant outer covering.

How does pollination take place?

In conifers, pollination is the transfer of pollen grains from the male cone to the female cone. Wind blows the pollen grains from the male cones. If a pollen grain falls near the opening of an ovule in the female cone, pollination can occur. The opening of the ovule is called the **micropyle** (MI kruh pile). The pollen grain gets trapped in a sticky drop of fluid that covers the micropyle. As the fluid evaporates, the pollen grain is pulled closer to the micropyle. Even though pollination has occurred, fertilization does not take place for at least a year. The pollen grain and the female gametophyte will mature during this time.

When are mature seeds released?

As the pollen grain matures, it produces a pollen tube that grows through the micropyle into the ovule. A sperm nucleus from the pollen grain moves through the pollen tube to the egg. If fertilization occurs, a zygote forms. The female gametophyte nourishes the zygote until it develops into an embryo with several cotyledons. The cotyledons will nourish the sporophyte after development begins. A seed coat forms around the ovule as the mature seed is produced. Mature seeds are released when the female cone opens.

3. Analyze What are the differences between male and female reproductive cones?

____

____

____

After You Read

Mini Glossary

megaspore: female spore formed by some plants that can develop into a female gametophyte

micropyle (MI kruh pile): the opening in the ovule through which the pollen tube enters

microspore: male spore formed by some plants that can develop into a male gametophyte, or pollen grain

protonema (proh tuh NEE muh): in mosses, a small green filament of haploid cells that develops from a spore; can develop into the gametophyte

vegetative reproduction: type of asexual reproduction in plants where a new plant is produced from existing plant organs or parts of organs

1. Read the terms and definitions in the Mini Glossary above. Then use the space below to write a brief paragraph describing vegetative reproduction.

2. Use the table below to review what you have learned about the life cycle of mosses, ferns, and conifers. Fill in the data table by entering "yes" or "no" in the appropriate space.

Trait	Moss	Fern	Conifer
Has alternation of generations			
Film of water needed for fertilization			
Dominant gametophyte			
Dominant sporophyte			
Produces seeds			
Produces sperm			
Produces pollen grains			
Produces egg			

3. Explain why mosses and ferns need a moist environment for sexual reproduction.

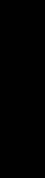

V oe Science Web site at **science.glencoe.com** to find your ıd learn more about the life cycles of mosses, ferns, and conifers.

Section 24.2 Flowers and Flowering

Before You Read

Do you associate certain flowers with particular seasons or days in the year? When you think of spring do you picture tulips? On the lines below identify the flowers and the particular seasons you associate with those flowers. After you read this section, you will understand why flowers bloom at certain times of the year.

Read to Learn

What is a flower?

The process of sexual reproduction in flowering plants takes place in a flower. The flower is a complex structure made up of several organs.

There are many sizes, shapes, and colors of flowers, yet all flowers share a basic structure. A flower's structure is genetically determined and is usually made up of four kinds of organs: petals, sepals, stamens, and pistils. You are probably most familiar with the petals. **Petals** are often the colorful structures at the top of a flower stem. The flower stem is called the peduncle. **Sepals** are green leaflike structures that encircle the peduncle just below the petals.

Inside the petals are the stamens. A **stamen** is the male reproductive organ of a flower. The **anther,** at the tip of the stamen, produces pollen that eventually contains sperm.

Attached to the peduncle, in the center of the flower, is one or more pistils. The **pistil** is the female reproductive organ of the flower. The bottom portion of the pistil is the ovary. The **ovary** is a structure with one or more ovules. Each ovule usually contains one egg. The female gametophyte develops inside the ovule.

A flower that has all four organs—sepals, petals, stamens, and pistils—is called a complete flower. A flower that lacks one or more organs is called an incomplete flower. Walnut trees have separate male and female flowers. The male flowers have stamens but no pistils; the female flowers have pistils but no stamens. The male and female flowers of walnut trees are incomplete flowers.

Mark the Text **Identify Details** As you read about a flower in the text, highlight each part of the flower illustrated on page 284.

Section
24.2 Flowers and Flowering, *continued*

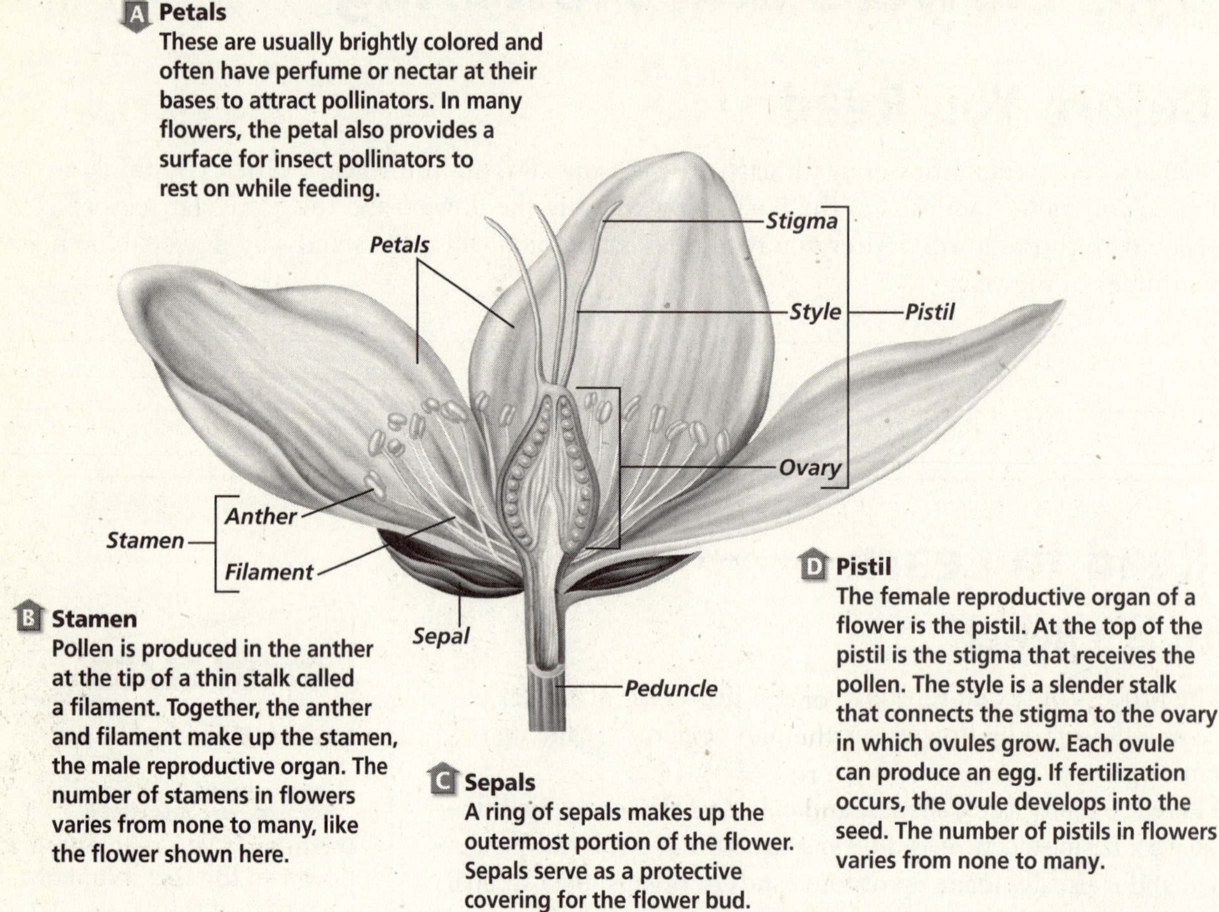

A Petals
These are usually brightly colored and often have perfume or nectar at their bases to attract pollinators. In many flowers, the petal also provides a surface for insect pollinators to rest on while feeding.
Petals
Stigma
Style
Pistil
Ovary
Stamen
Anther
Filament
Sepal
Peduncle
B Stamen
Pollen is produced in the anther at the tip of a thin stalk called a filament. Together, the anther and filament make up the stamen, the male reproductive organ. The number of stamens in flowers varies from none to many, like the flower shown here.
C Sepals
A ring of sepals makes up the outermost portion of the flower. Sepals serve as a protective covering for the flower bud.
D Pistil
The female reproductive organ of a flower is the pistil. At the top of the pistil is the stigma that receives the pollen. The style is a slender stalk that connects the stigma to the ovary in which ovules grow. Each ovule can produce an egg. If fertilization occurs, the ovule develops into the seed. The number of pistils in flowers varies from none to many.

The flowers of sweet corn and grasses have other adaptations. Their flowers do not have petals and are adapted for pollination by wind rather than by animals.

Photoperiodism

In summer, days are long and nights are short. In winter, days are short and nights are long. The length of daylight and darkness each day affects plant growth and the timing of flower production. For example, some chrysanthemum plants produce flowers only during the fall, when the length of daylight is getting shorter and the length of darkness is getting longer each day. A grower who wants to produce chrysanthemum flowers in the middle of summer must artificially increase the amount of darkness. The grower can do this by placing a black cloth around the plants before the sun sets, and then removing it later the following morning. The response of flowering plants to daily daylight-darkness conditions is called **photoperiodism.**

What is the critical period?

Plant biologists first thought that the length of daylight controlled flowering. Now they know it is the length of darkness that controls flowering. Each plant species has specific daylight-darkness conditions that will make flowering start, known as the critical period. Plants can be placed in one of four categories depending on the critical period they require to produce flowers. Plants are short-day plants, long-day plants, day-neutral plants, or intermediate-day plants.

A **short-day plant** flowers when the number of daylight hours is shorter than that of its critical period. Short-day plants usually flower sometime during late summer, fall, winter, or spring. Examples of short-day plants include asters, poinsettias, strawberries, ragweed, and pansies.

A **long-day plant** flowers when the number of daylight hours is longer than that of its critical period. Long-day plants flower in summer. They also will flower if lighted continuously. Carnations, petunias, potatoes, lettuce, spinach, and wheat are long-day plants.

As long as the proper growing conditions exist, **day-neutral plants** will flower over a range in the number of daylight hours. Examples of day-neutral plants are roses, cucumbers, cotton, and many tropical plants, as well as dandelions and many other weeds.

An intermediate-day plant will not flower if the days are longer or shorter than its critical period. Sugarcane and several grasses are in this category. ✓

How does photoperiodism benefit plants?

Photoperiodism is a physiological adaptation. All the plants of a particular species flower at the same time, ensuring that there is a large population of pollinators. This is important because pollination is a critical event in the life cycles of all flowering plants. ✓

Think it Over

1. **Analyze** Which controls flowering? (Circle your choice.)
   a. length of daylight
   b. length of darkness

Reading Check

2. What are the four different categories of flowering plants within photoperiodism?

____

____

____

____

Reading Check

3. How does photoperiodism ensure pollination?

ensuring that there is a large population

After You Read

Mini Glossary

anther: pollen-producing structure located at the tip of a flower's stamen

day-neutral plants: plants that flower over a range in the number of daylight hours

long-day plants: plants that flower when the number of daylight hours is longer than its critical period

ovary: in plants, the bottom portion of a flower's pistil that contains one or more ovules, each containing one egg

petals: leaflike flower organs, usually brightly colored structures at the top of a stem

photoperiodism: flowering plant response to differences in the length of night and day

pistil: female reproductive organ of a flower

sepals: leaflike, usually green structures that encircle the top of a flower stem below the petals

short-day plant: a plant that flowers when the number of daylight hours is shorter than its critical period

stamen: male reproductive organ of a flower consisting of an anther and a filament

1. Read the terms and definitions in the Mini Glossary above. Highlight four terms that deal with photoperiodism. Then circle the terms for the four organs that make up flowers.
2. Use the text and illustrations to list the organs and structures of a flower in the diagram below.

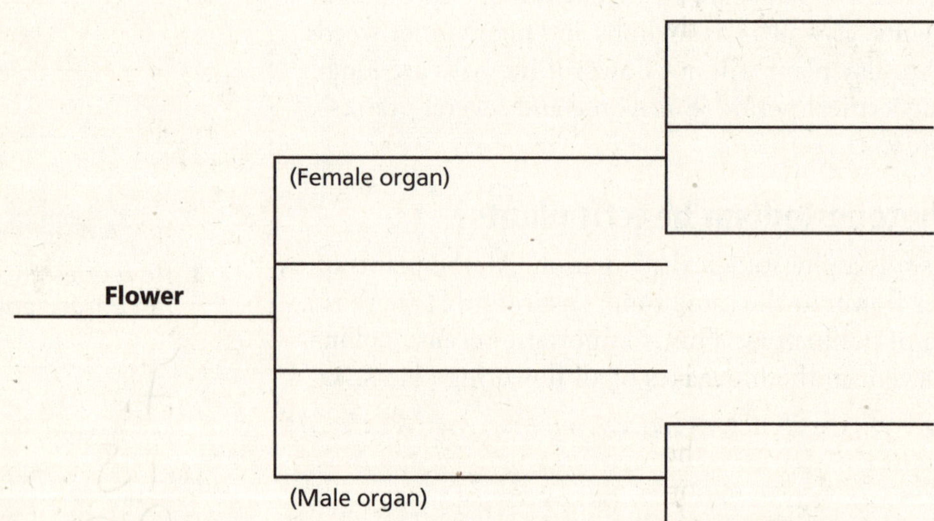

Visit the Glencoe Science Web site at **science.glencoe.com** to find your biology book and learn more about flowers and flowering.

Section 24.3

The Life Cycle of a Flowering Plant

Before You Read

Have you ever opened the kitchen pantry or refrigerator and seen a potato or onion with leaves or roots growing from it? Why do you think that happens sometimes? Write your thoughts on the lines below.

Read to Learn

The Life Cycle of an Anthophyte

The life cycle of flowering plants (anthophytes) and conifers is similar. In both, the sporophyte is the dominant generation, and the gametophyte generation develops within the sporophyte. However, in conifers the reproductive organs are in male and female cones. In the anthophytes, the reproductive organs are in the flowers. Anthophytes are the only plants that produce flowers and fruits.

Mark the Text **Identify Main Ideas** Highlight the main point in each paragraph. Study the points and come up with a way to state each main point in your own words.

What is the development of the female gametophyte?

Inside each ovary is one or more ovules. In the ovule, a cell undergoes meiosis and produces haploid megaspores. A megaspore is a female spore that eventually can become the female gametophyte. Four megaspores are produced, but three of them usually die. In most flowering plants, the megaspore's nucleus undergoes mitosis three times, producing eight haploid nuclei as illustrated in the figure to the right. Together, these eight nuclei make up the female gametophyte. Six of the nuclei develop cell walls. One of the six becomes the egg cell. The two remaining nuclei, called **polar**

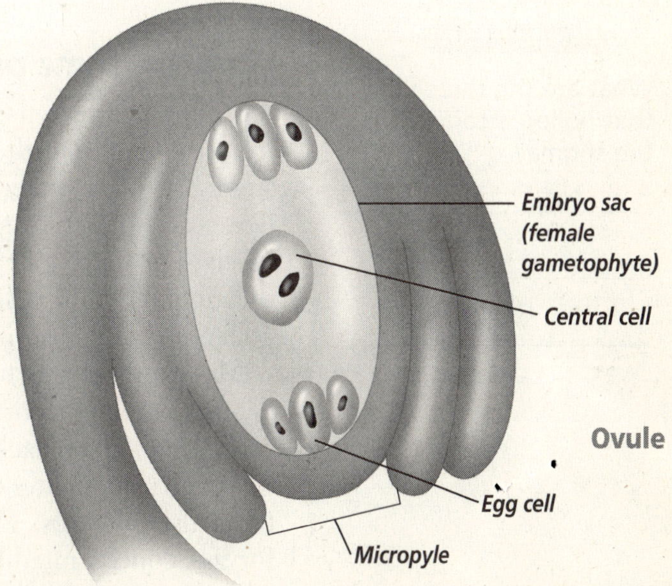

nuclei, are enclosed in one cell. This cell, called the central cell, is located at the center of the female gametophyte. The egg cell is near the micropyle or opening of the ovule. Eventually the other five cells break apart and disappear.

How does the male gametophyte develop?

Pollen sacs in the anther produce haploid microspores. The nucleus of each microspore undergoes mitosis, forming two nuclei. A thick, protective wall surrounds these two nuclei. This structure is the immature male gametophyte, or pollen grain. The nuclei within the pollen grain are the tube nucleus and the generative nucleus. When the pollen grains mature, the anther usually splits open.

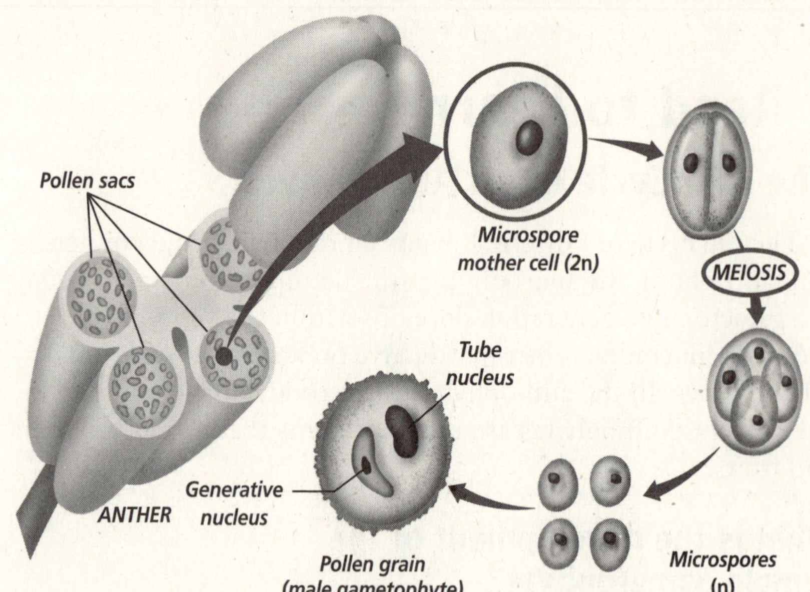

Reading Check

1. What are the three ways that pollen is carried to the stigma?

What are some pollination adaptations?

In flowering plants, pollination is the transfer of pollen grains from the anther to the stigma. Pollen grains can be carried to the stigma by wind, water, or animals. A high pollination rate means a better rate of plant reproduction. This means the pistil of a flower receives enough pollen of its own species to fertilize the egg in each ovule. Wind-pollinated plants produce large amounts of pollen to help guarantee pollination. ✓

Most anthophytes have adaptations that allow pollen grains to be deposited in the right place at the right time. For example, wind-pollinated flowers lack structures that might block wind currents.

Many anthopyhtes that are pollinated by animals produce nectar in their flowers. The animals that visit the flowers are looking for this concentrated liquid. Nectar is made up of proteins and sugars and collects in the cuplike area at the base of the petals. Insects and birds brush up against the anthers trying to get the nectar. The pollen that attaches to them is carried to another flower as they continue to look for nectar, resulting in pollination. Some insects also gather pollen to use as food. By producing nectar and attracting animal pollinators, animal-pollinated plants do not need to produce as much pollen as wind-pollinated plants. ✓

Some nectar-feeding pollinators are attracted to a flower by its color or scent or both. Butterflies and bees are attracted to bright, colorful flowers. Some of these flowers have markings that humans can't see but are easily seen by insects. Flowers that are pollinated by beetles and flies have a strong scent but are often dull in color.

Some flowers have structural adaptations that give them a better chance of being pollinated by another plant of the same species. This is called cross-pollination. It results in greater genetic variation than self-pollination. For example, the flowers of a certain species of orchid resemble female wasps. A male wasp visits the flower and attempts to mate with it. The male becomes covered with pollen, which is deposited on the next orchid he visits.

Reading Check

2. Why do animal-pollinated plants produce less pollen than wind-pollinated plants?

____

____

____

What are the steps in fertilization?

Animals or wind can bring a pollen grain to the stigma of the pistil, but that does not mean fertilization has occurred. Several events have to take place before fertilization occurs. Remember that inside each pollen grain are two haploid nuclei. One is the tube nucleus, the other is the generative nucleus. The tube nucleus directs the growth of the pollen tube down through the pistil to the ovary. The generative nucleus divides by mitosis, producing two sperm nuclei. The sperm nuclei move down the pollen tube to the micropyle.

Inside the ovule is the female gametophyte. One of the sperm unites with the egg cell forming a diploid zygote, which begins the new sporophyte generation. The other sperm nucleus fuses with the central cell, which contains the polar nuclei. This creates a cell with a triploid ($3n$) nucleus. This process, in which one sperm fertilizes the egg and the other sperm joins with the central cell, is called **double fertilization.** Double fertilization is unique to anthophytes. ✓

The triploid nucleus will divide many times, eventually forming the endosperm of the seed. The **endosperm** is food storage tissue that supports development of the embryo in anthophyte seeds.

Reading Check

3. What is double fertilization?

____

____

____

Section
24.3 The Life Cycle of a Flowering Plant, *continued*

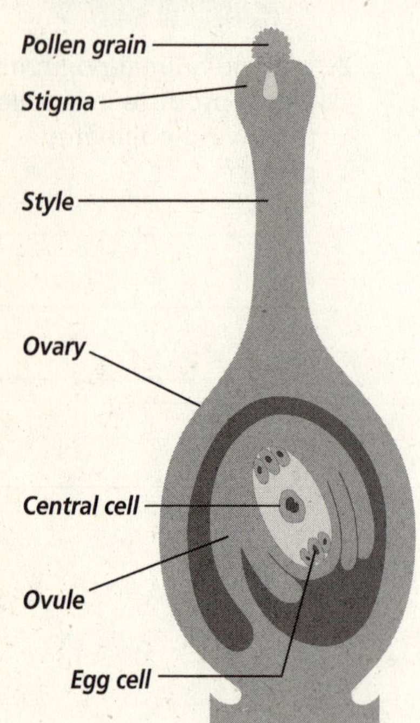

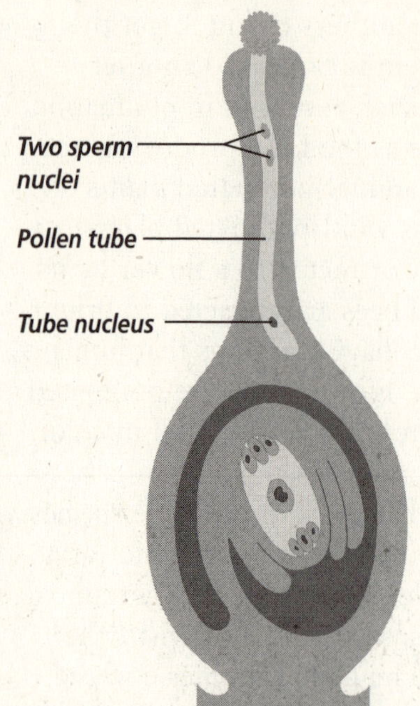

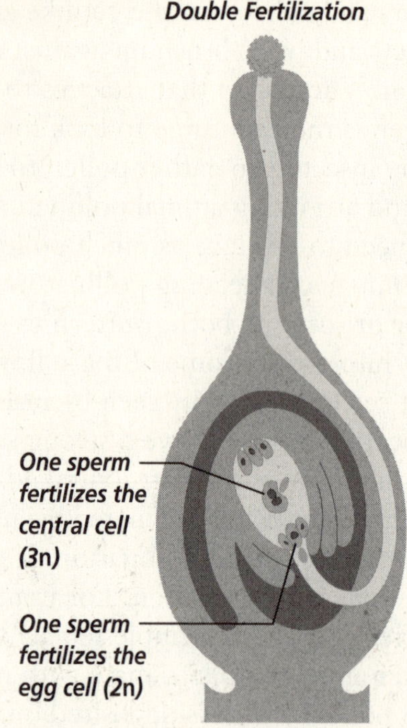

Many flower ovaries contain more than one ovule. Each ovule can become a seed as long as at least one pollen grain lands on the stigma for each ovule in the ovary. Think of the hundreds of seeds in a watermelon. Each seed is the result of a pollen grain pollinating a flower, followed by the process of double fertilization.

Seeds and Fruits

How do seeds form?

Once fertilization takes place, seed development begins. Inside the ovule, the zygote divides and develops into the embryo plant. While this is happening, the triploid central cell develops into the seed's endosperm. The wall of the ovule becomes the seed coat. The seed coat helps with seed dispersal and helps protect the embryo until it begins growing. The embryo inside a seed is the next sporophyte generation. The formation of seeds and the fruits that enclose them help ensure the survival of the next generation. ✓

Reading Check

4. What are the functions of the seed coat?

___

___

___

How does fruit form?

As seeds develop, the surrounding ovary grows larger and becomes the fruit. Sometimes other flower organs become part of the fruit. A fruit is a structure that contains the seeds of an anthophyte.

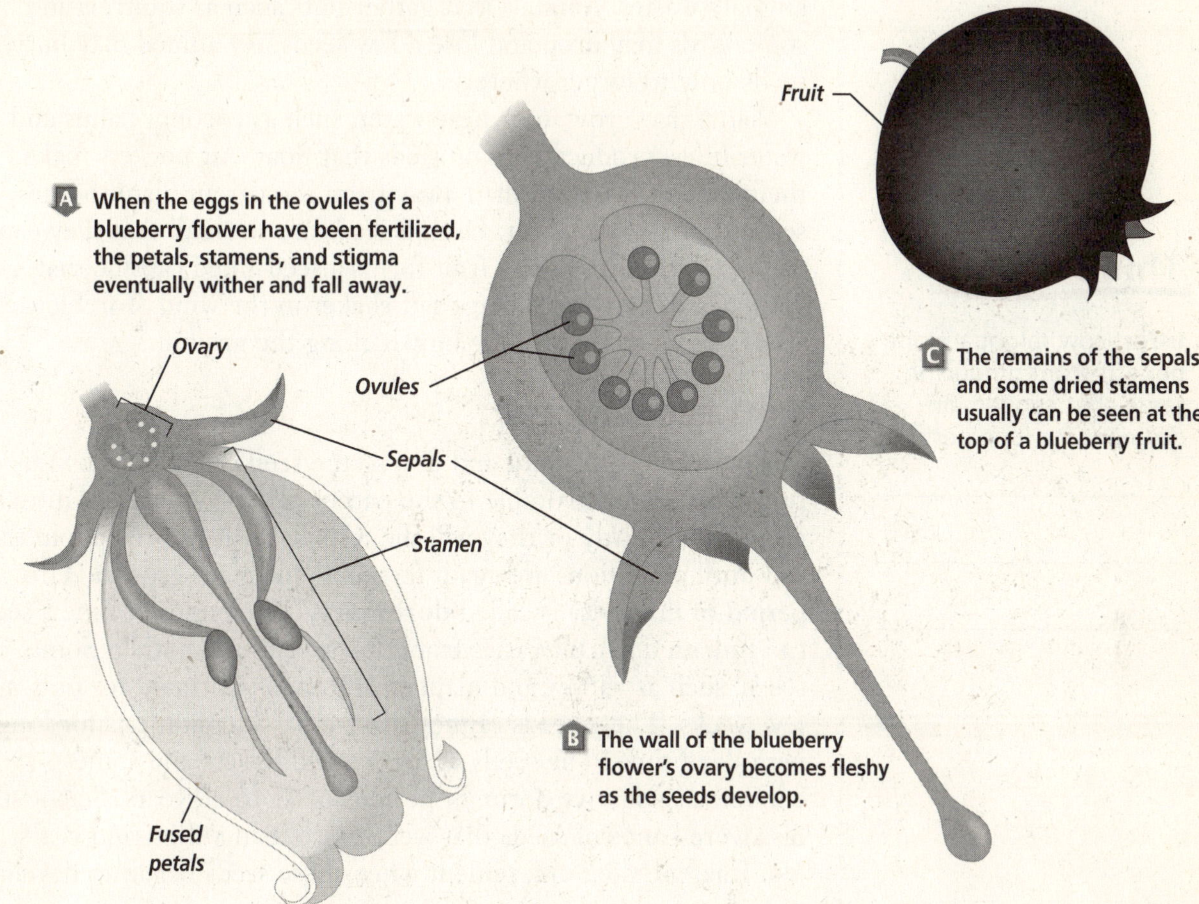

A When the eggs in the ovules of a blueberry flower have been fertilized, the petals, stamens, and stigma eventually wither and fall away.
Ovary
Ovules
Sepals
Stamen
Fused petals
B The wall of the blueberry flower's ovary becomes fleshy as the seeds develop.
Fruit
C The remains of the sepals and some dried stamens usually can be seen at the top of a blueberry fruit.

Each flowering plant species has different flowers and fruit that can be used for identification. Some plants develop fleshy fruits such as apples, grapes, melons, tomatoes, and cucumbers. Other plants develop dry fruits such as peanuts, sunflower "seeds," and walnuts. In dry fruits, the ovary around the seeds hardens as the fruit matures. Some plants that we call vegetables or grains are fruits. Did you know that green peppers are actually fruits?

What are some methods of seed dispersal?

A fruit protects the seeds within it. A fruit can help scatter seeds away from parent plants and into new habitats. This scattering, or dispersal, is important because it reduces competition for sunlight, soil, and water between the parent plant and offspring. Animals carry fruit away from the parent plant. While eating the fruit, they may spit the seeds out. Seeds that are eaten usually pass unharmed through the digestive system, then are deposited in the

Section
24.3 The Life Cycle of a Flowering Plant, *continued*

animal's wastes. Animals that gather nuts such as squirrels and some birds may drop and lose a few seeds. An animal may bury seeds only to forget where.

Plants that grow in or near water, such as coconut palms and water lilies, produce fruit or seeds that float. Air pockets make them able to float and drift away from the parent plant. Some seeds stick to animal fur. Orchid seeds are so small that they can become airborne. Poppy fruit forms a seed-filled capsule that sprinkles its tiny seeds like a salt shaker in the wind. Tumbleweed seeds scatter as the plant is blown along the ground.

5. Infer How might a plant benefit from producing seeds that can remain dormant for long periods?

____

____

____

What is dormancy?

The seed coat is hard and dry, so the seed can survive conditions that are unfavorable to the parent plant. Some seeds must germinate quickly or they will die. Other seeds can remain in the soil until conditions are again favorable for plant growth. This period of inactivity is called **dormancy.** The length of time a seed can remain dormant varies from one species to another. Some seeds, such as willow and maple, can remain dormant for only a few weeks. These seeds cannot survive harsh conditions for long periods of time. The seeds of desert wildflowers and some conifers can survive dormant periods of 15 to 20 years. Scientists discovered ancient seeds that were more than a thousand years old. Imagine their amazement when these seeds began to develop.

What is necessary for germination?

Dormancy ends when the seed is ready to germinate. **Germination** is the beginning of the development of the embryo into a new plant. Dormancy usually ends when water softens the seed coat, oxygen is present, and temperatures are favorable.

Water starts the chemical reaction process in the embryo. The chemical reaction process in living things is called metabolism. Once metabolism has begun, the seed must continue to receive water or it will die.

Many seeds germinate best at temperatures between 25°C and 30°C. Most seeds won't germinate if the temperature is below 0°C or above 45°C.

Some seeds need specific conditions for germination. The seeds of some conifers and wildflowers, such as lupines and gentians, will not germinate unless they have been exposed to fire. Apple seeds need a period of freezing temperatures. Coconut seeds have to soak in salt water. Some seeds germinate better if they pass through the acid environment in an animal's digestive system.

Section 24.3 The Life Cycle of a Flowering Plant, *continued*

Once the seed coat has been softened by water, the embryo starts to come out of the seed. The first part of the embryo to appear is the embryonic root called the **radicle** (RA dih kul). The radicle grows down into the soil and develops into a root. The part closest to the seed is called the **hypocotyl** (HI poh kah tul).

The part of the plant that first appears above the soil is often an arched portion of the hypocotyl. As the hypocotyl grows, it straightens upward. As it straightens, it brings the cotyledons (if it is a dicot), and the plant's first leaves with it. In monocots, the cotyledon remains below the soil surface. As the plant grows the leaves turn green, and produce food through photosynthesis. ✓

Reading Check

6. When does photosynthesis begin in a new plant?

____

____

____

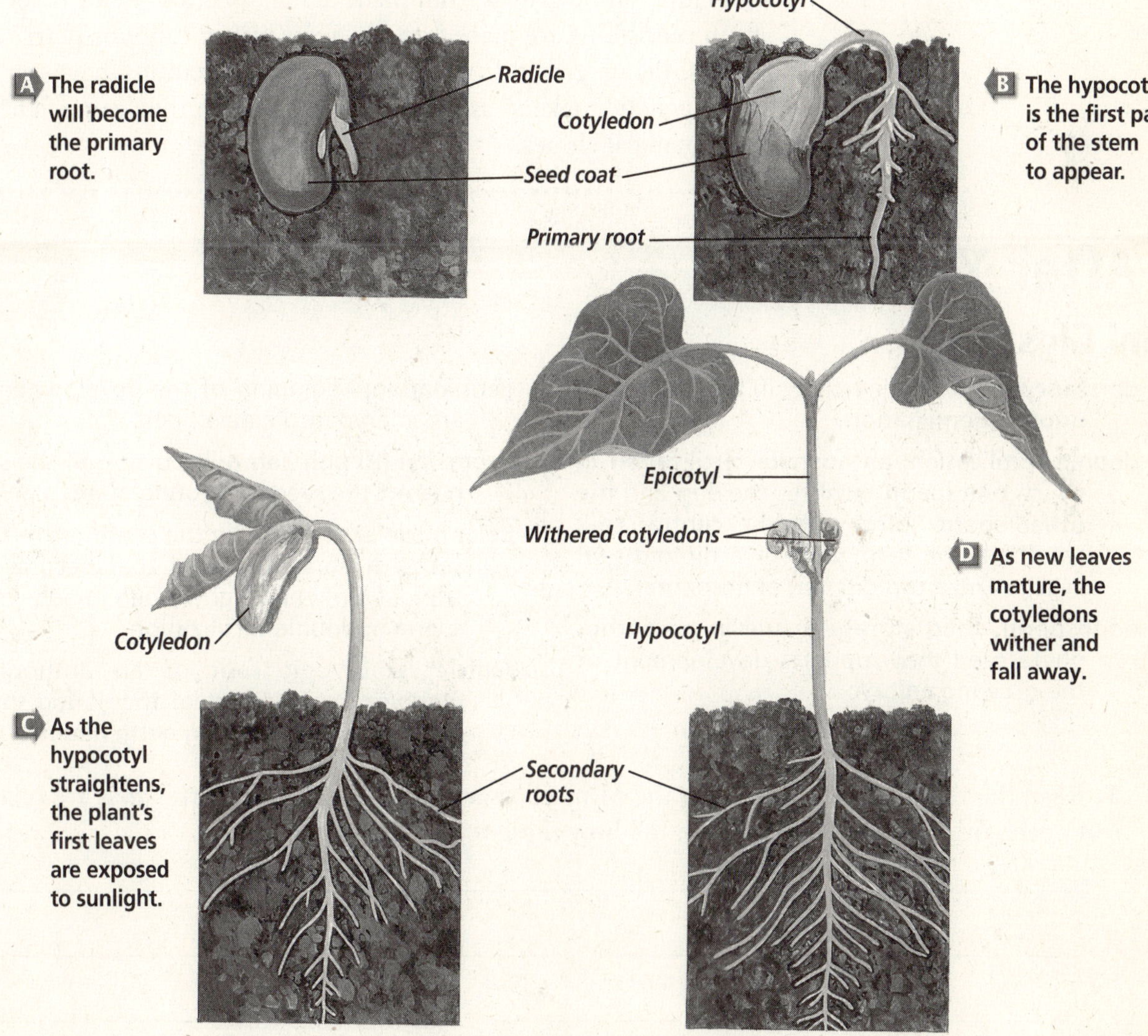
A The radicle will become the primary root.
Radicle
Seed coat
Hypocotyl
Cotyledon
Primary root
B The hypocotyl is the first part of the stem to appear.
Epicotyl
Withered cotyledons
Hypocotyl
D As new leaves mature, the cotyledons wither and fall away.
Cotyledon
C As the hypocotyl straightens, the plant's first leaves are exposed to sunlight.
Secondary roots

Reading Check

7. Producing a new plant from part of another plant is called (Circle your choice.)

a. photoperiodism.
b. vegetative reproduction.
c. germination.

How do plants reproduce without seeds?

The roots, stems, and leaves of plants are called vegetative structures. When these structures are used to produce a new plant, it is called vegetative reproduction. Vegetative reproduction is common among anthophytes. For example, potatoes are actually modified stems called tubers. Each bud or "eye" on a potato can produce a new plant. Farmers cut potato tubers into pieces and then plant them.

Using vegetative reproduction to grow many plants from one plant is called vegetative propagation. Some plants, such as begonias, can be propagated by planting cuttings. Cuttings are pieces of stem or leaf that have been cut off a plant.

Plants can be grown from plant tissue. Pieces of plant tissue from meristems are placed on nutrients in test tubes or petri dishes. Plants grown from cuttings and tissue culture have the same genetic makeup as the plants from which they came. They are botanical clones.

After You Read

Mini Glossary

dormancy: period of inactivity in a mature seed prior to germination

double fertilization: anthophyte fertilization in which one sperm fertilizes the egg and the other sperm joins with the central cell; results in the formation of a diploid ($2n$) zygote and a triploid ($3n$) endosperm

endosperm: food storage tissue in an anthophyte seed that supports development of the growing embryo

germination: beginning of the development of an embryo into a new plant

hypocotyl (HI poh kah tul): portion of the stem nearest the seed in a young plant

polar nuclei: two nuclei in the center of the egg sac of a flowering plant that become the triploid ($3n$) endosperm when joined with a sperm in double fertilization

radicle: embryonic root of an anthophyte embryo; the first part of the young sporophyte to emerge during germination

1. Read the terms and definitions in the Mini Glossary above. Circle two terms that are related to each other. On the lines below, tell how these terms are related.

Section 24.3 The Life Cycle of a Flowering Plant, *continued*

2. Place a check mark to indicate whether the organ or structure is involved in the various activities. You may place a check mark in more than one column.

	Pollination	Fruit Formation	Seed Production	Seed Dispersal
Ovary				
Pollen grain				
Nectar				
Endosperm				
Seed coat				
Fleshy fruits				

Visit the Glencoe Science Web site at **science.glencoe.com** to find your biology book and learn more about the life cycle of a flowering plant.

Section 25.1 Typical Animal Characteristics

Before You Read

Think about all the animals you are familiar with. They may be pets, animals in nature, or captive animals such as in a circus or zoo. This section explains what all animals have in common. On the lines below, list the characteristics you know about that all animals share.

Read to Learn

Mark the Text **Identify Details** As you read this section, highlight the sentences that provide information about sessile organisms.

Think it Over

1. **Analyze** What is one characteristic all animals share? (Circle your choice.)
   a. They are autotrophs.
   b. They spend the same amount of energy to obtain food.
   c. They are heterotrophs.

Characteristics of Animals

All animals have several characteristics in common. Animals are eukaryotic, multicellular organisms. They have ways of moving that help them reproduce, get food, and protect themselves. Most animals have specialized cells. These cells form tissues and organs, such as nerves and muscles.

How do animals obtain food?

One characteristic animals share is that they are heterotrophic. That means that they must consume or eat other organisms to get their energy and nutrients. All animals depend either directly or indirectly on autotrophs for food. Remember autotrophs are organisms that make their own food.

Scientists hypothesize that animals first evolved in water. Although water is denser and contains less oxygen than air, it usually has more food suspended, or floating in it. Some animals that live in water, such as barnacles and oysters, do not move from place to place. They have adaptations that allow them to capture food from their environment. Organisms that permanently attach to a surface are called **sessile** (SE sul). They do not use much energy to obtain their food. Some animals that live in water, such as corals and sponges, move only during the early stages of their lives. They hatch from fertilized eggs into free-swimming larval forms. As adults, most of them are sessile and attach to rocks or other objects.

There is very little suspended food in the air. Because of this, land animals need to use more oxygen and energy to find food.

How do animals digest food?

Animals are heterotrophs that ingest, or take in, their food. After they ingest it, they must digest it. In some animals, digestion takes place within individual cells. In other animals, digestion takes place in an internal cavity. Some of the food that an animal ingests and digests is stored as fat or glycogen, which will be used when other food is not available.

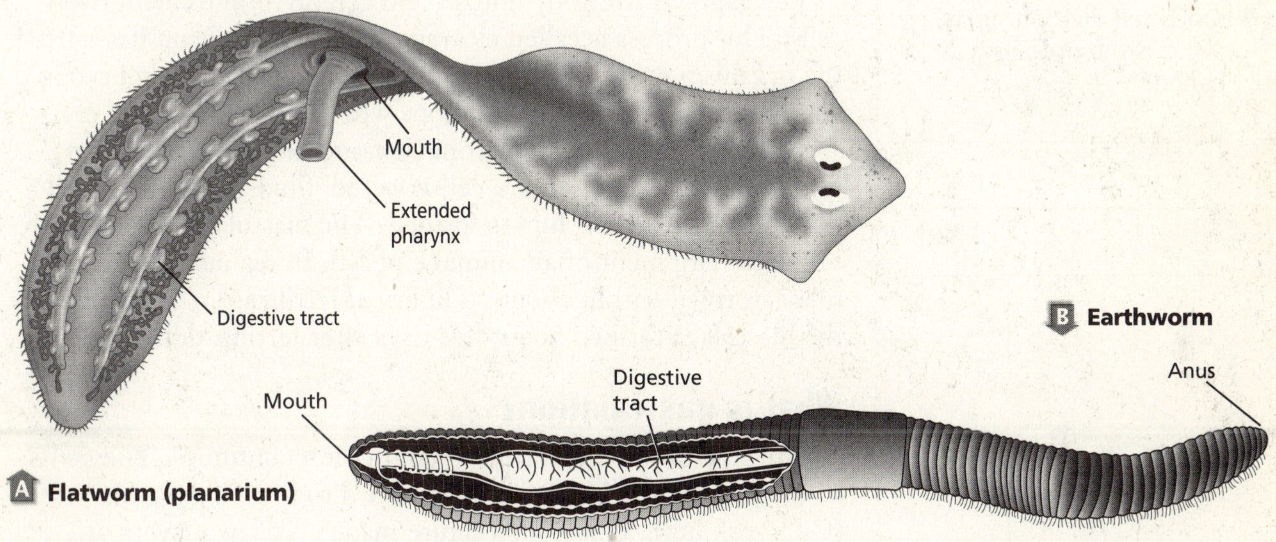

Examine the digestive tracts of a flatworm (planarium) and an earthworm in the illustration above. You will notice that there is only one opening to the flatworm's digestive tract, a pharynx. The earthworm has a digestive tract with two openings. There is a mouth at one end and an anus at the other.

What are the functions of some animal cells?

Most animal cells carry out different functions. Animals have specialized cells that allow them to sense and find food and mates. They also have specialized cells that help them to identify and protect themselves from predators.

Development of Animals

Most animals develop from a fertilized egg cell called a zygote. How does a zygote develop into the many different kinds of cells that make up a snail, a fish, or a human? After fertilization, the zygotes of different animal species all have similar, genetically determined stages of development. ✓

Think it Over

2. Analyze Why do land animals use more energy than aquatic animals to find food?

____

____

✓ Reading Check

3. Most animals develop from a fertilized cell called a (Circle your choice.)

a. sperm cell.
b. haploid cell.
c. zygote.

Reading Check

4. Once cell division starts, what is the organism called?

Reading Check

5. What do ectoderm cells develop into?

What happens during fertilization?

Most animals reproduce sexually. Male animals produce sperm cells and female animals produce egg cells. Fertilization occurs when a sperm cell penetrates an egg cell, forming a new cell called a zygote. Fertilization may occur inside or outside of the body.

How does cell division occur?

The zygote divides by mitosis and cell division to form two cells. This process is called cleavage. Once cell division has started, the organism is called an embryo. Remember that an embryo is an organism at an early stage of growth and development.

The two cells that result from cleavage then divide to form four cells, and so on, until a cell-covered, fluid-filled ball called a **blastula** (BLAS chuh luh) is formed. The blastula is formed early in the development of an animal embryo. In sea urchins, the blastula is formed within about 10 hours of fertilization. In humans, the blastula is formed about five days after fertilization.

What is gastrulation?

After the blastula is formed, cell division continues. The cells on one side of the blastula move inward to form a **gastrula** (GAS truh luh). This is a structure made up of two layers of cells with an opening at one end.

The way the gastrula forms can be compared to the way a potter makes a bowl from a ball of clay. By pushing in on one side of the clay ball, the potter forms a cavity that becomes the interior of the bowl. Similarly, the cells at one end of the blastula move inward, forming a cavity lined with a second layer of cells. The layer of cells on the outer surface of the gastrula is called the **ectoderm.** The layer of cells lining the inner surface of the gastrula is called the **endoderm.** The cells continue to grow and divide. Eventually the ectoderm cells develop into the skin and nervous tissue of the animal. The endoderm cells develop into the lining of the digestive tract and into organs that aid digestion.

How is the mesoderm formed?

In some animals, the development of the gastrula progresses until a layer of cells called the mesoderm forms as shown in the illustration on page 299. Mesoderm is found in the middle of the embryo. The term "meso" means middle.

The **mesoderm** (MEZ uh durm) is the third cell layer found in the developing embryo between the ectoderm and the endoderm. The mesoderm cells develop into the muscles, circulatory system,

Section **25.1** **Typical Animal Characteristics,** *continued*

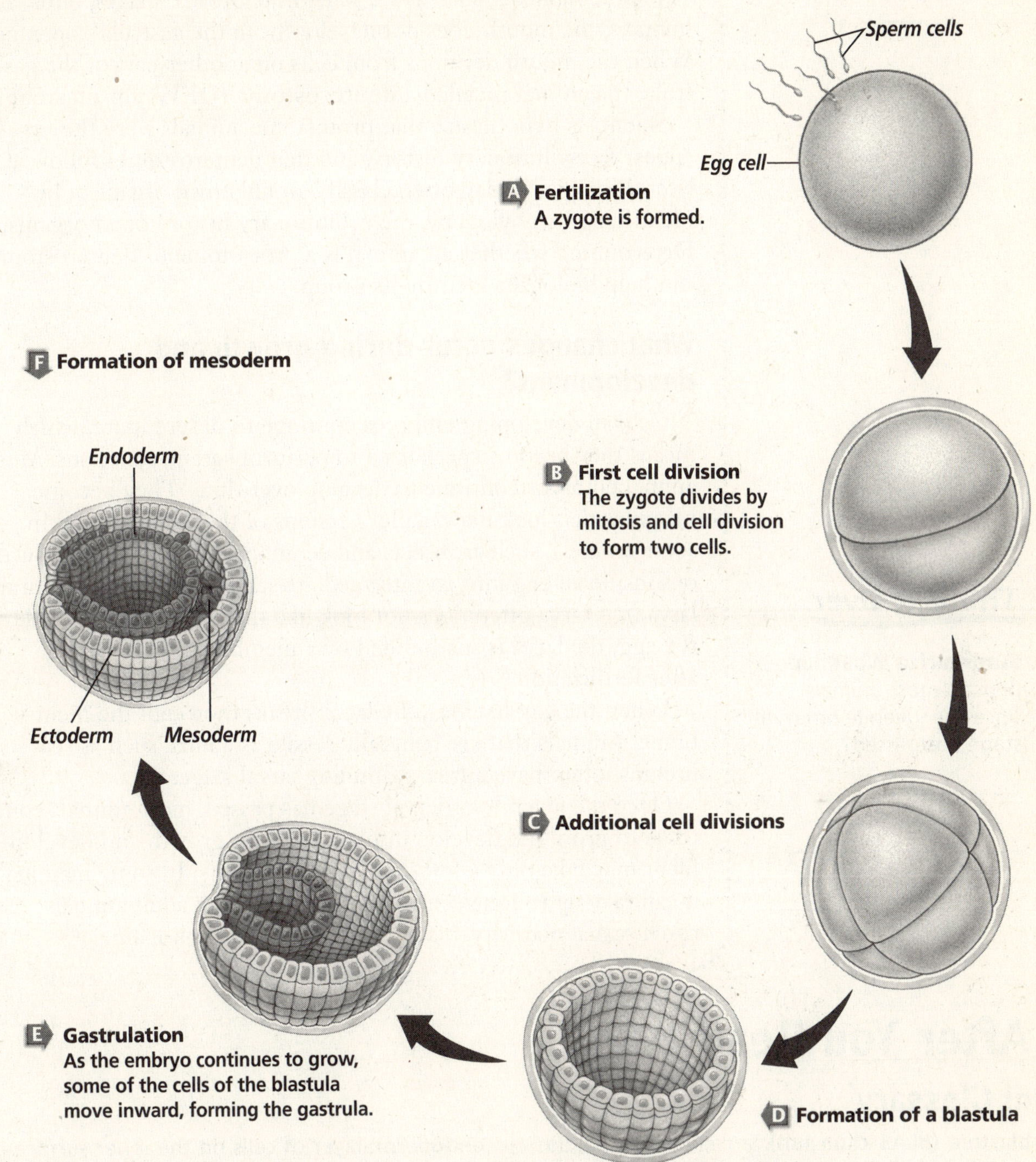
Sperm cells
Egg cell
A Fertilization
A zygote is formed.
B First cell division
The zygote divides by mitosis and cell division to form two cells.
C Additional cell divisions
D Formation of a blastula
E Gastrulation
As the embryo continues to grow, some of the cells of the blastula move inward, forming the gastrula.
F Formation of mesoderm
Endoderm
Ectoderm
Mesoderm

excretory system, and, in some animals, the respiratory system. When the opening in the gastrula develops into the mouth, the animal is called a **protostome** (PROH tuh stohm). Snails, earthworms, and insects are examples of protostomes.

In other animals, such as sea stars, fishes, toads, snakes, birds, and humans, the mouth does not develop from the gastrula's opening. When the mouth develops from cells on another part of the gastrula, the animal is called a **deuterostome** (DEW tihr uh stohm).

Scientists hypothesize that protostome animals were the first to appear in evolutionary history, and that deuterostomes followed later. Biologists today often classify an unknown organism by identifying its phylogeny, the evolutionary history of an organism. Determining whether an animal is a protostome or deuterostome can help biologists identify its group.

What changes occur during growth and development?

Cells in developing embryos continue to differentiate, which means they become specialized to perform certain functions. Most animal embryos continue to develop over time. They become juveniles that look like smaller versions of the adult animal. In some animals, such as insects and echinoderms, the embryo develops inside an egg into an intermediate stage called a larva (plural, larvae). A larva often does not look like the adult animal. Inside the egg, the larva is surrounded by a membrane formed right after fertilization.

When the egg hatches, the larva breaks thorough the membrane. Animals that are generally sessile as adults, such as sea urchins, often have a free-swimming larval stage.

Once the juvenile or larval stage has passed, most animals continue to grow and develop into adults. This growth and development may take only a few days in some insects. In some mammals it can take up to fourteen years. Eventually the adult animals reach sexual maturity, mate, and the cycle begins again.

Think it Over

6. **Summarize** What happens to most animals once the juvenile or larval stage has passed?

____

____

____

After You Read

Mini Glossary

blastula (BLAS chuh luh): a cell-covered, fluid-filled ball formed in the early development of an animal embryo

deuterostome (DEW tihr uh stohm): animal whose mouth develops from cells other than those at the opening of the gastrula

ectoderm: layer of cells on the outer surface of the gastrula; eventually develops into the skin and nervous tissue of an animal

endoderm: layer of cells on the inner surface of the gastrula; will eventually develop into the lining of the animal's digestive tract and organs associated with digestion

gastrula (GAS truh luh): animal embryo development stage where cells on one side of the blastula move inward forming a cavity of two or three layers of cells with an opening at one end

Section
25.1 Typical Animal Characteristics, *continued*

mesoderm (MEZ uh durm): middle cell layer in the gastrula, between the ectoderm and the endoderm; develops into the muscles, circulatory system, excretory system, and in some animals, the respiratory system

protostome (PROH tuh stohm): animal with a mouth that develops from the opening in the gastrula

sessile (SE sul): organism that permanently attaches to a surface

1. Read the terms and definitions in the Mini Glossary. Circle the three terms that refer to the layers of the gastrula. Then write the terms on the lines provided.

____

____

____

2. Write one phrase provided below in each division of the pyramid to show the stages of animal development. Be sure to list the stages in the correct order.

Embryo forms
Blastula forms
Mitosis occurs
Zygote forms
Cell division continues

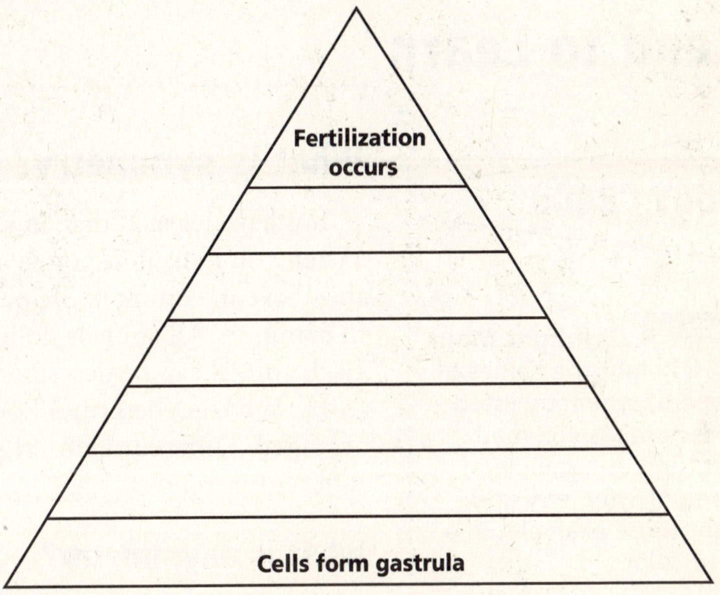

Visit the Glencoe Science Web site at **science.glencoe.com** to find your biology book and learn more about typical animal characteristics.

Section 25.2 Body Plans and Adaptations

Before You Read

In this section, you will learn about the arrangements of body structures in animals. Different animals have different types of body structures. Imagine that you are standing and someone drops a string from the middle of your forehead to the ground. Think about each side of your body. What limbs and facial features appear on both sides of the string? List them below.

Read to Learn

Mark the Text **Highlight Main Ideas** Highlight the names of each type of symmetry listed in this section. As you read the text, find an example of each type of symmetry and highlight those examples in a different color.

1. What is the common characteristic of sessile organisms?

What is symmetry?

You have learned that animals share certain characteristics. When you look at a sponge and a leopard, it's hard to see what they have in common. Shape is not something all animals have in common. All animals do have some kind of shape, however. Each animal can be described in terms of symmetry. **Symmetry** (SIH muh tree) describes how an animal's body structures are arranged. Different kinds of symmetry allow animals to move in different ways.

What is asymmetry?

Many sponges have irregularly shaped bodies. An animal that is irregular in shape has no symmetry. It has an asymmetrical body plan. Animals with no symmetry are often sessile organisms, meaning they do not move from place to place. Most adult sponges are sessile organisms. ✓

The bodies of most sponges consist of two layers of cells. Unlike all other animals, a sponge's embryonic development does not include the formation of an endoderm and mesoderm, or the gastrula stage. Fossil sponges first appeared in rocks dating back more than 700 million years. They represent one of the oldest groups of animals on Earth. This is evidence that their two-layer body plan suits their aquatic environment.

Section
25.2 Body Plans and Adaptations, *continued*

What is radial symmetry?

An animal with **radial** (RAY dee uhl) **symmetry** can be divided along any plane, through a central axis, into almost equal halves. Radial symmetry is an adaptation that enables an animal to detect and capture prey coming toward it from any direction. Hydra are examples of animals with radial symmetry.

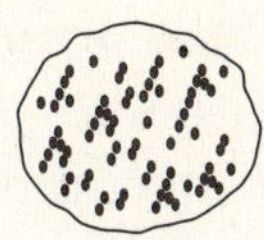

Sponge
Asymmetrical

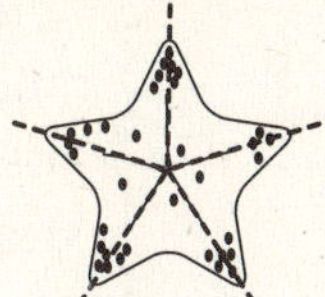

Sea star
Radial symmetry

Human
Bilateral symmetry

What is bilateral symmetry?

An organism with **bilateral** (bi LA tuh rul) **symmetry** can be divided down its length, along one plane, into similar right and left halves. Butterflies have bilateral symmetry. In animals with bilateral symmetry, the **anterior,** or head end, often has sensory organs. The **posterior** of these animals is the tail end. The **dorsal** (DOR sul), or upper surface, also looks different from the **ventral** (VEN trul), or lower surface. In animals that are upright, or nearly so, the back is on the dorsal surface and the belly is on the ventral surface. Animals with bilateral symmetry can find food and mates and avoid predators because they have sensory organs and good muscular control. ✓

✓ Reading Check

2. What type of symmetry does a butterfly have?

____

____

____

Bilateral Symmetry and Body Plans

Animals that are bilaterally symmetrical also share other important characteristics. All bilaterally symmetrical animals developed from three embryonic cell layers—ectoderm, endoderm, and mesoderm. Some bilaterally symmetrical animals also have fluid-filled spaces inside their bodies called body cavities. The internal organs are found inside the body cavities. These cavities make it possible for animals to grow larger because they allow for the efficient circulation and transport of fluids, and support for organs and the organ systems. ✓

✓ Reading Check

3. What are the names of the three embryonic cell layers that are present in bilaterally symmetrical animals?

____

____

____

What are acoelomates?

Animals that develop from these same three cell layers but do not have body cavities are called **acoelomate** (ay SEE lum ate) animals. They have a digestive tract that extends throughout the body. Acoelomate animals may have been the first group of

Reading Check

4. What group of animals may have been the first to develop organs?

animals in which organs evolved. Flatworms are bilaterally symmetrical animals with solid, compact bodies. Like other acoelomate animals, the organs of flatworms are surrounded in the solid tissues of their bodies. A flattened body and branched digestive tract allow for the diffusion of nutrients, water, and oxygen to supply all body cells and to eliminate wastes.

What are pseudocoelomates?

The roundworm is another animal with bilateral symmetry. Unlike the flatworm, the body of a roundworm has a space that develops between the endoderm and mesoderm. It is called a **pseudocoelom** (soo duh SEE lum)—a fluid-filled body cavity partly lined with mesoderm.

Pseudocoelomates have a one-way digestive tract that has regions with specific functions. The mouth takes in food, the breakdown and absorption of food occurs in the middle section, and the anus expels wastes.

What are coelomates?

A **coelom** (SEE lum) is a fluid-filled space that is completely surrounded by mesoderm. The body cavity of an earthworm develops from a coelom. Humans, insects, fishes, and many other animals have a coelomate body plan. In fact, the greatest diversity of animals is found among the coelomates.

In coelomate animals, specialized organs and organ systems develop in the coelom. The digestive tract and other internal organs are attached by double layers of mesoderm and are suspended within the coelom. Like the pseudocoelom, the coelom cushions and protects the internal organs. It provides room for them to grow and move independently within an animal's body.

Animal Protection and Support

Over time, the development of body cavities resulted in a greater diversity of animal species. These different species became adapted to life in different environments. Some animals, such as mollusks, evolved hard shells that protected their soft bodies. Sponges and some other animals evolved hardened spicules, small, needlelike structures, between their cells that provided support.

Section 25.2 Body Plans and Adaptations, *continued*

Some animals developed exoskeletons. An **exoskeleton** is a hard covering on the outside of the body. Exoskeletons provide a framework for support, protect soft body tissues, prevent water loss, and provide protection from predators. An exoskeleton is secreted, or formed, by the epidermis and extends into the body where it provides a place for muscle attachment. As an animal grows, it secretes a new exoskeleton and sheds the old one.

Invertebrate

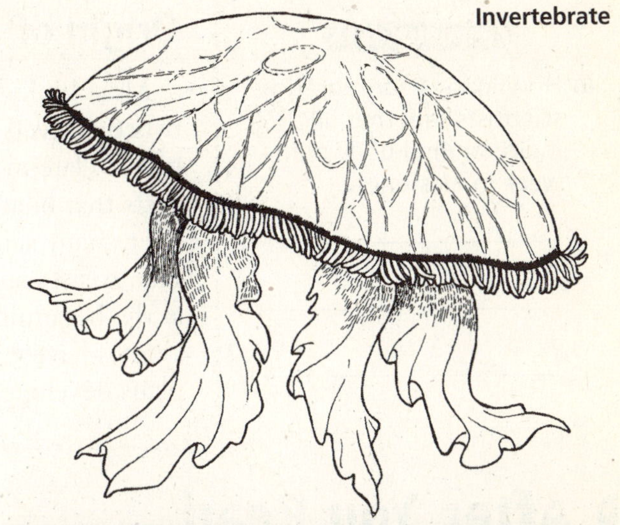

Exoskeletons are often found in invertebrates. **Invertebrates** are animals that do not have a backbone. Crabs, spiders, grasshoppers, dragonflies, and beetles are examples of invertebrates that have exoskeletons.

Other animals have evolved different structures to give them support and protection. Invertebrates, such as sea urchins and sea stars, have an internal skeleton called an **endoskeleton,** which is covered by layers of cells. It provides support for an animal's body. The endoskeleton protects internal organs and provides an internal brace for muscles to pull against. Endoskeletons may be made of one of the following: ✓

Vertebrate

Endoskeleton Substance	Example
Calcium carbonate	Sea stars
Cartilage	Sharks
Bone	Reptiles, birds, amphibians, mammals, bony fishes

A **vertebrate** is an animal with an endoskeleton and a backbone. All vertebrates are bilaterally symmetrical. Examples of vertebrates include fishes, amphibians, reptiles, birds, and mammals.

✓ Reading Check

5. What three substances may the endoskeleton be made of?

____

____

____

Reading Check

6. How many years ago do scientists feel that all major animal body plans were in existence?

Origin of Animals

Most biologists agree that animals probably evolved from protists that lived in groups in water. Scientists trace this evolution back to late in the Precambrian Period. Although evidence suggests that bilaterally symmetrical animals might have appeared later, many scientists agree that all the major animal body plans that exist today were already in existence 543 million years ago—at the beginning of the Cambrian Period. Since then, many new species have evolved but all known species have variations of body plans developed during the Cambrian Period.

After You Read

Mini Glossary

acoelomate (ay SEE lum ate): an animal with no body cavities

anterior: head end of bilateral animals where sensory organs are often located

bilateral (bi LA tuh rul) symmetry: animals with a body plan that can be divided down its length into two similar right and left halves

coelom (SEE lum): fluid-filled body cavity completely surrounded by mesoderm

dorsal (DOR sul): upper surface of bilaterally symmetric animals

endoskeleton: internal skeleton; provides support, protects internal organs, and acts as an internal brace for muscles to pull against

exoskeleton: hard covering on the outside of some animals, including spiders and mollusks; provides a framework for support, protects soft body tissues, and provides a place for muscle attachment

invertebrate: animal that does not have a backbone

posterior: tail end of bilaterally symmetric animals

pseudocoelom (soo duh SEE lum): fluid-filled body cavity partly lined with mesoderm

radial (RAY dee uhl) symmetry: an animal's body plan that can be divided along any plane, through a central axis, into roughly equal halves

symmetry (SIH muh tree): a term that describes the arrangement of an animal's body structures

ventral (VEN trul): lower surface of bilaterally symmetric animals

vertebrate: an animal with an endoskeleton and a backbone

1. Read the terms and definitions in the Mini Glossary above. On the lines below, write a sentence using at least two of the terms from the Mini Glossary.

Section 25.2 Body Plans and Adaptations, *continued*

2. Use the Venn diagram to compare and contrast acoelomate animals with pseudocoelomate animals. List characteristics of acoelomates in the left oval. List characteristics of pseudocoelomates in the right oval. List characteristics that both have where the ovals overlap.

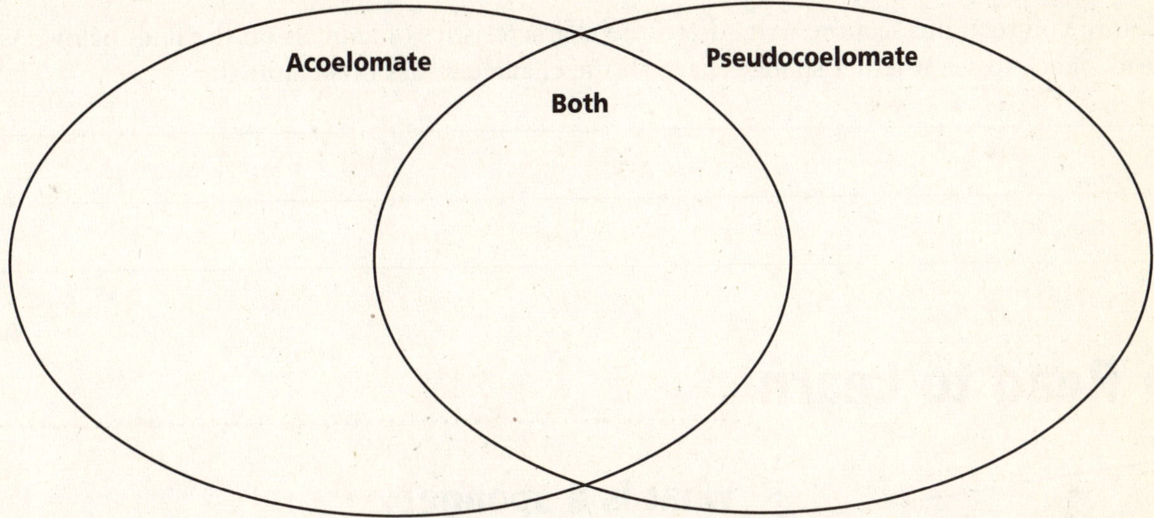

Visit the Glencoe Science Web site at **science.glencoe.com** to find your biology book and learn more about body plans and adaptations.

Name Date Class

Section 26.1 Sponges

Before You Read

Before you read this section, write down the characteristics of animals on the lines below. As you read, check to see whether sponges have all the characteristics of an animal.

Read to Learn

STUDY COACH

Create a Quiz After you have read this section, create a quiz based on what you have learned. After you have written the quiz questions, be sure to answer them.

What is a sponge?

Sponges are asymmetrical animals that live in water. Sponges can be brightly colored, shaped like balls or branches, smaller than a quarter, or larger than a door. Sponges are invertebrates, which means they do not have backbones. There are more than 5000 species of sponges. Most live in marine or saltwater environments, but about 150 species live in freshwater.

Sponges are mainly sessile organisms. Remember that sessile animals permanently attach to a surface. Adult sponges do not move in search of food. They get their food by a process called filter feeding. In **filter feeding,** an organism feeds by filtering small particles of food from water that passes by or through some part of the organism.

How are sponge cells organized?

Like all animals, sponges are multicellular. Their cells are differentiated to perform functions that help the sponges survive. The illustration on page 309 shows some of the different cells and their functions. Unlike most animals, sponges do not have tissues, organs, or organ systems.

For some sponge species, a living sponge can be torn apart and the cells would still be alive but separate from each other. Over a period of several weeks, the cells would come together, reorganize themselves, and form new sponges.

Many biologists hypothesize that sponges evolved from protists. An earlier chapter discussed protists that live together in colonies. The ancestors of sponges might have been colonial protists.

Section 26.1 Sponges, *continued*

A Osculum
Water and wastes are expelled through the osculum, the large opening at the top of the sponge.

B Epithelial-like cells
These cells are thin and flat. They contract in response to touch or to irritating chemical. In so doing, they close pores in the sponge.

C Collar cells
Lining the interior of the sponge are collar cells. Each collar cell has a flagellum that whips back and forth, drawing water into the sponge.

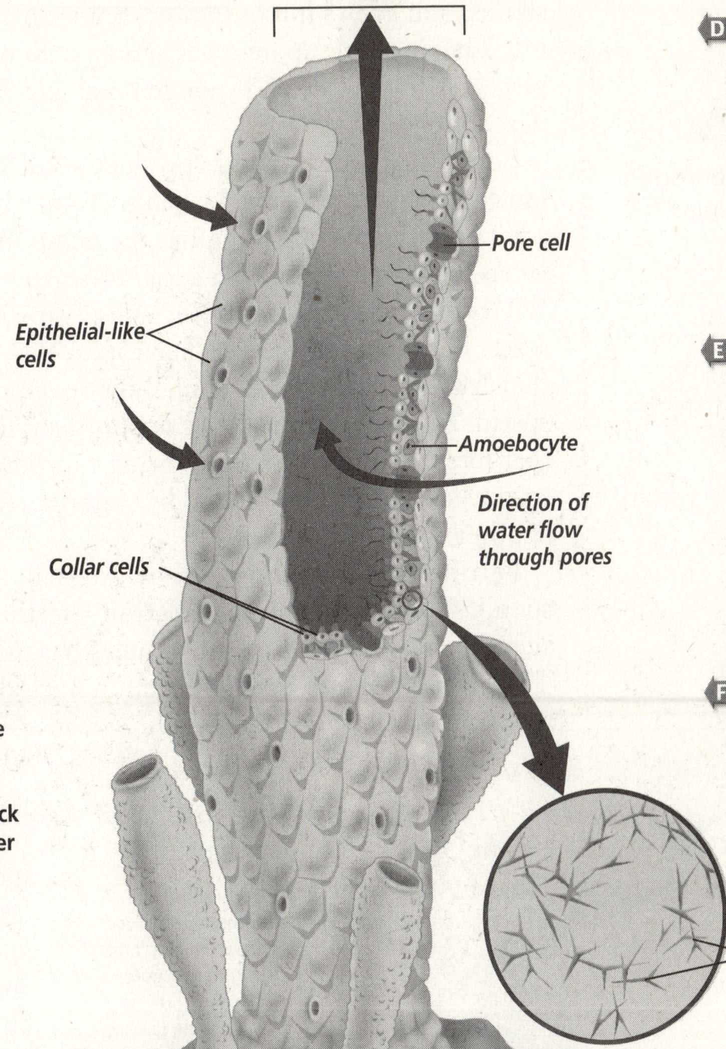

D Pore cell
Surrounding each pore is a pore cell. Pore cells allow water carrying food and oxygen into the sponge's body.

E Amoebocytes
Located between the two cell layers of a sponge, amoebocytes carry nutrients to other cells, aid in reproduction, and produce chemicals that help make up the spicules of sponges.

F Spicules
These small, needlelike structures located between the cell layers of a sponge form the hard support systems of sponges.

Sponges show a major change in the evolution of animals—the change from a unicellular life to a division of labor among groups of organized cells. ✓

How do sponges reproduce?

Sponges reproduce asexually and sexually. Asexual reproduction depends on the species of sponge. Asexual reproduction can be by budding, fragmentation, or the formation of gemmules. A bud is

✓ Reading Check

1. What organisms might be the ancestors of sponges?

____

____

____

an external growth on a sponge. If a bud drops off, it floats away, settles, and grows into a sponge. Sometimes buds do not break off. Then a single sponge becomes a colony of sponges. Often, pieces or fragments of a sponge break off. They can grow into new sponges.

Some freshwater sponges produce seedlike particles, called gemmules. They are produced when the water temperature cools. The adult sponges die in winter, but the gemmules survive. In spring when the water warms, the gemmules grow into new sponges.

Most sponges reproduce sexually. Some sponges are separate sexes, but most sponges are hermaphrodites. A **hermaphrodite** (hur MAF ruh dite) is an animal that can produce both eggs and sperm. In sessile animals, hermaphrodism increases the chances that fertilization will occur. During reproduction, sperm released by one sponge can be carried by water currents to fertilize another sponge. ✓

Reading Check

2. How does hermaphrodism benefit sessile animals?

____

____

____

Fertilization in sponges can be external or internal. A few sponges have **external fertilization**—fertilization that occurs outside the animal's body. Most sponges have **internal fertilization.** This means the eggs inside the animal's body are fertilized by sperm from another sponge. The sperm are carried into the sponge by water. Fertilization occurs and the result is the development of free-swimming larvae. The larvae settle and grow into adult sponges. A sponge is able to move about only during its larval stage.

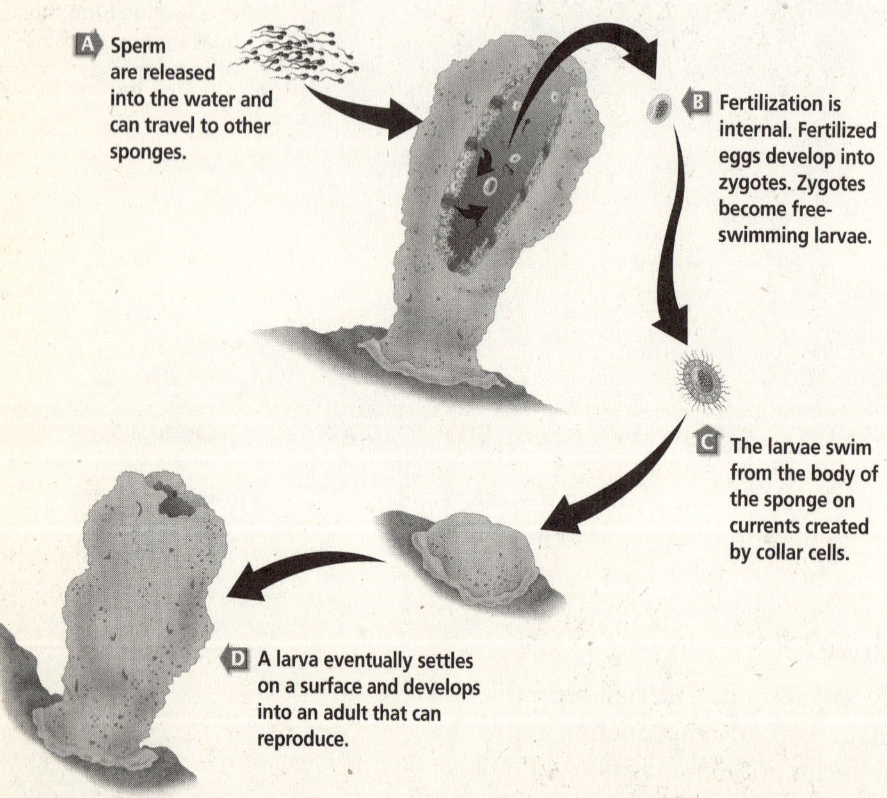

A Sperm are released into the water and can travel to other sponges.
B Fertilization is internal. Fertilized eggs develop into zygotes. Zygotes become free-swimming larvae.
C The larvae swim from the body of the sponge on currents created by collar cells.
D A larva eventually settles on a surface and develops into an adult that can reproduce.

What is the internal structure of a sponge?

Sponges are soft-bodied invertebrates. They have an internal structure that gives them support and can help protect them from predators. Some sponges have hard, sharp spicules located between the cell layers. Spicules may be made of glasslike material or of calcium carbonate. Some species have thousands of tiny, sharp, needlelike spicules that make them difficult for animals to eat. Other sponges have an internal framework made of silica or spongin, a fibrous proteinlike material. Sponges can be classified according to the shape of the spicules and frameworks and what they are made of. ✓

Some sponges contain chemicals that are toxic to fishes and other predators. Scientists are studying sponge toxins to identify those that might be used as medicines.

✓Reading Check

3. How are sponges classified?

After You Read

Mini Glossary

external fertilization: fertilization that occurs outside the animal's body

filter feeding: method in which food particles are filtered from water as it passes by or through some part of the organism

hermaphrodite (hur MAF ruh dite): an animal that can produce both eggs and sperm

internal fertilization: fertilization that occurs inside the animal's body

1. Read the terms and definitions in the Mini Glossary above. On the lines below, explain why filter feeding is important for sessile animals.

2. Write a fact about sponges on the lines under each heading.

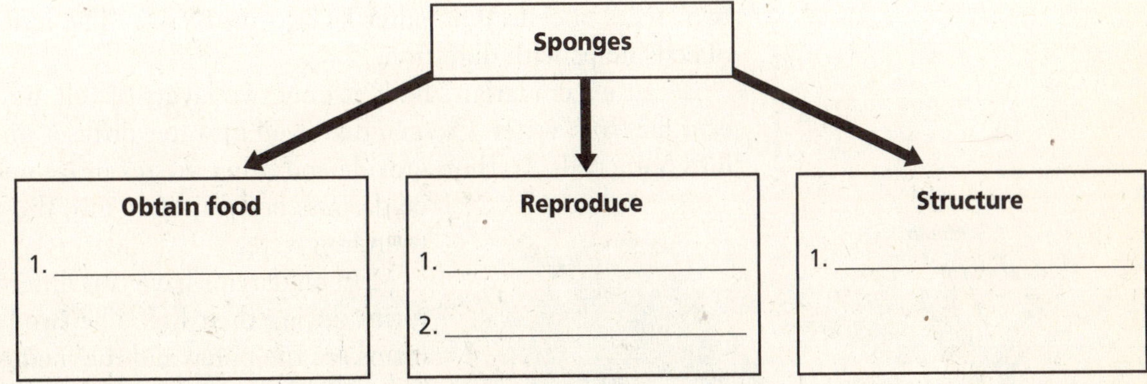

Visit the Glencoe Science Web site at **science.glencoe.com** to find your biology book and learn more about sponges.

Name Date Class

Section 26.2 Cnidarians

Before You Read

Imagine that you are an aquatic animal. Now imagine that for most of your life you had to remain in exactly the same place. What adaptations would you need to survive? List your thoughts on the lines below.

____

____

____

Read to Learn

STUDY COACH

Mark the Text **Identify Details** As you read this section, highlight facts about cnidarians in one color. Highlight facts that apply to polyps in a different color. Highlight facts about medusae in a third color.

What is a cnidarian?

Cnidarians (ni DARE ee uns) are a group of invertebrates made up of more than 9000 species of jellyfish, corals, sea anemones, and hydras. They can be found worldwide and almost all live in marine biomes.

What is the body structure of cnidarians?

All cnidarians have the same basic body structure. Their bodies are radially symmetrical. This means that they can be divided along any plane, through a central axis, into roughly equal halves. Cnidarians have one body opening. Their bodies are made up of two layers of cells. The outer layer is protective. The inner layer of cells helps with digestion.

Because a cnidarian's body is only two layers of cell, no cell is ever far from water. Oxygen dissolved in water diffuses directly into body cells. Carbon dioxide and other wastes move out of a cnidarian's body directly into the surrounding water.

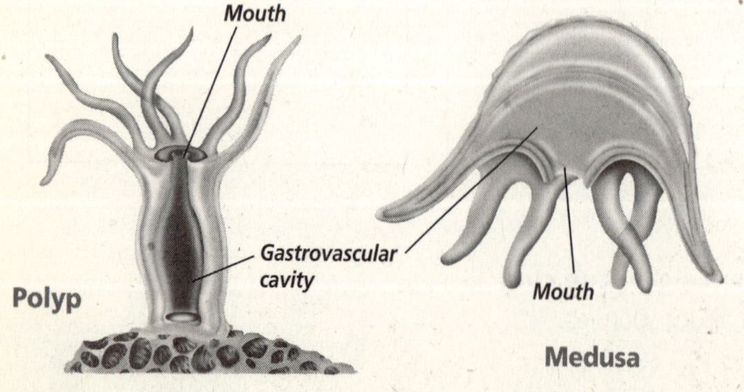

Most cnidarians have two different body forms during their life. The two body forms are the polyp and the medusa. A **polyp** (PAH lup) has a tube-shaped body with a mouth surrounded by tentacles. A **medusa** (mih DEW suh; plural, medusae) has an umbrella-shaped body, called a bell. The tentacles hang down. Its mouth is on the underside of the bell.

Section
26.2 Cnidarians, *continued*

In cnidarians, one body form is usually easier to observe than the other. For example, in jellyfishes, the medusa is the body form you observe. The jellyfish polyp is small and not easily seen. In hydras, the polyp is the form that is easier to see. A hydra's medusa form is small and delicate. Corals and sea anemones have only polyp forms.

How do cnidarians reproduce?

All cnidarians have the ability to reproduce sexually and asexually. Sexual reproduction usually occurs in the medusa stage. If there is no medusa stage, then the polyp can reproduce sexually.

The illustration of the jellyfish below shows how the sexual medusa stage alternates with the asexual polyp stage. Male

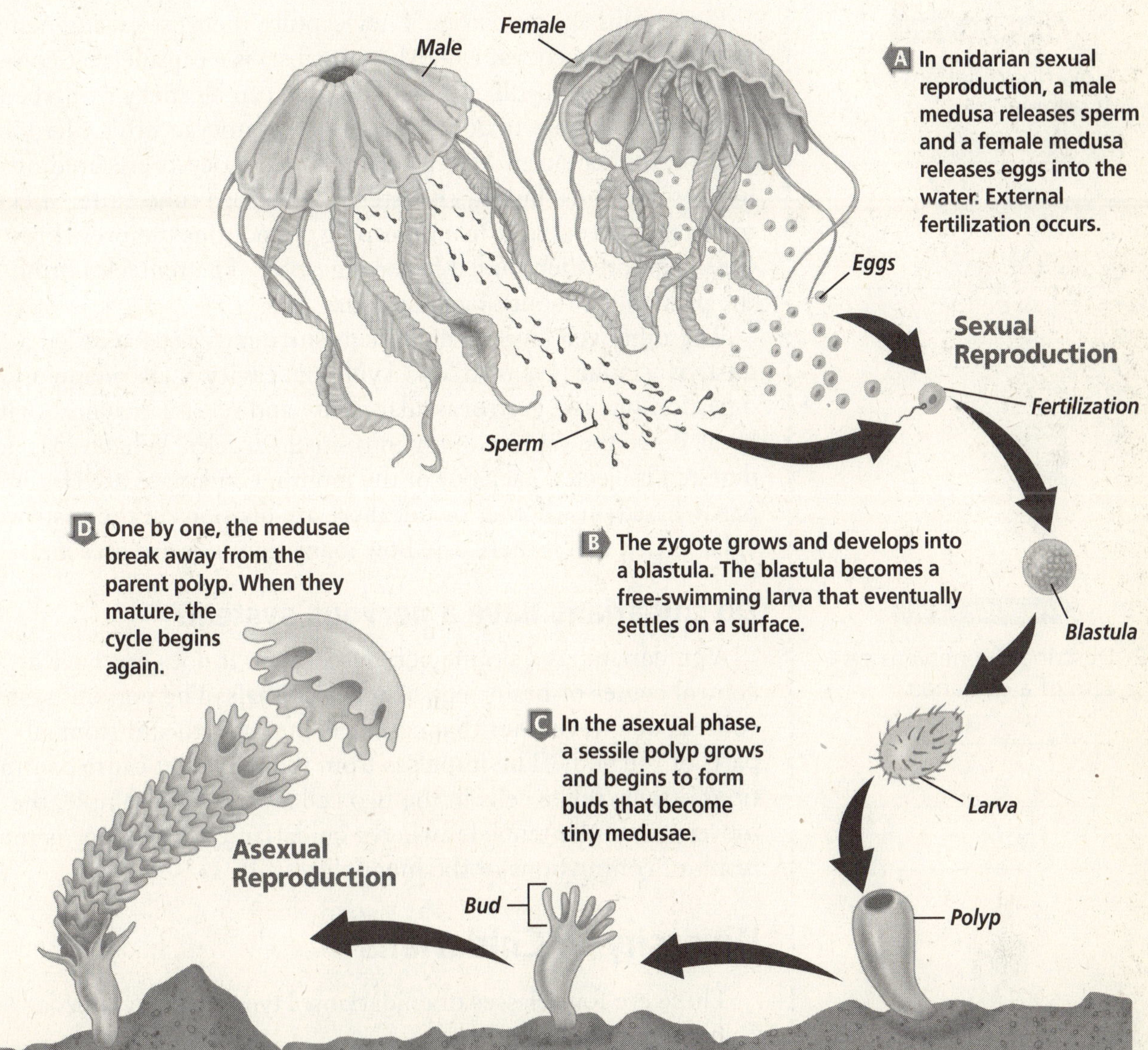

Female
Male
A In cnidarian sexual reproduction, a male medusa releases sperm and a female medusa releases eggs into the water. External fertilization occurs.
Eggs
Sexual Reproduction
Fertilization
Sperm
D One by one, the medusae break away from the parent polyp. When they mature, the cycle begins again.
B The zygote grows and develops into a blastula. The blastula becomes a free-swimming larva that eventually settles on a surface.
Blastula
C In the asexual phase, a sessile polyp grows and begins to form buds that become tiny medusae.
Larva
Asexual Reproduction
Bud
Polyp

medusae release sperm, and female medusae release eggs into the water. External fertilization occurs. Fertilization results in zygotes. Zygotes develop into embryos, and then into larvae. The free-swimming larvae settle and grow into polyps that reproduce asexually to produce new medusae. This may sound similar to alternation of generations in plant life cycles. However, in plants, the generations alternate between haploid and diploid. In cnidarians, medusae and polyps are diploid animals.

Asexual reproduction can occur in either the polyp or the medusa stage. Polyps reproduce asexually by budding. Cnidarians that remain in the polyp stage, such as corals and sea anemones, can reproduce sexually.

How do cnidarians digest food?

Reading Check

1. How do cnidarians capture prey?

______

______

______

Cnidarians are predators. They capture their prey using nematocysts. A **nematocyst** (nih MA tuh sihst) is a capsule that contains a coiled, threadlike tube. The tube can be sticky or barbed. It also may contain toxic substances. Nematocysts are located in cells on the tentacles. When touched, nematocysts are fired off like toy popguns, but much faster. The barbed tube either sticks to the prey, keeping it from escaping, or poisons the prey. Prey organisms are then pulled in for digestion. The tentacles bring the prey to the mouth by contracting. ✓

The inner cell layer of cnidarians surrounds a space called a **gastrovascular** (gas troh VAS kyuh lur) **cavity.** Cells adapted for digestion line the gastrovascular cavity and release enzymes that break down the captured prey into small particles. Whatever is not digested is ejected back out of the mouth. Cnidarians are classified partly based on whether or not there are divisions in the gastrovascular cavity, and if there are, how many divisions are present.

Do cnidarians have a nervous system?

Reading Check

2. Describe the nervous system of a cnidarian.

______

______

______

A cnidarian has a simple nervous system. It does not have a control center or brain such as other animals. The nervous system consists of a **nerve net** that conducts impulses to and from all parts of the body. The impulses from the nerve net cause contractions of musclelike cells in the two cell layers. For example, the movement of the tentacles when a cnidarian captures prey is the result of contractions of the musclelike cells. ✓

Diversity of Cnidarians

There are four classes of cnidarians: Hydrozoa, Scyphozoa, Cubozoa, and Anthozoa.

What are the hydrozoans?

The class Hydrozoa includes two groups—the hydroids, such as the hydras, and the siphonophores, including the Portuguese man-of-war. All hydrozoans have open gastrovascular cavities with no internal divisions. Most hydroids are marine animals. They form branching polyp colonies by budding. They are found attached to pilings, seashells, and other surfaces. The siphonophores include floating colonies that drift about on the ocean's surface.

Consider the Portuguese man-of-war. It looks so much like a single organism that it is hard to see how it is actually a closely associated group of animals. One individual forms a large, gas-filled float. Other polyps hanging from the float have functions, such as feeding and reproduction. The polyps all function together for the survival of the colony.

What are the scyphozoans?

Jellyfishes are beautiful and fragile. The medusa form of the jellyfishes is the dominant stage. Like other cnidarians, scyphozoans have musclelike cells in their outer cell layer that can contract. When these cells contract together, the bell contracts, which moves the animal through the water. The gastrovascular cavity of scyphozoans has four internal divisions. ✓

Jellyfishes can be found in arctic to tropical waters. They have been seen at depths of more than 3000 m. Swimmers should avoid jellyfishes because of their painful stings.

What are the anthozoans?

Anthozoans are cnidarians that exhibit only the polyp form. All anthozoans have many incomplete divisions in their gastrovascular cavities.

Sea anemones are anthozoans that live as individual animals. They are thought to live for centuries. They can be found in tropical, temperate, and arctic seas. Some tropical sea anemones may be more than a meter in diameter.

Corals are anthozoans that live in colonies. They are found in warm ocean waters. Corals form protective shelters around their soft bodies. The shelter is made of calcium carbonate. Colonies of many coral species build beautiful coral reefs that provide food and shelter for many marine species. When a coral polyp dies, its shelter is left behind, adding to the coral reef's structure. The living portion of a coral reef is a thin, fragile layer that grows on top of the shelters left behind by previous generations. Coral reefs

✓ Reading Check

3. What is the dominant form of jellyfishes?

____

____

____

Think it Over

4. Analyze Which anthozoans live in colonies and which live as individuals?

____

____

____

form slowly. It took thousands of years to form the reefs found today. Corals that form reefs are known as hard corals. Other corals, known as soft corals, do not build such structures.

A coral polyp extends its tentacles to feed. Corals thrive in nutrient-poor water because they have a symbiotic relationship with protists. Remember that symbiosis is a relationship in which both organisms benefit. The protists are called zooxanthellae (zoh oh zan THEH lee). Using photosynthesis, the protists produce oxygen and food that the corals use, while using carbon dioxide and waste materials provided by the corals. The protists are primarily responsible for the bright colors found in coral reefs. The zooxanthellae are able to swim, and sometimes leave the corals. When they do, the corals often die. ✓

Reading Check

5. What organism helps corals survive?

______

______

______

Origins of Sponges and Cnidarians

Sponges and cnidarians evolved from a common ancestor early in geologic time. The earliest sponge fossils date from the Precambrian Period, about 650 million years ago. The earliest cnidarian fossils are about 630 million years old. Scientists infer that sponges and cnidarians evolved from protists because, today, flagellated protists resemble collar cells of sponges. The larval form of cnidarians also resembles protists.

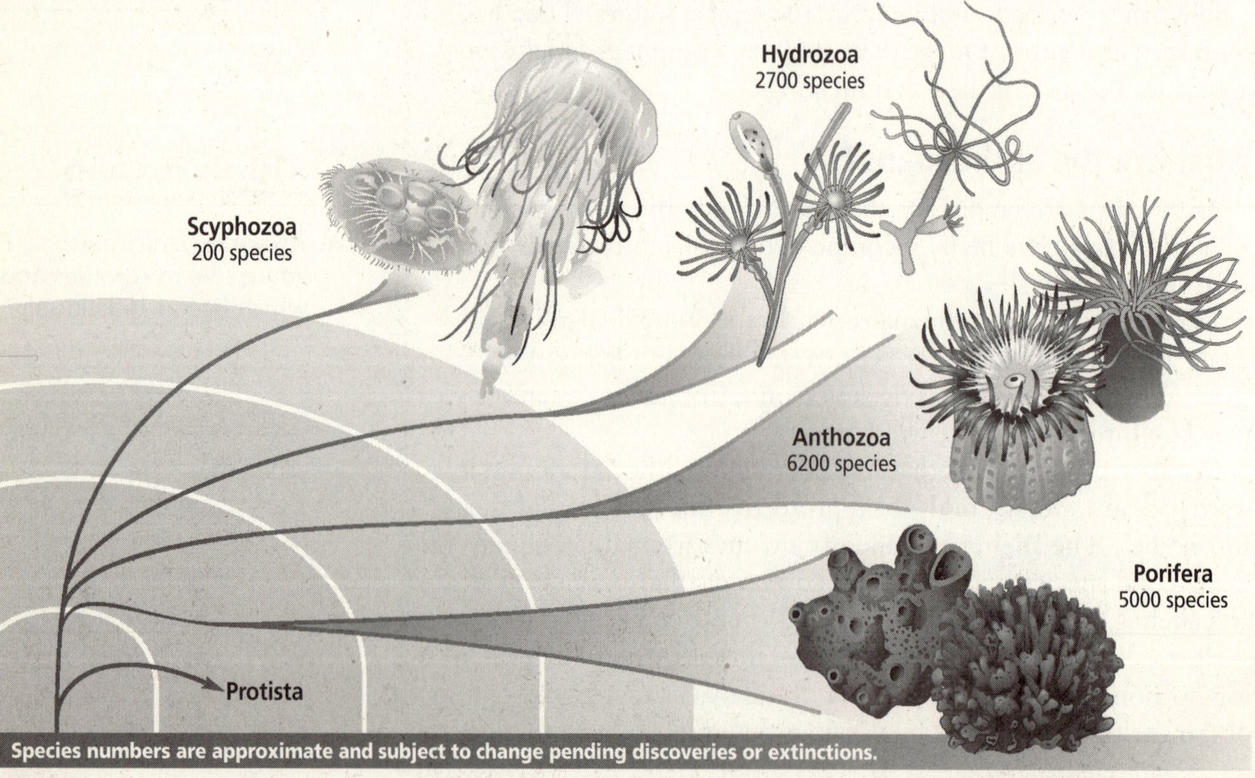

After You Read

Mini Glossary

gastrovascular (gas troh VAS kyuh lur) cavity: in cnidarians, a large cavity in which digestion takes place

medusa (mih DEW suh): a cnidarian body form that is umbrella-shaped with tentacles that hang down

nematocyst (nih MA tuh sihst): in cnidarians, a capsule that contains a coiled, threadlike tube that may be sticky, barbed, or contain poisons; used in capturing prey

nerve net: simple netlike nervous system in cnidarians that conducts nerve impulses from all parts of the cnidarian's body

polyp (PAH lup): a cnidarian body form that is tubelike with a mouth surrounded by tentacles

1. Read the terms and definitions in the Mini Glossary above. Highlight two terms that deal with cnidarian body forms. Then use each term in a sentence that illustrates a fact about cnidarians. Write your sentences on the lines provided.

____

____

____

____

2. Match the concepts in Column 1 to the examples in Column 2.

	Column 1	Column 2
_____	1. All cnidarians display two basic body forms.	a. The bell contracts
_____	2. Cnidarians sting their prey.	b. Nematocysts
_____	3. A cnidarian body is only two layers of cells.	c. Oxygen diffuses directly into cells
_____	4. The nerve net conducts impulses.	d. Medusa and polyp
_____	5. Coral polyps live with zooxanthellae.	e. Symbiosis

Visit the Glencoe Science Web site at **science.glencoe.com** to find your biology book and learn more about cnidarians.

Name Date Class

Section 26.3 Flatworms

Before You Read

From this point on, most of the animals you will learn about are bilaterally symmetrical. An organism with bilateral symmetry can be divided down its length into similar right and left halves. As a review, write on the lines below some of the other physical characteristics of bilateral symmetry.

____

____

____

Read to Learn

Create a Quiz After you have read this section, create a quiz based on what you have learned. After you have written the quiz questions, be sure to answer them.

What is a flatworm?

Flatworms belong to the phylum Platyhelminthes (pla tee HEL min theez). Flatworms are the least complex worms. The name *flatworm* accurately describes their appearance. They range in size from 1 mm up to several meters. There are approximately 14 500 species of flatworms. They are found in marine and freshwater environments as well as moist habitats on land.

Tapeworms and flukes are the most well known of the flatworms. Both are parasites and cause disease in humans and animals. Parasitic flatworms are discussed at the end of this section.

Planarians are free-living flatworms. That means they do not require another organism in order to survive. Planarians are the most commonly studied flatworms.

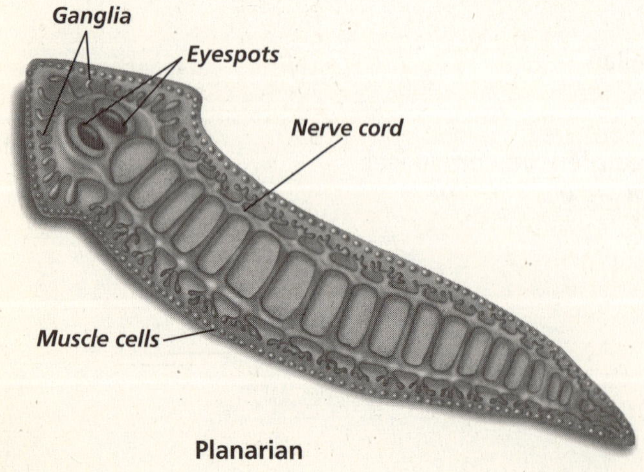

Planarian

What type of nervous system do flatworms have?

A planarian is bilaterally symmetrical. Most of its nervous system is located in the head. The nervous system helps the planarian respond to stimuli in its environment. Some flatworms have a nerve net; others have the beginning of a central nervous system. The illustration at left shows the nervous system of a planarian.

The nervous system includes two nerve cords along each side of the body. At the head, eyespots can detect the presence or absence of light. Sensory cells detect chemicals and movement in water. On each nerve cord, near the head, is a small swelling called a ganglion (plural, ganglia). The ganglion receives messages from the eyespots and sensory cells. The ganglion then communicates with the rest of the body along the nerve cords. Messages from the nerve cords trigger responses in a planarian's muscle cells. ✓

Reading Check

1. What do eyespots do?

How do planarians reproduce?

Planarians are hermaphrodites, meaning they produce both eggs and sperm. During reproduction, individual planarians exchange sperm, which travel along special tubes to reach the eggs. Fertilization happens inside the animals' bodies. The resulting zygotes are released into the water. The zygotes are in capsules and then hatch into tiny planarians.

Planarians also can reproduce asexually. If a planarian is damaged, it has the ability to regenerate, or regrow, new body parts. **Regeneration** is the replacement or regrowth of missing body parts. Missing body parts are replaced through cell division. If a planarian is cut in half horizontally, the section containing the head will grow a new tail, and the tail section will grow a new head. Since a planarian that is cut into two pieces may grow into two new organisms, scientists consider this a form of asexual reproduction.

What do planarians eat?

A planarian feeds on dead or slow-moving organisms. It extends a tubelike, muscular organ out of its mouth as shown in the illustration on page 320. The organ is called a **pharynx** (FAHR inx). Food particles are sucked into the digestive tract, where they are broken up. Cells lining the digestive tract obtain food by endocytosis. Remember that endocytosis is a process in which a cell surrounds materials with a portion of the cell's plasma membrane and then releases the contents inside the cell. Food is then digested in individual cells.

Think it Over

2. **Infer** Why do parasites not require complex nervous or muscular tissue?

How do parasitic flatworms obtain nutrients?

Parasitic flatworms live inside the bodies of their hosts. A parasite is an organism that lives on or in another organism. Parasites depend upon the host organism for nutrients. Parasitic flatworms have mouthparts with hooks that keep them firmly attached inside their hosts. Parasitic flatworms such as tapeworms are surrounded by nutrients. They do not need to move to find their food. Parasitic flatworms do not have complex nervous or muscular tissue.

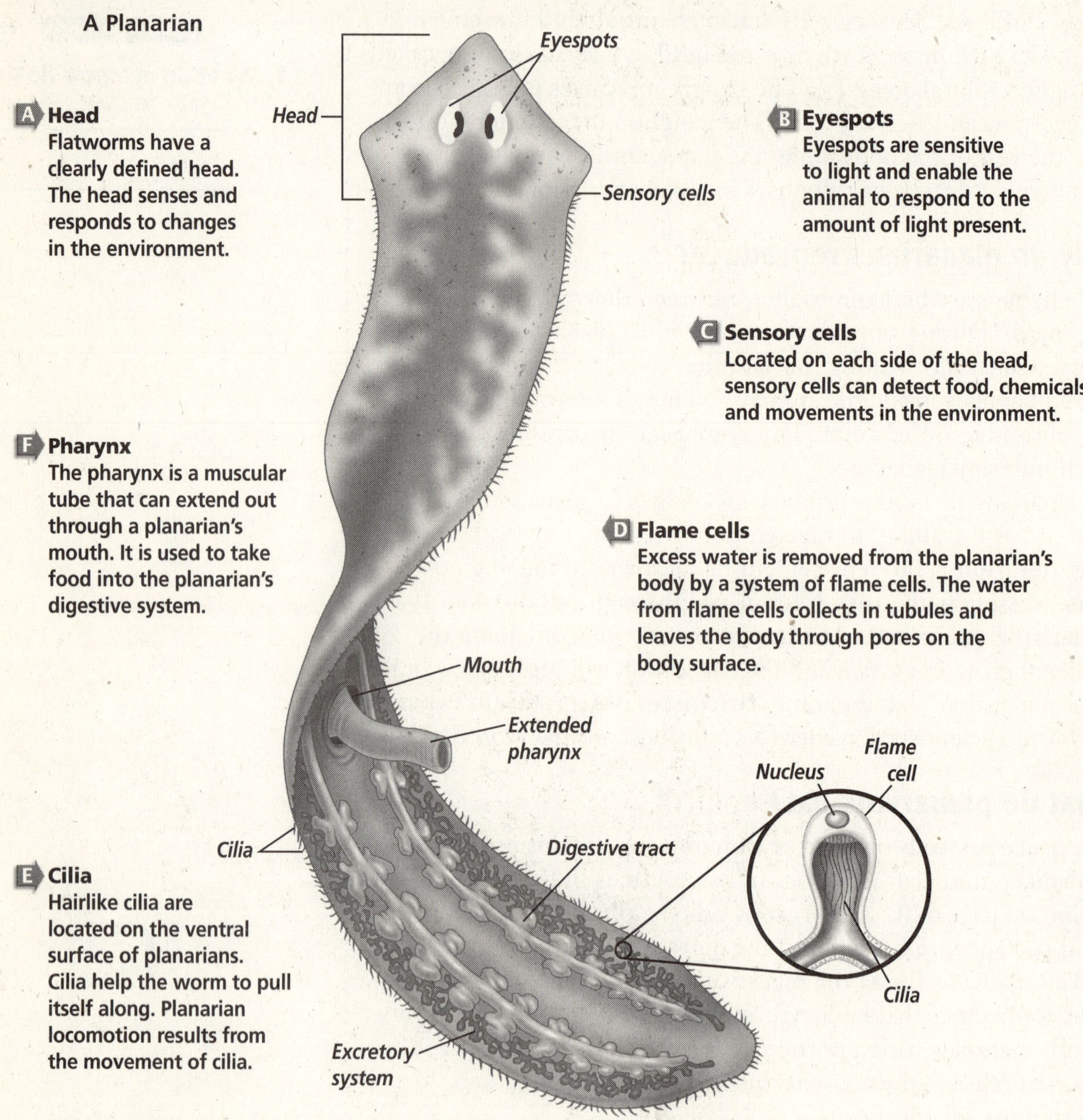

A Planarian
Eyespots
Head
Sensory cells
Mouth
Extended pharynx
Flame cell
Nucleus
Cilia
Digestive tract
Cilia
Excretory system
A Head
Flatworms have a clearly defined head. The head senses and responds to changes in the environment.
B Eyespots
Eyespots are sensitive to light and enable the animal to respond to the amount of light present.
C Sensory cells
Located on each side of the head, sensory cells can detect food, chemicals, and movements in the environment.
D Flame cells
Excess water is removed from the planarian's body by a system of flame cells. The water from flame cells collects in tubules and leaves the body through pores on the body surface.
E Cilia
Hairlike cilia are located on the ventral surface of planarians. Cilia help the worm to pull itself along. Planarian locomotion results from the movement of cilia.
F Pharynx
The pharynx is a muscular tube that can extend out through a planarian's mouth. It is used to take food into the planarian's digestive system.

The knob-shaped head of a tapeworm is called a **scolex** (SKOH leks). The tapeworm's body is made of detachable, individual sections called proglottids. A **proglottid** (proh GLAH tihd) contains muscles, nerves, flame cells, and male and female reproductive organs. Some adult tapeworms that live in animal intestines can be more than 10 m in length and consist of 2000 proglottids.

What are flukes?

A fluke is a parasitic flatworm that spends part of its life in the internal organs of a vertebrate, such as a human. The fluke feeds on cells, blood, and other fluids of the host organism. Flukes have complex life cycles. They can require more than one host. The host can be a vertebrate or invertebrate. For example, blood flukes of the genus *Schistosoma* cause a disease in humans known as schistosomiasis. The disease is common where rice is grown. Farmers plant and harvest rice in standing water. Refer to the illustration below to understand the life cycle. Blood flukes are common where the secondary host, snails, also are found.

Reading Check

3. Flukes only use one host in their life cycle. (Circle your choice.)
   a. True
   b. False

The Life Cycle of the Fluke

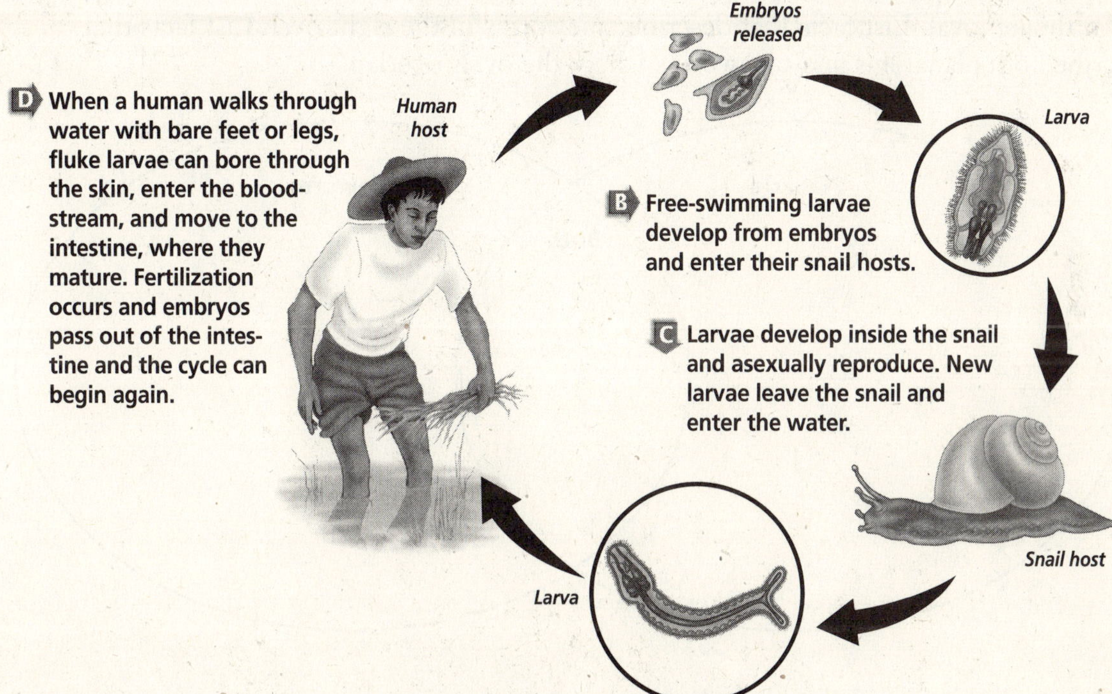

A Adult flukes are about 1 cm long and live in the veins of the human digestive tract. Fluke embryos that are encased in a protective capsule pass out of the body with human wastes. If they reach freshwater, they hatch.
Embryos released
Larva
B Free-swimming larvae develop from embryos and enter their snail hosts.
C Larvae develop inside the snail and asexually reproduce. New larvae leave the snail and enter the water.
Snail host
Larva
Human host
D When a human walks through water with bare feet or legs, fluke larvae can bore through the skin, enter the bloodstream, and move to the intestine, where they mature. Fertilization occurs and embryos pass out of the intestine and the cycle can begin again.

After You Read

Mini Glossary

pharynx (FAHR inx): in planarians, the tubelike, muscular organ that extends from the mouth; aids in feeding and digestion

proglottid (proh GLAH tihd): a section of a tapeworm that contains muscles, nerves, flame cells, and reproductive organs

regeneration: replacement or regrowth of missing body parts

scolex (SKOH leks): knob-shaped head of a tapeworm

1. Read the terms and definitions in the Mini Glossary above. In your own words explain why regeneration is considered asexual reproduction. Write your explanation on the lines provided.

2. Use the Venn diagram to compare planarians to tapeworms. List facts that describe planarians in the left oval. List facts that describe tapeworms in the right oval. List facts that describe both planarians and tapeworms where the ovals overlap.

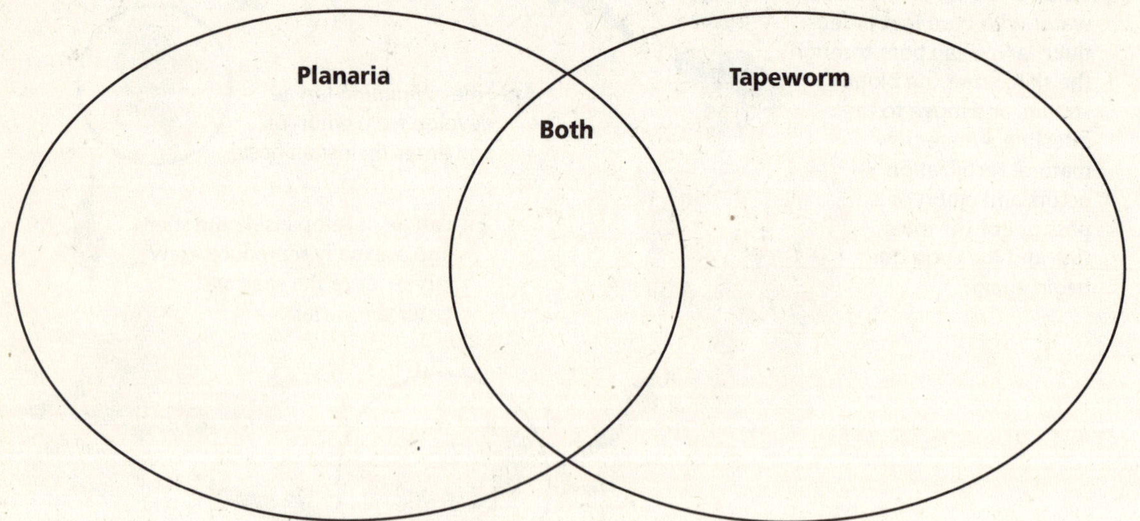

Visit the Glencoe Science Web site at **science.glencoe.com** to find your biology book and learn more about flatworms.

Section 26.4 Roundworms

Before You Read

Health care professionals often say that one of the best ways to prevent illness is to frequently wash your hands. Think of reasons why washing your hands helps prevent illness. Write those reasons on the lines below. After you read this section, list one disease from this section that can be prevented by frequent hand washing. You will have to infer the information from your reading.

Read to Learn

What is a roundworm?

Roundworms belong to the phylum Nematoda. They live in soil, inside animals, in freshwater, and in marine environments. There are more than 12 000 species of roundworms. Some roundworm species are free-living, but many are parasitic. Nearly all plant and animal species are affected by parasitic roundworms.

Roundworms are tapered at both ends. They have a thick outer covering that protects them in harsh environments. They shed the outer covering four times as they grow. Roundworms look like tiny, wriggling bits of thread. They do not have circular muscles but they do have lengthwise muscles. As one muscle contracts, another muscle relaxes. Roundworms move in a thrashing fashion due to the alternating contraction and relaxation of the muscles.

Make Flash Cards For each page, think of two questions a teacher might ask on a test. Write the question on one side of the flash card. Then write the answer on the other side. Quiz yourself until you know the answers.

Roundworms have a pseudocoelom, a fluid-filled body cavity partly lined with mesoderm. They are the simplest animals with a tubelike digestive system. Unlike flatworms, roundworms have two body openings—a mouth and an anus. The free-living species have well-developed sense organs, such as eyespots. These are less-developed in the parasitic forms.

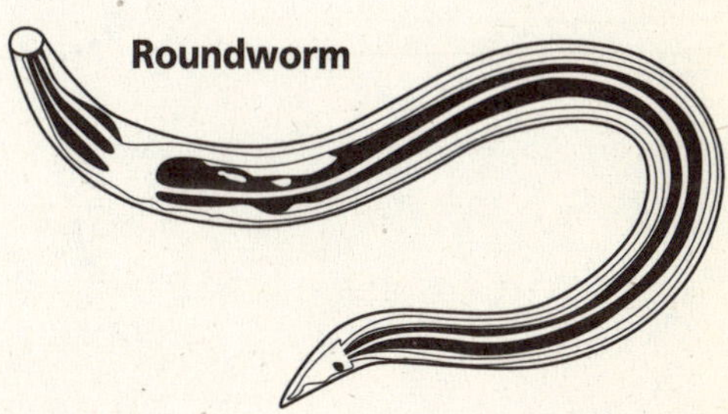

Roundworm

Think it Over

1. **Infer** Why are children more likely than adults to be infected by certain types of roundworms?

Diversity of Roundworms

About half of roundworm species are parasites. There are about 50 species that infect humans. Worldwide, the most common roundworm infection is from *Ascaris* (ASS kuh ris). It is more common in tropical and subtropical areas. Children become infected more often than adults. Eggs of *Ascaris* are found in soil. They enter the human body through the mouth. The eggs hatch in the intestines. *Ascaris* moves into the bloodstream and then into the lungs. They are coughed up, swallowed, and the cycle begins again.

Pinworms are the most common roundworm parasite in the United States. Children are infected more than adults. Pinworms are highly contagious because eggs can survive on surfaces up to two weeks. The life cycle begins when live eggs are ingested. They mature inside the intestinal tract of the host. Female pinworms exit through the host's anus—usually as the host sleeps—and lay eggs on nearby skin. These eggs fall onto bedding or nearby surfaces.

Trichinella causes a disease called **trichinosis** (trih keh NOH sis). The roundworm enters through the mouth if an individual eats infected pork or wild game that is raw or undercooked.

Hookworm infections are common in humans in warm climates where they walk on contaminated soil in bare feet. Hookworms cause people to feel weak and tired due to blood loss.

Reading Check

2. Other than humans, what organisms can roundworms infect?

Can roundworms infect other organisms?

There are about 1200 species of nematodes (roundworm parasites) that cause disease in plants. They are particularly attracted to plant roots, causing a slow decline of the plant. Nematodes can infect fungi and form symbiotic associations with bacteria. Nematodes have been used to control pests. Instead of using chemical pesticides, nematodes can be introduced to kill weevils that damage plants.

After You Read

Mini Glossary

trichinosis (trih ken NOH sis): a disease caused by the roundworm *Trichinella* that can be ingested in raw or undercooked pork, pork products, or wild game

1. Read the term and definition in the Mini Glossary above. Use the term in a sentence that explains how a person might contract this disease. Write your sentence on the lines below.

____

____

____

2. Use information from this section to fill in the table below.

Name of animal	Examples	One or two openings in body?	Do they live on land or in water?	Do they have sensory organs?	Are any species parasitic?
Roundworms	*Ascaris*				

Visit the Glencoe Science Web site at **science.glencoe.com** to find your biology book and learn more about roundworms.

Section 27.1 Mollusks

Before You Read

Snails and earthworms live in different places and have different habits. However, both move efficiently, if somewhat slowly. What questions do you have about the movement and body systems of these two animals? Write your questions on the lines below. After you have read this section, answer the questions that you have written.

____

____

____

Read to Learn

Mark the Text **Compare and Contrast** As you read this section, use one color to highlight the ways in which all mollusks are the same. Use another color to highlight ways in which mollusks differ from each other.

Reading Check

1. What five characteristics do mollusks have in common?

____

____

____

____

____

What is a Mollusk?

Snails, slugs, squid, octopuses, and some other animals that live in shells either in the ocean or on the beach are mollusks. Members of the phylum Mollusca, or mollusks, include both slow-moving slugs and fast, jet-propelled squid. Some mollusks, like snails and slugs, live in damp areas on land. They can often be found moving over leaves on a forest floor. Some mollusks, including oysters, live most of their lives attached to the ocean floor. Oysters also may attach themselves to the underwater parts of docks or boats. Other mollusks, including the octopus, swim easily in the ocean.

Some mollusks have a shell. Other mollusks, including slugs and squids, do not have a hard outer covering. All mollusks have bilateral symmetry, a coelom, a digestive tract with two openings, a muscular foot, and a mantle. A coelom is a fluid-filled body cavity that is completely surrounded by the mesoderm. The **mantle** (MAN tuhl) is the membrane that surrounds the internal organs of the mollusk. In mollusks that have shells, the mantle secretes the shell. Some mollusks, like snails, have adapted to life on land.

Although mollusks look different from one another on the outside, they are similar inside. Refer to the figure at the right. Compare the similarities and differences in the structures of a snail and a squid. Notice that both the snail and the squid have a mantle, shell, head, and gut. The foot area of the squid has been modified into tentacles and arms.

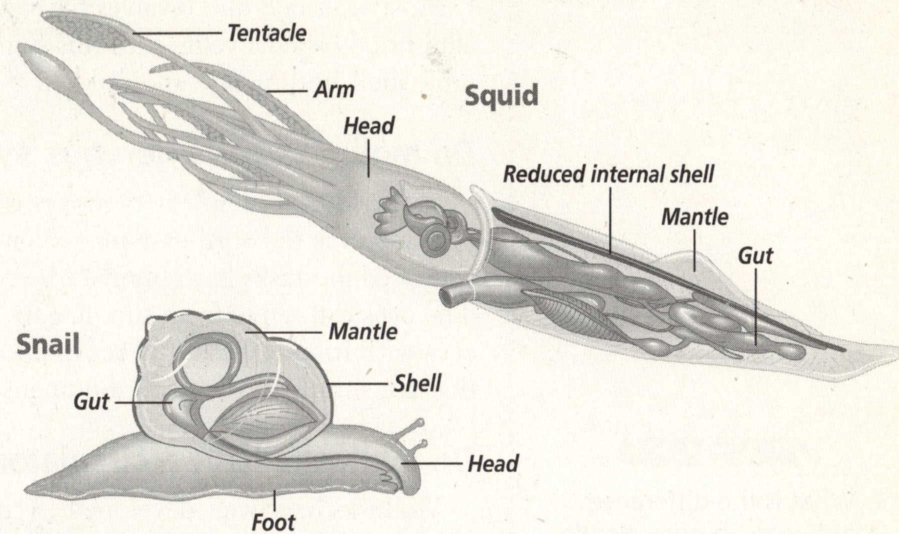

How do mollusks obtain food?

Snails and many other mollusks use a **radula** (RA juh luh) to obtain food. A radula is a structure located in the mouth of mollusks. Similar to a rough file, the radula is a tonguelike organ with rows of teeth. Mollusks use their radulas to drill, scrape, grate, and even to cut food. Octopuses and squids capture food with their tentacles. They use their radulas to tear up the food they have caught. Some mollusks are grazers. Others, including bivalves, are filter feeders. They do not have radulas. Instead, they filter their food from the water.

How do mollusks reproduce?

Mollusks reproduce sexually and most mollusks have separate sexes. For most mollusks that live in water, eggs and sperm are released at the same time. External fertilization then takes place. Many mollusks that live on land are hermaphrodites. Hermaphrodites are plants or animals that have both female and male reproductive organs. Many gastropods, the largest class of mollusks, produce both eggs and sperm, and fertilization takes place within the animal. Some bivalves also are hermaphrodites, producing both sperm and eggs.

Mollusks have different appearances as adults but they develop in similar ways. In one larval stage, most mollusks resemble spinning tops with tufts of cilia. Cilia are tiny hairlike structures that beat in order to produce movement. Most of these larvae will swim freely in the water until they settle down on the ocean floor. They will spend their adult lives on the ocean floor.

Think it Over

2. **Infer** Which of the following best explains why bivalves are filter feeders? (Circle your choice.)
   a. They do not have a mantle.
   b. They do not have a radula.
   c. They swim freely in the ocean.

Most sea snails and bivalves have another stage before reaching adulthood called a veliger. In this stage, the beginnings of the foot, shell, and mantle are visible.

Do mollusks have nervous systems?

Mollusks have simple nervous systems. The function of the nervous system is to coordinate movement and behavior. The more advanced mollusks have brains. Most mollusks have paired eyes. The eyes can range from simple cups that detect light to complex eyes with irises, pupils, and retinas. Octopuses have complex eyes that are similar to the eyes of humans.

Reading Check

3. What is the difference between an open circulatory system and a closed circulatory system?

____

____

____

Do mollusks have a circulatory system?

Mollusks have well-developed circulatory systems that include either a two- or three-chambered heart. In most mollusks, the heart pumps blood through an open circulatory system. In an **open circulatory system,** blood moves through vessels and into open spaces around the body organs. In an open circulatory system, body organs are directly exposed to blood that contains nutrients and oxygen. The blood removes metabolic waste from the organs. Other mollusks, like octopuses, have closed circulatory systems. In a **closed circulatory system,** blood moves through the body, but the blood is entirely enclosed in the blood vessels. The blood moves nutrients and oxygen through the closed blood vessels. A closed system provides an efficient means for gas exchanges within the animal. ✓

Reading Check

4. Why do mollusks have gills?

____

____

____

Respiration and excretion in mollusks

Most mollusks have respiratory structures called gills. Gills are specialized parts of the mantle. They are a system of tiny strands that contain a rich supply of blood for transporting gases. Gills increase the surface area where carbon dioxide and oxygen are exchanged. In snails and slugs that live on land, the mantle cavity appears to have become a primitive lung. ✓

Reading Check

5. What is the purpose of the nephridia?

____

____

____

The excretory structures are called nephridia. **Nephridia** (nih FRIH dee uh) are organs that remove metabolic wastes from the animal's body. Mollusks have one or two nephridia that collect waste from the coelom. The coelom is located around the heart only. Wastes pass from the coelom into the mantle cavity and are expelled from the body by the pumping of the gills. Remember that the gills are respiratory structures. ✓

Diversity of Mollusks

There are many kinds of mollusks. Three classes, Gastropoda, Bivalvia, and Cephalopoda include the most common and the best-known species.

Which mollusks belong to the class Gastropoda?

The largest class of mollusks is Gastropoda, or gastropods. Gastropods are stomach-footed mollusks. The name gastropod comes from the way the animal's large foot is positioned under its body. Most gastropods, such as snails, abalones, conches, periwinkles, whelks, limpets, cowries, and cones have a shell. They can be found in freshwater, in saltwater, or in moist land habitats. Gastropods with shells may be plant eaters, predators, or parasites.

Other gastropods, like slugs, do not have shells. They have a thick layer of mucus that protects their bodies. Nudibranches are colorful sea slugs. Nudibranches are protected from predators in another way. When these sea slugs feed on jellyfish, the poisonous nematocysts of the jellyfish are taken into the tissues of the sea slug. When a fish tries to eat the sea slug, the nematocysts are discharged into the predator and the predator is repelled. The bright colors of nudibranches warn predators of the danger before they attack.

Which mollusks belong to the class Bivalvia?

Clams, oysters, and scallops belong to class Bivalvia. Bivalves have two shells. The two shells are connected with a ligament, called a valve. The valve works like a hinge. Strong muscles allow the valves to open and close over the soft body of the bivalve. Most bivalve mollusks live in saltwater, but a few species live in freshwater. Some bivalves are tiny, measuring less than 1 mm in length. Others, such as the tropical giant clam, can be as large as 1.5 m long. Bivalves do not have a distinct head. They use their large, muscular foot for digging and hiding in the sand.

Bivalves do not have radula, the rough tonguelike organ of many mollusks. They are filter feeders. They obtain their food by filtering small particles of food from the water in which they live. Bivalve mollusks have several features that help them filter feed. They have gill cilia that beat so water can be drawn through the shells. Water and particles in the water move over the gills. Food is trapped in the bivalve's mucus. The cilia that line the gills push

Reading Check

6. What protects the bodies of gastropods that do not have shells?

____

____

____

Reading Check

7. What connects the two shells of a bivalve?

____

____

____

food particles to the mouth. The cilia also sort out food from large particles and sediment. These and other items that are rejected are carried to the bivalve's mantle. These rejected particles are pushed out. Rejected particles also may be carried to the foot where they are eliminated from the animal's body.

Which mollusks belong to the class Cephalopoda?

Cephalopoda means head-footed. Cephalopods live in oceans. Octopus, squid, cuttlefish, and the chambered nautilus are cephalopods. The only cephalopod that has a shell is the chambered nautilus. Some species, such as the cuttlefish, have reduced internal shells, but they do not have external shells. Scientists consider the cephalopods to have the most complex structure. Scientists also think the cephalopods are the most recently evolved mollusk.

In the cephalopods, the foot has evolved into tentacles with suckers, hooks, or sticky adhesive structures. Cephalopods swim or walk over the ocean floor to catch their food. They catch food in their tentacles, bring it to their mouths, and bite it with strong, beaklike jaws. Like many other mollusks, they have radulas. They use their radulas to tear food and pull it into their mouths. ✓

Cephalopods, like bivalves, have siphons that push water out. They can expel water in any direction, and they can move quickly by jet propulsion. Using the force of expelled water, squids can travel up to 20 m per second. Squids and octopuses use jet propulsion to escape danger. They also can release a dark fluid, or "ink," that darkens the water around them. This "ink" helps to confuse their predators because the cephalopod can no longer be easily seen.

✓ Reading Check

8. What body structure do cephalopods have in place of a foot?

____

____

____

After You Read

Mini Glossary

closed circulatory system: blood moves through the body, but it is enclosed entirely in a series of blood vessels

mantle (MAN tuhl): a membrane that surrounds the internal organs of the mollusk

nephridia (nih FRIH dee uh): organs that remove metabolic wastes from an animal's body

open circulatory system: blood moves through vessels and also into open spaces around the body organs

radula (RA juh luh): a tonguelike organ with rows of teeth

Section 27.1 Mollusks, *continued*

1. Read the terms and definitions in the Mini Glossary on page 330. Then on the lines below, explain how each term functions in a mollusk.

2. Use the text and the diagram below to identify mollusks according to their classes and list an identifying characteristic of each mollusk.

Mollusk	Class	Identifying Characteristics
Snail		
Squid		
Clam		
Sea slug		
Oyster		
Octopus		

Visit the Glencoe Science Web site at **science.glencoe.com** to find your biology book and learn more about mollusks.

Name Date Class

Section 27.2 Segmented Worms

Before You Read

Have you ever watched a segmented worm crawl into the soil? Was it burrowing away from you for safety or for other reasons? Why do you think worms are segmented? Why do they burrow into the ground? Write your thoughts on the lines below.

____

____

____

Read to Learn

Mark the Text **Identify Main Ideas** Highlight the main point in each paragraph. Read the main points carefully and restate each in your own words. Use your words to create an outline of this section.

1. What is the basic body plan of a segmented worm?

____

____

____

What is a segmented worm?

Segmented worms include leeches, bristleworms, and earthworms. They are classified in the phylum Annelida. Segmented worms are bilaterally symmetrical, the same on both sides. Like mollusks and other animals you have studied, they have a coelom. Segmented worms also have two openings on their bodies. One opening is for taking in food and the other opening is for releasing wastes. Some segmented worms have a larval stage similar to larval stages of certain mollusks. This similar larval stage suggests that mollusks and segmented worms may have a common ancestor.

The basic body plan of a segmented worm is a tube within a tube. The internal tube, which is suspended within the coelom, is the digestive tract. The worm takes food in through the mouth, which is the opening in the front, or anterior, end of its body. The worm releases its waste through the anus, an opening at the back, or posterior, end of its body. ✓

Most segmented worms have tiny bristles called **setae** (SEE tee) on each segment. These bristles help the worms move. Using the setae, the worm anchors its body in the soil. Each segment then helps move the animal along its path.

You can find segmented worms almost everywhere except in the frozen soil of the polar regions and in the dry sand and soil of the deserts. There are about 15 000 species of segmented worms that live in soil, freshwater, and saltwater. Earthworms are just one of the many species of segmented worms that live on our planet.

Why is segmentation important?

The body of a segmented worm is cylindrical, long and round. The body is divided into ringed segments. The giant earthworm of Australia can grow to more than 3 m long and the ringed segments are easily seen. The segmentation on the outside of a worm is repeated inside the worm. Internally, each segment is separated from others by a body partition. Each segment has its own muscles. By using these separated muscles, a worm can shorten and lengthen its body to move.

Segmentation also allows for specialization of body tissues. Groups of segments work together for a particular purpose or function. Certain segments have adaptations or modifications for sensing surroundings and for reproduction. ✓

Reading Check

2. What are the two primary advantages of segmentation?

____

____

____

What kind of nervous system does a segmented worm have?

Segmented worms have simple nervous systems. Organs in the front segments are able to sense the environment. Other sensory organs in the front segments detect light. Some segmented worms have eyes with lenses and retinas. Some species have a brain in the front segment. Nerve cords connect the brain to nerve centers that are found in each segment of the worm. These nerve centers are called ganglia.

What type of circulatory system do segmented worms have?

Segmented worms have a closed circulatory system. As you learned in the first section of this chapter, a closed circulatory system means that blood flows through closed vessels. Blood that carries oxygen to the body cells also carries carbon dioxide away from body cells. In a segmented worm, blood flows to all parts of the worm's body. Segmented worms must live in water or in wet areas on land because they exchange oxygen and carbon dioxide directly through their moist skin.

Do segmented worms have a digestive system?

Segmented worms have a complete internal digestive tract that runs from the front of the worm to the end of the worm. When a worm eats, food and soil that go into the mouth eventually pass into the **gizzard.** A muscular sac and hard particles in the gizzard help grind the soil and the food before they are passed into the

✓Reading Check

3. What is the function of the gizzard?

intestine. Material that cannot be digested and solid wastes pass out of the worm's body through the anus. The anus is the opening in the worm's body that is located at the posterior end. Like mollusks, segmented worms also have nephridia as shown in the illustration below. In a segmented worm, there are two nephridia in almost every segment. The nephridia collect wastes and move them through the coelom and out of the worm's body. ✓

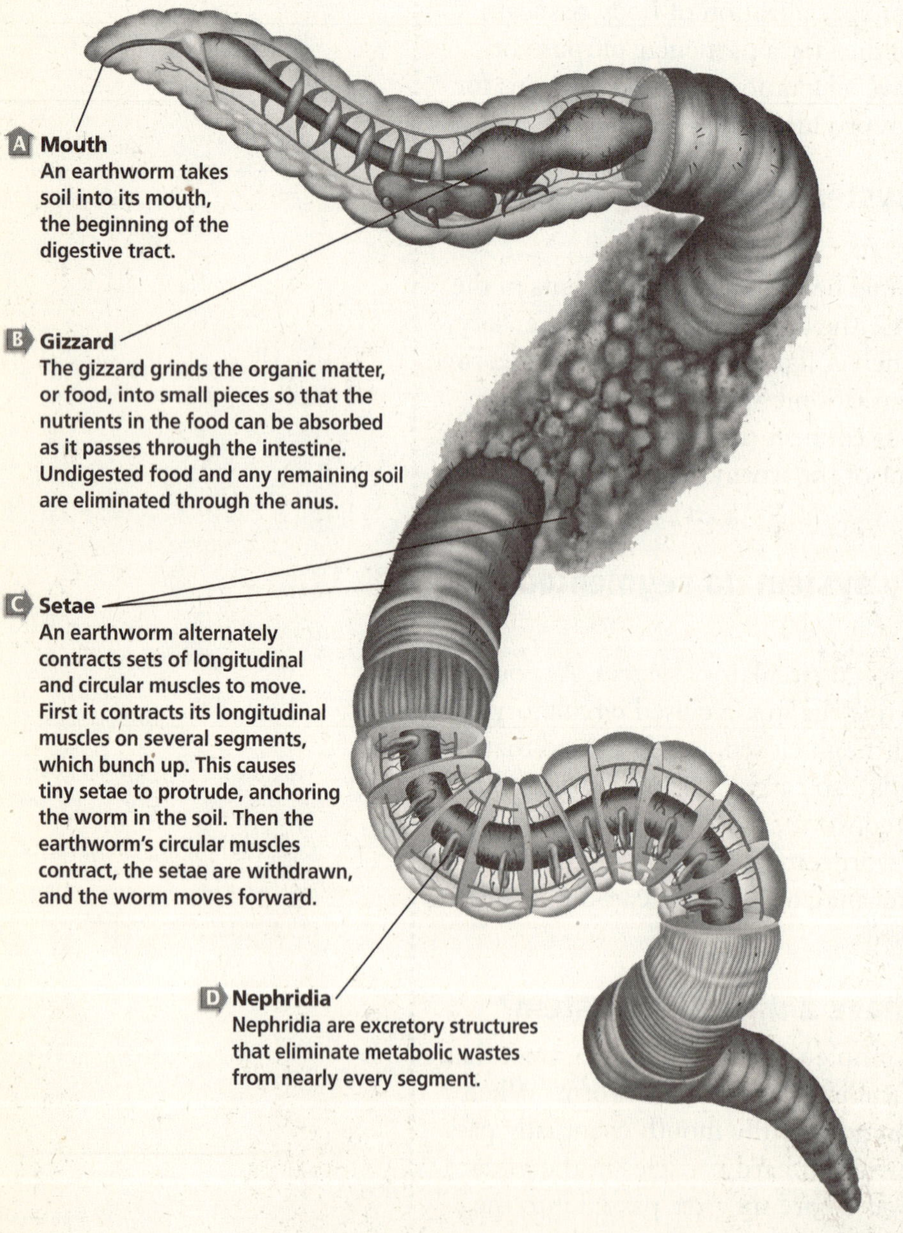

A Mouth
An earthworm takes soil into its mouth, the beginning of the digestive tract.

B Gizzard
The gizzard grinds the organic matter, or food, into small pieces so that the nutrients in the food can be absorbed as it passes through the intestine. Undigested food and any remaining soil are eliminated through the anus.

C Setae
An earthworm alternately contracts sets of longitudinal and circular muscles to move. First it contracts its longitudinal muscles on several segments, which bunch up. This causes tiny setae to protrude, anchoring the worm in the soil. Then the earthworm's circular muscles contract, the setae are withdrawn, and the worm moves forward.

D Nephridia
Nephridia are excretory structures that eliminate metabolic wastes from nearly every segment.

How do segmented worms reproduce?

Some segmented worms, including earthworms and leeches, are hermaphrodites. Hermaphrodites produce both sperm and eggs. When worms mate, two worms exchange sperm. Each worm forms a capsule for the eggs and the sperm. The sperm fertilize the eggs in this capsule. The capsule slips off the worm, and the capsule stays behind in the soil. Two to three weeks later, young worms emerge from the eggs.

Bristleworms and other closely related species have separate sexes. They reproduce sexually. Eggs and sperm are usually released into the seawater where they live. Fertilization takes place in the water. Bristleworm larvae hatch in the sea. They become part of the plankton. When a larva begins to develop segments, the worm will settle to the bottom of the ocean.

Diversity of Segmented Worms

There are three classes of segmented worms in the phylum Annelida. Earthworms belong to the class Oligochaeta. Bristleworms and their relatives belong to the class Polychaeta. Leeches belong to the class Hirudinea.

What are earthworms?

Earthworms are probably the best-known Annelids because they can be seen easily by most people. Earthworms belong to the class Oligochaeta. Earthworms have anterior and posterior sections, but lack distinct heads. They have only a few setae on each of their segments. As you can see in the illustration, earthworms have a brain, a pharynx, an esophagus, blood vessels, and nephridia. Notice that a single segment contains muscle layers, blood vessels, the esophagus, nephridia, and the ventral nerve cord. ✓

Earthworms improve garden soil. As they eat, they create new spaces for air and water to flow through the soil. As soil passes through the worm's digestive tract, nutrients are extracted. The undigested materials pass out of the worm's digestive tract. These wastes, called castings, help fertilize the soil.

What are bristleworms and their relatives?

Bristleworms and their relatives, including fanworms, lug worms, plumed worms, and sea mice, belong to the class Polychaeta. Most live in the oceans. Most body segments of these worms have many setae. Their class name Polychaete means "many bristles."

Most body segments of Polychaetes have pairs of appendages, similar to tiny limbs. The appendages are called parapodia. Polychaetes use parapodia for swimming or crawling over corals and the bottom of the sea. The parapodia also help in gas exchange. Polychaetes have heads with well-developed sense organs, including eyes.

What are leeches?

Leeches belong to the class Hirudinea. Leeches are segmented worms with flattened bodies. Usually they do not have setae. Most leeches live in freshwater streams or in rivers. Many species of leeches are parasites that suck blood or other body fluids from animals called hosts. Host animals include ducks, turtles, fishes, and humans. Leeches have front and rear suckers than enable them to attach themselves to their hosts.

✓ Reading Check

4. Why are earthworms the best known of the Annelids?

____

____

____

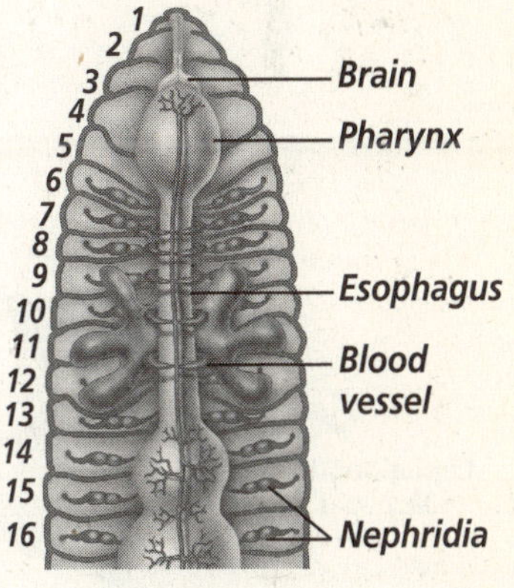

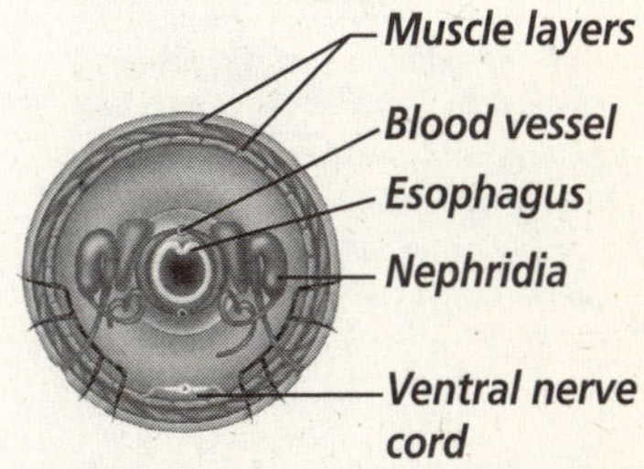

Reading Check

5. Why isn't the bite of a leech painful to its host?

____

____

____

The bite of a leech is not painful. The saliva of the leech contains chemicals that act as an anesthetic, a pain killer. Other chemicals stop the host's blood from clotting. A leech can ingest two to fives times its own weight in a single meal. Once a leech has finished eating, it lets go and drops off its host. The leech may not need to eat again for several months.

Origins of Mollusks and Segmented Worms

Fossil records show that there were great numbers of mollusks 500 million years ago. Fossils of gastropods, bivalves, and cephalopods, the three classes of mollusks, have been found in Precambrian deposits. Some species of mollusks, such as the chambered nautilus, have not changed much from their ancestors that lived long ago. Because mollusks lived so long ago, scientists use fossil mollusks to help determine the ages of rocks.

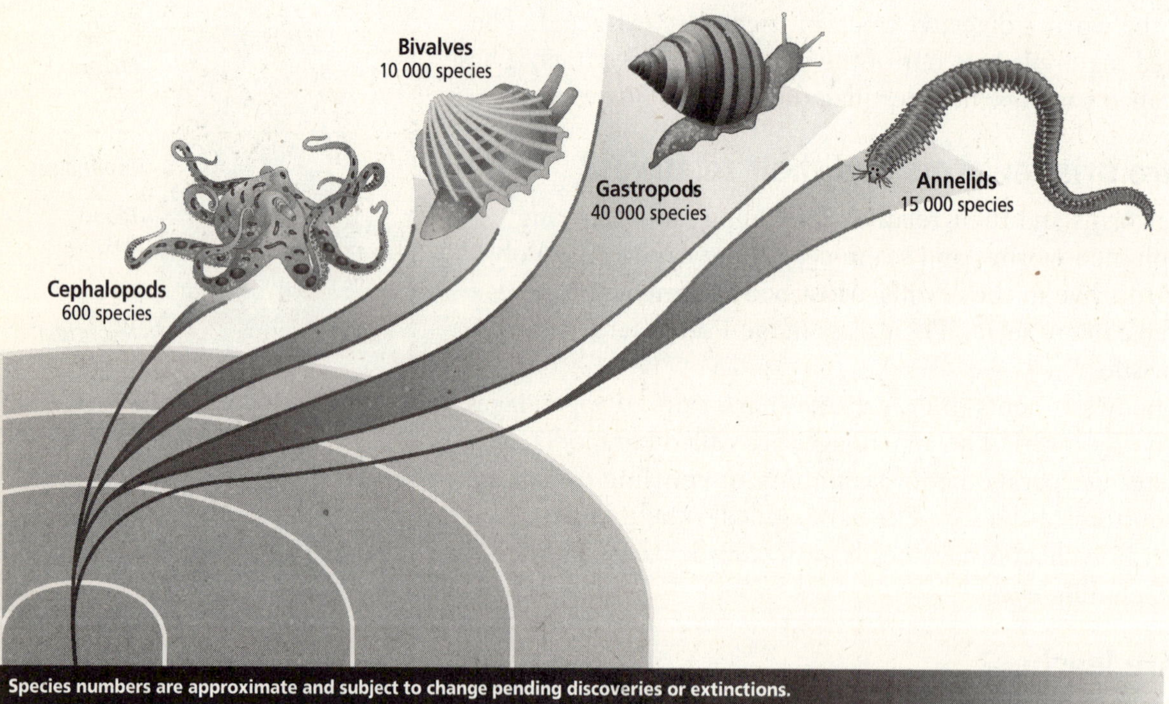

Annelids, segmented worms, probably evolved from the sea. They may have originated from the larvae of ancestral flatworms. The fossil records for segmented worms are much more limited than for mollusks because segmented worms have almost no hard body parts. Little remains from their soft bodies in the fossil records. Tubes constructed by Polychaetes are the most common fossils for this phylum. Some of these tubes appear in the fossil record as early as 540 million years ago. ✓

✓ Reading Check

6. Why is the fossil record of segmented worms more limited than the fossil record of mollusks?

____

____

____

After You Read

Mini Glossary

gizzard: part of the worm's digestive tract that consists of a muscular sac and hard particles that help grind soil and food before they pass into the worm's intestine

setae (SEE tee): tiny bristles on each segment of a segmented worm that help the worm move

1. Read the terms and definitions in the Mini Glossary above. Then, using both terms, describe how a segmented worm moves and digests its food.

____

____

____

2. Complete the chart below to identify the various types of segmented worms discussed in this section. Identify two characteristics of the Annelid in each class.

Type of Annelid	Class	Characteristics
	Oligochaeta	**1.**______________________ **2.**______________________
	Polychaeta	**1.**______________________ **2.**______________________
	Hirudinea	**1.**______________________ **2.**______________________

Visit the Glencoe Science Web site at **science.glencoe.com** to find your biology book and learn more about Annelids.

Name Date Class

Section 28.1 Characteristics of Arthropods

Before You Read

Have you seen a drawing or picture of a knight in armor? What is the purpose of armor? On the lines below list the advantages and disadvantages of wearing a heavy suit of armor. As you read this section about arthropods, notice the similarities between a knight's armor and a feature of arthropods.

____

____

____

Read to Learn

Mark the Text **Locate Information** Underline every heading in the reading that asks a question. Then, highlight the answers to those questions as you find them.

1. What are the advantages of jointed appendages?

____

____

____

What is an arthropod?

A typical arthropod is a segmented, coelomate invertebrate animal. A coelom is a fluid-filled cavity completely surrounded by mesoderm. Arthropods have bilateral symmetry, an exoskeleton, and jointed structures called appendages. An exoskeleton is a hard, thick outer covering. An **appendage** (uh PEN dihj) is any structure, such as a leg or an antenna that grows out of the body of an animal. In arthropods, appendages are adapted for a variety of purposes. There are appendages for walking, sensing, feeding, and mating.

Arthropods are the earliest known invertebrates to have jointed appendages. Joints allow more flexibility in animals that have hard, rigid exoskeletons. Joints allow powerful movements of appendages. Joints allow appendages to be used in many different ways. For example, the second pair of appendages in spiders is used for mating. In scorpions, the second pair is used for seizing prey. ✓

How does an exoskeleton provide protection?

Arthropods as a group have been successful partly because they have exoskeletons. The exoskeleton is made of protein and chitin (KI tun). In some species, the exoskeleton is a continuous covering over most of the body. In other species, it is made of separate plates held together by hinges. The exoskeleton protects and supports internal tissue and provides a place for the attachment of muscles. In many species that live on land, a waxy layer on the exoskeleton protects against water loss. In aquatic species, the exoskeleton may be reinforced with calcium carbonate.

Section 28.1 Characteristics of Arthropods, *continued*

What are the disadvantages of an exoskeleton?

One disadvantage of an exoskeleton is that it is a relatively heavy structure. The larger an arthropod is, the thicker and heavier its exoskeleton must be to support its larger muscles. The weight of the exoskeleton limits the size of arthropods. Some land dwelling and flying arthropods have adapted by having thinner, lighter exoskeletons. A thinner exoskeleton offers less protection but allows the animal to fly or jump. ✓

Another disadvantage is that the exoskeleton does not grow. An exoskeleton must be shed periodically. Shedding the old exoskeleton is called **molting.** Before an arthropod molts, a new, soft exoskeleton forms under the old one. When the new exoskeleton is ready, the animal contracts muscles and takes in air or water. This causes the animal's body to swell. The old exoskeleton splits open and the animal sheds it. The new exoskeleton is soft. Before it hardens, the animal puffs up as a result of increased blood circulation to all parts of the body. Many insects and spiders also increase in size by taking in air. The new exoskeleton hardens in a larger size, allowing some room for the animal to continue growing. ✓

Most arthropods molt four to seven times in their lives before they become adults. During molting, they are at risk from predators. While the new exoskeleton is soft, arthropods cannot protect themselves from danger. Many species hide or remain motionless for a few hours or days until the new exoskeleton hardens.

How is fusion related to movement and protection?

Most arthropods are segmented but they do not have as many segments as worms. In most groups of arthropods, segments have become fused into three body sections—head, thorax, and abdomen. In other groups, even these segments may be fused. Some arthropods have a head, but the thorax and abdomen are fused. Others have a head fused with a thorax, and an abdomen. A fused head and thorax is called a **cephalothorax** (se fuh luh THOR aks). ✓

Fusion of the body segments is related to movement and protection. Less fusion means more movement, but less protection. Beetles and some other arthropods that have separate head and thorax sections are more flexible than those with fused sections. Shrimps, lobsters, and crayfishes have a cephalothorax, which protects the animal but limits its movement.

✓ Reading Check

2. What limits the size of arthropods?

____

____

____

✓ Reading Check

3. What happens before an animal molts?

____

____

____

✓ Reading Check

4. Name the three body sections of most arthropod groups.

____

____

____

Section 28.1 Characteristics of Arthropods, *continued*

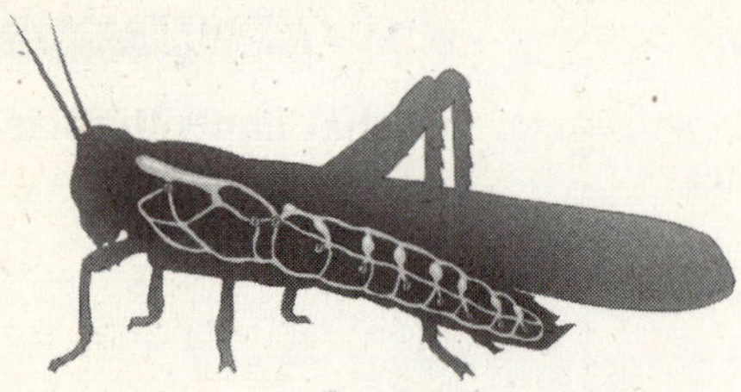

A Tracheal tubes are inside the body, thereby reducing water loss through the respiratory surface while carrying air close to each cell.

B Gills, with their large surface area, enable a large amount of blood-rich tissue to be exposed to water containing oxygen.

C Book lungs are folded membranes that increase the surface area of blood-rich tissue exposed to air.

What kinds of respiratory structures do arthropods have?

Arthropods are quick, active animals. They crawl, run, climb, dig, swim, and fly. In order to have oxygen delivered quickly to the cells, they need efficient respiratory structures.

Three types of respiratory structures have evolved in arthropods: gills, tracheal tubes, and book lungs. Aquatic arthropods use gills to get oxygen from the water and release carbon dioxide into the water. Land arthropods either have a system of tracheal tubes or book lungs. Most insects have **tracheal** (TRAY kee ul) **tubes,** a network of hollow air passages that carry air throughout the body. Muscle activity helps pump the air through the tracheal tubes. Air enters and leaves the tracheal tubes through openings on the thorax and abdomen called **spiracles** (SPIHR ih kulz).

Most spiders and their relatives have **book lungs,** air-filled chambers that contain leaflike plates. The folded membranes increase the surface area of tissue exposed to air. The stacked plates of the book lungs are arranged like pages in a book.

What senses do arthropods have?

Arthropods use antennae to detect movement, sound, and chemicals with great sensitivity. Antennae are stalklike structures that allow arthropods to detect changes in the environment. Antennae also are used for sound and odor communication. Ants and other arthropods are able to communicate with each other by pheromones. **Pheromones** (FER uh mohnz) are chemical odor signals given off by animals. For an ant to communicate the presence of food to another ant, the first ant releases a pheromone. Antennae sense the odor, which signals to the second ant that food is present. Pheromones signal animals to engage in a variety of behaviors. Some are used in scent trails. Others are important for mating behavior. ✓

✓ Reading Check

5. How do ants communicate?

____

____

____

Section 28.1 Characteristics of Arthropods, *continued*

Accurate vision is important to arthropods. Most arthropods have one pair of large compound eyes and three to eight simple eyes. A **simple eye** is a visual structure with only one lens that is used for detecting light. A **compound eye** is a visual structure with many lenses. Each lens registers light from a tiny portion of the field of view. The total image that is formed is made up of thousands of parts. The multiple lenses of a flying arthropod, such as a dragonfly, help it to analyze a fast-changing landscape during flight. ✓

Arthropods have well-developed nervous systems. The nervous system processes information coming in from the sense organs. The nervous system consists of a double ventral nerve cord, an anterior brain, and several ganglia. Arthropods have ganglia that have become fused. These ganglia act as control centers for the body section in which they are located.

✓ Reading Check

6. What is the difference between a simple eye and a compound eye?

____

____

____

What are other arthropod body systems?

Arthropod blood is pumped by a heart in an open circulatory system. Vessels carry blood away from the heart. The blood flows out of the vessels, bathing the tissues of the body. Blood returns to the heart through open body spaces.

Arthropods have complete digestive system. They have a mouth, stomach, intestine, anus, and various glands that produce digestive enzymes. The mouthparts of most arthropods include one pair of jaws called **mandibles** (MAN duh bulz). The mandibles are adapted for holding, chewing, sucking, or biting the different foods eaten by arthropods.

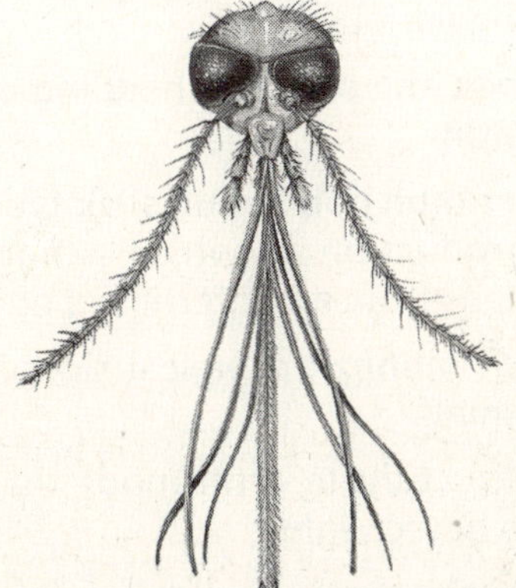

A Sand flies and other insects that feed by drawing blood have piercing blades or needlelike mouthparts.

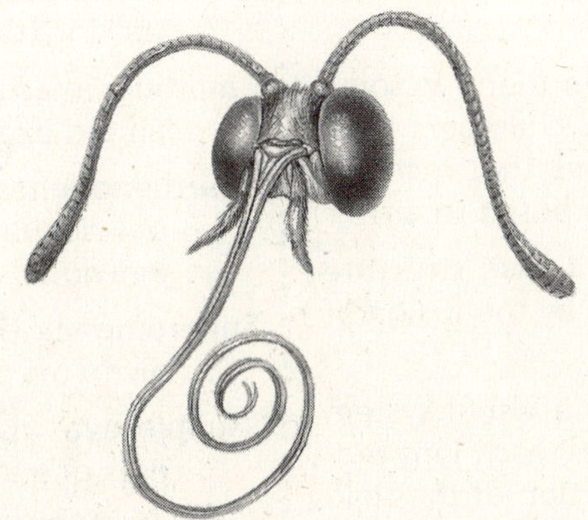

B The rolled-up sucking tube of moths and butterflies can reach nectar at the bases of long, tubular flowers.

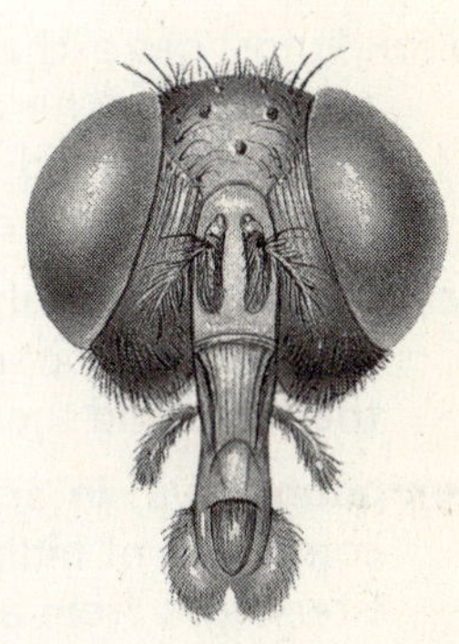

C The sponging tongue of the housefly has an opening between its two lobes through which food is lapped.

Most arthropods that live on land get rid of wastes through **Malpighian** (mal PIH gee un) **tubules.** In insects, the tubules are all located in the abdomen. Malpighian tubules are attached to and empty directly into the intestine.

The muscular system is well-developed in arthropods. In human limbs, muscles attach to bones. In arthropod limbs, the muscles attach to the inner surface of the exoskeleton. An arthropod muscle attaches to the exoskeleton on both sides of the joint.

How do arthropods reproduce?

Most arthropod species have separate males and females and reproduce sexually. For land species, fertilization is usually internal. For aquatic species, fertilization is usually external. A few species, such as barnacles, are hermaphrodites, animals with both male and female reproductive organs. Some species, such as bees, ants, aphids, and wasps, exhibit parthenogenesis. **Parthenogenesis** (par thuh noh JE nuh sus) is a form of asexual reproduction in which a new individual develops from an unfertilized egg. ✓

Reproductive diversity has contributed to the success of arthropods. There are more arthropod species than all other animal species combined.

Reading Check

7. Name three arthropod systems discussed in this section.

____

____

____

After You Read

Mini Glossary

appendage (uh PEN dihj): any structure, such as a leg or an antenna, that grows out of an animal's body

book lungs: gas exchange system found in some arthropods where air-filled chambers have plates of folded membranes that increase the surface area of tissue exposed to air

cephalothorax (se fuh luh THOR aks): structure in some arthropods formed by the fusion of the head and thorax

compound eye: in arthropods, a visual system composed of multiple lenses; each lens registers light from a small portion of the field of view, creating an image composed of thousands of parts

Malpighian (mal PIH gee un) tubule: in arthropods, tubules located in the abdomen that are attached to and empty waste into the intestine

mandibles (MAN duh bulz): in most arthropods, mouthparts adapted for holding, chewing, sucking, or biting various foods

molting: in arthropods, the periodic shedding of an old exoskeleton

parthenogenesis (par thuh noh JE nuh sus): type of asexual reproduction in which a new individual develops from an unfertilized egg

pheromones (FER uh mohnz): chemical signals given off by animals

simple eye: visual structure in arthropods that uses one lens to detect light

spiracles (SPIHR ih kulz): in arthropods, openings on the thorax and abdomen through which air enters the tracheal tubes

tracheal (TRAY kee ul) tubes: hollow passages in some arthropods that transport air throughout the body

Section 28.1 Characteristics of Arthropods, *continued*

1. Read the terms and definitions in the Mini Glossary on page 342. Then, circle a term that describes a type of structural adaptation. On the lines below, write a sentence using the term.

2. List the advantages and disadvantages of an exoskeleton in the web diagram below.

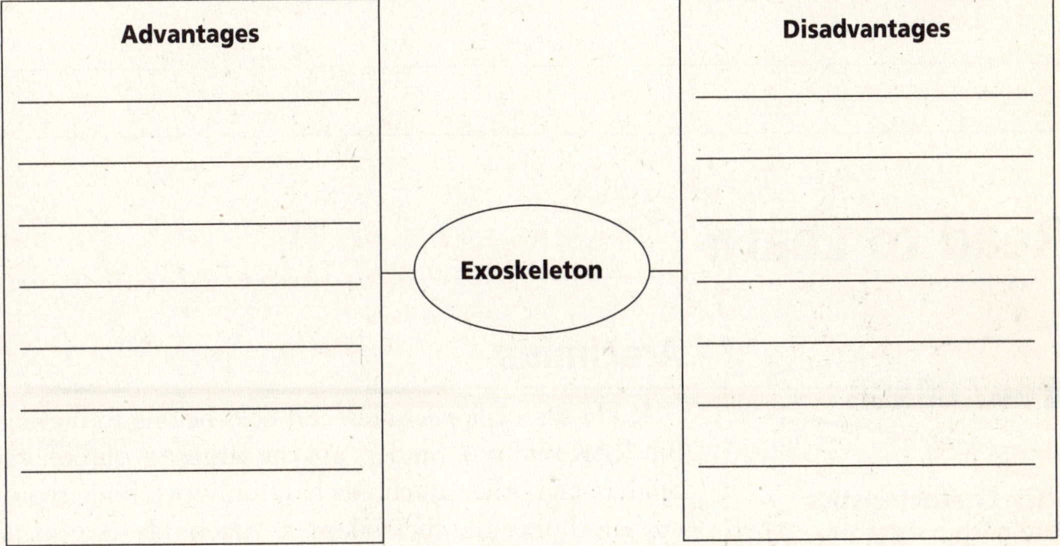

Visit the Glencoe Science Web site at **science.glencoe.com** to find your biology book and learn more about characteristics of arthropods.

Name Date Class

Section 28.2 Diversity of Arthropods

Before You Read

Spiders are arthropods. They show many of the characteristics of arthropods, but they also are different from the arthropods you have learned about. On the lines below list what you have observed about the behavior or appearance of spiders and how they are different from the arthropods you already know about.

____

____

____

Read to Learn

Identify Characteristics Point to each part of the spider's body in the illustration on page 345 as you read about it.

Arachnids

Spiders, scorpions, mites, and ticks belong to the class Arachnida (uh RAK nud uh). Spiders are the largest group of arachnids. Spiders and other arachnids have only two body regions—the cephalothorax and the abdomen. Arachnids have six pairs of jointed appendages.

The first pair of appendages, called **chelicerae** (chih LIH suh ree), is located near the mouth. Chelicerae are often modified into pincers or fangs. Pincers hold food. Fangs inject prey with poison. Spiders do not have mandibles for chewing. Using a process of extracellular digestion, digestive enzymes from the spider's mouth turn the internal organs of the prey into liquid. The spider then sucks up the liquefied food.

The second pair of appendages is called the **pedipalps** (PE dih palpz). Pedipalps are adapted for handling food and sensing. In male spiders, pedipalps carry sperm during reproduction. The four remaining appendages in arachnids are adapted as legs. Arachnids do not have antennae.

Most people think of webs when they think of spiders. Although all spiders spin silk, not all make webs. Spider silk is secreted by silk glands in the abdomen. As silk is secreted, it is spun into thread by structures called **spinnerets,** located at the rear of the spider.

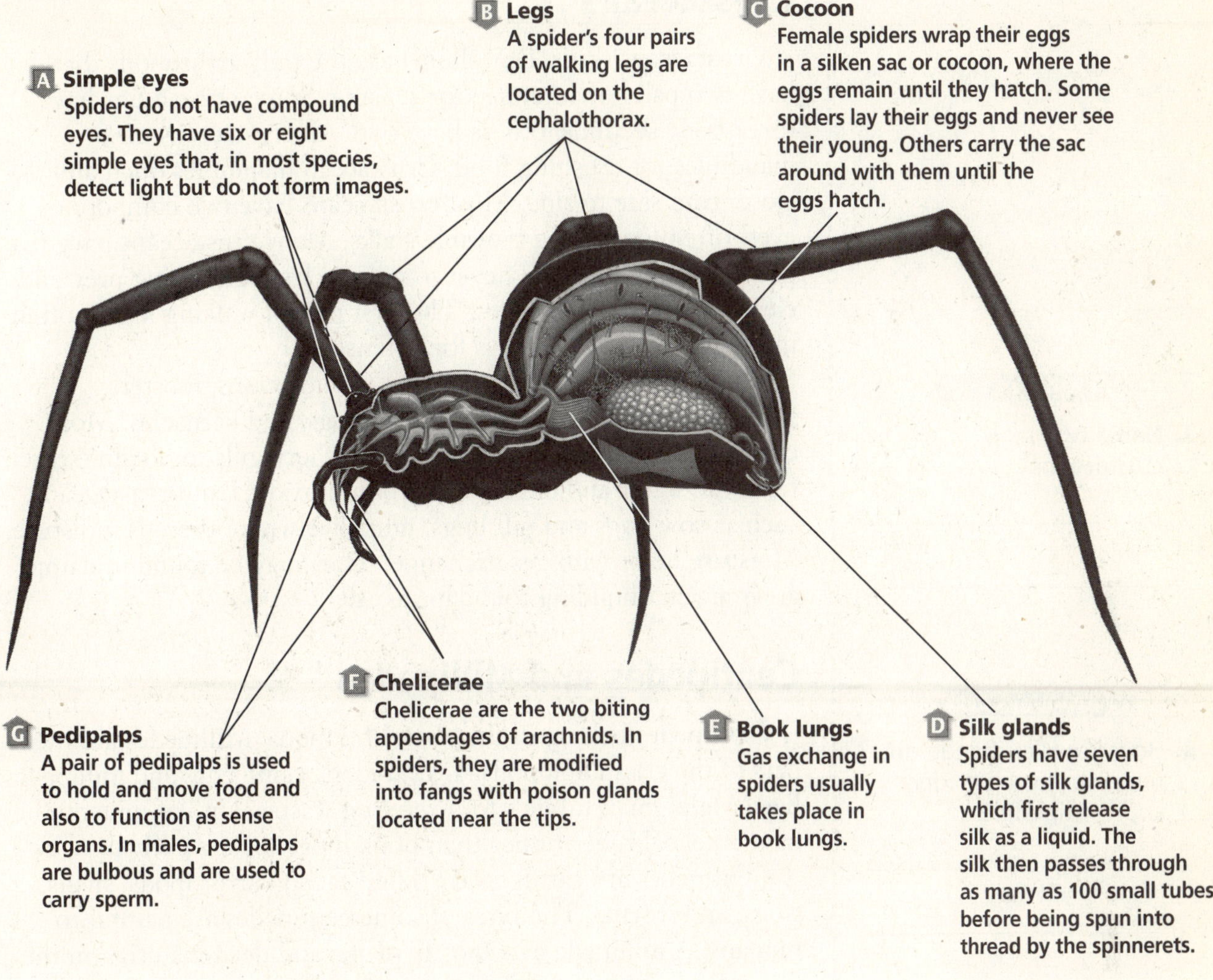

A Simple eyes
Spiders do not have compound eyes. They have six or eight simple eyes that, in most species, detect light but do not form images.
B Legs
A spider's four pairs of walking legs are located on the cephalothorax.
C Cocoon
Female spiders wrap their eggs in a silken sac or cocoon, where the eggs remain until they hatch. Some spiders lay their eggs and never see their young. Others carry the sac around with them until the eggs hatch.
G Pedipalps
A pair of pedipalps is used to hold and move food and also to function as sense organs. In males, pedipalps are bulbous and are used to carry sperm.
F Chelicerae
Chelicerae are the two biting appendages of arachnids. In spiders, they are modified into fangs with poison glands located near the tips.
E Book lungs
Gas exchange in spiders usually takes place in book lungs.
D Silk glands
Spiders have seven types of silk glands, which first release silk as a liquid. The silk then passes through as many as 100 small tubes before being spun into thread by the spinnerets.

What other animals are arachnids?

Arachnida also includes ticks, mites, and scorpions. Ticks and mites have only one body section. The head, thorax, and abdomen are completely fused. Ticks feed on blood from reptiles, birds, and mammals. They are small but can enlarge up to three times their normal size after feeding. ✓

Mites feed on fungi, plants, and animals. They are so small that they usually cannot be seen without magnification. The bite of a mite can irritate the skin. Both mites and ticks can spread disease.

Scorpions are easy to recognize because of their many body segments and large pincers. They have long tails with stingers at the tips. They use the poison in their stingers to paralyze prey. Scorpions live in warm, dry climates. They eat insects and spiders.

Reading Check

1. What animals, in addition to spiders, are included in Arachnida?

____

____

____

Section 28.2 Diversity of Arthropods, *continued*

Crustaceans

Crustaceans (krus TAY shuns) are the only arthropods that have two pairs of antennae for sensing. Some crustaceans have three body sections, others have two. All crustaceans have mandibles for crushing food. Crustacean mandibles open and close from side to side. Most crustaceans have two compound eyes, often located on movable stalks. Many crustaceans have five pairs of walking legs. They also use the legs for seizing prey and cleaning other appendages. The first pair of walking legs is often modified into strong claws for defense.

Members of the class Crustacea include crabs, lobsters, shrimps, crayfishes, water fleas, pill bugs, and barnacles. Most crustaceans live in water. They use feathery gills to obtain oxygen from the water and to release carbon dioxide. Land crustaceans, such as sow bugs and pill bugs, must live where there is moisture. Moisture helps with gas exchange. They may be found in damp areas around building foundations. ✓

Reading Check

2. Name two land crustaceans.

______

______

Centipedes and Millipedes

Centipedes belong to the class Chilopoda. Millipedes are members of the class Diplopoda. Like spiders, centipedes and millipedes have Malpighian tubules for excreting wastes. Unlike spiders, they have tracheal tubes rather than book lungs for gas exchange. ✓

Centipedes are carnivorous. They eat soil arthropods, snails, slugs, and worms. The bites of some centipedes are painful to humans. A millipede eats mostly plants and dead material on the forest floor. Millipedes do not bite, but they can spray foul-smelling fluids from their stink glands. Millipedes walk with a slow, graceful motion.

Reading Check

3. How is gas exchange different between spiders and centipedes?

______

______

______

Horseshoe Crabs: Living Fossils

Horseshoe crabs are members of the class Merostomata. One of the three living genera, *Limulus*, is found along the east coast of North America. Horseshoe crabs are considered to be living fossils. *Limulus* has remained relatively unchanged since the Triassic Period about 220 million years ago. ✓

Horseshoe crabs are heavily protected by a large exoskeleton. They live in deep coastal waters. Horseshoe crabs search on sandy or muddy ocean bottoms for algae, annelids, and mollusks. These arthropods migrate to shallow water in the spring, then mate at night during high tide.

Reading Check

4. Why are horseshoe crabs considered to be living fossils?

______

______

______

Insects

Flies, grasshoppers, lice, butterflies, bees, and beetles are a few of the members of the class Insecta. Insects have three body segments and six legs. There are more species of insects than all other classes of animals combined.

How do insects reproduce?

Insects usually mate once in their lifetime. The eggs are usually fertilized internally. Some insects reproduce from unfertilized eggs, a process called parthenogenesis. In aphids, parthenogenesis produces all-female generations. Most insects lay a large number of eggs. This increases the chances that some offspring will survive long enough to reproduce. Many female insects are equipped with an appendage that can pierce into wood or the surface of the ground. The female insect lays eggs in the hole. ✓

Reading Check

5. How do insects increase the chances that offspring will survive long enough to reproduce?

____

____

____

What happens during metamorphosis?

After eggs are laid, the embryos develop and the eggs hatch. In some wingless insects, such as silverfish, the eggs hatch into small forms that look like tiny adults. These insects molt several times until they reach adult size. Many other species of insects go through a series of major changes in body structure as they develop. Often, the adult insect does not resemble the juvenile form of the insect. This series of changes, controlled by chemical substances in the animal, is called **metamorphosis** (me tuh MOR fuh sus). ✓

Reading Check

6. What is metamorphosis?

____

____

____

Insects that undergo metamorphosis usually go through four stages on their way to adulthood: egg, larva, pupa, and adult. The **larva** is the free-living, wormlike stage of an insect, often called a caterpillar. As the larva eats and grows, it molts several times.

The **pupa** (PYEW puh) stage of insects is a period of reorganization. Tissues and organs of the larva are broken down and replaced by adult tissues. Usually the insect does not move or feed during the pupa stage. After a period of time, a fully formed adult emerges from the pupa.

The series of changes that occur as an insect goes through the egg, larva, pupa, and adult stages is called complete metamorphosis. In winged insects that undergo complete metamorphosis, the wings do not appear until the adult stage. More than 90 percent of all insects undergo complete metamorphosis.

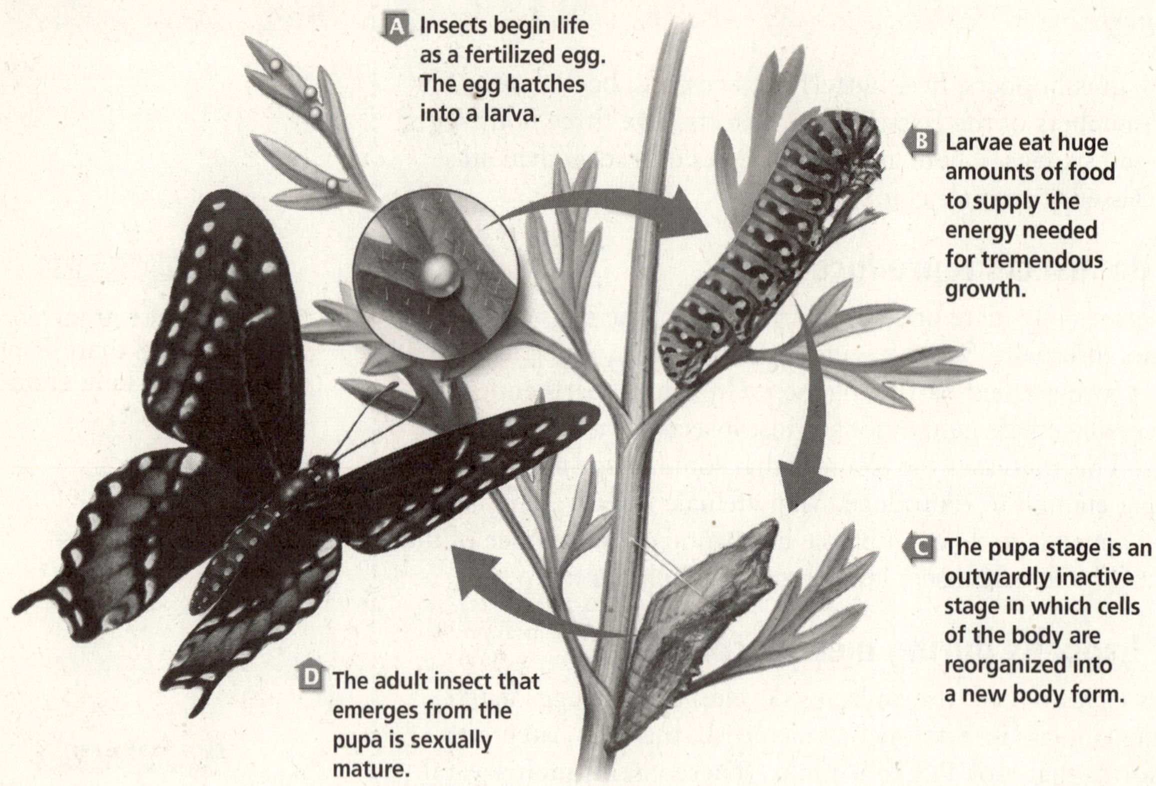

A Insects begin life as a fertilized egg. The egg hatches into a larva.
B Larvae eat huge amounts of food to supply the energy needed for tremendous growth.
C The pupa stage is an outwardly inactive stage in which cells of the body are reorganized into a new body form.
D The adult insect that emerges from the pupa is sexually mature.

The advantage of complete metamorphosis is that larva do not compete with adults for the same food. For example, butterfly larvae (caterpillars) feed on leaves. Adult butterflies feed on nectar from flowers. The complete metamorphosis of a butterfly is shown above.

What is incomplete metamorphosis?

Some insect species undergo a gradual or incomplete metamorphosis. The insect only goes through three stages of development. The three stages are egg, nymph, and adult. A **nymph,** which hatches from an egg, has the same general appearance as the adult but is smaller. Nymphs may lack certain appendages, or have appendages not seen in adults. A nymph cannot reproduce. With each nymph molt, it looks more like the adult. Wings begin to form and an internal reproductive system develops. Gradually, the nymph becomes an adult. Grasshoppers and cockroaches are insects that undergo incomplete metamorphosis. ✓

Reading Check

7. What are the differences between a nymph and an adult?

____

____

____

Origins of Arthropods

Arthropods live successfully on every surface of Earth. Their ability to survive in just about every habitat is unequaled in the animal kingdom. The success of arthropods is due in part to their varied life cycles, high reproductive output, and structural adaptations, such as small size, hard exoskeleton, and jointed appendages.

Reading Check

8. Why have arthropods been successful in so many habitats?

How did arthropods evolve?

Arthropods most likely evolved from an ancestor of the annelids. As arthropods evolved, body segments became fused. They adapted for certain functions such as locomotion, feeding, and sensing the environment. Segments in arthropods are more complex than in annelids. Arthropods have more developed nerve tissue and sensory organs such as eyes.

The exoskeletons of arthropods provide protection for their soft bodies. The circular muscles of annelids do not exist in arthropods. Muscles in arthropods are arranged in bands. The muscles are associated with particular segments and appendages. Because arthropods have many hard parts, much is known about their evolutionary history. Trilobites were once an important group of ancient arthropods, but they have been extinct for 248 million years.

After You Read

Mini Glossary

chelicerae (chih LIH suh ree): first pair of an arachnid's six pair of appendages; located near the mouth, they are often modified into pincers or fangs

larva: in insects, the free-living, wormlike stage of metamorphosis, often called a caterpillar

metamorphosis (met uh MOR fuh sus): in insects, a series of chemically-controlled changes in body structure from juvenile to adult

nymph: stage of incomplete metamorphosis where an insect hatching from an egg has the same general appearance as the adult insect but is smaller and sexually immature

pedipalps (PE dih palpz): second pair of an arachnid's six pair of appendages that are often adapted for handling food and sensing

pupa (PYEW puh): stage of insect metamorphosis where tissues and organs are broken down and replaced by adult tissues; larva emerges from pupa as mature adult

spinnerets: glands that spin silk into thread, located at the rear of a spider

1. Read the terms and definitions in the Mini Glossary above. Circle three terms that relate to metamorphosis. Then, choose one of these terms and use it correctly in a sentence..

Section 28.2 Diversity of Arthropods, *continued*

2. Fill in the blanks below to list the stages of metamorphosis.

Metamorphosis

Incomplete ____egg____ → ____________ → ____________

Complete ____________ → ____________ → ____________ → ____adult____

3. Write on the lines below, the stage at which molting occurs.

Incomplete metamorphosis: ______________________________

Complete metamorphosis: ______________________________

Visit the Glencoe Science Web site at **science.glencoe.com** to find your biology book and learn more about diversity of arthropods.

Name Date Class

Section 29.1 Echinoderms

Before You Read

Have you ever seen a sea star, a sea urchin, or a sand dollar? What do you know about them? Write two questions you have about them on the lines below. Answer the questions after you have completed the Read to Learn section.

Read to Learn

What is an echinoderm?

Echinoderms belong to the phylum Echinodermata. Echinoderms (ih KI nuh durmz), which means spiny skinned, can be found in oceans all over the world. Echinoderms move using hundreds of suction-cup-tipped appendages. The skin of an echinoderm is covered with tiny, jawlike pincers. These characteristics distinguish echinoderms from other animals.

What kinds of skeletons do echinoderms have?

Echinoderms have hard, spiny, or bumpy endoskeletons. An endoskeleton is an internal skeleton. The skeleton of all echinoderms is mostly calcium carbonate, the compound that makes up limestone. A thin epidermis, or outer layer of skin, covers the endoskeleton. A sea urchin has long, pointed spines. Sea stars, sometimes called starfishes, may not look spiny, but they are. If you look closely at a sea star, you will see that its long arms are covered with short, rounded spines. The arms of a sea star are called **rays.** The skin of a sea cucumber is made of soft tissue that does not appear spiny. The small platelike structures that cover a sea cucumber's skin have replaced true spines. ✓

Mark the Text **Create a Chart** Highlight the text that describes the body parts or characteristics that make an echinoderm unique. In a different color, highlight the body parts or characteristics that make one echinoderm species different from another. Use the highlighted text to create a chart that shows these unique features.

1. What are sea stars' arms called?

Some of the spines found on the skin of sea stars and sea urchins have been modified into pincerlike appendages. These appendages are called a **pedicellariae** (PEH dih sih LAHR ee ay; singular, pedicellaria). An echinoderm uses its pedicellariae to protect itself. It also uses the pedicellariae to clean the surface of its body.

How do echinoderms benefit from radial symmetry?

As you will recall, radial symmetry means that the parts of the animal's body are arranged regularly around a central axis. Radial symmetry is an advantage to animals that move slowly or that do not move freely from place to place. Radial symmetry allows echinoderms to sense possible food or predators that might be nearby. Radial symmetry also allows echinoderms to sense other conditions of their environment from all directions.

Reading Check

2. How do echinoderms use a water vascular system?

____

____

____

What is a water vascular system?

Echinoderms have a water vascular system. A **water vascular system** is a hydraulic system that operates under water pressure. Water comes in and goes out of a sea star through the water vascular system. This system allows sea stars to move, to catch their food, to exchange gases, and to excrete wastes. ✓

As the water goes through the sea star, it passes through the madreporite. The **madreporite** (mah druh POHR ite) is a disk-shaped opening on the upper surface of the animal. It functions like a strainer that fits into a kitchen sink drain and keeps large bits of material from going into the drain pipes. The madreporite prevents large particles from entering the echinoderm's body.

Sea stars and other echinoderms have tube feet. **Tube feet** are hollow tubes with thin walls. They look like tiny medicine droppers. The end of a tube foot works like a tiny suction cup. The round, muscular structure is called an **ampulla** (AM pew lah). It looks like the bulb of the dropper. The ampullae contract and relax, creating a strong suction action. The total suction action of ampullae in tube feet is so strong that a sea star can open a clam shell. Each tube foot works independently of the other tube feet. Echinoderms move by pushing out and then pulling in their tube feet.

Tube feet also carry out gas exchange and excretion. Gases are exchanged and some wastes are excreted by diffusion through the thin walls of the tube feet. ✓

Reading Check

3. What purposes do tube feet serve?

What does an echinoderm eat?

All echinoderms have a mouth, stomach, and intestines. The way echinoderms catch their food differs from one species to another. Sea stars are carnivores, or meat-eaters. They eat worms and mollusks such as clams and oysters. Most sea urchins are herbivores. They eat algae. Brittle stars, sea lilies, and sea cucumbers feed on dead and decaying matter on the ocean floor.

Do echinoderms have a nervous system?

An echinoderm has a simple nervous system. It consists of a nerve ring that surrounds the mouth. Echinoderms do not have heads or brains. Nerves extend from the nerve ring down into each ray. Nerves in the rays are called radial nerves. The radial nerves branch out into a network of nerves. This nerve network provides sensory information to the animal. ✓

Reading Check

4. What kind of nervous system does an echinoderm have?

Most echinoderms do not have sensory organs. They have cells that detect touch and light. Sea stars, however, do have sensory organs. A sea star's body is composed of long rays that extend from the animal's central disk. On the underside of each arm or ray is an eyespot. Eyespots consist of a cluster of light-detecting cells. They help sea stars detect the intensity of light. Most sea stars use their eyespots to help them move toward light. The tube feet of sea stars have chemical receptors. When a sea star senses a chemical signal from a prey animal, it moves in the direction of the ray that most strongly senses the chemical.

Do echinoderm larvae have radial or bilateral symmetry?

The larval stages of echinoderms have bilateral symmetry. The larvae are free swimming, and they go through metamorphosis. During metamorphosis, the larvae undergo many changes both in their body parts and in their symmetry. Remember that the adult forms of echinoderms have radial symmetry.

✓Reading Check

5. How many classes and how many species of echinoderms exist today?

____

____

____

Are echinoderms protostomes or deuterostomes?

A protostome is an animal with a mouth that develops from the opening in the gastrula. A deuterostome is an animal with a mouth that develops from cells elsewhere on the gastrula. Echinoderms are deuterostomes. Echinoderms have a close relationship to chordates because chordates also are deuterostomes.

Diversity of Echinoderms

About 6000 species of echinoderms exist today. About 1500 of these belong to the class Asteroidea (AS tuh ROY dee uh). Sea stars belong to this class. The six classes of echinoderms and an example of a species in each class are as follows: ✓

Class	Example of Species in Class
Asteroidea (AS tuh ROY dee uh)	Sea stars
Ophiuroidea (OH fee uh ROY dee uh)	Brittle stars
Echinoidea (eh kihn OY dee uh)	Sea urchins
Holothuroidea (HOH loh thuh ROY dee uh)	Sea cucumbers
Crinoidea (cry NOY dee uh)	Sea lilies
Concentricycloidea (kon sen tri sy CLOY dee uh)	Sea daisies

What are sea stars?

Sea stars are probably the best-known echinoderm. Most sea stars have only five rays. Some sea stars, however, have more than 40 rays. Sea stars have all the common characteristics of echinoderms. They have an endoskeleton made of calcium carbonate plates. The sea star can change from a rigid structure to a flexible one by contracting or relaxing the muscles that connect the plates.

As shown in the illustration on page 355, they have tube feet with ampullae and use eyespots on each ray to distinguish between light and dark. Sea stars have a stomach. In order to eat, a sea star pushes its stomach out of its mouth and spreads its stomach over the food. Enzymes secreted by the digestive gland turn solid food into liquid. The stomach absorbs this liquid. The sea star then pulls its stomach back into its body. The sea star has an anus for ridding its body of waste. It also has pedicellariae that it uses to keep the surface of its body clean.

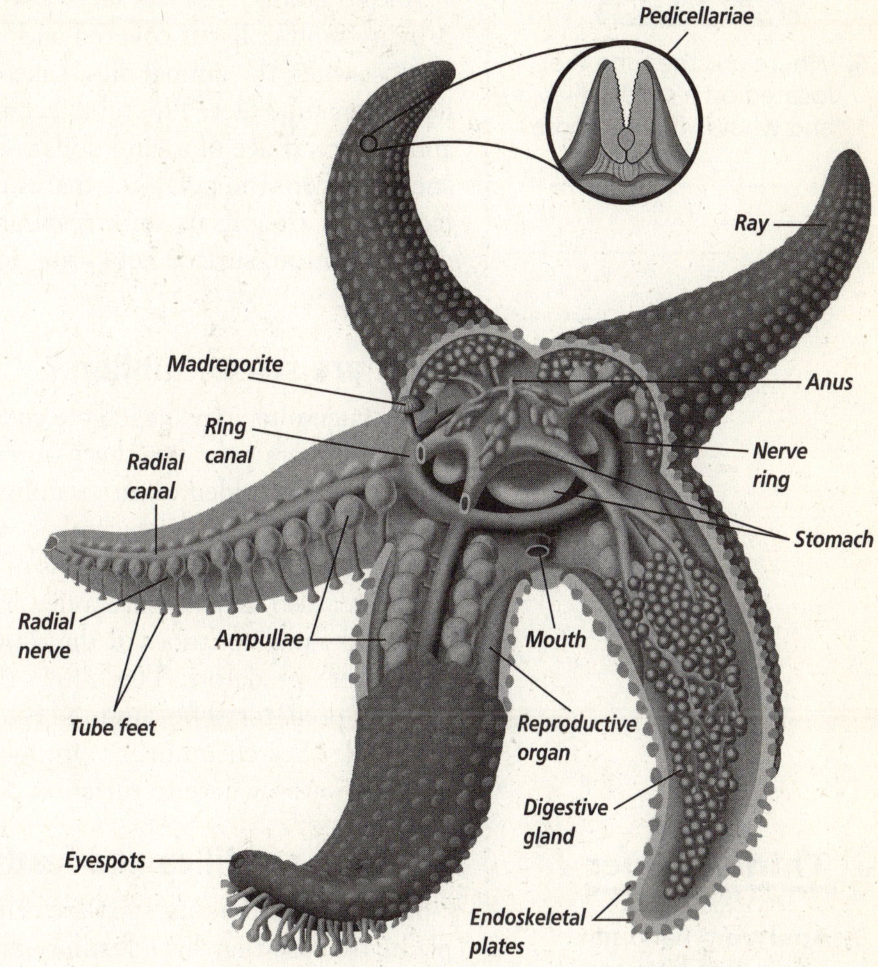

What are brittle stars?

Brittle stars are fragile. If you pick up a brittle star, parts of its rays will break off in your hand. That is how this star got its name. If a predator tries to attack a brittle star, rays will break off and the brittle star can escape. New rays will develop to replace those that break off.

Brittle stars do not use their tube feet for movement. They move with a snakelike, slithering motion of their flexible rays. The brittle star uses its tube feet to move bits of food along the rays and into its mouth. The mouth is located in the star's central disk.

What are sea urchins and sand dollars?

Sea urchins and sand dollars do not have rays. Their shapes resemble globes or disks, and they are covered with spines. Sea urchins often burrow into rocks to protect themselves from predators and from rough water. A sea urchin resembles a pincushion because of its long, pointed spines. The spines help protect the sea urchin from predators. Sea urchins have long and slender tube feet that help them move.

Reading Check

6. Where are the tube feet located on a sand dollar and what is their purpose?

____

____

____

A sand dollar has a flat surface with a petal-like flower pattern. A living sand dollar is covered with tiny, hair-like spines that are lost when the animal dies. Like other echinoderms, a sand dollar has tube feet. The tube feet are found on both the upper and lower surface of a sand dollar. The feet on the upper surface stick out from the petal-like marks on the sand dollar's top surface. They are gills used for respiration. Tube feet on the sand dollar's bottom surface help bring food particles into the sand dollar's mouth.

What are sea cucumbers?

Sea cucumbers and vegetable cucumbers have a similar shape. Sea cucumbers have a leathery outer covering that allows them the flexibility needed to move along the ocean floor. When a predator threatens a sea cucumber, it can protect itself in two ways. It can expel a tangled, sticky mass of tubes through the anus. A sea cucumber also may rupture, releasing some of its internal organs. The sticky mass of tubes or the rupturing organs confuse a sea cucumber's predators. This allows the sea cucumber to move away from its predator. The ruptured internal organs regenerate in a few weeks. Sea cucumbers reproduce by shedding sperm and eggs into the water, where fertilization occurs.

Think it Over

7. Analyze What distinguishes sea lilies and feather stars from other echinoderms? (Circle your choice.)

a. They regenerate.
b. They are sessile.
c. They have a larval stage.

What are sea lilies and feather stars?

Sea lilies and feather stars are echinoderms. They resemble plants because they have feathery rays. Sea lilies are the only stationary, or sessile, echinoderms. Feather stars are sessile only in the larval stage. The adult feather star uses its feathery arms to swim from place to place. Sea lilies and feather stars feed by capturing downward-drifting particles with their feathery rays.

What are sea daisies?

Two species of sea daisies were discovered in 1986 in deep waters off New Zealand. Sea daisies are flat, disk-shaped echinoderms less than 1 cm in diameter. Sea daisies have tube feet. The feet are located around the edge of the disk, not along radial lines.

Origins of Echinoderms

The earliest echinoderms may have been bilaterally symmetrical as adults. They may have been attached to the ocean floor by a kind of stalk. Another view suggests that echinoderms swam freely in the oceans.

Most invertebrates show protostome development. Echinoderms are the only major group of deuterostome invertebrates. Deuterostome development appears mainly in chordates. For this reason some biologists suggest that echinoderms are the closest invertebrate relatives of the chordates.

Because the endoskeletons are made of calcium carbonate, echinoderms easily turn into fossils. There is a good fossil record of the phylum. Echinoderms, as a group, date from the Paleozoic Era. More than 13 000 fossil species have been identified.

After You Read

Mini Glossary

ampulla (AM pew lah): the round, muscular structure on a tube foot

madreporite (mah druh POHR ite): a disk-shaped opening on the upper surface of an echinoderm's body; operates like a sieve or strainer to keep large particles out of the animal

pedicellariae (PEH dih sih LAHR ee ay): pincerlike appendages found on sea stars and sea urchins; used for protection and for cleaning the body

rays: the long tapering arms of an echinoderm, especially a sea star

tube foot: a hollow, thin-walled tube that ends in a suction cup

water vascular system: the system that allows water to enter and leave the system of a sea star; works as a hydraulic system that operates under water pressure

1. Read the terms and definitions in the Mini Glossary above. Use the space below to write a brief paragraph describing how the parts of an echinoderm work together.

2. Complete the table below for any three echinoderms that you have read about. List at least one characteristic for each echinoderm. Then list at least one way that each is similar to and different from the others.

Echinoderm	Class	Characteristics	Similarities	Differences

Visit the Glencoe Science Web site at **science.glencoe.com** to find your biology book and learn more about echinoderms.

Section 29.2 Invertebrate Chordates

Before You Read

You are familiar with vertebrates. Cats, dogs, canaries, and humans are vertebrates. Some animals are invertebrates. What do you think is the primary way in which vertebrates and invertebrates differ? Write your answer on the lines below.

Read to Learn

What is an invertebrate chordate?

You are probably most familiar with the vertebrate chordates. They have backbones and include animals such as birds, fishes, and mammals. The phylum Chordata (kor DAH tuh) also includes invertebrates. The three subphyla are listed below.

Subphyla	Members of Subphyla
Urochordata	Tunicates
Cephalochordata	Lancelets
Vertebrata	Fishes, reptiles, amphibians, birds, mammals

Create a Quiz After you have read this section, create a quiz question for each paragraph. After you have written the quiz questions, be sure to answer them.

This section examines tunicates and lancelets, invertebrate chordates that do not have backbones. Invertebrate chordates do not look much like fishes, reptiles, or humans. However, all of these animals have the characteristics common to all chordates. At some time during their development, they all have a notochord, a dorsal hollow nerve cord, and pharyngeal pouches. In addition, at some time during their development, all chordates have a postanal tail. All chordates have bilateral symmetry, a well-developed coelom, and segmentation.

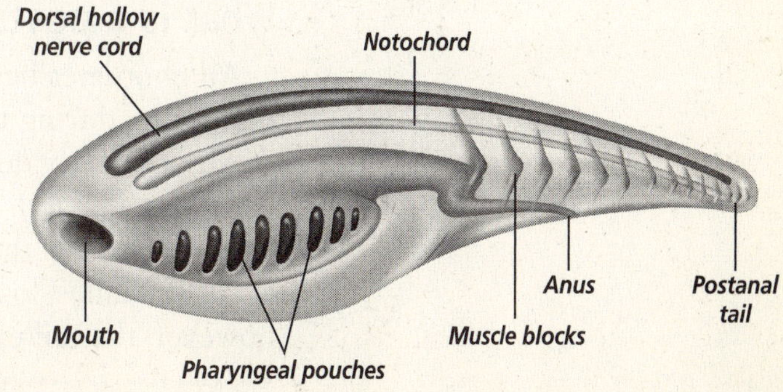

What is a notochord?

The embryos of all chordates have a notochord. The **notochord** (NOH tuh kord) is a long, semirigid structure that resembles a rod. The notochord is made of large, fluid-filled cells. These cells are held within stiff, fibrous tissues.

1. Where is the notochord located?

The notochord is located between the digestive system and the dorsal hollow nerve cord. Invertebrate chordates may keep their notochords in their adult stages. In vertebrate chordates, the notochord is replaced by a backbone. Invertebrate chordates do not develop a backbone.

The notochord develops on what will be the dorsal side of the embryo. It develops after the formation of a gastrula from mesoderm. The gastrula is the embryo development stage in animals where cells on one side of the blastula move inward, forming a cavity of two or three layers of cells with an opening at one end. The mesoderm is the middle cell layer in the gastrula, between the ectoderm and the endoderm. The notochord holds internal muscles in place. This makes it possible for invertebrate chordates to move their bodies quickly.

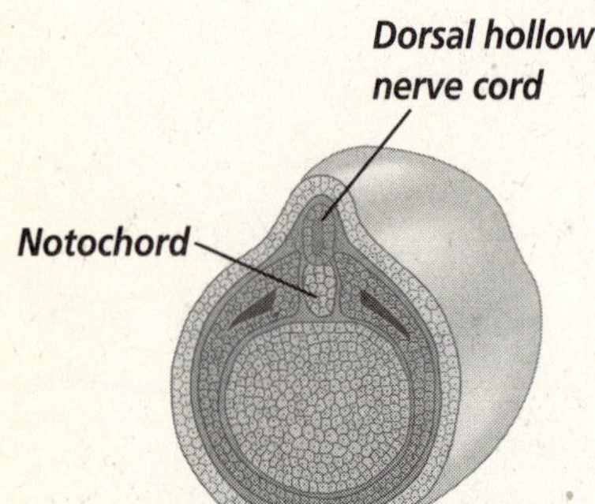

What is a dorsal hollow nerve cord?

All chordates have a dorsal hollow nerve cord. In chordates, the **dorsal hollow nerve cord** develops from a plate of ectoderm that rolls itself into a hollow tube. This tube is made of cells that surround a fluid-filled canal. The tube lies above the notochord.

In most adult chordates, the cells in the posterior or back portion of the dorsal hollow nerve cord develop into the spinal cord. The cells in the anterior or front portion develop into the brain. A pair of nerves connects the nerve cord to each block of muscles.

What is a pharyngeal pouch?

All chordates have pharyngeal pouches. The **pharyngeal pouches** of a chordate embryo are paired openings that are located in the pharynx. The pharynx is located behind the mouth. Many chordates have these pouches only during their embryonic stages. In chordates that live in the water, pharyngeal pouches develop openings called gill slits. The gill slits filter food and gas exchange occurs as water flows through them. In chordates that live on land, the pharyngeal pouches develop into other structures.

What is a postanal tail?

All chordates have a postanal tail. In some chordates the tail disappears during the early developmental stages. For example, during the early development of the human embryo, the embryo has a postanal tail. The tail disappears as development continues. In most animals that have tails, the digestive system extends to the tip of the tail. This is where the anus is located. In chordates, however, the tail extends beyond the anus.

Muscle blocks help the tail move. Muscle blocks are modified body segments that consist of stacked layers of muscle. Muscle blocks are held in place by the notochord. The notochord gives the muscles a firm structure to pull against. As a result, chordates generally have stronger muscles than members of other phyla.

Think it Over

2. **Analyze** Why do chordates have stronger muscles than members of other phyla?

____

____

____

How do homeotic genes control development?

Homeotic genes outline body organization. They also direct the development of tissues and organs in the embryo. Scientists have studied chordate homeotic genes. These studies have helped scientists understand the development of chordates and the relationship between invertebrate and vertebrate chordates.

Diversity of Invertebrate Chordates

The invertebrate chordates belong to two subphyla in the phylum chordata. One subphylum is called Urochordata. Tunicates (TEW nuh kaytz), or sea squirts, belong to this subphylum. The other subphylum is called Cephalochordata. Lancelets belong to this subphylum.

What are tunicates?

Tunicates are members of the subphylum Urochordata. Adult tunicates, or sea squirts, do not seem to have any of the common chordate features. In the larval stage, a sea squirt has a tail that makes it look similar to a tadpole. Sea squirt larvae do not feed. After they hatch, the larvae are free swimming. They soon settle and attach themselves with a sucker to boats, rocks, and the ocean floor. Many adult sea squirts secrete a tunic, a tough covering, or sac, made of cellulose. It surrounds their bodies. The figure at right shows the parts of a tunicate.

Sometimes tunicates form a group or a colony. A colony of tunicates sometimes secretes one big tunic that will have one opening to the outside.

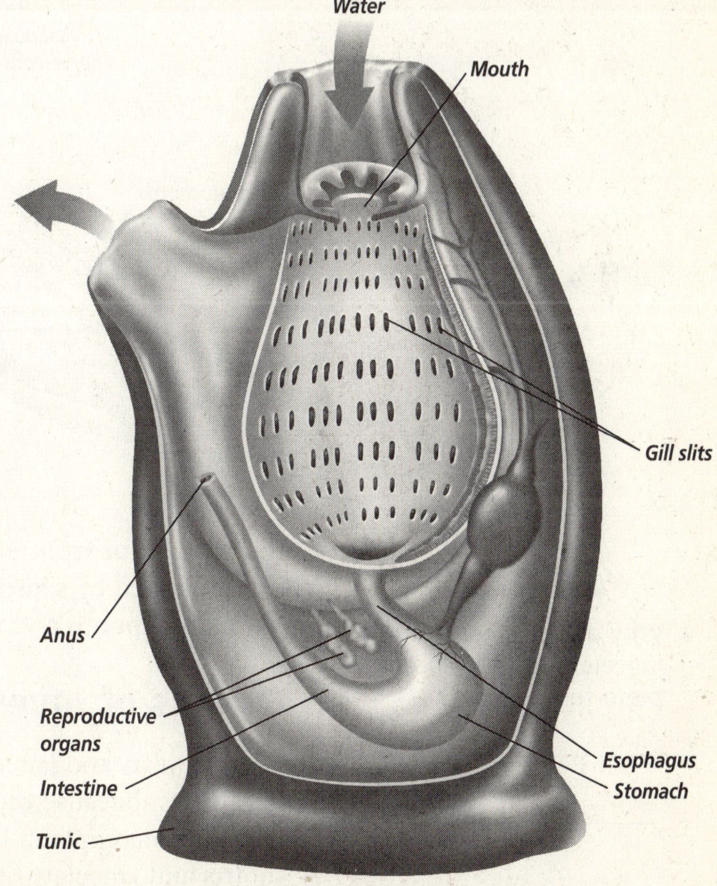

Reading Check

3. How do we know that sea squirts are chordates?

The gill slits in adult sea squirts show that they are chordates. The adult animals are small, tube-shaped animals. They may be microscopic in size, or they may be several centimeters long. If you take an adult tunicate from its home, it may squirt water at you. This is how the tunicate got the name sea squirt.

What are lancelets?

Lancelets are similar to fishes. They are small, common animals that live in the sea. Lancelets belong to the subphylum Cephalochordata. Lancelets spend most of their time buried in the sand with only their heads sticking out. Like tunicates, lancelets are filter feeders. Unlike tunicates, lancelets keep all their chordate features for their entire lives.

Although lancelets look similar to fish, there are some differences. Lancelets have only one layer of skin. They do not have pigment or color. They do not have scales or a distinct head. Lancelets can sense light because they have light-sensitive cells on the anterior or front end of their bodies. They have a mouth that is surrounded by sensory tentacles. Lancelets have a hood that covers the mouth and these tentacles.

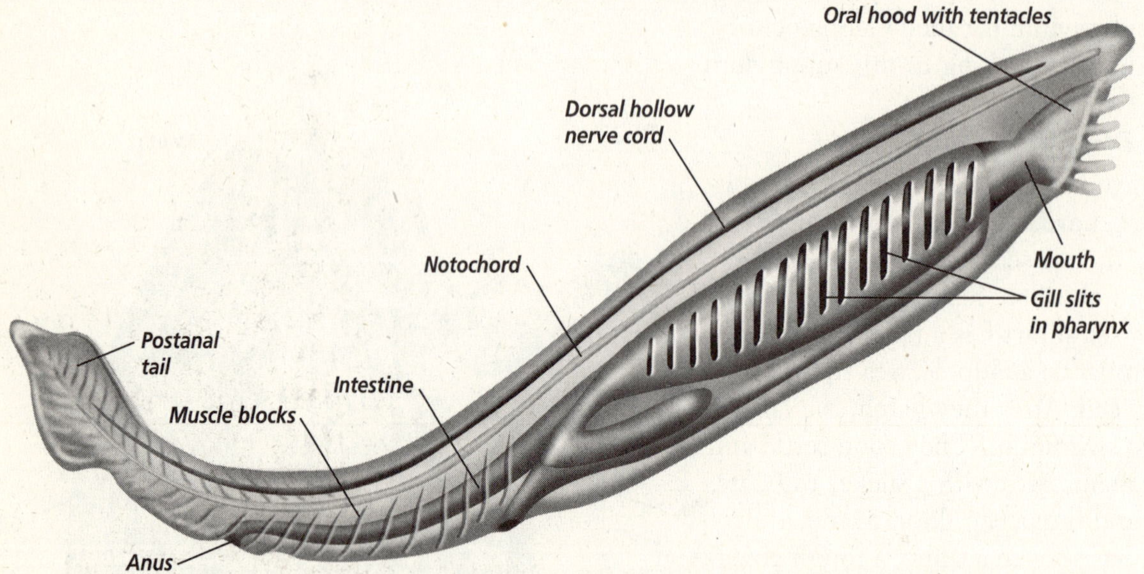

Reading Check

4. Why do sea squirts and lancelets have an incomplete fossil record?

Origins of Invertebrate Chordates

Sea squirts and lancelets do not have bones, shells, or other hard parts. Because they do not have hard parts, their fossil record is incomplete. Biologists do not know exactly where sea squirts and lancelets fit in the phylogeny of chordates.

Section 29.2 Invertebrate Chordates, *continued*

It is possible that echinoderms, invertebrate chordates, and vertebrates came from ancestral, stationary animals that caught food in their tentacles. Modern vertebrates may have come from free-swimming larval stages of ancestral invertebrate chordates. Scientists have recently found fossil forms of organisms that are similar to present-day lancelets. These fossil forms are in rocks that are 550 million years old. These fossils tell us that invertebrate chordates probably existed before vertebrate chordates.

After You Read

Mini Glossary

dorsal hollow nerve cord: develops from a plate of ectoderm that rolls into a hollow tube. In most adult chordates, this develops into the spinal cord.

notochord: a long, semirigid rodlike structure located between the dorsal hollow nerve cord and the digestive system

pharyngeal pouches: paired openings located in the pharynx, behind the mouth

1. Read the terms and their definitions in the Mini Glossary above. Use the space below to explain the importance of the dorsal hollow nerve cord.

2. In the diagram, fill in seven characteristics that all chordates have.

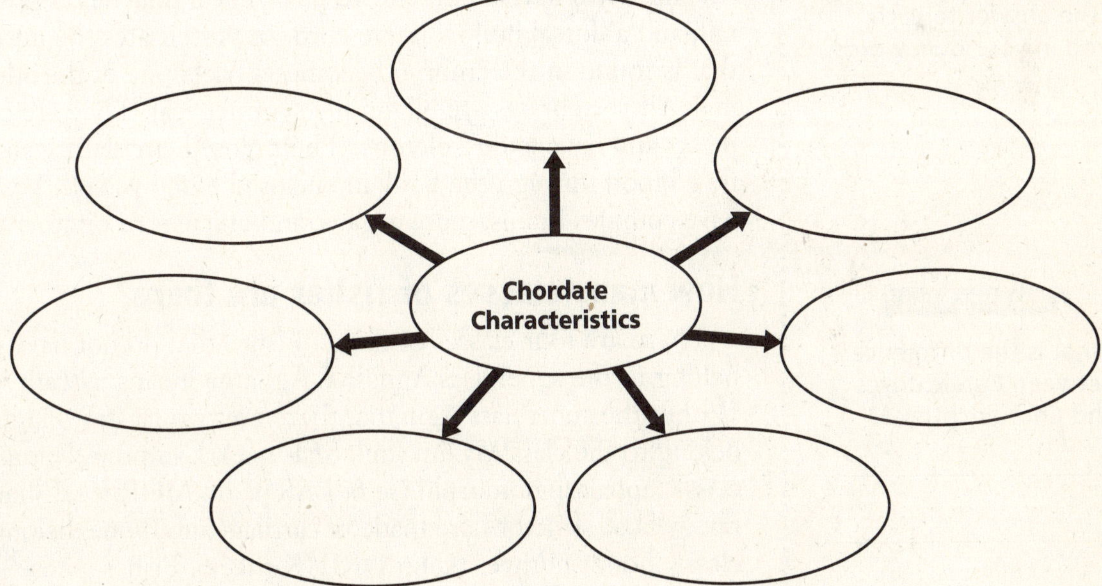

Visit the Glencoe Science Web site at **science.glencoe.com** to find your biology book and learn more about invertebrate chordates.

Name Date Class

Section 30.1 Fishes

Before You Read

Why do you think fish swim so easily and effortlessly in the water? On dry land, they cannot move efficiently, and they quickly die. On the lines below, explain why you think fish have adapted so well to their watery environment but have never adapted to a land environment.

Read to Learn

Mark the Text **Summarize** Highlight each of the question headings about fishes. After you have read the section, write the answer to each question in your own words.

What is a fish?

Fishes are vertebrates. This means that they have backbones. Fishes are members of the phylum Chordata. There are three subphyla in Chordata. The Urochordata include the tunicates or sea squirts. The Cephalochordata include the lancelets. Vertebrata are the vertebrates. Fishes, amphibians, reptiles, birds, and mammals are all vertebrates. Remember that vertebrates are chordates. All chordates have a notochord, pharyngeal pouches, a postanal tail, and a dorsal hollow nerve cord. In vertebrates the notochord that is found in the embryo becomes a backbone in the adult animal. All vertebrates are bilaterally symmetrical. They are coelomates and have endoskeletons. Their closed circulatory systems flow blood through the body in enclosed blood vessels. Vertebrates have complex brains, sense organs, and efficient respiratory systems.

1. What is the difference between Osteichthyes and Chondrichthyes?

How many classes of fishes are there?

There are four classes of fishes. Fishes that do not have jaws belong to the superclass Agnatha. Agnatha means *without jaws.* Within the superclass Agnatha, there are two classes. Hagfish belong to the class Myxini (mik SEE nee). Lampreys belong to the class Cephalaspidomorphi (se fa LAS pe do MOR fee). Sharks and rays, whose skeletons are made of cartilage, not bone, belong to the class Chondrichthyes (kahn DRIHK theez). Fishes whose skeletons are made of cartilage are called cartilaginous fishes. Cartilage is a tough, flexible material. Fishes with bone skeletons belong to the class called Osteichthyes (ahs tee IHK theez).

Fishes live in almost every kind of water environment on Earth. They live in salt water and freshwater. They live in shallow, warm water, as well as deep, cold water that has very little or no light.

How do fishes breathe?

Fishes breathe by using their gills. Notice in the figure to the right that gills are made up of filaments, which are feathery, thread-like structures. These filaments contain tiny blood vessels called capillaries. As a fish takes water in through its mouth, the water flows over the gills. The water goes out through slits in the side of the fish. Oxygen and carbon dioxide are exchanged through the tiny blood vessels, or capillaries, in the gill filaments.

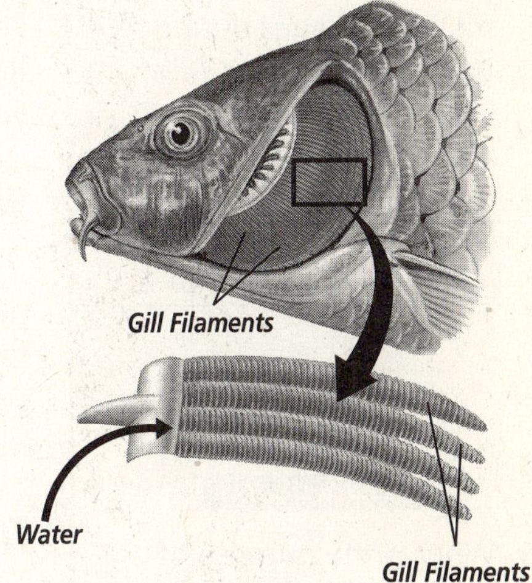

What kind of hearts do fishes have?

All fishes have two-chambered hearts. Blood that no longer has oxygen flows into one chamber of the heart from the body tissues. The second heart chamber pumps blood directly to the capillaries located in the fish's gills. Oxygen is picked up from the water passing over the gills. Carbon dioxide is released. The oxygen is carried from the gills to the fish's body tissues. Blood flows through a fish's body slowly. The flow of blood is shown in the figure to the right. Most of the heart's pumping action is used to push blood through the gills, not through the body tissues.

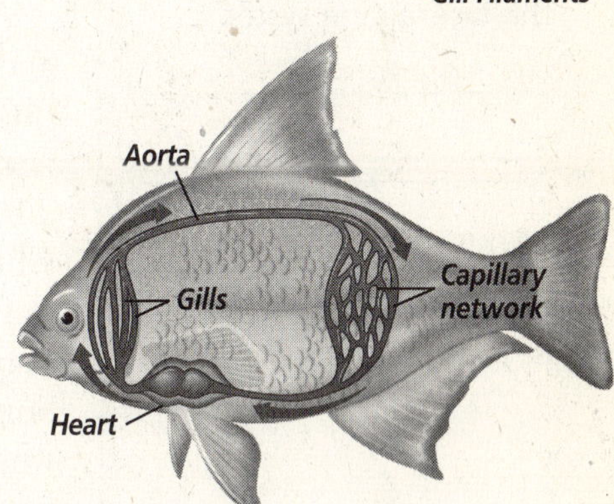

How do fishes reproduce?

All fishes reproduce sexually. The method of reproduction varies from one fish species to another. For most fishes, fertilization and development of the embryos is external. Eggs and sperm are released directly into the water. Some species of fishes leave their eggs and sperm in a more protected area, such as on plants floating in the water. Most fishes produce large numbers of small eggs at one time. Hagfishes produce small numbers of large eggs. ✓

Fishes whose skeletons are made of cartilage have internal fertilization. Skates leave fertilized eggs on the ocean floor. Some female sharks and rays carry developing young inside their bodies. These young fishes are well developed when they are born. This increases their chance for survival.

Most bony fishes have external fertilization and development. This type of external reproduction is called **spawning.** Fishes that spawn, such as cod, may produce as many as 9 million eggs. Most

Reading Check

2. What form of reproduction is common to all fishes?

______

______

______

Think it Over

3. **Analyze** Explain why most bony fishes produce millions of eggs during reproduction.

____

____

____

Reading Check

4. What is the purpose of the lateral line system?

____

____

____

of these fishes provide no care for their young after they spawn. Only a few of the young will live to be adults. In some bony fishes, such as guppies and mollies, fertilization and development take place internally. Some fishes, such as the mouth-brooding cichlids, stay with their young after they hatch. When young cichlids are threatened by predators, the parent fishes collect their young in their mouths to protect them.

What kind of fins do fishes have?

Fishes that belong to Chondrichthyes and Osteichthyes have paired fins. The **fins** are fan-shaped membranes. Fins are used for balance, swimming, and steering. Fins are attached to and supported by the endoskeleton. Fins foreshadowed the development of limbs for moving on land and wings for flying.

Do fishes have developed sensory systems?

All fishes have well-developed sensory systems. Cartilaginous and bony fishes have a lateral line system that helps them sense objects and changes in their environment. The **lateral line system** is a line of fluid-filled canals that runs along the sides of a fish. These canals make it possible for the fish to detect movement and vibrations in the water. ✓

Fishes have eyes that allow them to see objects. The eyes also help fishes see contrasts between light and dark in the water. Some fishes see well. Other fishes that live deep in the ocean where there is little or no light have little vision or their eyes do not function at all.

Some fishes have a developed sense of smell. These fishes can smell even small amounts of a chemical in the water. Sharks can follow a trail of blood for several hundred meters. This ability helps sharks find their prey.

What are fish scales?

Cartilaginous fishes and bony fishes have scales. Fish **scales** are thin bony plates formed from the skin. On some species, the scales are intermittent. This means that the fish body is not completely covered in a continuing pattern of scales. Some species have overlapping rows of scales. Scales have various shapes. They can be toothlike, shaped like diamonds, shaped like cones, or they may be round. Shark scales resemble teeth that are found in other vertebrates.

Do fishes have jaws?

The development of jaws in ancestral fishes was an important event. Jaws, such as those shown in the figure to the right, enable a fish to grasp and crush its food. Jaws allowed early fishes to eat a greater variety of organisms. This is one of the reasons that early fishes grew very large.

Sharks have jaws. They also have up to 20 rows of teeth that are continually replaced. The teeth of a shark point backward. This prevents prey from escaping from the shark's mouth.

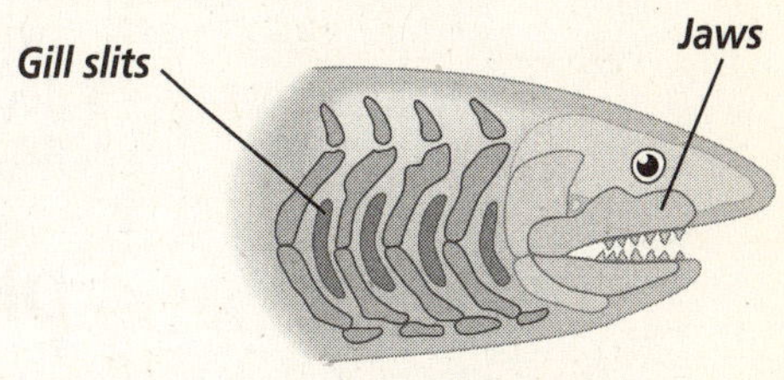

Fish with jaws

What kinds of skeletons do most fishes have?

Most fishes have bony skeletons. They belong to the class Osteichthyes, the bony fishes. The other two classes of fishes have skeletons made of cartilage, not bone. Bone is the hard, mineralized living tissue that makes up the endoskeleton of most vertebrates. There are many differences among species of bony fishes. They live in different habitats, eat different organisms, and have different shapes. Their bony skeletons allow them to adapt to a variety of water environments.

Are bony fishes flexible?

Bony fishes have a backbone composed of separate, hard segments. These segments are called vertebrae. The evolution of a backbone made of separate vertebrae in the vertebrate skeleton was important. Vertebrae provide the major support structure of the vertebrate skeleton. Because the vertebrae are separate, the animal is flexible. It can bend its back and its body. This is especially important for fish movement. When a fish swims, it continually flexes its backbone. Some fish can swim fast because of their flexible skeletons. ✓

Reading Check

5. What is the advantage of a segmented backbone?

____

____

____

What is a swim bladder?

A **swim bladder** is a thin-walled internal sac. It is located just below the backbone in most bony fishes. The swim bladder can be filled with mostly oxygen or nitrogenous gases that diffuse out of a fish's blood. If a fish has a swim bladder, it can control its depth by regulating the amount of gas in the bladder. The gas works like the gas in a blimp that adjusts the height of the blimp above the ground.

Some fishes have a special duct that attaches the swim bladder to the esophagus. Fishes use this duct to expel gases from the swim bladder. In fishes that do not have this duct, the swim bladder empties when the gases from the bladder diffuse back into the fish's blood.

Diversity of Fishes

Fishes can be tiny. The dwarf goby is less than 1 cm long. Fishes also can be huge. The whale shark can grow to 15 m—about the length of three cars.

What are Agnathans?

6. What is cartilage?

Hagfishes and lampreys are members of the superclass Agnathans. The skeleton of an Agnathan is made of cartilage, not bone. **Cartilage** is a tough, flexible material. Agnathans do not have jaws. A hagfish has a toothed mouth. It feeds on dead or dying fish. The hagfish can drill a hole into a fish and suck out the blood and the insides. Lampreys are parasites. Lampreys have sharp teeth and suckerlike mouths. They use their mouths to attack other fishes. They use their teeth to scrape away the flesh from their prey. Then they suck out the blood. ✓

Which fishes are cartilaginous?

Sharks, skates, and rays all have skeletons that are composed of cartilage, not bone. They are called cartilaginous fishes. They belong to the class Chondrichthyes. These fishes are similar to the sharks, skates, and rays that lived more than 100 000 years ago. Because of this similarity, sharks, skates, and rays are considered living fossils. Sharks are probably the best-known predator in the ocean.

Rays also are predators. They feed on or near the ocean floor. Rays have flat bodies and broad pectoral fins. A pectoral fin begins near the fish's chest. Rays flap their fins up and down slowly. This creates a gliding motion that rays use as they look for mollusks and crustaceans on the ocean floor. Some species of rays have sharp spines with poison glands on their tails. They use these glands to protect themselves. Other species of rays have organs that generate electricity so that the rays can stun or kill their prey and their predators.

Are there any subclasses of bony fishes?

There are two subclasses of bony fishes—lobe-finned fishes, including lungfishes, and ray-finned fishes. There are seven living species of lobe-finned fishes. Six species of lobe-finned fishes are lungfishes. They have both gills and lungs. The other species is the coelacanth (SEE luh kanth). Ray-finned fishes include catfish, perch, salmon, and cod. The fins of these fishes are fan-shaped membranes. Stiff spines called rays support the fins.

Origins of Fishes

Scientists have found fossils of fishes that existed during the late Cambrian Period, 500 million years ago. At that time, early jawless fishes called ostracoderms (OHS trah koh durmz) were the main vertebrates on Earth. Most ostracoderms became extinct at the end of the Devonian Period, about 354 million years ago. Present-day agnathans appear to be descendants of ostracoderms. ✓

Ostracoderms swam slowly over the seafloor. They had cartilaginous skeletons and heavy, bony, external armor. Shields of bone covered their heads and necks. In ancestral fishes, bone formed into plates, or shields, that protected the fishes. The development of bone in early vertebrates was important because bone provides a place for muscles to attach. This enabled the fishes to move more quickly and efficiently.

Ostracoderms may have been the common ancestor of all fishes. Modern cartilaginous and bony fishes evolved during the Devonian Period. Lobe-finned fishes, such as the coelacanths, are another ancient group. They appear in the fossil record about 395 million years ago. They had lobelike, fleshy fins, and they lived in deep places in the ocean. This makes them hard to find. The skeletal structure of fleshy fins is believed to be an ancestral trait of all animals that have four limbs, called tetrapods. The earliest tetrapods had gills and were aquatic.

Reading Check

7. When did most ostracoderms become extinct?

____

____

____

After You Read

Mini Glossary

cartilage: a tough, flexible material that forms the skeletons of two classes of fishes

fin: fan-shaped membrane used for balance, swimming, and steering

lateral line system: a line of fluid-filled canals that run along the sides of a fish that enable the fish to detect movement and vibrations in the water

scale: thin bony plates formed from the skin of fishes

spawning: external reproduction in some species of fishes; fishes that spawn may produce as many as 9 million eggs, and provide no care for their young after spawning

swim bladder: a thin-walled internal sac located just below the backbone in most bony fishes; the swim bladder can be filled with mostly oxygen or nitrogenous gases that diffuse out of a fish's blood

Section 30.1 Fishes, *continued*

1. Read the terms and definitions in the Mini Glossary on page 369. In the space below create a drawing of a shark or skate and label the parts, using at least five of the terms listed above.

2. Use the diagram below to describe the parts of a fish. Describe the parts listed in each oval.

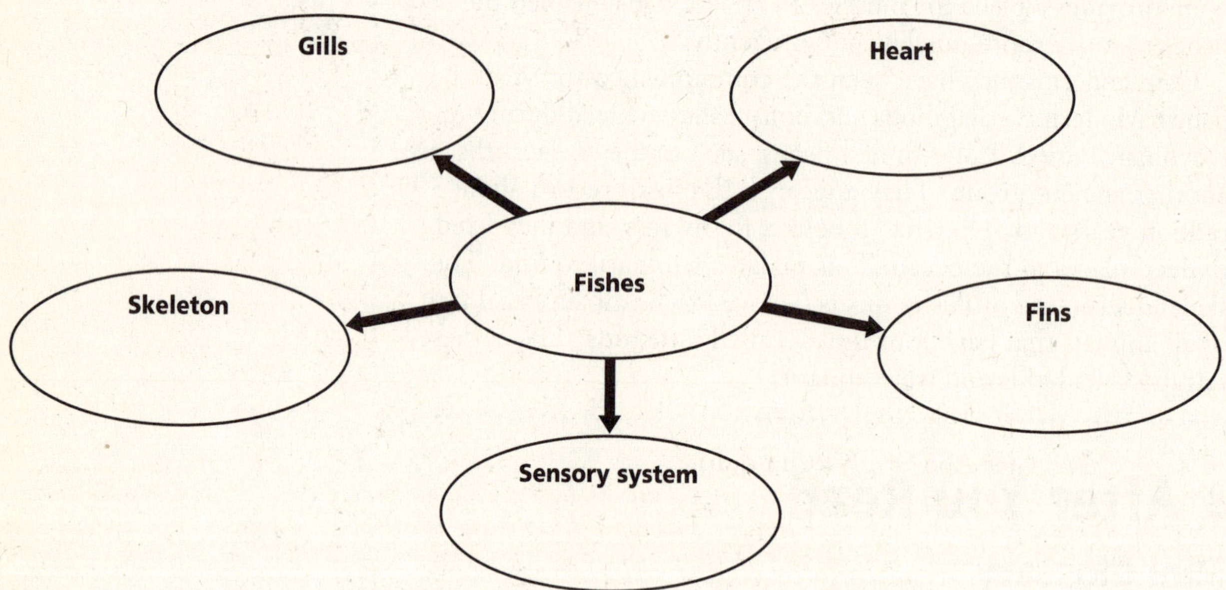

Visit the Glencoe Science Web site at **science.glencoe.com** to find your biology book and learn more about fishes.

Name Date Class

Section 30.2 Amphibians

Before You Read

You may have heard frogs croak or seen toads hopping near creeks or ponds. Frogs and toads are members of the same order and have a number of characteristics in common. On the lines below, explain what you have observed about frogs and toads.

____

____

____

____

Read to Learn

What is an amphibian?

Amphibia (am FIHB ee uh) means double life. Animals that belong to the class Amphibia change from aquatic to semiterrestrial during their life cycles. This is how they got the name amphibian. As larvae, almost all of these animals are completely aquatic. When most amphibians reach the adult stage, they breathe air. Amphibians can be found on land, but they must live near water or other moist areas. Therefore, these animals are called semiterrestrial. There are three orders in the class Amphibia. Salamanders and newts belong to the order Caudata (kaw DAH tuh). Frogs and toads belong to the order Anura (uh NUHR uh). Caecilians, amphibians that do not have legs, belong to the order Apoda (uh POH duh).

Amphibians have thin, moist skin. Most amphibians have four legs. Most adult amphibians can exist on land, but nearly all of them need water for reproduction. Fertilization in most amphibians is external, and water is needed to carry the sperm to the eggs. The eggs do not have a membrane or a shell to protect them. They must be laid in water or in other damp areas. ✓

What is an ectotherm?

An **ectotherm** (EK tuh thurm) is an animal that has a variable, or changing, body temperature. Amphibians cannot regulate their own body temperature or maintain their temperature at a stable

Mark the Text **Compare**

Highlight each question in this section about amphibians. Use these questions and the information in the text to write a comparison of amphibians and fishes.

Reading Check

1. Why do amphibians lay their eggs in water?

____

____

____

level as humans do. An ectotherm gets its heat from external sources. Amphibians are more common in areas that have warm temperatures all year round because they are ectotherms. Many biological processes require certain temperature ranges. Because of this, amphibians are dormant, or completely inactive, when they live in areas that are too hot or too cold for them during certain parts of the year. During these hot and cold times, many amphibians burrow into the mud and stay buried in the mud until the temperature changes.

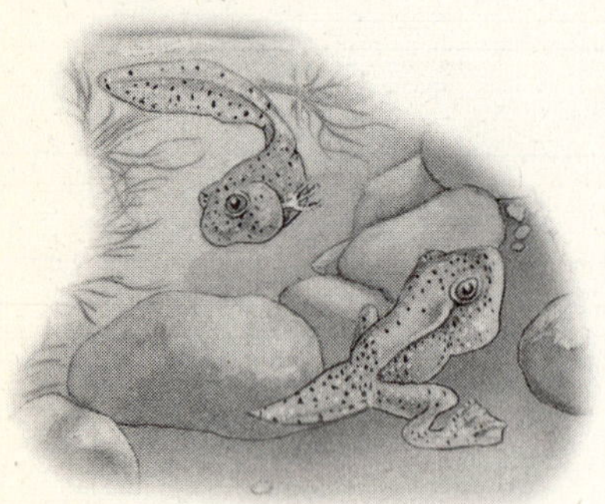

How does an amphibian change during metamorphosis?

Most amphibians go through metamorphosis. Fertilized frog and toad eggs hatch into tadpoles. This is the totally aquatic stage. Tadpoles have fins, gills, and a two-chambered heart like fishes. As tadpoles grow into adult frogs and toads, they develop legs, lungs, and a three-chambered heart. ✓

Young salamanders resemble adult salamanders. Most salamander larvae also are aquatic. In this stage they have gills, and they usually have a tail fin. Most adult salamanders do not have gills or fins. They breathe through their skin or with their lungs. Some salamanders do not have lungs. They can only breathe through their moist skin. Salamanders that live completely on land do not have a larval stage. When these young salamanders hatch, they look like small adults. Most salamanders have four legs, but some have only two front legs.

✓ Reading Check

2. What changes occur within the tadpole's body as it becomes an adult frog or toad?

____

____

____

What kind of hearts do amphibians have?

Amphibians have a three-chambered heart. When early amphibians began to walk, walking required a great deal of energy from food. It also required a large amount of oxygen for respiration. The evolution to a three-chambered heart helped to ensure that the amphibian's cells would receive enough oxygen. The change from a two- to a three-chambered heart was an important evolutionary change.

In the three-chambered heart, one chamber receives oxygen-rich blood from the lungs and the skin. Another chamber receives oxygen-poor blood from the body tissues. Blood from both of these two chambers goes into the third chamber of the heart. This third chamber pumps oxygen-rich blood into the body tissues. It also pumps oxygen-poor blood back to the lungs and the skin. This blood then picks up oxygen. The three-chambered heart mixes some oxygen-rich and some oxygen-poor blood.

Oxygen-rich and oxygen-poor blood also are mixed in blood vessels that lead away from the heart. This is why the skin in amphibians is more important than the lungs for gas exchange.

The skin of an amphibian must stay moist to exchange gases. This means that most amphibians have to live on the water's edge or in other, very moist areas. Some newts and some salamanders spend their entire lives in the water. Toads, which have a thicker skin, live primarily on land. They return to water to reproduce.

Amphibian Diversity

Because most amphibians depend on water to reproduce, they must live on the edges of ponds, streams, and rivers, or in areas that remain damp during parts of the year. Amphibian species are found worldwide.

How are frogs and toads different from other amphibians?

Frogs and toads are members of the order Anura. Frogs and toads do not have tails. Frogs have long back legs and smooth, moist skin. Toads have short legs and bumpy, dry skin. Frogs and toads have jaws and teeth. They are predators. They eat insects and worms. To defend themselves from predators, many frogs and toads secrete chemicals through their skin. ✓

3. Name two differences between frogs and toads.

____

____

Frogs and toads have vocal cords. **Vocal cords** are bands of tissue located in the throat. As air passes over the vocal cords, they vibrate. This makes molecules in the air vibrate, and a wide range of sounds is produced. In many male frogs, air passes over the vocal cords, and then goes into a pair of vocal sacs that are underneath the throat. This enlarges these sacs, making them look like small balloons.

Most frogs and toads spend part of their lives in water and part of their lives on land. They breathe through lungs or through their skins. Because they breathe through their skins, they are exposed to pollutants in the air, on land, or in the water.

How are salamanders different from other amphibians?

Salamanders belong to the order Caudata. Salamanders have long, slender bodies. They have necks and tails. Salamanders look like lizards, but they have smooth, moist skin. They do not have claws. Some salamanders spend their entire life cycle in water.

Others live on land in damp places. Salamanders can be small, a few centimeters long, or they can be up to 1.5 m long. When young salamanders hatch from eggs, they look like small adults. Salamanders are carnivores, or meat-eaters.

Think it Over

4. Analyze What characteristics do caecilians have that make them adapted for life as burrowers?

____

____

____

How are caecilians different than other amphibians?

Caecilians belong to the order Apoda. They do not have limbs. Caecilians are burrowing amphibians. Some have short tails; others have no tail at all. They have small eyes, but are often blind. Caecilians eat earthworms and other invertebrates that live in the soil. All caecilians have internal fertilization. They live primarily in tropical areas.

Origins of Amphibians

Tetrapods, animals with four legs, evolved over 360 million years ago. One type of tetrapod developed gills for breathing and a finned tail for swimming. These tetrapods may have used their limbs to move along the bottom of marshlands. Later fossils show the four limbs located further below the body. These limbs could lift the body off the ground. Amphibians probably arose as they developed the ability to breathe air through well-developed lungs. The ability to live on land depended on adaptations, or changes, that would provide support for the body, protect the membranes involved in respiration, and provide for efficient circulation.

Reading Check

5. Name two advantages and two disadvantages of living on land.

____

____

____

____

What were the benefits and challenges of living on land?

There were many benefits to living on land. There was a large food supply and good shelter. At the time, there were no predators. There was more oxygen on land than in the water. However, life on land was dangerous. Air temperature varied. When the body was out of the water, it was heavy and clumsy. Movement was more difficult. ✓

Amphibians first appeared about 360 million years ago. They probably evolved from an aquatic tetrapod around the middle of the Paleozoic Era. The climate on Earth was warm and wet. Because these early amphibians could breathe through their lungs, gills, or skin, they became the dominant vertebrates on the land.

After You Read

Mini Glossary

ectotherm (EK tuh thurm): an animal whose body temperature changes; an ectotherm gets its heat from external sources

vocal cord: band of tissue located in the throat; as air passes over the cord, it vibrates, and the molecules around the vocal cord tissues vibrate, making sounds

1. Read the terms and their definitions in the Mini Glossary above. Use the space below to write a brief paragraph explaining how these two terms relate to the life cycle of a frog.

2. Use the diagram below to show list the facts you have learned about the three orders of Amphibia. In the box labeled **Amphibians** list characteristics common to all amphibians.

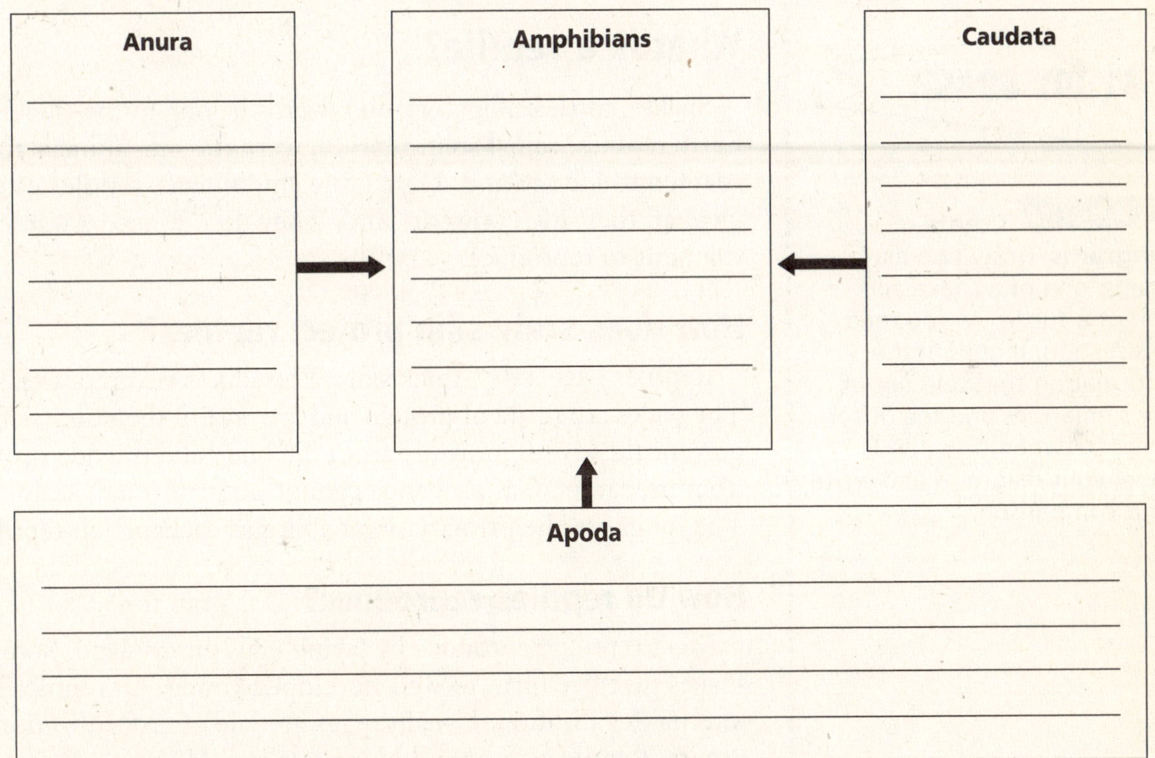

3. How are caecilians different from other amphibians? Write your response on the line below.

Visit the Glencoe Science Web site at **science.glencoe.com** to find your biology book and learn more about amphibians.

Section 31.1 Reptiles

Before You Read

Often lizards and snakes appear to be lazily enjoying a sunny day. At the same time, they may be watching for both prey and predators. Why do snakes and lizards stay in the warm sun if it means exposing themselves to danger? On the lines below, explain why you think lizards and snakes behave this way. After you read this section, check your explanation and update it to reflect what you have learned.

Read to Learn

STUDY COACH

Create Diagrams Draw two diagrams, one of a snake and one of a turtle. As you read this section, highlight the information that tells about the similarities and the differences between these two. Label your diagrams and write your comparison.

What is a reptile?

Snakes, turtles, alligators, and lizards belong to the class Reptilia. Early reptiles, called stem reptiles, were the first animals that adapted to life on land. Unlike the amphibians, reptiles can complete all their life cycles on land. They do not need a watery environment to reproduce.

How does scaly skin protect reptiles?

Reptiles have a dry, thick skin. The skin is covered with scales. The scales are made of protein and are part of the skin itself. Scales prevent the loss of moisture from the body and provide protection from predators. Gas exchange cannot occur through scaly skin. The lungs are the primary organs for gas exchange in reptiles.

How do reptiles reproduce?

Most reptiles reproduce by laying eggs on dry land. Some snakes give live birth to well-developed young. All reptiles have internal fertilization. Usually, eggs are laid after fertilization occurs. Reptile eggs are amniotic (am nee AH tihk). An **amniotic egg** provides nourishment, or food, for the embryo. An embryo is the earliest stage of growth and development of both plants and animals. The amniotic egg, shown in the figure on page 377, contains membranes that protect the embryo while it develops on land. An amniotic egg serves as the embryo's total life-support system. The evolution of the amniotic egg was the adaptation that enabled reptiles to reproduce on land rather than in water.

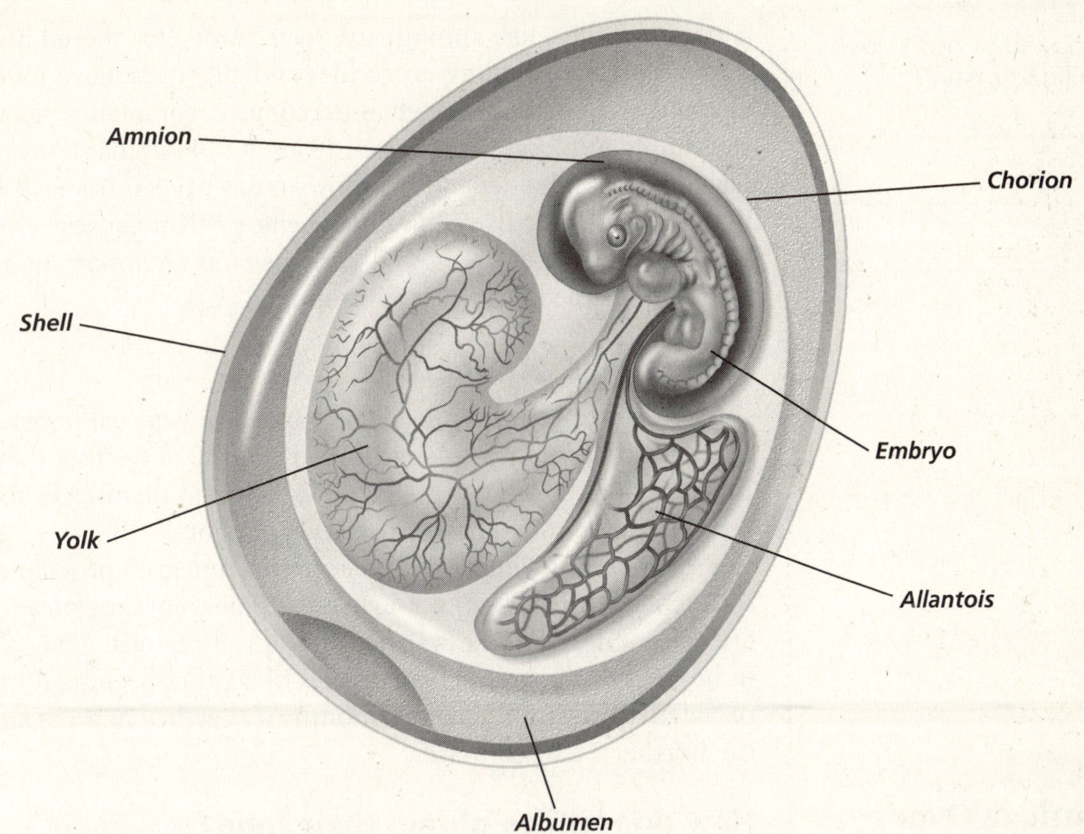

The embryos develop after the eggs are laid. Most reptiles lay their eggs under rocks, tree bark, or grass. Some reptiles dig holes or prepare a nest for the eggs. Most reptiles provide no care for the young, but female crocodiles may guard their nests from predators.

How have body changes helped reptiles?

The legs of amphibians are set at right angles to the body. The legs of early reptiles were placed more directly under the body. This under-the-body positioning provided better body support. It also made running and walking easier. Because reptiles with legs could run and walk, they could catch their prey more easily. They could also better avoid their predators. Reptiles with legs also have claws. Sharp claws help them catch food and protect themselves. Other evolutionary changes in the jaws and teeth of early reptiles helped them to use other resources found on land.

Reading Check

1. What kind of heart does an alligator have?

What kind of hearts do reptiles have?

Most reptiles, like amphibians, have three-chambered hearts. Some reptiles, including crocodiles and alligators, have four-chambered hearts. A four-chambered heart completely separates the supply of blood with oxygen from the blood that does not have oxygen. This separation allows more oxygen to reach body tissues. Land animals require more energy than aquatic animals. Delivering more oxygen to body tissues was an important adaptation that enabled reptiles to live on land.

Are reptiles ectotherms?

Reptiles are ectotherms. They depend on external heat to maintain their body temperatures within the range that they need for their bodies to function. Reptiles may sun themselves to get warm and then find shade when they get hot.

Because reptiles depend on their environment to provide warmth for their bodies, they do not live in extremely cold regions. Reptiles are commonly found in warmer or tropical regions. They also live in hot desert climates. Many species of reptiles become dormant, or inactive, during the colder months if they live in areas such as the northern United States.

Think it Over

2. Analyze Why are most turtles and tortoises herbivores?

How do reptiles obtain their food?

Some reptiles are herbivores, and some reptiles are carnivores. All reptiles, however, have adaptations that make it possible for them to find food. Most turtles and tortoises are too slow to be effective predators. Most of these species are herbivores, but the turtles and tortoises that are predators eat worms and mollusks. Snapping turtles are aggressive predators. They attack fish and amphibians and will even pull ducklings under the water.

Lizards mainly eat insects but some species of lizards are herbivores. The marine iguana of the Galápagos Islands eats marine algae. The largest lizard, the Komodo dragon, lives on several islands in Indonesia. The Komodo dragon is a strong predator and will attack humans. Although lizards, especially the large ones, may look slow or clumsy, they are able to run for short periods of time. They use this ability to catch their prey.

Snakes are also effective predators. Some species, including rattlesnakes, have poison fangs that they use to subdue or kill their prey. A constrictor wraps its body around its prey to prevent the prey from breathing.

How do reptiles use their sensory organs?

Reptiles have several sensory organs that help them detect danger. The sensory organs also help them detect food. The heads of some snakes have heat-sensitive organs. These organs or pits enable snakes to detect tiny changes in air temperature. These changes are brought about when a warm-bodied animal comes into a snake's environment.

Snakes and lizards also have a keen sense of smell. A snake will flick its tongue out. The tongue picks up molecules of air. The snake then pulls its tongue back into its mouth and moves its tongue into its Jacobson's organ. The **Jacobson's organ** is a pitlike structure located in the roof of the mouth in both snakes and lizards. Special cells in the Jacobson's organ help the animal identify and differentiate the smells found in the air molecules.

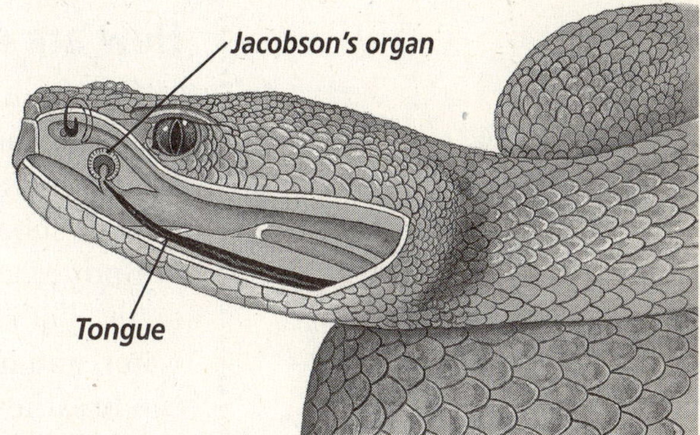

Diversity of Reptiles

There are four orders in the class Reptilia. Snakes and lizards belong to the order Squamata. Turtles belong to the order Chelonia. Crocodiles and alligators belong to the order Crocodilia. Tuataras, lizardlike reptiles, belong to the order Rhynochocephalia.

How are turtles different from other reptiles?

Turtles are the only reptiles that have a shell made up of two parts. The dorsal, or top, shell of the turtle's body is called the carapace. The ventral, or bottom, shell is called the plastron. The vertebrae and the expanded ribs of turtles are connected to the inside of the carapace. Most turtles have a two-layer shell. The inner layer is hard and bony. The outer layer is made of horny keratin. Keratin is a protein found in the exterior portion of the epidermis. Keratin helps to protect the living cells that are found in the interior epidermis. In a few species of turtles, the shell is made of tough, leathery skin. Most turtles can draw their limbs, their tails, and heads into their shells. They do this to protect themselves against predators. Turtles do not have teeth. They do have powerful jaws that have a beaklike structure. Turtles use their jaws to crush their food.

Not all turtles live on land. Some are aquatic. Turtles that live on land are called tortoises. Tortoises eat fruit, berries, and insects. The largest tortoises in the world live on the Galápagos Islands.

Some adult marine turtles swim long distances to lay their eggs. For example, green turtles travel from the coast of Brazil to Ascension Island in the Atlantic Ocean, more than 4000 km, to lay their eggs.

How are crocodiles different from other reptiles?

Crocodiles and alligators belong to the order Crocodilia. They are excellent, fast hunters. When they float in the water, only their eyes and nostrils remain above water. Crocodiles have long, slender snouts, while alligators have short, broad snouts.

Both crocodiles and alligators have powerful jaws and sharp teeth. They catch their prey in their jaws and teeth, drag it underwater, and hold it there until it drowns. Crocodiles and alligators can breathe air with their mouths full of food and water. This makes them especially efficient predators. The American alligator lives in freshwater. It can be found in the southeastern regions of the United States. The American alligator can grow up to 5 m long. The American crocodile lives only in salt water, and it can be found in the estuaries of southern Florida. Crocodiles, such as the Nile crocodile of Africa, can grow longer than alligators. ✓

Crocodiles and alligators do not migrate to reproduce. They lay their eggs in nests on the ground. They stay close to their nests and guard them from predators. Several species hold their newly hatched young in their mouths and carry the young alligators or crocodiles to the safety of the water.

Reading Check

3. How do alligators and crocodiles kill their prey?

____

____

____

How are lizards and snakes different from other reptiles?

Most lizards have four legs. Many species are adapted to hot, dry climates. Lizards are found in many places throughout the world. Some lizards live on the ground while others burrow. Some lizards live in trees, and some lizards are aquatic.

Snakes, unlike most vertebrates, do not have limbs. They do not have the bones to support limbs. Pythons and boas, however, do have bones of the pelvis. Snakes have many vertebrae, which permit them to move quickly over grass and rough land. Some snakes swim and climb trees.

Think it Over

4. Compare/Contrast How are snakes similar to and different from alligators?

____

____

____

____

____

Snakes usually kill their prey in one of three ways. Constrictors, including boas, pythons, and anacondas, wrap themselves around the prey and suffocate it. Venomous snakes use poison to paralyze or kill their prey. These include rattlesnakes, cobras, and vipers. When these snakes bite their prey, they inject poison from their venom glands. Most snakes are neither constrictors nor poisonous. They grab food with their mouths and swallow it whole. Snakes eat rodents, amphibians, insects, fishes, eggs, and other reptiles.

What are Rhynchocephalia?

There are two living species of this order. Both of the species are tuatara, and they are found only in New Zealand. Tuataras are the only survivors of a primitive group of reptiles. Tuataras have ancestral features including teeth that are fused to the edge of their jaws. Most of the other species of this order died out 100 million years ago.

Origins of Reptiles

The ancestors of snakes and lizards are traced to a group of early reptiles, called scaly reptiles. Scaly reptiles branched off from stem reptiles. Although the evolutionary history of turtles is not complete, scientists have suggested that they may also be descendants of stem reptiles. Dinosaurs and crocodiles are the third group to descend from stem reptiles. ✓

Scientists used to think that birds arose as a separate group from this third branch. Now there is fossil evidence that leads biologists to suggest that birds are the living descendants of the dinosaurs.

Reading Check

5. What ancestor do snakes, lizards, turtles, and crocodiles have in common?

____

____

____

Reptiles, *continued*

After You Read

Mini Glossary

amniotic egg: provides nourishment for the embryo; contains membranes that protect the egg while it develops on land

Jacobson's organ: a pitlike structure located in the roof of the mouth in both snakes and lizards; special cells in the organ help identify and differentiate smells

1. Read the terms and definitions in the Mini Glossary above. Then use the space below to explain why the development of an amniotic egg was critical to land animals.

____

____

____

2. Use the Venn diagram below to show the differences and similarities between reptiles that have limbs and those that do not.

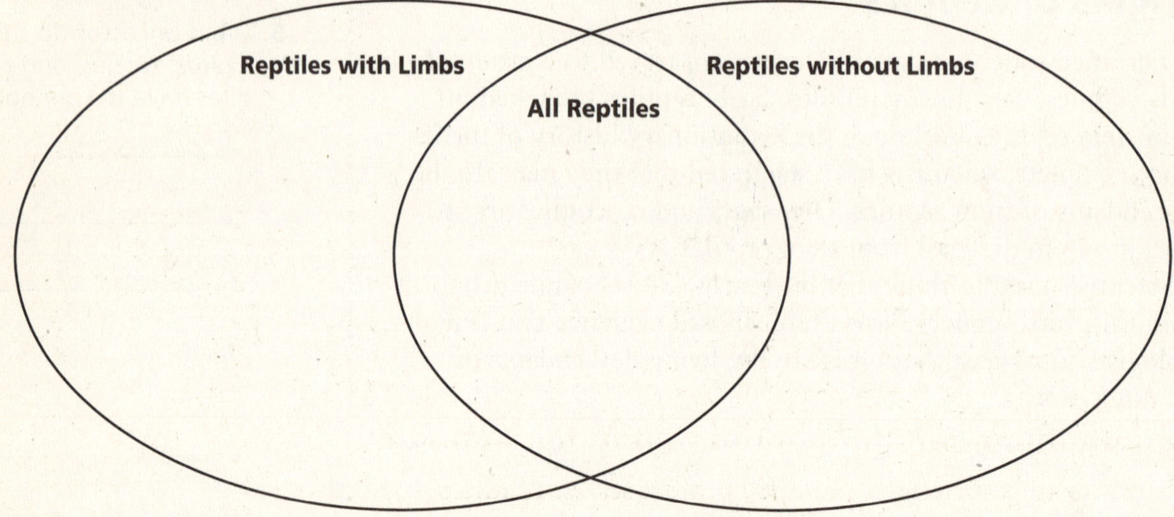

Visit the Glencoe Science Web site at **science.glencoe.com** to find your biology book and learn more about reptiles.

Name Date Class

Section 31.2 Birds

Before You Read

When you compare a bird and a reptile, there are many obvious differences. Yet they may be descended from a common ancestor. On the lines below, write ways in which birds and reptiles are similar and ways in which they differ.

____

____

____

Read to Learn

Mark the Text **Identify Details** Highlight the portions of text that provide information about bird flight.

What is a bird?

There are more than 8600 species of modern birds in the class Aves. Biologists sometimes refer to birds as feathered dinosaurs. Fossil evidence may indicate that birds evolved from small two-legged dinosaurs called theropods. Birds have clawed toes and protein scales on their feet. Fertilization is internal, and birds produce amniotic eggs that have shells. Amniotic eggs provide nourishment for the embryo developing inside. The shells offer some protection for the embryo. All birds have feathers and wings, but all birds do not fly. Birds live all around the world, including Antarctica, deserts, and tropical rain forests.

What are feathers?

Feathers are lightweight, modified protein scales. Feathers provide insulation and enable a bird to fly. Birds frequently run their bills or beaks through their feathers. This process, called preening, keeps the feathers in good condition for flight. When it preens, a bird also rubs oil from a gland located near the tail onto the feathers. Water birds must do this in order to waterproof their feathers. ✓

Bird feathers wear out and are replaced. Birds shed their old feathers and grow new ones in a process called molting. Most birds molt in late summer. They do not lose all their feathers at once and can continue to fly while they molt. Wing and tail feathers are usually lost in pairs. This enables a bird to maintain its balance while flying.

✓ Reading Check

1. Why do birds preen their feathers?

____

____

____

Reading Check

2. What two major adaptations were necessary so that birds could fly?

____

____

How do birds fly?

The front limbs of birds have changed into wings. Strong muscles that are used while a bird flies are attached to a large breastbone and to the upper bone of each wing. The breastbone is called the **sternum.** The sternum supports the thrust and power that the muscles produce when the wings move to get the bird off the ground. Feathers and wings are adaptations that allow birds to fly.

Why do birds need strong circulatory and respiratory systems?

Birds need high levels of energy to fly. Several factors are involved in maintaining these high energy levels. Birds' four-chambered hearts beat quickly. The rapid heartbeat moves oxygenated blood quickly through the body. The quick movement of blood keeps birds' cells supplied with the oxygen they need to produce energy.

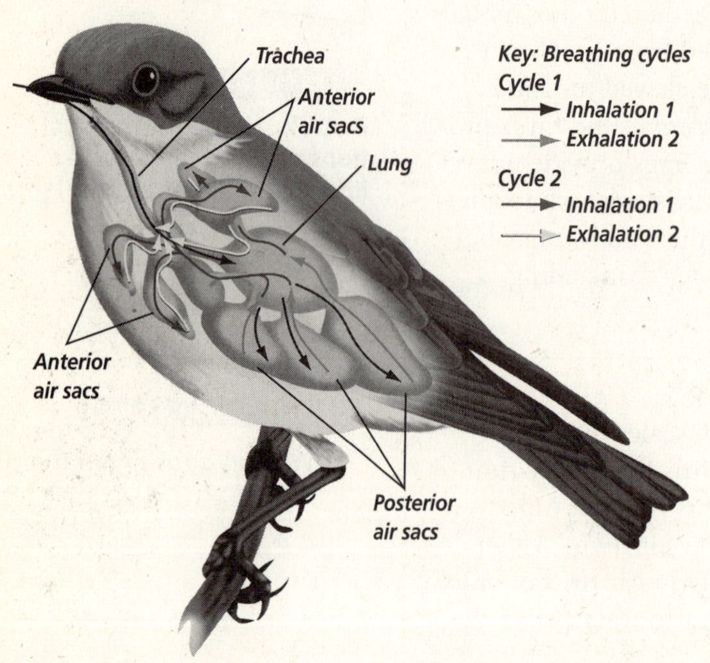

Birds have two breathing cycles. A bird's respiratory system brings oxygenated air to the lungs both when the bird inhales and when it exhales. The respiratory system consists of lungs and anterior, or front, and posterior, or back, air sacs. Air follows a one-way path in birds. When a bird inhales in Cycle 1, oxygenated air passes through its trachea and into the lungs. The trachea is located in the throat portion of the bird. Gas exchange occurs in the lungs. Most of the inhaled air passes directly into the posterior air sacs. When a bird exhales the deoxygenated air from the lungs, oxygenated air goes into the lungs from the posterior air sacs. At the next inhalation in Cycle 2, the deoxygenated air in the lungs flows into the anterior air sacs. Then, at the next exhalation, air flows from the anterior air sacs out of the trachea.

What is an endotherm?

Birds are endotherms. An **endotherm** is an animal that maintains a nearly constant body temperature. It does not depend on the environment to change its body temperature. Because they are endotherms, birds can maintain the high energy levels they need to fly. Birds have different ways to retain and release body heat to maintain a constant body temperature. When it is cold, feathers help reduce heat loss. Feathers fluff up and trap a layer of air around the bird's body. This limits the amount of heat that the bird loses. When it is hot, birds flatten their feathers and hold their wings away from their bodies to release heat. Birds also pant to increase respiratory heat loss.

Because birds are endotherms, they can live in all environments. Birds are found in the arctic regions and in the hot tropics. In order to maintain the high energy levels, birds and other endotherms must eat large amounts of food.

How do birds reproduce?

Birds, like reptiles, reproduce by internal fertilization and they lay amniotic eggs. Bird eggs have a hard exterior shell. Birds usually make a nest and lay their eggs in the nest. Nests may be made of bits of straw and twigs or may be a depression in the sand. Some birds build nests that they add to every year. Birds **incubate** or sit on their eggs to keep them warm. Birds turn the eggs in the nest so that the eggs will develop properly. In some species, both parents, male and female birds, take turns incubating the eggs. In others, only one parent incubates the eggs. Bird eggs do not all look the same. Often the species of bird can be identified by the color, size, and shape of an egg.

Think it Over

3. **Evaluate** What are the advantages of being an endotherm rather than an ectotherm?

____

____

____

Diversity of Birds

Reading Check

4. What adaptations help ptarmigans live in colder climates?

____

____

____

The basic form and structure of all birds are similar. They do show differences, or adaptations, depending on where they live and what they eat. Ptarmigans have feathered legs and feet, which serve as snowshoes in the winter, making it easier for them to walk in the snow. Penguins do not fly. Their wings and feet are modified for swimming. They have a thick layer of insulating fat on their bodies to help keep them warm. Owls have large eyes, a keen sense of hearing, and sharp claws, which make them successful predators of the night. They can swoop down on their unsuspecting prey. ✓

The shape of a bird's beak or bill indicates what kind of food the bird eats. Hummingbirds have long beaks used to obtain nectar from flowers. Hawks have curved beaks that tear apart their prey. Pelicans have large bills with pouches. They use the pouches as nets for capturing fish. A cardinal has a short, stout beak for cracking seeds.

Origins of Birds

Caudipteryx zoui

Scientists hypothesize that today's birds come from an evolutionary line of dinosaurs that did not become extinct. They evolved. *Archaeopteryx* is the earliest known bird in the fossil record. *Archaeopteryx* was about the size of a crow and had feathers and wings similar to a modern bird. It also had teeth, a long tail, and clawed front toes.

Fossil finds in China support the idea that birds evolved from a theropod dinosaur. It did not fly and it ran to capture its prey. It was about 1 m tall and had feathers. The feathers may have helped to insulate the animal or they may have been used for camouflage. Scientists suggest that feathers evolved before flight. The figure at the left pictures how a theropod dinosaur may have looked.

Modern birds and theropods have other features in common. Both have a sternum, a wishbone, shoulder blades, flexible wrists, and three fingers on each hand.

After You Read

Mini Glossary

endotherm: animal that maintains a nearly constant body temperature; it is not dependent upon the environmental temperature

feathers: lightweight, modified protein scales; provide insulation, enable a bird to fly

incubate: to sit on eggs to keep them warm

sternum: breastbone

1. Read the terms and their definitions in the Mini Glossary above. Then use the space below to write a brief paragraph explaining how birds fly and how they reproduce.

2. Use the diagram below to write the facts you have learned about each step of the respiratory system of birds.

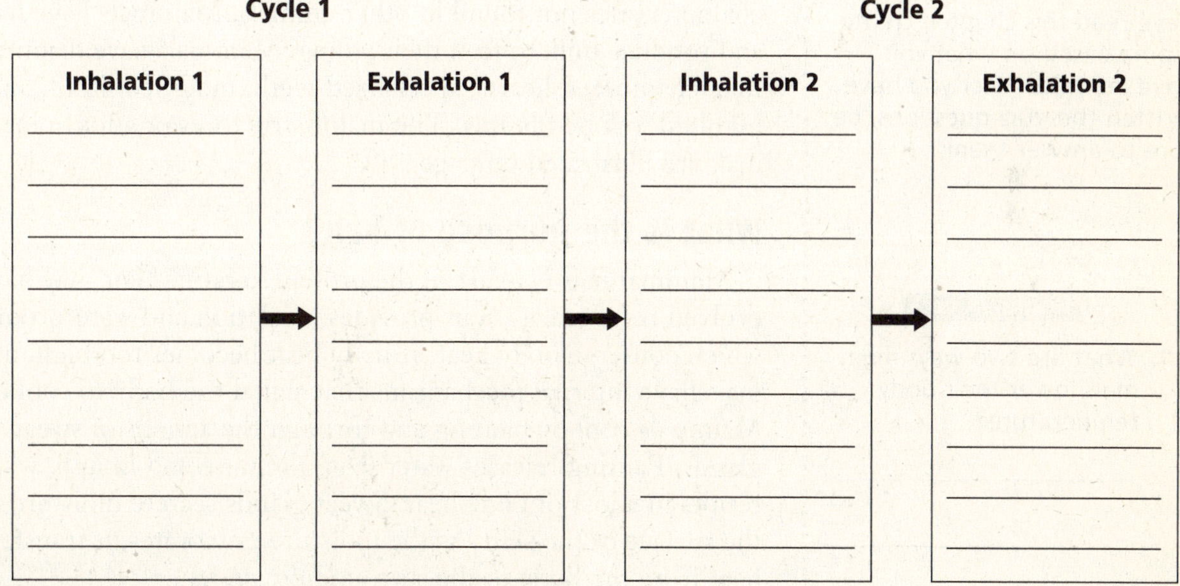

Visit the Glencoe Science Web site at **science.glencoe.com** to find your biology book and learn more about birds.

Name Date Class

Section 32.1 Mammal Characteristics

Before You Read

You may have had a dog that gave birth to puppies. Even the friendliest mother dog may growl when someone attempts to come near her pups. How do the puppies benefit from the mother dog's protective behavior?

Read to Learn

Create a Quiz After you have read this chapter, create a quiz based on what you have learned. After you have written the quiz questions, be sure to answer them.

1. What are two ways mammals lower their body temperature?

What is a mammal?

Mammals are endotherms. They have the ability to maintain a fairly constant body temperature. This enables mammals to live in almost every possible environment on Earth. Mammals have characteristics not found in other animals. Mammals have hair and produce milk to feed their young. Mammals have diaphragms, four-chambered hearts, specialized teeth, modified limbs, and highly developed brains. The major structures of a fox, a mammal, are illustrated on page 389.

What is the purpose of hair?

Mammal hair is made of the protein keratin. Hair may have evolved from scales. Hair provides insulation and waterproofing, which conserve body heat. If body heat becomes too high, mammals have internal mechanisms that signal the body to cool off. Mammals cool by panting and through the actions of sweat glands. Panting releases water from the nose and mouth, which results in a loss of body heat. Sweat glands secrete moisture onto the surface of the skin. As the moisture evaporates, it transfers heat from the body to the surrounding air.

How do mammals feed their young?

Mammals have several types of **glands,** which are groups of cells that secrete fluids. Glands produce saliva, sweat, oil, digestive enzymes, milk, and scent.

Mammals feed their young from **mammary** (MA muh ree) **glands.** These may be modified sweat glands, which produce

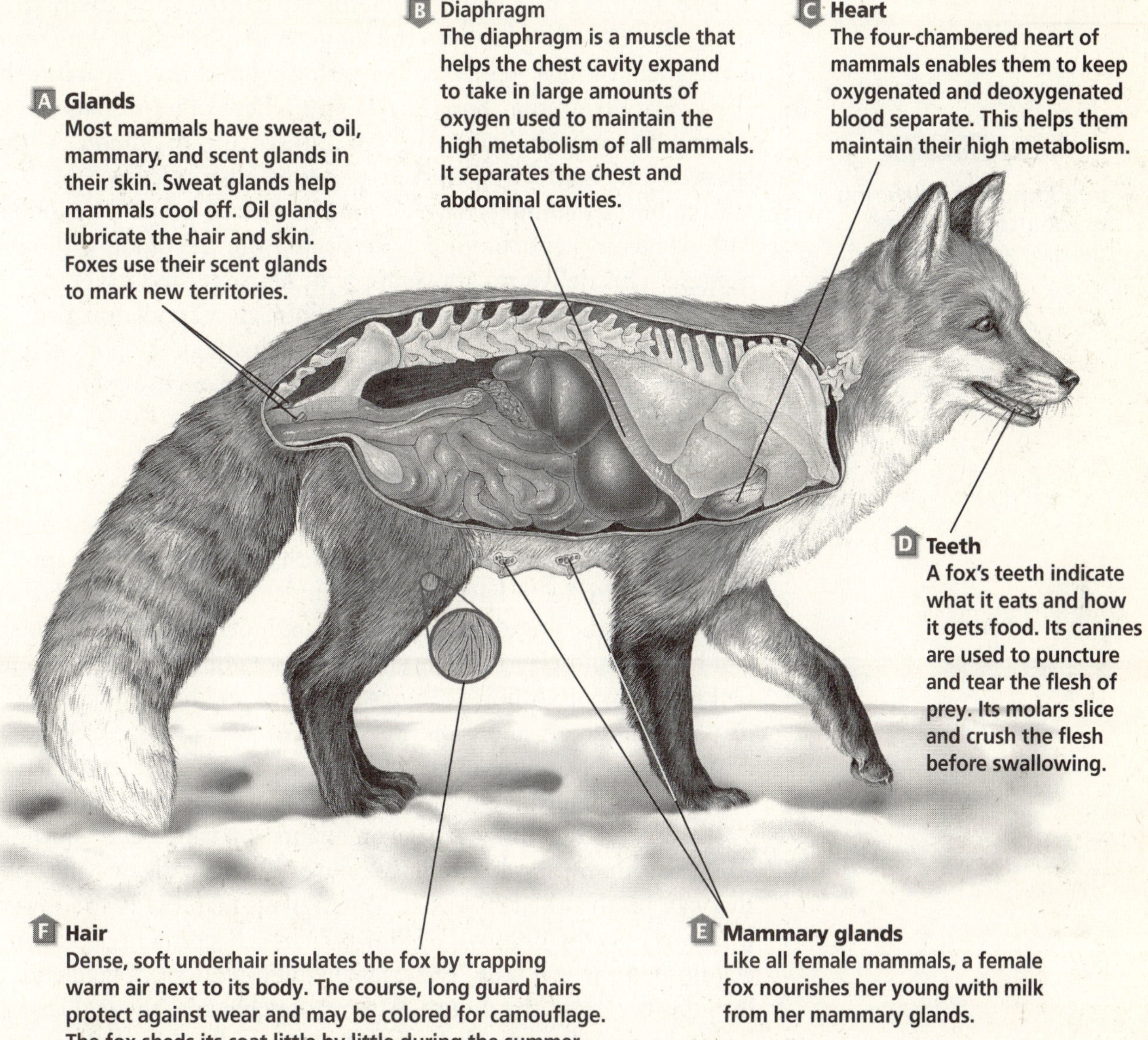

A Glands
Most mammals have sweat, oil, mammary, and scent glands in their skin. Sweat glands help mammals cool off. Oil glands lubricate the hair and skin. Foxes use their scent glands to mark new territories.
B Diaphragm
The diaphragm is a muscle that helps the chest cavity expand to take in large amounts of oxygen used to maintain the high metabolism of all mammals. It separates the chest and abdominal cavities.
C Heart
The four-chambered heart of mammals enables them to keep oxygenated and deoxygenated blood separate. This helps them maintain their high metabolism.
D Teeth
A fox's teeth indicate what it eats and how it gets food. Its canines are used to puncture and tear the flesh of prey. Its molars slice and crush the flesh before swallowing.
E Mammary glands
Like all female mammals, a female fox nourishes her young with milk from her mammary glands.
F Hair
Dense, soft underhair insulates the fox by trapping warm air next to its body. The course, long guard hairs protect against wear and may be colored for camouflage. The fox sheds its coat little by little during the summer.

and secrete milk. Milk is a liquid rich in fats, sugars, proteins, minerals, and vitamins. Mammals nurse their young until the young are able to digest and absorb nutrients from solid food.

How do the processes of respiration and circulation work in mammals?

Mammals need a high level of energy to maintain their endothermic metabolism. Remember that metabolism is all the chemical reactions that occur in an organism. The energy level is maintained when large amounts of nutrients and oxygen enter the body and reach the cells.

The diaphragm of mammals helps expand the chest cavity to allow the flow of oxygen into the lungs. A **diaphragm** (DI uh fram) is the sheet of muscle located beneath the lungs that separates the chest cavity from the abdominal cavity, where other organs are located. Once in the lungs, oxygen diffuses into the blood. As the chest cavity returns to its resting position, air is released.

Like birds, mammals have a four-chambered heart. The blood with oxygen is kept completely separate from the blood without oxygen. This delivers a good supply of nutrients and oxygen to cells, supporting endothermic metabolism. Circulation also removes waste products from cells and helps regulate body temperature. Blood helps keep a constant cellular environment, maintaining homeostasis. ✓

Reading Check

2. In addition to delivering oxygen to cells, what does circulation do?

______

______

______

Why do mammals have varied teeth?

Teeth are a distinguishing feature of most mammals. Although fishes and reptiles have teeth, their teeth are all about the same. Mammals with teeth have different kinds that are adapted to the type of food the animal eats. The pointed incisors of moles grasp and hold small prey. The chisel-like incisors of beavers are modified for gnawing. A lion's canines puncture and tear the flesh of its prey. Premolars and molars are used for slicing or shearing, crushing, and grinding. By examining the teeth of a mammal, a scientist can determine what kind of food it eats.

Many hoofed mammals have an adaptation called cud chewing. Cud chewing breaks down the cellulose in plant walls into nutrients that can be absorbed and used. When plant material is swallowed, it moves into the first two of four pouches. Bacteria break down cellulose in the cell wall. The partially digested food, called cud, is brought back into the mouth. After more chewing, the cud is swallowed again. When the food particles are small enough, they are passed on to the other stomach areas where digestion continues.

Reading Check

3. How do limb adaptations help mammals meet their energy needs?

______

______

______

What are the benefits of limb adaptations?

Mammals have several adaptations for gathering food to meet their energy needs. For example, primates use their opposable thumb to grasp objects, such as fruits and other foods. Mammals have other limb modifications. Moles have short powerful limbs with large claws that help them dig. Bats have long finger bones that support the flight membrane of their wings. ✓

Do mammals learn and remember?

One reason mammals are successful is that they protect their young, sometimes fiercely. They also teach their young survival skills. Mammals can accomplish complex behaviors, such as learning and remembering what they have learned.

Primates, including humans, are perhaps the most intelligent animals. For example, chimpanzees can use tools, operate machines, and use sign language to communicate with humans. The intelligence of mammals is a result of complex nervous systems and highly developed brains. The outer layer of a mammalian brain often is folded, forming ridges and grooves. These ridges and grooves increase the brain's active surface area.

Think it Over

4. **Infer** List animal behaviors that illustrate intelligence.

____

____

____

After You Read

Mini Glossary

diaphragm (DI uh fram): in mammals, the sheet of muscles located beneath the lungs that separates the chest from the abdominal cavity; expands the chest cavity, which increases the amount of oxygen entering the body

gland: in mammals, a cell or a group of cells that secretes fluids

mammary (MA muh ree) gland: modified gland in female mammals, which produces and secretes milk to feed their young

1. Read the terms and the definitions in the Mini Glossary above. Write a sentence using at least two of the terms correctly.

____

____

2. Match the concepts from Column 1 with the examples in Column 2.

	Column 1	Column 2
_____	1. Mammals maintain a fairly constant body temperature.	a. highly developed brain
_____	2. Mammals learn and remember.	b. opposable thumb
_____	3. Mammals regulate temperature by excreting moisture.	c. endotherms
_____	4. Scientists can determine what a mammal eats by examining teeth.	d. sweat glands
_____	5. Mammals have modified limbs.	e. specialized teeth

Visit the Glencoe Science Web site at **science.glencoe.com** to find your biology book and learn more about mammal characteristics.

Name Date Class

Section
32.2 Diversity of Mammals

Before You Read

You may have seen kangaroos at the zoo. Write down all the information you know about kangaroos on the lines below. After you read this section, add the subclass to which kangaroos belong.

____

____

____

Read to Learn

Mark the Text **Identify Main Points** Highlight each subclass of mammal as it is introduced in this section. Circle the distinguishing features of each subclass.

1. How has gestation helped mammals to succeed?

____

____

____

Mammal Classification

Scientists place mammals into one of three subclasses based on their method of reproduction. The first subclass you will study is placental mammals. About 90 percent of all mammals are placental. **Placental mammals** give birth to young that have developed inside the mother's uterus. Birth occurs when the body systems of the young are fully functional and they can live outside their mother's body. The **uterus** (YEWT uh rus) is a hollow, muscular organ in which offspring develop. The young are nourished inside the uterus through an organ called the **placenta** (pluh SEN tuh). The placenta develops during pregnancy. Nutrients and oxygen pass through the placenta to the developing embryo. Wastes from the embryo are removed through the placenta.

The time during which placental mammals develop inside the uterus is called **gestation** (jeh STAY shun). The length of gestation varies from species to species. Developing inside the mother's body is an adaptation that has helped mammals succeed. This is because the offspring are protected from predators and the environment during the early stages of development. ✓

How is reproduction different in marsupials?

Marsupials make up the second subclass of mammals. A **marsupial** (mar SEW pee uhl) is a mammal in which the young have a short period of development within the mother's body, followed by a period of development inside a pouch made of skin and hair on the outside of the mother's body.

Section 32.2 Diversity of Mammals, *continued*

There is only one North American marsupial, the opossum. Most marsupials are found in Australia and surrounding islands. The theory of plate tectonics explains why most marsupials are found in Australia. Scientists have found fossil marsupials on continents that once made up Gondwana (a prehistoric super continent made up of Australia, South America, Africa, India, and Antarctica). These fossils support the idea that marsupials originated in South America, moved across Antarctica, and populated Australia before Gondwana broke up.

What effect have placental mammals had on marsupials?

Ancestors of today's marsupials did not have to share the landmass that became Australia with the competitive placental mammals. Placental mammals evolved in other places. Marsupials were able to spread and fill niches similar to those occupied by placental mammals in other parts of the world. For example, the giant anteater of Mexico, a placental mammal, has a long, sticky tongue that it uses to collect ants and termites from their nests. The numbat of Australia, a marsupial, fills the same niche. The numbat has a long, sticky tongue that it uses to eat termites and ants. Since the introduction of placental mammals, such as sheep and rabbits, to Australia, many native marsupial species have become threatened, endangered, or even extinct. Remember that endangered means that the number of individuals falls so low that extinction is possible.

Which mammals lay eggs?

A mammal that reproduces by laying eggs is called a **monotreme** (MA nuh treem). Only three species of monotremes are alive today. Monotremes are found only in Australia, Tasmania, and New Guinea. ✓

The platypus is a mostly aquatic monotreme with a broad, flat tail much like that of a beaver. Its rubbery snout looks like the bill of a duck. The platypus has webbed front feet for swimming. All four feet have sharp claws for digging and burrowing. Much of the body is covered with thick, brown hair. Like all mammals, the platypus has mammary glands.

There are two species of spiny anteaters, or echidnas, in the monotreme subclass. The spiny anteater has coarse, brown hair. Its back and sides are covered with sharp spines that it can raise to defend itself when threatened by enemies. From its mouth, the anteater extends its long, sticky tongue to catch insects.

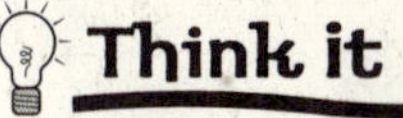

2. **Infer** How does the presence of placental mammals threaten marsupials in Australia?

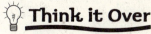

Reading Check

3. Where are monotremes found?

Section 32.2 **Diversity of Mammals,** *continued*

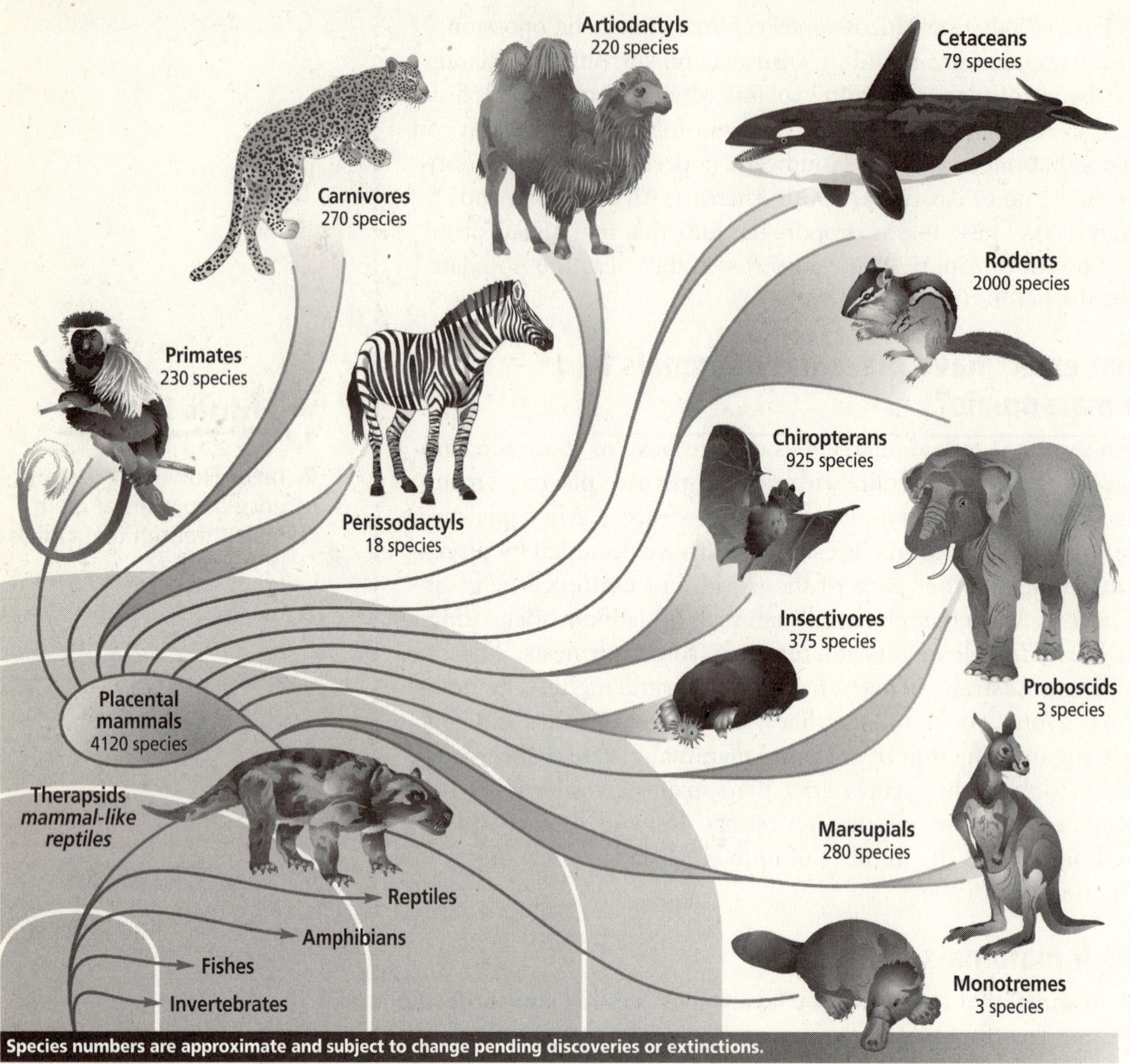

Origins of Mammals

The first placental mammals appeared in the fossil record about 125 million years ago. The oldest placental mammal fossil is *Eomaia*, a group of mouse-sized animals. Scientists trace the origins of placental mammals to a group of reptilian ancestors called therapsids. **Therapsids** (ther AP sidz) had features of both reptiles and mammals. They existed between 270 and 180 million years ago. ✓

Reading Check

4. Why do scientists suggest that therapsids are the ancestors of mammals?

____

____

____

Section
32.2 Diversity of Mammals, *continued*

The mass extinction of the dinosaurs at the end of the Mesozoic Era, the breaking apart of Pangaea (all the continents formed a single land mass), and changes in climate opened up new niches for early mammals to fill. Flowering plants appeared at the end of this era. They provided new living areas, food sources, and shelter. Some mammals that moved into the drier grasslands became fast-running grazers, browsers, and predators. The Cenozoic Era is sometimes called the golden age of mammals because of the dramatic increase in their numbers and diversity.

Reading Check

5. How did the development of flowering plants benefit mammals?

After You Read

Mini Glossary

gestation (jeh STAY shun): time during which placental animals develop inside the uterus

marsupial (mar SEW pee uhl): subclass of mammals in which young develop for a short period in the uterus and complete their development outside of the mother's body inside a pouch made of skin and hair

monotreme (MAHN uh treem): subclass of mammals that have hair and mammary glands but reproduce by laying eggs

placenta (pluh SEN tuh): organ that provides food and oxygen to and removes wastes from young inside the uterus of placental mammals

placental mammal: mammals that give birth to young that have developed inside the mother's uterus until their body systems are fully functional and they can live independently of their mother's body

therapsids (ther AP sidz): reptilian ancestors of mammals that had features of both reptiles and mammals

uterus (YEWT uh rus): in females, the hollow, muscular organ in which offspring of placental mammals develop

1. Read the terms and their definitions in the Mini Glossary above. Circle the subclass of mammals that has the fewest species. Highlight the subclass of mammals that contains the most mammals. Then use the term **gestation** in a sentence on the lines below.

Section 32.2 Diversity of Mammals, *continued*

2. Use the diagram to compare and contrast the three subclasses of mammals. List characteristics of each group in the appropriate box.

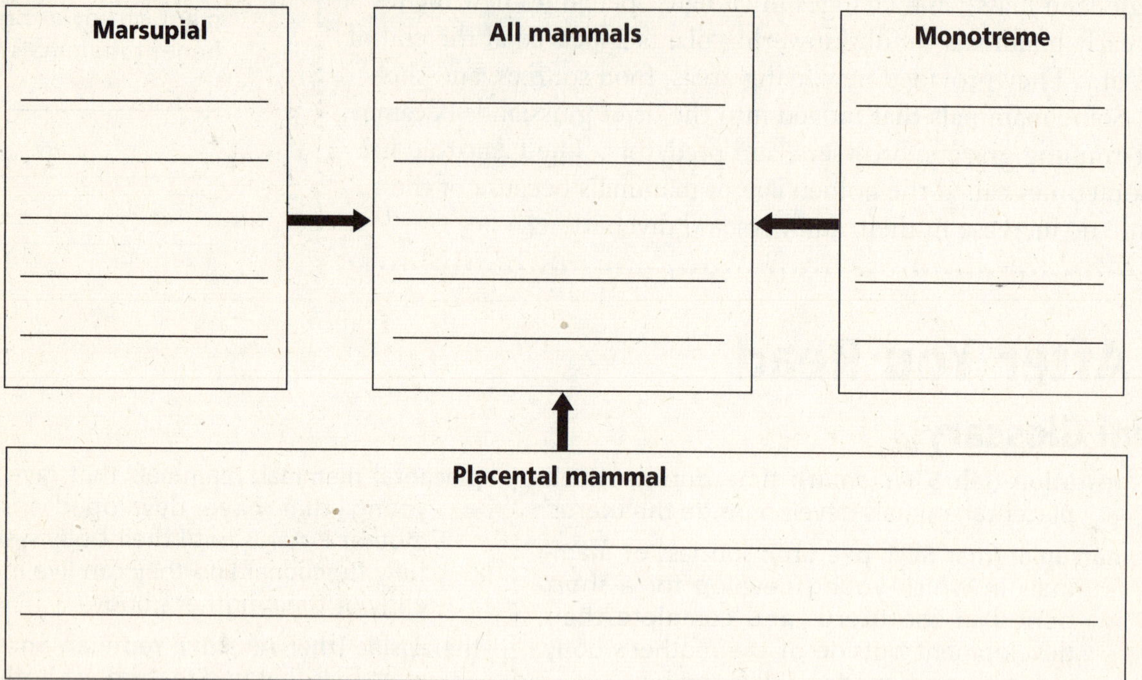

Visit the Glencoe Science Web site at **science.glencoe.com** to find your biology book and learn more about the diversity of mammals.

Name Date Class

Section 33.1 Innate Behavior

Before You Read

You may live in an area where you can observe animals migrating. Have you seen flocks of birds flying overhead in spring or fall? Perhaps you have seen salmon returning upstream. How do animals know when to migrate? Write your thoughts on the lines below.

____

____

____

Read to Learn

Mark the Text **Identify Main Ideas** This section introduces innate behavior. Skim the section and highlight three important facts about innate behavior.

What is behavior?

A peacock displaying his colorful tail, a whale spending the winter months in the ocean off the coast of southern California, and a lizard seeking shade from the hot desert sun are all examples of animal behavior. **Behavior** is anything an animal does in response to a stimulus. A stimulus is an environmental change that directly influences the activity of an organism. The presence of a peahen stimulates the peacock to open its tail feathers. A change in the length of daylight hours may cause the whale to leave its summertime arctic habitat. Heat stimulates the lizard to seek shade.

Inherited Behavior

Inheritance plays an important role in the ways animals behave. You would not expect to see a hummingbird tunnel underground or a mouse fly. Yet, why does a mouse run away when a cat appears? Why does a hummingbird fly south for the winter? These behaviors are genetically programmed. An animal's genetic makeup determines how that animal reacts to certain stimuli.

Does natural selection favor certain behaviors?

Often, a behavior exhibited by an animal species is the result of natural selection. A variety of behaviors among individuals affects their ability to survive and reproduce. Individuals with behavior that makes them more successful at surviving and reproducing usually produce more offspring. These offspring inherit the genetic basis for the successful behavior. Individuals with less successful behavior produce fewer offspring or none at all. ✓

Reading Check

1. How does natural selection favor certain behaviors?

____

____

____

Inherited behavior of animals is called **innate** (ih NAYT) **behavior.** A toad captures prey by flipping out its tongue. To capture prey, a toad must first be able to detect and follow the prey's movement. Toads have "insect detector" cells in the retinas of their eyes. As an insect moves across the toad's line of sight, the "insect detector" cells signal the brain, causing an innate response; the toad's tongue flips out. This is an innate behavior known as a fixed-action pattern. A fixed-action pattern is an unchangeable behavior pattern that, once begun, continues until completed.

What is the basis of innate behavior?

Reading Check

2. What does innate behavior include?

Scientists have found that an animal's hormonal balance and its nervous system affect how sensitive the individual is to certain stimuli. The sense organs responsible for sight, sound, touch, and odor identification are especially important. In fire ant colonies, a single gene influences the acceptance or rejection of the ant queen, thereby controlling the colony's social structure. Innate behavior includes fixed-action response, automatic response, and instincts. ✓

Automatic Responses

What happens if something is thrown at your face? Your first reaction is to blink and jerk back your head. Even if a protective clear shield is placed in front of you, you cannot stop yourself from behaving this way. This reaction is called a reflex, the simplest form of innate behavior. A **reflex** (REE fleks) is a simple, automatic response to a stimulus that involves no conscious control. If you accidentally touch a hot stove, you will automatically jerk your hand away. Before you even have time to think about it, the reflex movement saves your body from serious injury.

Reading Check

3. What controls the fight-or-flight response?

Another automatic response, called fight-or-flight response, has adaptive value. Think about a time when you were suddenly scared. Your heart began to beat faster. Your skin got cold and clammy and your breathing rate increased. You were having a fight-or-flight response. A **fight-or-flight response** mobilizes the body for greater activity. Your body is being prepared to either fight or run from the danger. A fight-or-flight response is automatic and controlled by hormones and the nervous system. ✓

Instinctive Behavior

The fixed-action response of the toad capturing prey, the reflex response to a hot stove, and the fight-or-flight response are quick, automatic responses to stimuli. Some behaviors, however, take a longer time because they involve more complex actions. An **instinct** (IHN stingt) is a complex pattern of innate behavior. Instinctive behavior begins when the animal recognizes a stimulus and continues until all parts of the behavior are completed. For example, a female greylag goose instinctively retrieves an egg that she sees has rolled out of the nest.

Much of an animal's courtship behavior is instinctive. **Courtship behavior** is the behavior that males and females of a species carry out before mating. Like other instinctive behaviors, courtship has evolved through natural selection. Courtship behavior helps members recognize other members of the same species. That is important for the survival of the species. In courtship, behavior ensures that members of the same species find each other and mate. Such behavior has adaptive value for the species. For example, different species of fireflies can be seen at dusk flashing distinct light patterns. Female fireflies of one species respond only to those males flashing the species-correct patterns. ✓

Some courtship behaviors prevent females from killing males before they have had the opportunity to mate. For example, in some spider species, the male is smaller than the female and risks being eaten if he gets close to her. Before mating, the male in some species presents the female with an object, such as an insect wrapped in a silk web. While the female is unwrapping and eating the insect, the male is able to mate with her without being attacked. Sometimes, after mating, the female eats the male anyway.

✓Reading Check

4. How has courtship behavior evolved?

____

____

____

Can instinctive behavior reduce aggression?

A **territory** is a physical space an animal defends against other members of its species. It may contain the animal's breeding area, feeding area, potential mates, or all three. Animals that have territories will defend their space. They will drive away other individuals of the same species. For example, a male sea lion patrols the area of beach where his harem of female sea lions is located. He does not bother a neighboring male that has a harem of his own. Both males have marked their territories and each respects the boundaries. However, if a young, unattached male tries to enter the sea lion's territory, the owner of the territory will attack and drive the intruder away.

Setting up territories reduces conflicts, controls population growth, and provides for efficient use of environmental resources. When animals space themselves out, they don't compete for the same resources within a limited space. This behavior improves survival rates. If the male has selected an appropriate site and the young survive, they may have inherited his ability to select an appropriate territory. Territorial behavior has survival value, not only for individuals, but also for the species. ✓

Reading Check

5. What are the benefits of territorial behavior?

____

____

____

Pheromones are chemicals that communicate information among individuals of the same species. Many animals produce pheromones to mark territorial boundaries. For example, wolf urine contains pheromones that warn other wolves to stay away. Pheromones work day and night, and they work whether or not the animals that made the marks are present.

What is the purpose of aggressive behavior?

Animals sometimes act aggressively. **Aggressive behavior** is used to intimidate another animal of the same species. Animals fight or threaten one another in order to defend their young, their territory, or another resource, such as food. Aggressive behavior includes bird calling, teeth baring, or growling. It is a message to "keep away".

Animals of the same species rarely fight to the death. The fights are usually symbolic. Why does aggressive behavior rarely result in serious injury? It may be that the defeated individual shows signs of submission. These signs stop further aggression by the victor.

What is a dominance hierarchy?

Sometimes, aggressive behavior among several individuals results in a grouping in which there are different levels of dominant and submissive animals. A **dominance hierarchy** (DAH muh nunts • HI rar kee) is a form of social ranking in which some individuals are more subordinate than others. Usually one animal is the top-ranking, dominant animal. This animal may lead others to food, water, and shelter. A dominant male often sires most or all

Think it Over

6. Analyze The formation of a dominance hierarchy is (Circle your choice.)

a. a learned behavior.
b. innate behavior.
c. both.

of the offspring. There might be several levels in the hierarchy. Individuals in each level are subordinate to the one above. The ability to form a dominance hierarchy is innate. However, the position each animal assumes may be learned. You may have heard the term *pecking order*. It describes a dominance hierarchy formed by chickens. The top-ranking chicken can peck any other chicken. The chicken lowest in the hierarchy is pecked by all the other chickens in the group.

What are some behavioral cues?

Sometimes behavior is a response to internal biological rhythms. Behavior based on a 24-hour day/night cycle is one example. Many animals, humans included, sleep at night and are awake during the day. Other animals, such as owls, have the opposite pattern. They sleep during the day and are awake at night. A 24-hour, light regulated, sleep/wake cycle of behavior is called a **circadian** (sur KAY dee uhn) **rhythm.** Circadian rhythms keep you alert during the day and help you relax at night. Even if you forget to set your alarm clock, they may wake you. Circadian rhythms are controlled by genes. They are also influenced by factors such as jet lag and shift work.

Rhythms also occur on a yearly or seasonal cycle. Migration, for example, occurs on a seasonal cycle. **Migration** is the instinctive, seasonal movement of animals. In North America about two-thirds of bird species fly south in the fall. There is food available in areas such as South America. The birds fly north in the spring to areas where they breed during the summer. Whales migrate seasonally too. Scientists hypothesize that change in day length stimulates the onset of migration in the same way that it controls the flowering of plants. Butterflies, salmon, and caribou are just a few of the animals that make seasonal migrations. ✓

Migration requires remarkable strength and endurance. The arctic tern migrates between the arctic circle and the Antarctic, a one-way flight of almost 18 000 km.

Animals navigate in a variety of ways including:

- using the positions of the sun and stars
- using geographic clues such as mountain ranges
- using Earth's magnetic field ✓

✓ Reading Check

7. What might stimulate the onset of migration?

____

____

____

✓ Reading Check

8. How do animals navigate during migration?

____

____

____

It is possible that some animals migrate in response to colder temperatures and shorter days, as well as hormones. Young animals may learn when and where to migrate by following their parents.

What happens to animals that do not migrate?

It is easy to see why some animals migrate from a colder place to a warmer place, yet most animals do not migrate. The ways in which many animals cope with winter is another example of instinctive behavior.

In preparation for winter, some animals store food in burrows and nests. Other animals survive winter by undergoing changes in their bodies that reduce the need for energy. Many mammals, some birds, and a few other types of animals go into a deep sleep during the cold winter months. This period of inactivity is called hibernation. **Hibernation** (hi bur NAY shun) is a state in which the body temperature drops. Oxygen use decreases and the breathing rate falls to a few breaths per minute. Hibernation conserves energy. Animals that hibernate typically eat large amounts of food to build up body fat before entering hibernation. The fat provides fuel for the animal's body.

What about an animal that lives in a climate that is hot year-round? Some animals respond to heat in a way that is similar to hibernation. **Estivation** (es tuh VAY shun) is a state of reduced metabolism that occurs in animals living in conditions of intense heat. Desert animals appear to estivate in response to lack of food or periods of drought. On the other hand, Australian long-necked turtles will estivate even when they are kept in a laboratory with constant food and water. That means that estivation is an innate behavior that depends on internal and external cues. ✓

✓Reading Check

9. What is estivation?

______

______

______

After You Read

Mini Glossary

aggressive behavior: innate behavior used to intimidate another animal of the same species in order to defend young, territory, or resources

behavior: anything an animal does in response to a stimulus in its environment

circadian (sur KAY dee uhn) rhythm: innate behavior based on the 24-hour cycle of the day; light regulated; may determine when an animal sleeps or wakes

courtship behavior: an instinctive behavior that males and females of a species carry out before mating

dominance hierarchy (DAH muh nunts • HI rar kee): innate behavior by which some animals form social ranking within a group in which some individuals are more subordinate than others; usually has one top-ranking individual

Section 33.1 Innate Behavior, *continued*

estivation (es tuh VAY shun): state of reduced metabolism that occurs in animals living in conditions of intense heat

fight-or-flight response: automatic response controlled by hormones that prepares the body to either fight or run from danger

hibernation (hi bur NAY shun): state of reduced metabolism occurring in animals that sleep during parts of cold winter months; an animal's temperature drops, oxygen consumption decreases, and breathing rate declines

innate (ih NAYT) behavior: an inherited behavior in animals; includes automatic responses and instinctive behaviors

instinct (IHN stingt): complex innate behavior patterns that begin when an animal recognizes a stimulus and continue until all parts of the behavior have been performed

migration: instinctive seasonal movements of animals from place to place

reflex (REE fleks): simple, automatic response in an animal that involves no conscious control; usually acts to protect an animal from serious injury

territory: physical space an animal defends against other members of its species; may contain an animal's breeding area, feeding area, potential mates, or all three

1. Read the terms and their definitions in the Mini Glossary. On the line below explain how a reflex is different from an instinct.

____

2. Complete the web diagram by using the following concepts in the appropriate box: **reflex, courtship, hibernation, fight-or-flight, territoriality,** and **migration.**

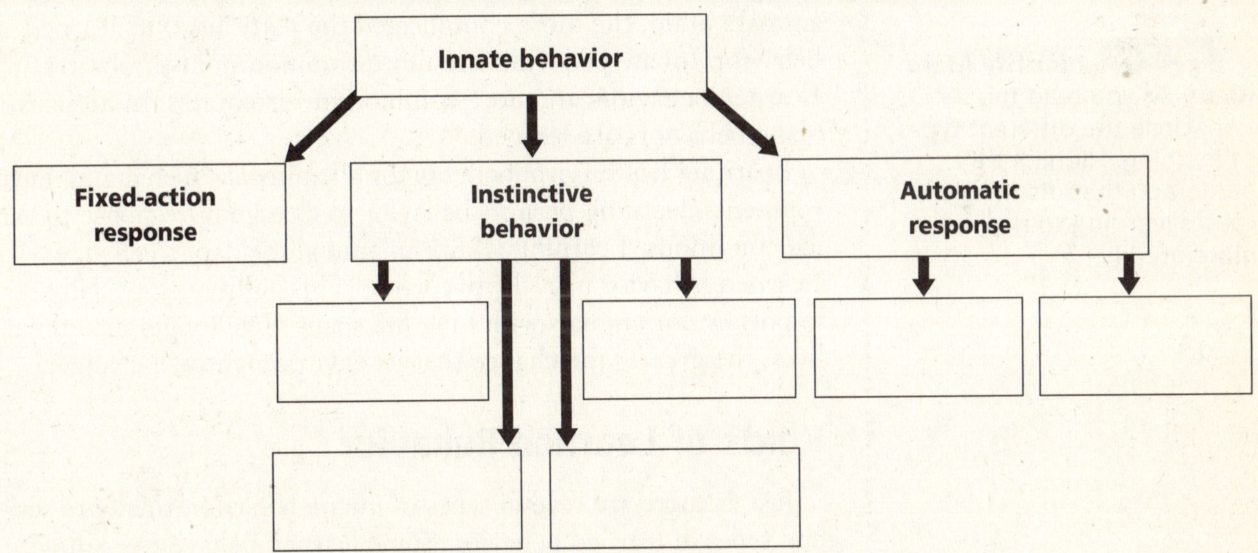

3. How is innate behavior an advantage in a species where the mother leaves once the young have hatched?

____

Visit the Glencoe Science Web site at **science.glencoe.com** to find your biology book and learn more about innate behavior.

Section 33.2 Learned Behavior

Before You Read

Have you ever seen a police officer patrolling on horseback? In many U.S. cities, horses play an important role in law enforcement. This is despite the fact that young horses often are afraid of cars, noisy streets, and sudden movements. After a while, the horses become used to the typical sights and sounds of the city and adjust to their work environment. How are horses able to do this? Write your thoughts on the lines below.

Read to Learn

Mark the Text **Identify Main Ideas** As you read this section, circle the different types of learning. Highlight the paragraph that discusses what causes learning to happen more quickly.

What is learned behavior?

Learning, or learned behavior, takes place when behavior changes through practice or experience. The more complex an animal's brain, the more complicated the patterns of its learned behavior. Innate behaviors are more common in invertebrates. Learned behaviors are more common in vertebrates. In humans, many behaviors are learned.

Learning has survival benefits for all animals. In changing environments, learning permits behavior to change in response to varied conditions. Learning allows an animal to adapt to change. Learned behavior has adaptive value. This ability is especially important for animals with long life spans. The longer an animal lives, the greater the chance that its environment will change.

Kinds of Learned Behavior

Just as there are several types of innate behavior, there are several types of learned behavior. Some learned behavior is simple and some is complex.

What is habituation?

Horses normally shy away from an object that suddenly appears from the trees or bushes, yet after awhile they disregard noisy cars with honking horns that speed by their pasture. This lack of response is called habituation. **Habituation** (huh bit choo AY shun) occurs when an animal is repeatedly given a stimulus. However, the stimulus is not associated with any punishment or reward. An animal has become habituated to a stimulus when it stops responding to the stimulus. ✓

What is imprinting?

You may have seen young ducklings following their mother. This behavior is the result of imprinting. **Imprinting** is a form of learning in which an animal forms a social attachment to another object. This occurs at a specific, critical time in the animal's life. Many kinds of animals do not innately know how to recognize members of their own species. Instead, they learn how to do this early in life. Imprinting takes place only during a specific period of the animal's life. It is usually irreversible. For example, birds that leave the nest immediately after hatching, such as geese, imprint on their mother. They learn to recognize her within a day of hatching. Imprinting also occurs in ducks. Ducklings quickly learn to recognize and follow the first highly visible moving object they see. Normally that object is the ducklings' mother. Learning to recognize their mother and following her helps ducklings survive. Their mother means that food and protection will be nearby.

Do animals learn by trial and error?

You may remember learning how to ride a bike. You probably tried it several times before you were able to do it successfully. Some animal abilities are acquired the same way. For example, nest building may be a learning experience. The first time a jackdaw builds a nest, it uses grass, bits of glass, stones, empty cans, old lightbulbs, and anything else it can find. With experience, the bird finds that grasses and twigs make better nests than lightbulbs and empty cans. The animal has used **trial-and-error learning** in which an animal receives a reward for making a particular response. When an animal tries one solution and then another in the course of obtaining a reward, in this case a suitable nest, it is learning by trial-and-error.

Learning happens more quickly if there is a reason to learn or be successful. **Motivation** is an internal need that causes an animal to act. In most animals, motivation often involves satisfying a physical need such as hunger or thirst. If an animal is not motivated, it will not learn. Usually, animals that are not hungry will not respond to a food reward. ✓

✓ Reading Check

1. How do you know when an animal has become habituated to a stimulus?

____

____

____

Think it Over

2. Infer What would happen if the first highly visible moving object that a newly hatched group of ducklings saw was a human?

____

____

____

✓ Reading Check

3. What usually motivates an animal to act?

____

____

____

Do animals learn by association?

Suppose you have a new kitten. Each time it smells the aroma of cat food in the can you are opening, it begins to meow. After a few weeks, the sound of the can opener attracts the kitten, causing it to meow. The kitten has become conditioned to respond to a stimulus other than the smell of food. **Classical conditioning** is learning by association. You can see a well-known example of an early experiment in classical conditioning in the illustration below.

A Pavlov noted that dogs salivate when they smell food. Responding to the smell of food is a reflex, an example of innate behavior.

B By ringing a bell each time he presented food to a dog, Pavlov established an association between the food and the ringing bell.

C Eventually, the dog salivated at the sound of the bell alone. The dog had been conditioned to respond to a stimulus that it did not normally associate with food.

What is the most complex type of learning?

In a classic study of animal behavior, a chimpanzee was given two bamboo poles. Neither of the poles was long enough to reach some fruit placed outside its cage. The chimpanzee connected the two shorter poles to make one longer pole. The chimpanzee solved the problem of how to reach the fruit. This type of learning is called insight. **Insight** is learning in which an animal uses previous experience to respond to a new situation. It is the most complex type of learning.

Much of human learning is based on insight. When you were a baby you learned a great deal by trial-and-error. As you grew older, you relied more on insight. Solving math problems is an example of insight. Most likely your first math experience was learning to count. Based on your understanding of numbers, you then learned to add, subtract, multiply, and divide. Years later, you continue to solve math problems based on your past experiences. When you encounter a problem or a situation you have never experienced before, you use insight to solve it. ✓

Reading Check

4. What is the most complex type of learning?

____

____

____

The Role of Communication

When you think about interactions among animals, you realize that some sort of communication has taken place. **Communication** is an exchange of information that results in a change of behavior. Black-headed gulls visually communicate their availability for mating with instinctive courtship behavior. The pat on the head from a dog's owner after the dog fetches a stick signals a job well done.

Do most animals communicate?

Animals have several ways to communicate. They signal each other by sounds, sights, touches, or smells. Sounds vibrate in all directions. They can be heard a long way from their sources. Sounds such as songs, roars, and calls communicate a lot of information quickly. For example, the song of a male cricket tells his location, his sex, and his social status. Communication by sound usually varies according to species, so the male cricket also communicates his species. ✓

Reading Check

5. List four ways animals communicate.

____

____

____

____

Signals that involve odors may be spread over a wide area and carry a general message. Ants leave odor trails that are followed by other members of their nest. These odors are specific to each ant species. As you know, pheromones such as those used by moths may be used to attract mates. Because only small amounts of pheromones are needed, other animals, especially predators, may not be able to detect the odors.

Some communication combines innate and learned behavior. In some species of songbirds, males automatically sing when they reach sexual maturity. Their songs are specific to their species, and singing is an innate behavior. Sometimes members of the same species that live in different regions learn variations of the song. They learn to sing with a regional dialect. In other species, birds raised in isolation never learn to sing their species song.

Reading Check

6. How do humans benefit from language?

Can animals use language?

Language, the use of symbols to represent ideas, is present primarily in animals with complex nervous systems, memory, and insight. Humans, with the help of spoken and written language, can benefit from what other people and cultures have learned. Humans do not have to experience everything for themselves. People can use accumulated knowledge to build new knowledge.

After You Read

Mini Glossary

classical conditioning: learning by association

communication: exchange of information that results in a change of behavior

habituation (huh bit choo AY shun): learned behavior that occurs when an animal is repeatedly given a stimulus not associated with any punishment or reward

imprinting: learned behavior in which an animal, at a specific critical time of its life, forms a social attachment to another object; usually occurs early in life and allows an animal to recognize its mother and others of its species

insight: type of learning in which an animal uses previous experiences to respond to a new situation

language: use of symbols to represent ideas; usually present in animals with complex nervous systems, memory, and insight

motivation: internal need that causes an animal to act and that is necessary for learning to take place; often involves hunger or thirst

trial-and-error learning: type of learning in which an animal receives a reward for making a particular response

1. Read the terms and their definitions in the Mini Glossary above. Select two key terms that describe a type of learning and provide an example of each learning type.

2. Match the terms with the correct statements. Put the letter of the term in Column 2 on the line in front of the statement it matches in Column 1.

Column 1	Column 2
_____ 1. Takes place when behavior changes through practice or experience.	a. adaptive value
_____ 2. Learning has survival benefits for animals in changing environments.	b. insight
_____ 3. This is the most complex type of learning.	c. language
_____ 4. Communication that enables humans to benefit from what others have learned without having to experience it directly.	d. learning

Visit the Glencoe Science Web site at **science.glencoe.com** to find your biology book and learn more about learned behavior.

Name Date Class

Section 34.1 Skin: The Body's Protection

Before You Read

Your entire body is covered by skin. Skin is an important organ. What are the functions of your skin? List your ideas on the lines below. After you have read this section, add other functions to your list.

____

____

____

Read to Learn

Structure and Functions of the Integumentary System

Skin is your body's largest organ. It is also the main organ of the integumentary (inh TE gyuh MEN tuh ree) system. Hair, nails, and some glands are also part of the system. Skin covers our bodies. It is composed of layers of the four types of body tissues. The four types of tissues are epithelial, connective, muscle, and nervous. Epithelial tissue is found in the outer layer of the skin. It covers body surfaces. Connective tissue consists of both tough and flexible protein fibers. Connective tissue holds your body together. Muscle tissues interact with hairs on the skin to respond to stimuli, such as cold and fright. Nervous tissue helps humans sense external stimuli, such as pain or pressure. ✓

What is the epidermis?

The **epidermis** is the outermost layer of the skin. It has two parts—the exterior, or outside, part and the interior, or inside, part. The exterior layer of the epidermis consists of 25 to 30 layers of dead, flattened cells. These cells are continually shed. Although the cells are dead, they serve an important function. They contain a protein called **keratin** (KER uh tin). Keratin helps protect the living cell layers underneath from exposure to bacteria, heat, and chemicals.

The interior layer of the epidermis contains living cells. The living cells continually divide so that they can replace the dead cells. As new cells are pushed toward the skin's surface, the nuclei

Mark the Text **Identify Specific Ideas** As you read though this section, highlight the text each time that you read about ways in which the skin helps to protect the body.

Reading Check

1. What are the four types of body tissues that make up the skin?

____

____

____

____

Think it Over

2. Explain Why is melanin important to the body?

Reading Check

3. What is the purpose of fat deposits in the skin?

in the cells degenerate, and the cells die. Then these cells are shed. This process takes about 28 days. So, every four weeks, all of the cells of the epidermis are replaced by new cells.

Some of these cells in the interior layer contain melanin. **Melanin** is a pigment that colors the skin. Differences in skin color are due to the amount of melanin produced by the cells. Melanin helps protect the underlying body cells from solar radiation, or sun damage, by absorbing ultraviolet light. If ultraviolet light damages cells, skin cancer could develop.

The epidermis on the fingers and palms of your hands and on the toes and soles of your feet contains ridges and grooves that are formed before birth. These ridges increase friction, which improves the skin's grip. Each person has a unique pattern of ridges and grooves. Because these patterns are unique, footprints and fingerprints can be used to identify individuals.

What is the dermis?

The second principal layer of the skin is the dermis. The **dermis** is the inner, thicker portion of the skin. The thickness of the dermis changes from body part to body part. The thickness depends on how that body part is used.

As shown in the illustration on page 411, the dermis contains structures such as blood vessels, nerves, nerve endings, hair follicles, sweat glands, and oil glands. Underneath the dermis, the skin is attached to the underlying tissues by the subcutaneous layer. This layer consists of fat and connective tissue. Fat deposits help the body absorb impacts, retain heat, and store food. ✓

Hair is another structure of the integumentary system. It grows out of **hair follicles,** narrow, hollow openings in the dermis. The primary function of hair is to protect the skin from injury and damage from the sun. Hair also provides an insulating layer of air just above the surface of the skin. As hair follicles develop, they are supplied with blood vessels and nerves. These follicles become attached to muscle tissue. Most hair follicles have an oil gland. Oil prevents hair from drying out and keeps the skin soft. Oil also helps prevent the growth of certain bacteria. When oil and dead cells block the opening of the hair follicle, pimples may form.

What are the functions of the integumentary system?

One function of skin is to help maintain homeostasis. Homeostasis is the regulation of an organism's internal environment to maintain conditions suitable for its survival. Skin helps

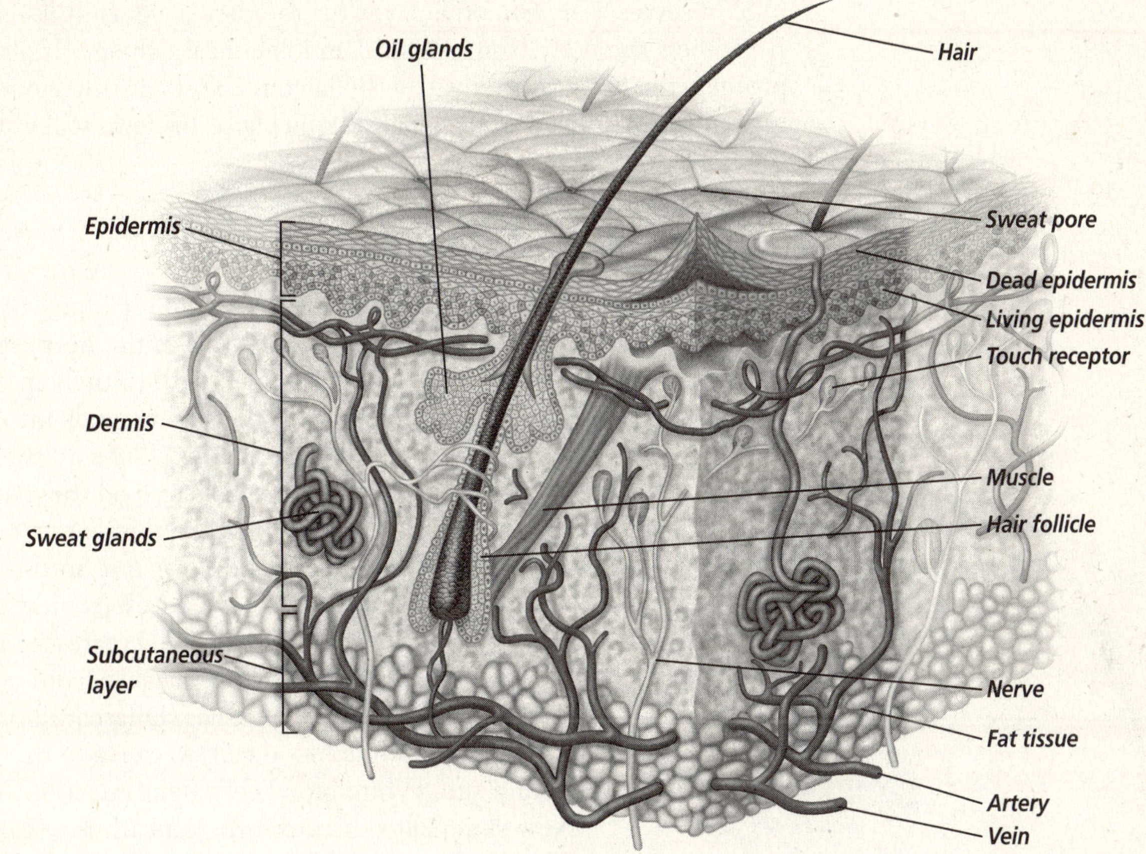

regulate your internal body temperature. When your temperature rises, blood vessels in the dermis dilate. This dilation causes increased blood flow. Body heat is transferred from the blood vessels to the surface of the skin. From the skin, the heat is lost by radiation. When you are cold, the blood vessels in the skin contract, and the body conserves heat.

Glands in the integumentary system help cool the body. When the body heats up, glands in the dermis produce sweat. The wet skin helps reduce body temperature. As the sweat evaporates, the water changes from liquid to vapor. Heat is lost, and the body cools itself. ✓

The skin also is a sense organ. Nerve cells in the dermis receive stimuli from outside the body. Nerve cells provide information about pressure, pain, and temperature to the brain.

Skin helps produce essential vitamins. When exposed to ultraviolet light, or sunlight, skin cells produce vitamin D. Vitamin D helps the body absorb calcium into the bloodstream. Because too much sunlight can damage the skin, people may need to take vitamin D supplements or eat foods that are enriched with this vitamin.

✓ Reading Check

4. How does the body cool itself?

____

____

____

Skin serves as a protective layer for the tissues beneath it. It protects the body from physical and chemical damage. It also protects the body from invasion by bacteria. Cuts or other openings in the skin need to be repaired quickly or bacteria will enter the body.

Skin Injury and Healing

It does not take the skin long to heal after a minor injury or wound. If the skin receives a minor scrape, cells in the deepest layer of the epidermis divide. The cells quickly fill in the gap on the skin. If the injury to the skin extends into the dermis, bleeding usually occurs. The skin then goes through a series of stages to heal the damaged tissue. The body's first reaction is to close the break in the skin. Blood flows out onto the skin until a clot forms. A scab develops on the skin to close the wound. The scab creates a barrier that prevents bacteria on the skin from reaching the underlying tissues. Dilated blood vessels allow white blood cells to move to the wound site. White blood cells fight infections. New skin cells begin to form beneath the scab. These cells eventually push the scab off, and new skin can be seen. If a wound to the skin is large, dense connective tissue used to close the wound may leave a scar. ✓

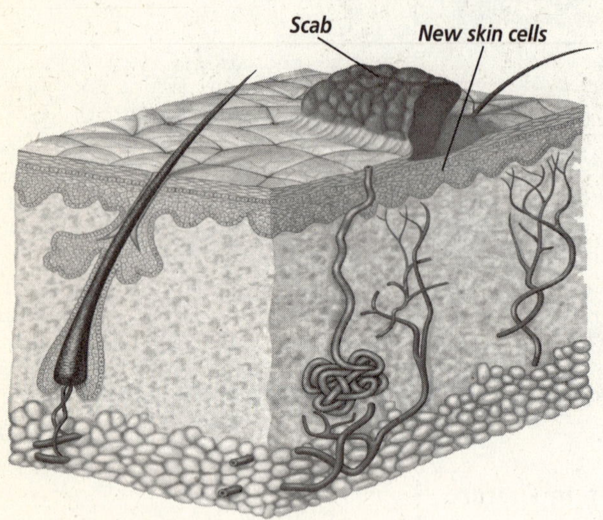

✓ Reading Check

5. What purpose does a scab serve?

____

____

____

How are burns rated?

Burns result from exposure to the sun, contact with chemicals, or contact with hot objects. Burns are rated according to how severe they are.

A first-degree burn, such as a mild sunburn, results in the death of epidermal cells. When the skin receives a first-degree burn, the skin turns red, and you feel mild pain. A first-degree burn heals in about a week. It will not leave a scar. A second-degree burn damages the skin cells of both the epidermis and the dermis. A second-degree burn can result in blisters and scars. The most severe burn is a third-degree burn. A third-degree burn destroys both the epidermis and the dermis. With a third-degree burn, the skin loses its function. The cells will not be replaced by new cells. Skin grafts may be required to replace the lost skin.

How does skin change?

As people get older, their skin changes. It becomes drier because the glands produce smaller amounts of skin oils. Skin oils are a mixture of fats, cholesterol, proteins, and inorganic salts. Wrinkles appear as the skin loses its elasticity. These changes happen more quickly when the skin has been repeatedly exposed to ultraviolet sun rays.

After You Read

Mini Glossary

dermis: the inner, thicker portion of the skin

epidermis: the outermost layer of the skin

hair follicle: narrow, hollow openings in the dermis; as hair follicles develop, they are supplied with blood vessels and nerves

keratin (KER uh tun): a protein that helps protect the living cell layers in the interior layer of the epidermis from exposure to bacteria, heat, and chemicals

melanin: a pigment that colors the skin

1. Read the terms and definitions in the Mini Glossary above. Use the space below to describe the purpose of keratin and melanin in the epidermis.

2. Use the Venn diagram below to compare and contrast the epidermis with the dermis.

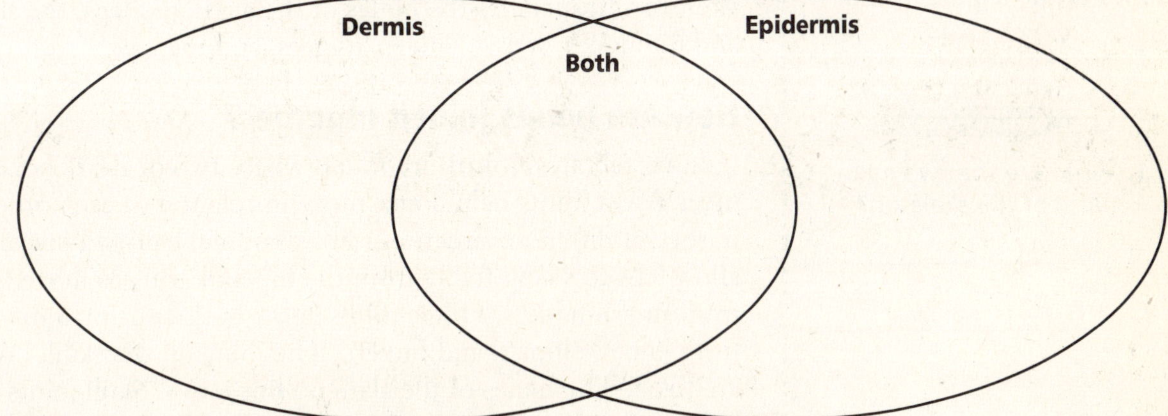

3. What is the purpose of hair?

Visit the Glencoe Science Web site at **science.glencoe.com** to find your biology book and learn more about skin, the body's protection.

Section 34.2 Bones: The Body's Support

Before You Read

Consider how much you have grown since you were born. As an infant, you had a completely developed skeleton. However, your bones are living tissue and they grow. Write a short paragraph on the lines below explaining how you think your bones have changed as you have grown.

____

____

____

____

Read to Learn

Create a Drawing Draw a picture of a human leg, from the hip to the toes. Label the drawing to explain how the bones of the leg are joined.

1. What are the two main parts of the skeleton?

____

____

Skeletal System Structure

The adult human skeleton contains about 206 bones as shown in the illustration on page 415. The skeleton has two main parts. The **axial skeleton** includes the skull and the bones that support it. These bones include the vertebral column, the ribs, and the sternum. The sternum is the breastbone. The other main part of the human skeleton is the **appendicular** (a pen DI kyuh lur) **skeleton.** It includes the bones of the arms and legs, the shoulder and hip bones, wrists, ankles, fingers, and toes. ✓

How are bones joined together?

In vertebrates, **joints** are found where two or more bones meet. Most joints help bones move in relation to each other, and in several different directions. For example, ball-and-socket joints allow legs to swing freely from the hip and arms to move freely from the shoulders. Hinge joints allow back-and-forth movement from knees, elbows, and fingers. The joints in the skull, however, are fixed. The bones of the skull do not move. Skull joints are held together by bone that has grown together, or by fibrous cartilage. Recall that cartilage is not bone. It is a tough, flexible material that makes up portions of the skeletons of bony animals.

Section 34.2 Bones: The Body's Support, *continued*

Joints are often held together by ligaments. A **ligament** is a tough band of connective tissue that attaches one bone to another. Joints with large ranges of motion, such as the knee, usually have more ligaments surrounding them. In movable joints, the ends of the bones are covered in cartilage. This layer of cartilage allows for smooth movement between the bones. In some joints,

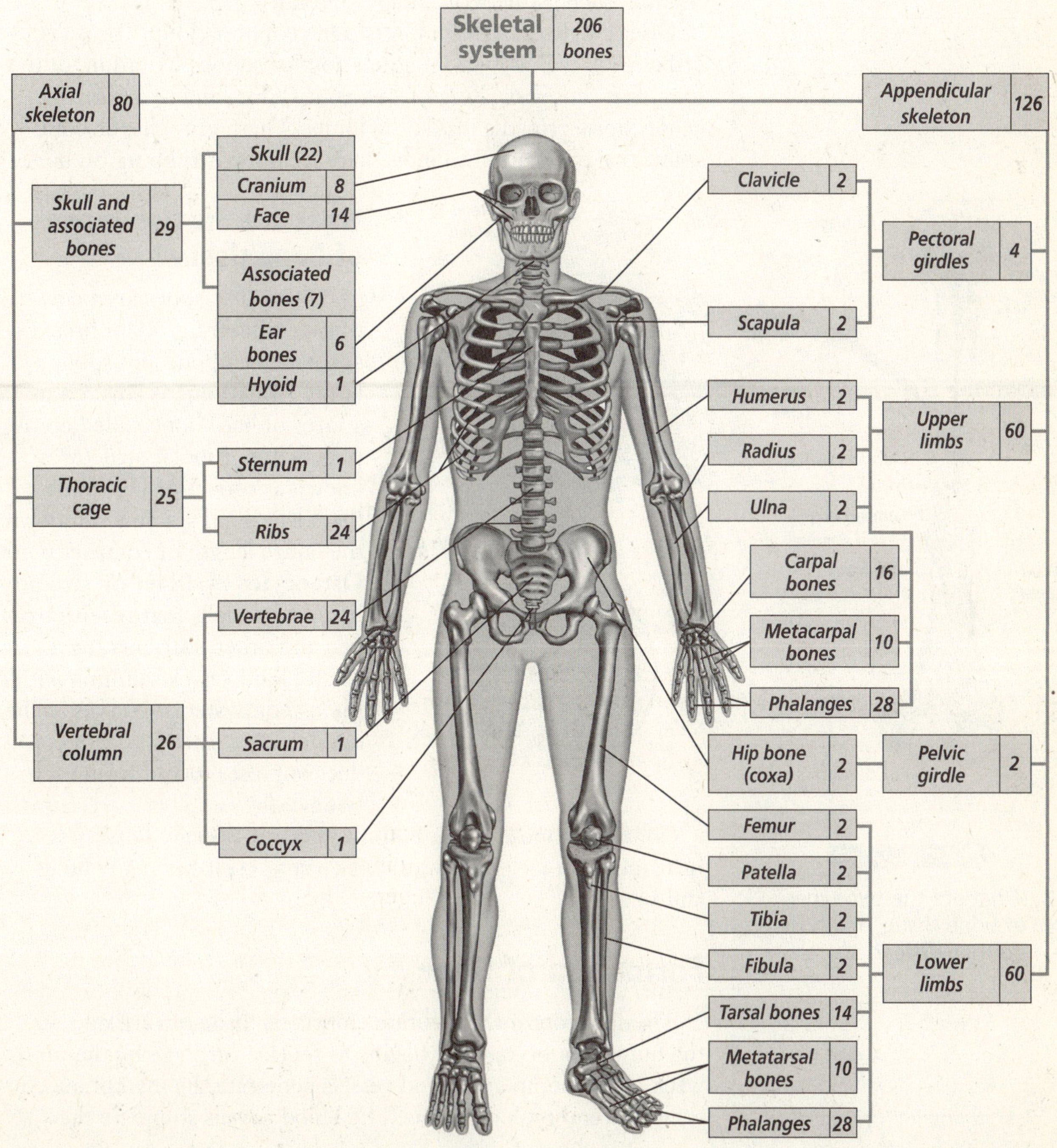

Think it Over

2. Compare/Contrast What is the difference between a ligament and a tendon?

including the shoulder and the knee, there are fluid-filled sacs located on the outside of joints. These sacs are called **bursae.** The bursae decrease friction and keep bones and tendons from rubbing against each other. **Tendons** are thick bands of connective tissue that attach muscles to bones.

When a joint is twisted with force, an injury called a sprain can result. Sprains usually occur in joints that have a wide range of motion, such as the wrist, ankle, and knee.

Diseases also can harm joints. One common joint disease is arthritis. Arthritis is an inflammation, or serious irritation, of the joint that causes swelling or deformity. One kind of arthritis causes bony growths inside the joints. These growths, or bone spurs, make it painful to move because bone is rubbing on bone.

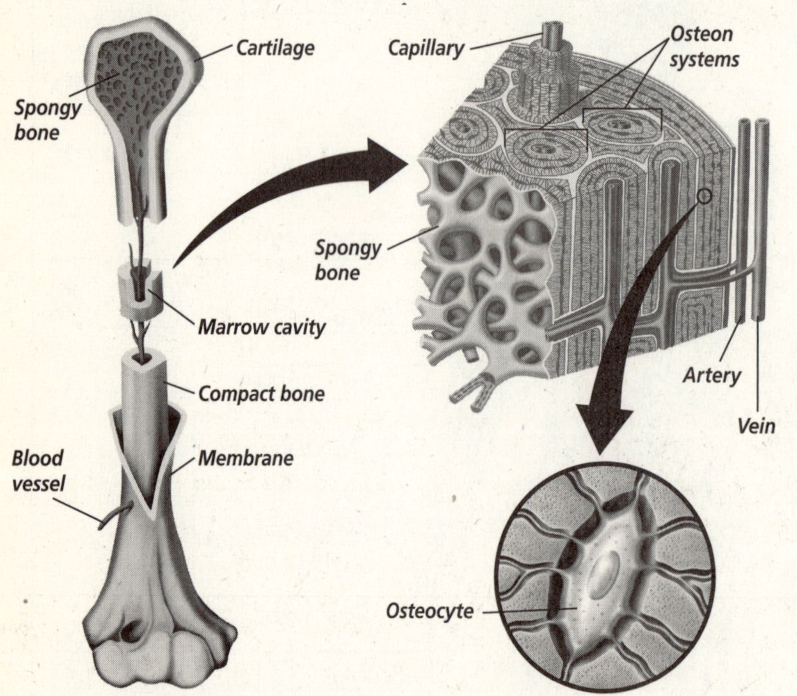

What are the two types of bone tissue?

Notice that bones are made of two different types of bone tissue: compact bone and spongy bone. Every bone is covered in a layer of hard bone called **compact bone.** Tubular structures known as osteon or Haversian (ha VER zhen) systems run down the entire length of compact bone. **Osteocytes** (AHS tee oh sitz) are living bone cells that receive oxygen and other nutrients from small blood vessels running within the osteon systems. Nerves in the canals of the osteon system conduct impulses to and from each bone cell.

Compact bone surrounds **spongy bone.** Spongy bone gets its name from its appearance. Like a sponge, it has many holes and spaces. ✓

Reading Check

3. What are the two types of bone tissue, and how do they differ?

Formation of Bone

The skeleton of a vertebrate embryo is made of cartilage. In the human embryo, bone begins to replace cartilage by the ninth week of development. Blood vessels penetrate the membrane covering the embryo's cartilage. The blood vessels stimulate the embryo's cartilage cells to become potential bone cells. These potential bone cells are called **osteoblasts** (AHS tee oh blastz). Osteoblasts secrete a protein called collagen. Minerals from the bloodstream begin to deposit themselves in the collagen. Calcium salts and other ions harden the newly formed bone cells. These new living bone cells are osteocytes. ✓

The skeleton of an adult human is almost all bone. Cartilage is found only where flexibility is needed. Regions with cartilage include the nose tip, the external ears, discs between individual vertebrae, and movable joint linings.

Reading Check

4. What are osteoblasts?

___

___

___

How do bones grow?

Bones grow in both length and in diameter. In bones that end in cartilage, bone growth occurs at both ends of the bones. During the teen years, increased production of sex hormones causes the osteoblasts, the cells that form bone, to divide more rapidly. This results in a growth spurt. These hormones also cause the growth centers at the ends of the bones to slow production. As these cells begin to die, growth slows. After growth stops, bone-forming cells repair and maintain the bones.

Skeletal System Functions

The primary function of the skeleton is to provide a framework for the body tissues. The skeleton also protects internal organs, such as the heart, the lungs, and the brain.

The human skeleton allows for efficient movement. Muscles that move body parts need to be firmly attached to a strong structure that the muscles can pull against. The skeleton provides these attachment points.

Bones also produce blood cells. Red blood cells, white blood cells, and cell fragments that are needed for blood clotting are produced in the **red marrow** of a bone. Red marrow is found in the humerus, the femur, the sternum, the ribs, the vertebrae, and the pelvis. **Yellow marrow** is found in many other bones. Yellow marrow consists of stored fat. The stored fat can be used in times of need.

Reading Check

5. Why is calcium important in the human diet?

___

___

___

What other functions do bones have?

Your bones store minerals. Minerals stored in bones include calcium and phosphate. Calcium is needed to form strong, healthy bones. It is important to eat foods that are rich in calcium. These foods include milk, yogurt, cheese, lettuce, spinach, and other leafy vegetables. ✓

What are some common bone injuries and diseases?

As people get older, their bones age and change. A disease called osteoporosis (ahs tee oh puh ROH sus) involves a loss of bone volume and minerals. These losses cause the bones to become more porous and brittle. Osteoporosis is most common in older women because they produce less estrogen, a hormone that helps bones form.

When bones break, a doctor can move them back into position. The doctor will put a cast or a splint over the bone to hold it in place until the bone tissue regrows.

After You Read

Mini Glossary

appendicular (a pen DI kyuh lur) skeleton: includes the bones of the arms and legs, the shoulder and hip bones, wrists, ankles, fingers, and toes

axial skeleton: a part of the skeleton that includes the skull and the bones that support it, such as the vertebral column, ribs, and sternum

bursa: fluid-filled sac located on the outside of some joints; sacs decrease friction, and help bones and tendons move smoothly together

compact bone: hard bone composed of repeating units of osteon systems; this type of bone tissue covers or surrounds every bone

joint: found where two or more bones meet; most joints help bones move in relation to each other

ligament: a tough band of connective tissue that attaches one bone to another

osteoblast (AHS tee oh blast): potential bone cell

osteocyte (AHS tee oh sit): living bone cell that receives oxygen and nutrients from small blood vessels running within the osteon systems

red marrow: the production site for red blood cells, white blood cells, and cell fragments that are needed for blood clotting

spongy bone: has many holes and spaces and is surrounded by compact bone

tendon: thick band of connective tissue that connects muscles to bones

yellow marrow: consists of stored fat; found in many bones

1. Read the terms and their definitions in the Mini Glossary above. Use the space below to write a brief paragraph describing bones and bone structure. Use at least six of the terms from the glossary in your paragraph.

____

____

____

Section 34.2 Bones: The Body's Support, *continued*

2. Place the sentences below in the correct sequence to show the development of bone cells in a human.

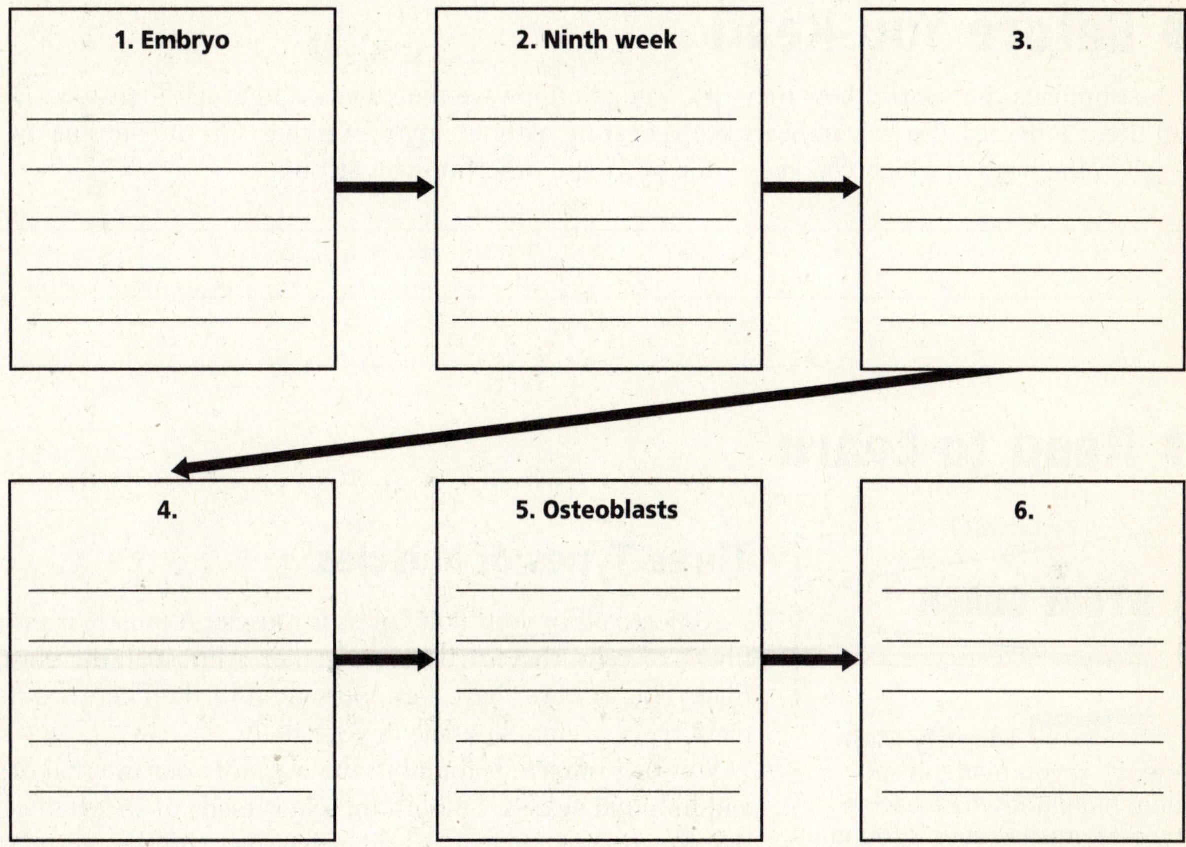

a. Bone begins to replace cartilage.
b. Osteoblasts secrete collagen.
c. Embryonic cartilaginous skeleton forms.
d. Blood vessels in the embryo penetrate the membrane covering the cartilage.
e. Minerals in the bloodstream are deposited in living bone cells.
f. This stimulates embryonic cells to become osteoblasts.

3. What is the difference between yellow and red marrow?

Visit the Glencoe Science Web site at **science.glencoe.com** to find your biology book and learn more about bones, the body's support.

Name Date Class

Section
34.3 Muscles for Locomotion

Before You Read

Most humans can control how they run, walk, jump, wave their arms, and touch their toes. During all these activities, the human heart keeps beating without any prompting. On the lines below, explain the ways in which you keep your heart and other muscles strong.

____

____

____

Read to Learn

Mark the Text **Identify Main Ideas** As you read this section, highlight where each type of muscle tissue is found. In another color, highlight what each muscle tissue does.

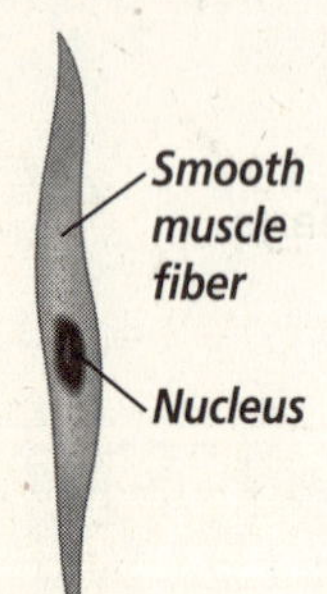

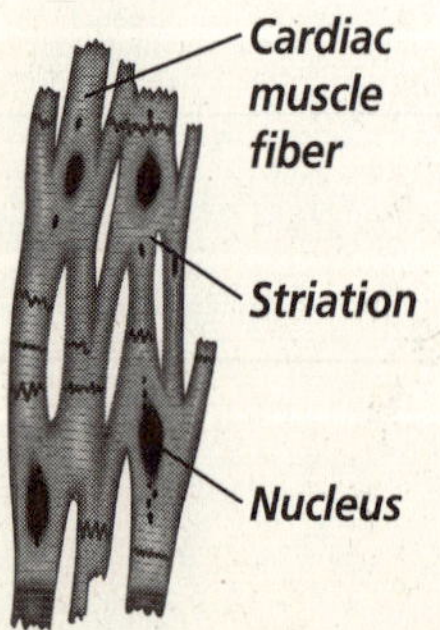

Three Types of Muscles

Almost half of your body mass is muscle. A muscle is groups of fibers, or cells, that are bound together. Almost all the muscle fibers you will ever have were present at birth. There are three main types of muscle tissue in your body.

Smooth muscle is found in the walls of your internal organs and in blood vessels. Smooth muscle is made of sheets of cells that line organs, such as the digestive tract and the reproductive tract. The most common job of smooth muscle is to squeeze. A smooth muscle applies pressure on the space inside the tube or organ it surrounds. This pressure moves material through the organ. For example, food moves through the digestive tract because smooth muscles squeeze the material as it moves through the tract. Gametes move through the reproductive system because they are squeezed by smooth muscle. You do not consciously control a smooth muscle and its contractions. Therefore, smooth muscle is considered an **involuntary muscle.** It contracts by itself.

Cardiac muscle is also an involuntary muscle. Cardiac muscle makes up your heart muscle. Cardiac muscle fibers are connected, forming a network that helps the heart muscle contract efficiently. Cardiac muscle can generate and conduct electrical impulses. These impulses are necessary for the regular, rhythmic contractions of the heart—your heartbeat. Cardiac muscle is found only in the heart.

Name Date Class

Section 34.3 Muscles for Locomotion, *continued*

Skeletal muscle is the third type of muscle tissue. Skeletal muscle is attached to and moves your bones. Most of the muscles in your body are skeletal muscles. You can control their contractions. When you want to move your arm or your leg, the muscles are under your control. A muscle that contracts under conscious control is called a **voluntary muscle.** Skeletal muscles are voluntary muscles.

Skeletal muscle fiber
Nucleus
Striation

Skeletal Muscle Contraction

Movement occurs because muscles can contract and relax. Most of your skeletal muscles work in opposing pairs. When one muscle contracts, another relaxes. When you bend your arm, the biceps muscle, which is located on the front of your arm, contracts. The muscle on the back of your arm, the triceps, relaxes. When you straighten your arm, the biceps relaxes, and the triceps contracts.

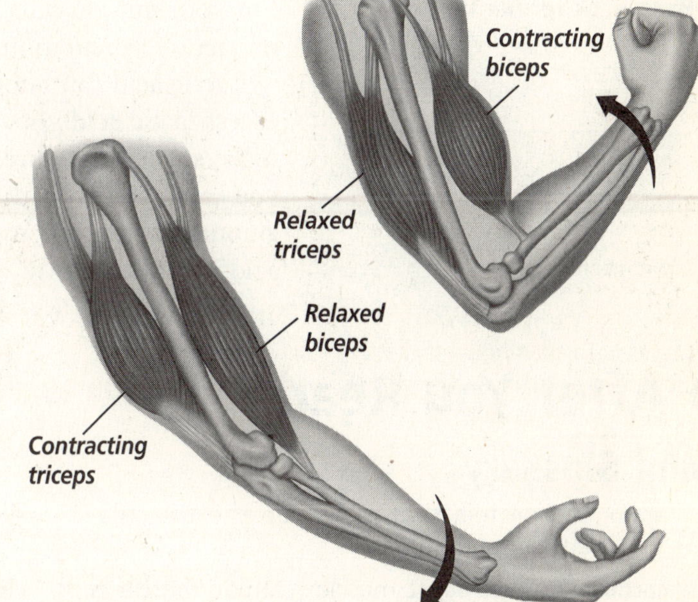

Muscle tissue is made up of muscle fibers. Muscle fibers are long muscle cells that are connected. Each fiber is made up of smaller units called **myofibrils** (mi oh FI brulz). Myofibrils consist of even smaller protein filaments that can be either thick or thin. The thicker filaments are made of the protein **myosin.** The thinner filaments are made of the protein **actin.** Each myofibril can be divided into sections called **sarcomeres** (SAR kuh meerz). Sarcomeres are the functional units of muscles.

One of the best explanations for how muscle contraction occurs is called the **sliding filament theory.** The sliding filament theory states that when a muscle receives a signal from a nerve, the actin filaments in each sarcomere slide toward each other. This shortens the sarcomeres in a fiber. This shortening causes the muscle to contract. The myosin filaments do not move. ✓

Reading Check

1. What is the sliding filament theory?

____

____

____

Muscle Strength and Exercise

Muscle strength does not depend upon the number of fibers in a muscle. The number of fibers in each muscle was fixed before

Reading Check

2. Does muscle strength depend upon the number of fibers in a muscle or the thickness of the fibers?

____

____

____

Reading Check

3. Is lactic fermentation an aerobic or anaerobic process?

____

birth. Muscle strength depends on the thickness of the fibers. It also depends upon how many of the fibers contract at one time. Regular exercise stresses the muscle fibers. This stress increases the size of the fibers. When you exercise regularly, your muscle fibers increase in diameter by adding myofibrils. ✓

Remember that ATP is the energy-storing molecule in cells. ATP is produced during cellular respiration. Muscle cells are continually supplied with ATP from both aerobic and anaerobic processes. When adequate oxygen is delivered to muscle cells, the aerobic respiration process dominates. This occurs when a muscle is resting or during moderate activity.

During vigorous activity, your muscles may not be able to get oxygen fast enough to sustain aerobic respiration and produce adequate ATP. The amount of available ATP becomes limited. For your muscle cells to get the energy they need, they must rely on lactic fermentation, an anaerobic process. ✓

Lactic acid can build up in muscle cells as you exercise. As the excess lactic acid goes into the bloodstream, the blood becomes more acidic, rapid breathing begins, and cramping may occur. As you catch your breath after vigorous activity or exercise, adequate amounts of oxygen are supplied to your muscles and the lactic acid is broken down.

After You Read

Mini Glossary

actin: the protein that makes up the thin filaments of myofibrils

cardiac muscle: heart muscle; cardiac muscle is an involuntary muscle

involuntary muscle: contracts by itself, not by conscious control

myofibril (mi oh FI brul): small units of muscle fiber; myofibrils are made up of small protein filaments that can be either thick or thin

myosin: the protein that makes up the thick filaments of myofibrils

sarcomere (SAR kuh meer): the functional unit of muscle; myofibrils can be divided into sections, called sarcomeres

skeletal muscle: attached to and moves bones; contracts under conscious control

sliding filament theory: theory that states that when actin filaments receive a signal, the actin filaments in each sarcomere slide toward each other, shortening the sarcomeres in a fiber and causing the muscle to contract

smooth muscle: made of sheets of cells that line organs; the most common job of smooth muscle is to squeeze

voluntary muscle: a muscle that contracts under conscious control

Section

34.3 Muscles for Locomotion, *continued*

1. Read the terms and their definitions in the Mini Glossary on page 422. Use the space below to describe the purpose and relationship of **myofibrils, myosin, actin,** and **sarcomeres.**

2. Use the cause and effect diagram to show what happens when you exercise moderately and when you exercise vigorously.

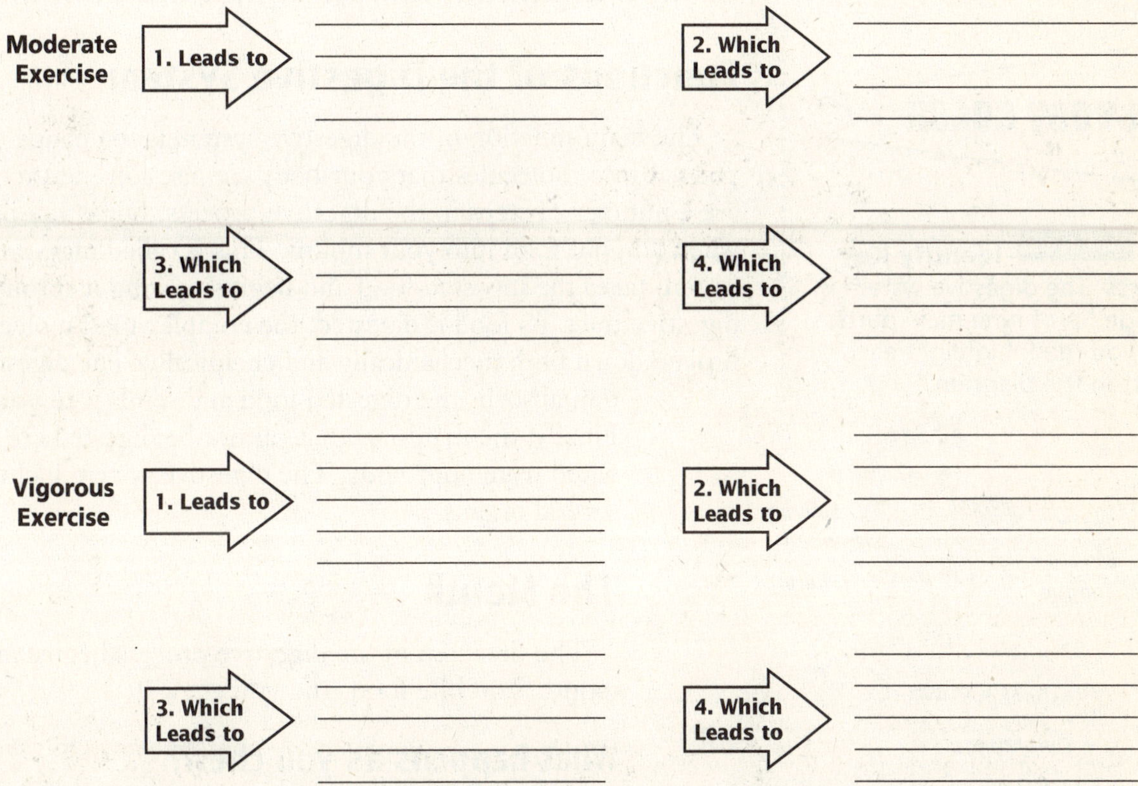

Visit the Glencoe Science Web site at **science.glencoe.com** to find your biology book and learn more about muscles for locomotion.

Name Date Class

Section 35.1 Following Digestion of a Meal

Before You Read

Your stomach growls, indicating hunger. You eat your lunch, and the process of digestion begins. Digestion is complex. What roles do you think the mouth and the stomach play in digestion? Write your thoughts on the lines below. After you have read this section, add any new information you learned.

Read to Learn

Mark the Text **Identify Key Parts** The digestive system is made up of nine main parts. As you read, highlight each part in the diagram.

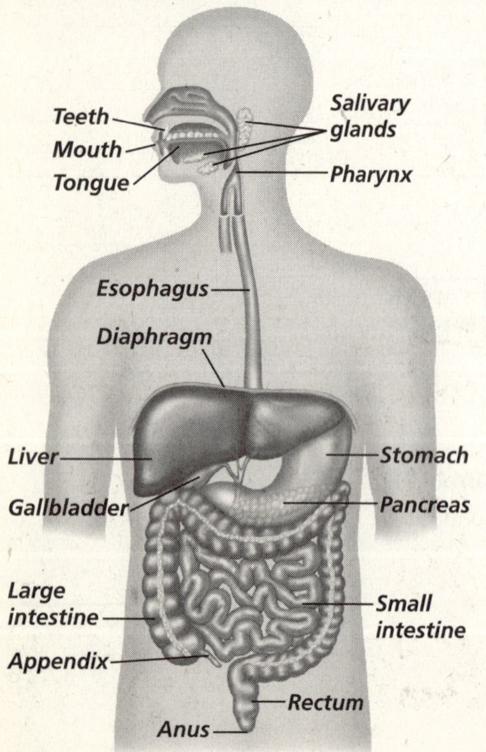

Functions of the Digestive System

The main function of the digestive system is to change the food you eat into molecules that your body can use for energy. There are a number of steps in the digestive process. Digestion begins when you put food into your mouth. This is called ingestion. The system takes the ingested food and begins moving it through the digestive tract. As food is digested, the complex food molecules are broken down both mechanically and chemically. The digestive system absorbs the digested food and sends it to your cells. Finally, the materials that cannot be digested are eliminated from your body. The digestive system includes several organs.

The Mouth

The first step in the digestive process begins in your mouth. You bite food, and you chew it.

What happens as you chew?

As you chew food, your tongue moves the food around in your mouth. Your tongue helps move the food between your teeth. Chewing is a form of mechanical digestion. It is the physical process of breaking food down into smaller pieces. Mechanical digestion prepares the food particles for chemical digestion. Chemical digestion is the process of changing food on a molecular level. This change occurs because of the actions of enzymes in your digestive system.

Section 35.1 **Following Digestion of a Meal,** *continued*

Where does chemical digestion begin?

Chemical digestion begins in the mouth. Salivary glands in your mouth secrete saliva. Saliva contains a digestive enzyme called amylase. **Amylase** breaks down the starches in food into smaller molecules. Many of the nutrients in the food you eat contain starches, large molecules known as polysaccharides. The polysaccharides are broken down into di- or monosaccharides. In the stomach, which is a muscular, pouchlike enlargement in the digestive tract, amylase digests the swallowed starches for about 30 minutes.

Think it Over

1. **Compare/Contrast** What is the difference between mechanical digestion and chemical digestion?

____

____

____

What happens after you have swallowed your food?

Once food is chewed, the tongue shapes it into a ball. The tongue moves this ball of chewed food into the back of the mouth. The food is swallowed. Swallowing food forces it from the mouth into the throat. Food then moves from the mouth into the esophagus. The **esophagus** is a muscular tube that connects the mouth to the stomach. Food moves down the esophagus by peristalsis. **Peristalsis** (per uh STAHL sus) is a series of involuntary, smooth muscle contractions along the walls of the digestive tract. The contractions occur in waves called peristaltic waves. First, circular muscles relax and longitudinal muscles contract. Then, circular muscles contract and longitudinal muscles relax. Since smooth muscles are involuntary, you do not consciously control these contractions.

When you swallow, food enters the esophagus. Usually, a flap of cartilage called the **epiglottis** (ep uh GLAH tus) closes over the opening to the respiratory tract as you swallow. This prevents food from entering the respiratory tract. If you talk while swallowing, the epiglottis may open, and food can enter the respiratory tract. The body responds by choking and coughing, forcing the food out of the respiratory tube and back into the throat.

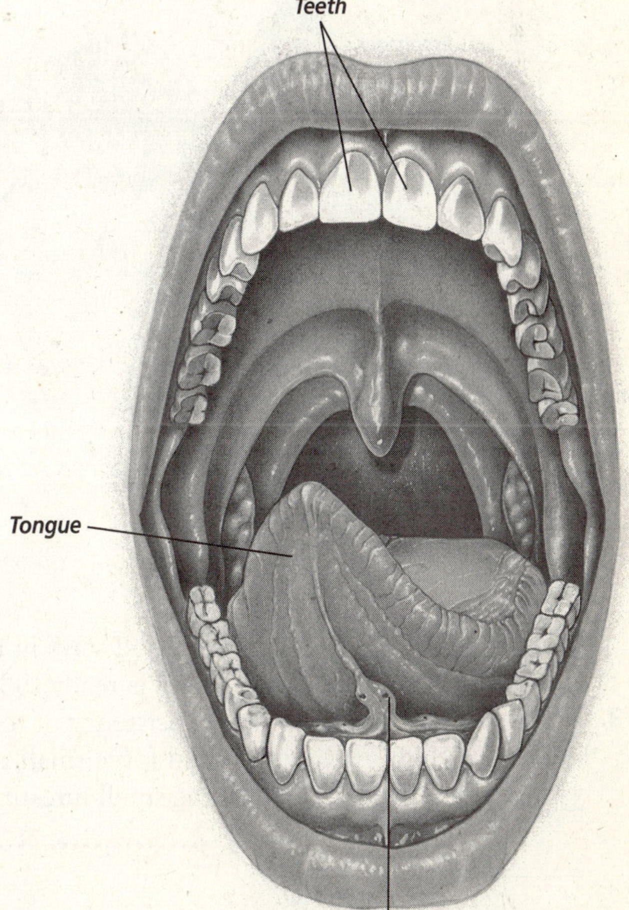

Reading Check

2. What types of digestion take place in the stomach?

____

____

____

The Stomach

When chewed food reaches the end of the esophagus, it enters the stomach. The **stomach** is a muscular, pouchlike enlargement of the digestive tract. Both mechanical and chemical digestion take place in the stomach. ✓

How do muscles in the stomach break down food?

The stomach contains three layers of involuntary muscles. They lie across each other, and they are located within the stomach's walls. When these muscles contract, they physically break down swallowed food into smaller pieces. As the muscles continue to work on the pieces of food, the pieces are mixed with digestive juices produced by the stomach.

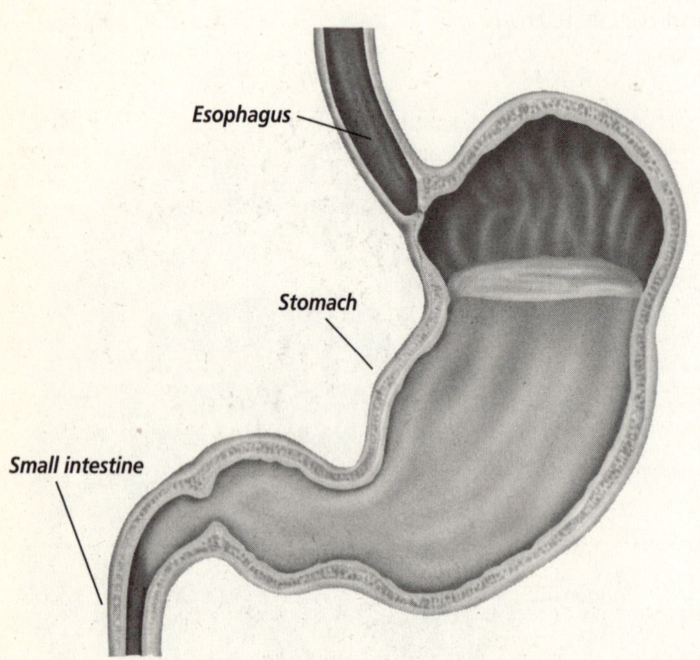

How do chemicals in the stomach break down food?

The inner lining of the stomach contains millions of glands. These glands secrete a mixture of chemicals called gastric juice. Gastric juice contains pepsin and hydrochloric acid. **Pepsin** is an enzyme that begins the chemical digestion of proteins in food. Pepsin works best in an acidic environment. This environment is provided by hydrochloric acid. ✓

How is the stomach lining protected from powerful digestive enzymes and strong acids? The stomach lining secretes mucus. This mucus forms a protective layer between the stomach lining and the acidic environment of the stomach.

Food stays in the stomach for about two to four hours. When food is ready to leave the stomach, its consistency is similar to the consistency of tomato soup. Peristaltic waves become stronger and force small amounts of the liquid out of the stomach and into the small intestine.

Reading Check

3. What role does pepsin play in digestion?

____

____

____

The Small Intestine

The **small intestine** is a muscular tube about 6 m long. It is called *small* because it has a narrow diameter. Its diameter is only about 2.5 cm. Food digestion is completed in the small intestine. Muscle contractions continue to help break down the food mechanically. Carbohydrates and proteins undergo additional chemical digestion. The pancreas and the liver secrete enzymes that break down the food substances even further. ✓

What is the purpose of the duodenum?

The first 25 cm of the small intestine is called the duodenum (doo ah DEE num). Most of the enzymes and chemicals that work in the duodenum enter it through ducts that collect juices from the pancreas, the liver, and the gallbladder. Food does not pass into these three organs, but they all help in the digestion process.

How does the pancreas help in the digestion process?

The **pancreas** is a soft, flattened gland that secretes both digestive enzymes and hormones. The enzymes that the pancreas secretes break down carbohydrates, proteins, and fats. Alkaline pancreatic juices also help to neutralize the acidity of the liquid food in the small intestine. This stops any further action of pepsin.

How does the liver help in the digestion process?

The **liver** is a large, complex organ that has many functions. It produces bile. **Bile** is a chemical substance used in digestion that breaks down fats mechanically. Bile breaks large drops of fat into smaller droplets. After the liver makes bile, the bile is stored in the gallbladder. The **gallbladder** is a small organ located just under the liver. Bile passes from the gallbladder into the duodenum.

How is food absorbed?

After it leaves your stomach, liquid food stays in the small intestine for three to five hours. The food moves slowly through the small intestine by peristalsis. As digested food moves through the small intestine, it passes over thousand of villi. A **villus** (plural, villi) is a tiny, fingerlike structure. Villi are projections on the

✓ Reading Check

4. If the small intestine is 6 m long, why is it called *small*?

____

____

____

lining of the small intestine that help absorb digested food. Because villi increase the surface area of the small intestine, they allow the body to absorb more food from the small intestine.

Digested food in the small intestine is in the form of small molecules. These small molecules can be absorbed into the cells of the villi. The food molecules diffuse into the blood vessels of the villus and enter the body's bloodstream. Villi are the link between the digestive system and the circulatory system.

The Large Intestine

The material that cannot be digested in the small intestine passes into the large intestine. The **large intestine** is a muscular tube that is also called the colon. The large intestine, or colon, is only about 1.5 m long, but it is about 6.5 cm in diameter. The large intestine is much wider than the small intestine. The appendix is a tubelike extension off the large intestine. It seems to serve no purpose in human digestion.

What does bacteria in the large intestine do?

The human body does not waste water. The walls of the large intestine absorb water and salts from the indigestible material. A more solid material remains in the large intestine. Anaerobic bacteria in the large intestine produce some B vitamins and vitamin K. Both these vitamins are absorbed as needed by the body. Other bacteria in the large intestine stop harmful bacteria from colonizing. This helps to reduce the risk of infections in the intestines.

Reading Check

5. What are three purposes of the large intestine?

___

___

___

How are wastes eliminated?

After 18 to 24 hours in the large intestine, indigestible material, now called feces, reaches the rectum. The **rectum** is the last part of the digestive system. Feces are eliminated from the rectum through the anus.

After You Read

Mini Glossary

amylase: breaks down the starches in food into smaller molecules

bile: a chemical that helps break down large drops of fats into smaller droplets

epiglottis (ep uh GLAH tus): a flap of cartilage that closes over the opening of the respiratory tract during swallowing

esophagus: a muscular tube that connects the mouth to the stomach

gallbladder: a small organ located just under the liver; bile passes from the gallbladder into the duodenum

large intestine: a muscular tube, also called the colon, that holds material that cannot be digested in the small intestine

liver: a large, complex organ that produces bile

pancreas: a soft, flattened gland that secretes both digestive enzymes and hormones

pepsin: an enzyme that begins the chemical digestion of protein in food; works best in an acidic environment

peristalsis (per uh STAHL sus): a series of smooth muscle contractions along the walls of the digestive tract

rectum: the last part of the digestive system; feces are eliminated from the rectum through the anus

small intestine: a muscular tube about 6 m long and about 2.5 cm in diameter; food digestion is completed in the small intestine

stomach: a muscular, pouchlike enlargement of the digestive tract

villus: a tiny, fingerlike structure that is a projection on the lining of the small intestine; works in the absorption of digested food

1. Read the terms and their definitions in the Mini Glossary above. Use them to create and label your own diagram of the digestive system on a separate sheet of paper.
2. Complete the sequencing diagram to show the body parts involved in the passage of food from the mouth to the anus.

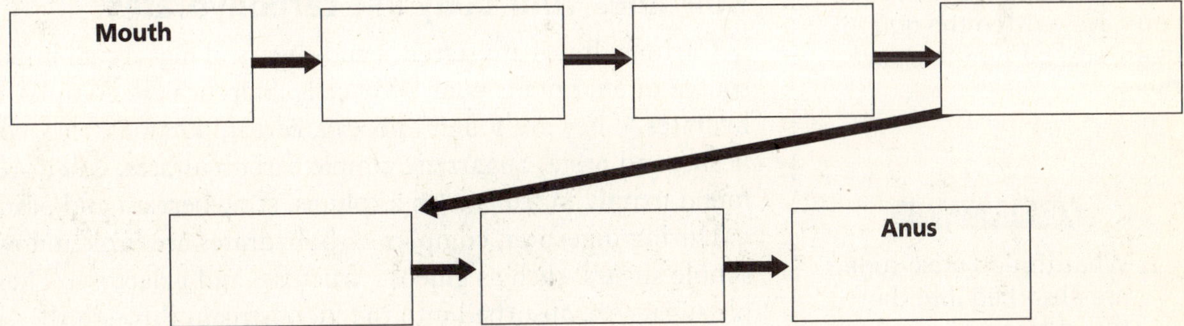

Visit the Glencoe Science Web site at **science.glencoe.com** to find your biology book and learn more about following digestion of a meal.

Name Date Class

Section 35.2 Nutrition

Before You Read

Consider the following two lunches—a tuna salad, an apple, and a glass of milk, or a cheeseburger, french fries, and a soda. Which would you choose? Which might be the healthier choice? On the lines below, explain your answer to this question.

Read to Learn

Mark the Text **Create a Quiz** After you have read this section, create a quiz based on what you have learned. After you have written the quiz questions, be sure to answer them.

Reading Check

1. What three simple sugars are absorbed into the bloodstream?

The Vital Nutrients

Six basic kinds of nutrients can be found in foods. They are carbohydrates, fats, proteins, minerals, vitamins, and water. These nutrients are essential for your body to function properly. Using the food pyramid, shown on page 430, to shape your diet can help you get the essential nutrients your body needs.

How does your body use carbohydrates?

Carbohydrates are starches and sugars. They are an important source of energy for your body cells. Starches are complex carbohydrates. They are found in bread, cereal, potatoes, rice, corn, beans, and pasta. Sugars are simple carbohydrates. They are found mainly in fruits such as plums, strawberries, and oranges.

During digestion, complex carbohydrates are broken down into simple sugars, such as glucose, fructose, and galactose. These simple sugars are absorbed into the bloodstream through the villi of the small intestine. These sugar molecules circulate through the blood to fuel body functions. Some sugar is carried to the liver where it is stored as glycogen. ✓

Cellulose, another complex carbohydrate, is found in all plant cell walls. Even though the human body cannot digest cellulose (also known as fiber), cellulose is important in the diet. It helps in the elimination of wastes. Sources of cellulose include bran, beans, and lettuce.

USE SPARINGLY

Includes butter, oils, salad dressings and soft drinks

Fat

Sugar

2–3 SERVINGS

2–3 SERVINGS

3–5 SERVINGS

2–4 SERVINGS

6–11 SERVINGS

How does your body use fats?

Fats are an essential nutrient. They provide energy for your body. Fats also are essential building blocks of the cell membrane. Fats help synthesize hormones, protect body organs against injury, and insulate the body from cold.

Meats, nuts, and dairy products contain fats. Cooking oils are another source of dietary fat. In the digestive system, fats break down into fatty acids and glycerol. They are absorbed by the villi of the small intestine. Some fatty acids eventually end up in the liver. The liver converts them to glycogen or stores them as fat throughout your body.

Reading Check

2. What role do fats play in maintaining the human body?

____

____

____

How does your body use proteins?

The body uses proteins in many ways. Enzymes, antibodies, many hormones, and substances that help the blood clot are all proteins. Proteins form parts of your muscles. Many cell structures, including cell membranes, are formed of proteins.

During digestion, proteins are broken down into amino acids. The amino acids are absorbed by the small intestine, and they enter the bloodstream. They are carried through the bloodstream to the liver. The liver can change amino acids to fats or to glucose. Both fats and glucose can be used for energy. Your body uses amino acids for energy only if other energy sources have been used up. Most amino acids are absorbed by cells and used for protein synthesis. The human body needs 20 different amino acids for protein synthesis. The body can make only 12 of these, so the other 8 amino acids must come from the food that you eat. Sources of essential amino acids include meats, dried beans, whole grains, eggs, and dairy products.

How does your body use minerals and vitamins?

A **mineral** is an inorganic substance that serves as a building material for your body. Minerals take part in chemical reactions in your body. They make up about four percent of your total body weight. Most minerals are found in your skeleton. Minerals are not used as an energy source.

Vitamins are organic nutrients that help maintain growth and metabolism. Your body needs only small amounts of them. There are two main groups of vitamins. One group is water-soluble, which means that they dissolve in water. Water-soluble vitamins cannot be stored in the body and must be included regularly in the diet. The other main group of vitamins is fat-soluble. These vitamins dissolve in fats. Fat-soluble vitamins can be stored in the body's liver. Excessive amounts of fat-soluble vitamins in the liver, however, can be poisonous.

Vitamin D, a fat-soluble vitamin, is synthesized in your skin. The bacteria in your large intestine synthesize vitamin K and some B vitamins. The rest of the vitamins that your body needs must be taken in through your food. ✓

Reading Check

3. What is the difference between vitamins and minerals?

____

____

____

How does your body use water?

Between 45 and 75 percent of your body mass is water. Water is the most abundant substance in your body. Water helps chemical reactions take place in your body. It is necessary for breaking down foods during digestion. Water is a solvent. Oxygen and nutrients from food cannot enter your cells unless they have been first dissolved in water.

Water helps your body maintain its internal temperature because water absorbs and releases heat slowly. The body contains so much water, it takes a lot of added energy to raise your body temperature. Every day your body loses about 2.5 L of water through exhalation, sweat, and urine. Therefore, you need to drink enough water every day to replace the water that is lost.

Think it Over

4. Evaluate Why would teenagers need more Calories per day than adults?

____

____

____

Calories and Metabolism

A calorie is the amount of heat that is needed to raise the temperature of 1 mL of water by 1°C. The energy content of food is measured in Calories. The term **Calories,** written with a capital letter C, represents a kilocalorie, or 1000 calories. Some foods contain more Calories than other foods. 1 g of fat generally contains nine Calories. 1 g of carbohydrate or protein generally contains only four Calories.

The number of Calories needed each day depends on a person's metabolism. Metabolism is the rate at which energy is burned. A person's body mass, age, gender, and level of physical activity also affect how much energy is used. Males usually need more Calories per day than females. Teenagers use more Calories than adults, and active people use more Calories than inactive people.

What is the relationship between Calories and health?

If a person takes in more Calories than his or her body can metabolize, or burn, the extra energy is stored as body fat. The person gains weight. If a person eats fewer Calories than the body can metabolize, some of the body's stored energy is used. That person will lose weight.

Many Americans are overweight. Being overweight increases the risk of developing health problems including high blood pressure, diabetes, and heart disease. Underweight people can have health problems, too. These can include anemia, fatigue, and a decreased ability to fight off infections and diseases.

After You Read

Mini Glossary

Calorie: unit of heat used to measure energy content of food, each Calorie represents a kilocalorie, or 1000 calories; a calorie is the amount of heat that is needed to raise the temperature of 1 mL of water by 1°C

mineral: an inorganic substance found mainly in the skeleton that serves as a building material for your body

vitamin: organic nutrients required in small amounts to maintain growth and metabolism

1. Read the terms and their definitions in the Mini Glossary above. On the lines below, write a sentence for each term that describes the functions of the term.

____

____

____

____

2. Use the web diagram below to list ways that each item helps a healthy body function.

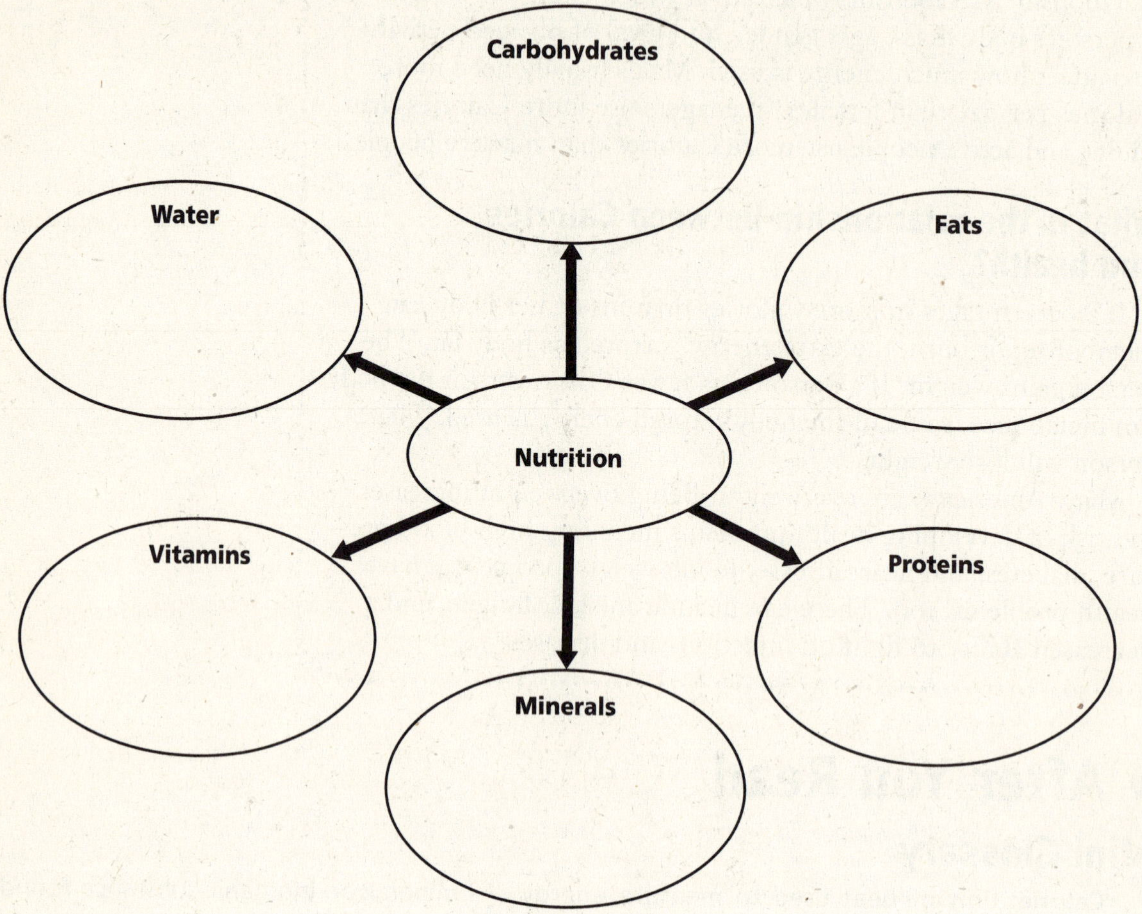

Nutrition

Carbohydrates

Fats

Proteins

Minerals

Vitamins

Water

Visit the Glencoe Science Web site at **science.glencoe.com** to find your biology book and learn more about nutrition.

Section 35.3 The Endocrine System

Before You Read

The endocrine system releases chemicals called hormones directly into the bloodstream. These hormones control many of the body's functions. One important hormone is the human growth hormone. What role do you think this hormone serves in helping the body function? When do you think this hormone is most important to human development? Write your response on the lines below.

Read to Learn

Mark the Text **Identify Main Ideas** As you read, highlight the glands of the endocrine system. Use another color to highlight what these glands do.

Control of the Body

Internal control of the body is directed by two systems. In this section you will learn about the endocrine system. Later, you will learn about the other system, the nervous system. The functions of all body systems are controlled by the interaction between the nervous system and the endocrine system. As you will recall, in mammals, a gland is a cell or a group of cells that secretes fluid. The endocrine system is made up of a series of glands, called endocrine glands. **Endocrine glands** release chemicals directly into the bloodstream. These chemicals relay information to other parts of the body.

How do the nervous and endocrine systems interact?

The endocrine system and the nervous system work together much of the time. The two systems maintain homeostasis in the body. As you will recall, homeostasis is the ability of a living organism to maintain internal equilibrium or conditions that enable it to survive. Because there are two control systems within the human body, the nervous system and the endocrine system, coordination is needed between the two. The **hypothalamus** (hi poh THA luh mus) is the part of the brain that connects the endocrine system and the nervous system. The hypothalamus receives messages from other areas of the brain. It also receives messages from the internal organs. When a change in homeostasis is detected, the hypothalamus stimulates the pituitary gland.

Think it Over

1. **Evaluate** What is the relationship between the hypothalamus and the pituitary gland?

____

____

____

The **pituitary gland** (pih TEW uh ter ee) is the main gland of the endocrine system.

The pituitary gland is located in the skull, just beneath the hypothalamus. The hypothalamus controls the pituitary gland. The two are connected by nerves and by blood vessels. When the hypothalamus receives messages, the pituitary gland releases its own chemicals, or it stimulates other glands to release their chemicals. The pituitary gland controls endocrine glands, including the thyroid gland, the adrenal glands, and glands associated with reproduction.

How do hormones travel?

The endocrine glands secrete chemicals called hormones into the bloodstream. Remember that a hormone is a chemical that is released in one part of an organism that affects another part of the organism. Hormones carry information to other cells in the body. Hormones give these other cells instructions regarding metabolism, growth, development, and behavior. Once glands release the hormones, the hormones travel in the bloodstream. Hormones then attach themselves to target cells. **Target cells** have specific binding sites for hormones. These binding sites are located either on the plasma membranes or in the nuclei of these cells. The binding sites on target cells are called **receptors.**

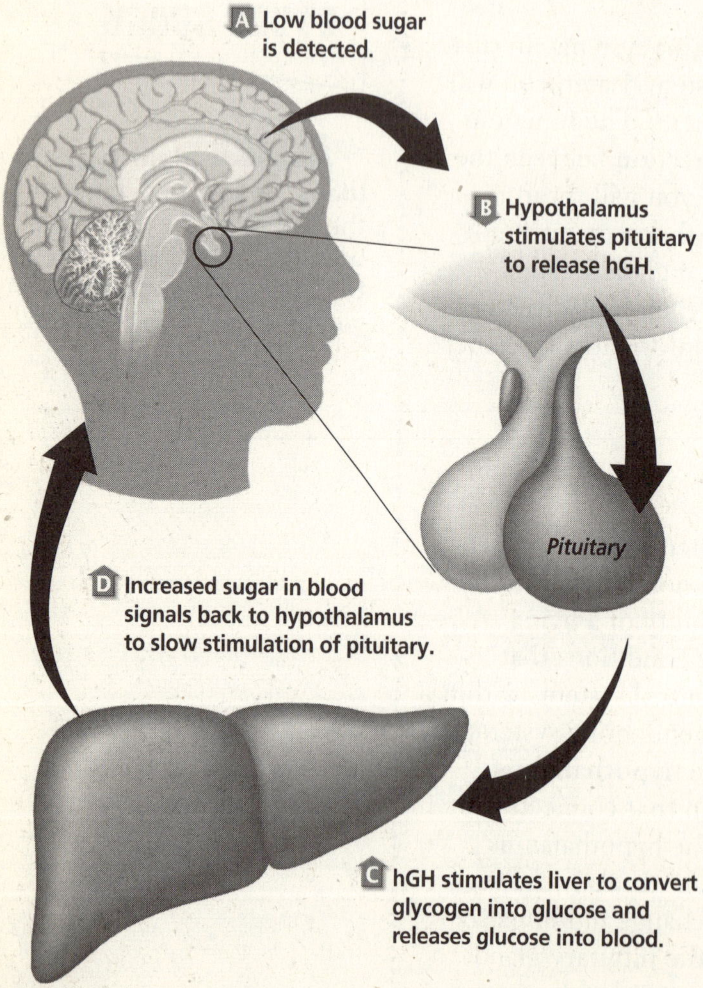

A Low blood sugar is detected.

B Hypothalamus stimulates pituitary to release hGH.

Pituitary

C hGH stimulates liver to convert glycogen into glucose and releases glucose into blood.

D Increased sugar in blood signals back to hypothalamus to slow stimulation of pituitary.

How does human growth hormone (hGH) work?

Human growth hormone, hGH, provides a good example of an endocrine system hormone. When your body is actively growing, blood glucose levels are slightly lowered because the growing cells use up the sugar. The hypothalamus detects this low blood glucose level. The hypothalamus then stimulates the production and release of hGH from the pituitary gland into the bloodstream. The hormone hGH binds to receptors on the plasma membranes of liver cells. This, in turn, stimulates the liver cells to release glucose into your blood. Your cells need this glucose to keep growing.

Negative Feedback Control

If homeostasis is disrupted, the body responds. The endocrine glands are stimulated. Endocrine glands can be stimulated by the nervous system, by changes in blood chemistry, or by other hormones. One type of internal feedback mechanism generally controls adjustments to the endocrine system. This is called a negative feedback system. In a **negative feedback system,** the hormones, or their effects, are fed back to suppress or slow the original signal. Once homeostasis is reached, the signal stops. The hormone is no longer released.

How does the feedback system work?

Most of the endocrine glands operate under a negative feedback system. A gland synthesizes and secretes its hormone. The hormone travels in the blood to the target cells. The needed response occurs in these target cells. Information concerning the hormone level or its effect on these target cells is fed back. The feedback is usually sent to the hypothalamus or the pituitary gland to regulate, or change, the gland's production of the hormone.

How do hormones controlled by the negative feedback system work?

Antidiuretic hormone, ADH, is one of the hormones that is controlled by a negative feedback system. If you have lost water because your body has been sweating, you will feel thirsty. You feel thirsty because the water content of your blood is reduced. The hypothalamus is able to sense the concentration of water in your blood. The hypothalamus determines that your body is dehydrated. It responds by stimulating the pituitary gland to release antidiuretic (AN tih di yuh reh tihk) hormone (ADH). ✓

Antidiuretic hormone (ADH) reduces the amount of water in your urine. The hormone binds to receptors in the kidney cells. In the kidneys, the hormone ADH promotes the reabsorption of water. ADH also reduces the amount of water that is excreted in urine. Information about the blood water levels is constantly fed back to the hypothalamus. The hypothalamus can then regulate the pituitary gland's release of ADH. If the body becomes overhydrated, or has too much water, the hypothalamus stops stimulating the release of ADH.

Reading Check

2. What does the hypothalamus do when you are dehydrated?

____

____

____

How do hormones that are controlled by a negative feedback system control blood glucose levels?

Another example of a negative feedback system involves the regulation of blood glucose levels. Unlike most other endocrine glands, the pancreas is not controlled by the pituitary gland. After you have finished eating a meal, your blood glucose levels are high. When the blood glucose levels are high, the pancreas releases the hormone insulin. Insulin signals the liver and muscle cells to take in glucose. This lowers the blood glucose levels. When the blood glucose levels become too low, the pancreas releases another hormone called glucagon. Glucagon signals cells in the liver to release stored glycogen as glucose.

Reading Check

3. What are the two groups of hormones?

____

____

Hormone Action

Once hormones are released by an endocrine gland, they travel to target cells, and they cause a change. There are two basic types of hormones. They are grouped according to how they act on their target cells. The two groups are steroid hormones and amino acid hormones. ✓

What are steroid hormones?

Hormones that are made from lipids, or fats, are called steroid hormones. Steroid hormones are lipid-soluble. As a result, they can diffuse freely into a cell through the cell's plasma membrane. The hormones bind to a hormone receptor inside the cell. The hormone, bound to its receptor, forms a hormone-receptor complex. The hormone-receptor complex travels to the cell nucleus. In the cell nucleus, this hormone-receptor complex starts the process for making specific messenger RNA (mRNA) molecules. The mRNA molecules move out to the cytoplasm. The mRNA molecules transport instructions from DNA in the nucleus for the synthesis of the required proteins while in the cytoplasm.

What are amino acid hormones?

The second group of hormones is made from amino acids. Remember that amino acids can be strung together in chains. Proteins are made from long chains of amino acids. Some hormones are short chains of amino acids while other hormones are long chains. Once amino acid hormones are secreted into the bloodstream, they bind to receptors. These receptors are embedded in the plasma membrane of the target cell. From the plasma membrane, they open ion channels in the membrane. Ions have electrical charges. These ion channels route signals down the surface of the membrane to activate enzymes within the cell. The enzymes change the behavior of other molecules inside the cell.

Adrenal Hormones and Stress

The adrenal glands help prepare your body for stressful, difficult situations. The **adrenal glands** are located on top of the kidneys. These glands consist of two parts—an inner portion and an outer portion. The outer portion secretes steroid hormones. These include glucocorticoids (glew ko KOR tuh koydz) and aldosterone (ahl DOS tuh rohn).

These steroid hormones cause an increase in available glucose. They also raise blood pressure. As a result, they help the body combat fear, very hot or very cold temperatures, bleeding, infection, disease, and other common anxieties.

The inner portion of the adrenal gland secretes two amino acid hormones. One is epinephrine (eh puh NEH frun). Epinephrine is often called adrenaline. The other hormone is called norepinephrine. Remember the fight-or-flight response? During this response, the hypothalamus relays impulses to the nervous system. The nervous system stimulates the adrenal glands to increase the output of both epinephrine and norepinephrine. These two hormones stimulate heart rate, blood pressure, and the rate of breathing. They increase the efficiency of muscle contractions and they also increase blood sugar levels. ✓

Reading Check

4. How are epinephrine and norepinephrine involved in the fight-or-flight response?

____

____

____

Thyroid and Parathyroid Hormones

The **thyroid gland** is located in the neck. This gland regulates metabolism, growth, and development. The main metabolic and growth hormone of the thyroid is thyroxine. Thyroxine affects the rate at which your body uses energy. It also determines how much food you need to eat.

The thyroid gland secretes calcitonin (kal suh TOH nun). This hormone regulates calcium levels in the blood. The body needs the mineral calcium for blood clotting, the formation of bones and teeth, and for normal nerve and muscle function. Calcitonin binds to the membranes of kidney cells. Calcitonin then causes the kidneys to excrete more calcium. Calcitonin also binds to bone-forming cells. It causes these bone-forming cells to increase calcium absorption and to make new bone.

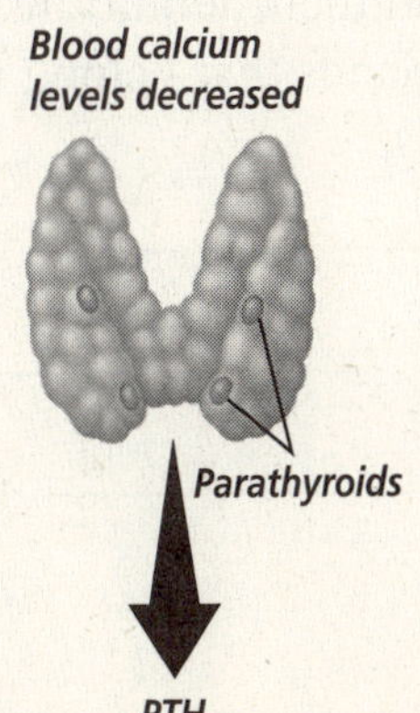

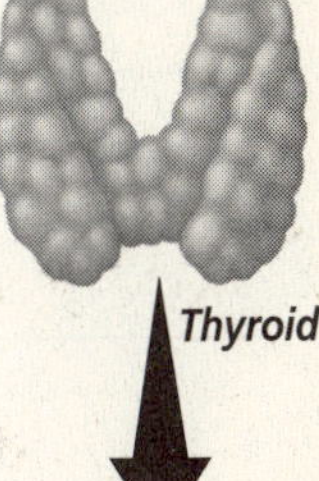

Reading Check

5. What three minerals are regulated by PTH?

____

____

____

Parathyroid hormone (PTH) is involved in mineral regulation. This hormone is produced by the **parathyroid glands.** Parathyroid glands are attached to the thyroid gland. The release of PTH leads to an increase in the rate that minerals are absorbed in the intestine. The three minerals affected are calcium, phosphate, and magnesium. PTH causes the release of calcium and phosphate from bone tissue. PTH also increases the rate at which the kidneys remove calcium and magnesium from urine and return these two minerals to the blood. Hormones associated with the endocrine system control many different body functions. Different hormones play more important roles during various stages of growth and development. They are the main biological influence on your behavior and on your development.

After You Read

Mini Glossary

adrenal gland: located on top of the kidneys; outer portion secretes steroid hormones; inner portion secretes amino acid hormones; amino acid hormones are responsible for the fight-or-flight response

endocrine glands: glands that release chemicals directly into the bloodstream; relay information to other parts of the body

hypothalamus: the part of the brain that connects the endocrine system and the nervous system

negative feedback system: type of internal feedback mechanism that generally controls adjustments to the endocrine system

parathyroid gland: gland attached to the thyroid gland that is involved in mineral regulation in the body

pituitary gland: the main gland of the endocrine system; controlled by the hypothalamus

receptors: the binding sites on target cells

target cells: cells to which hormones attach themselves; contain specific binding sites either on the plasma membranes, or in the nuclei

thyroid gland: gland located in the neck that regulates metabolism, growth, and development

1. Read the terms and their definitions in the Mini Glossary above. Use the lines below to list and describe the glands that belong to the endocrine system. Use your own words to describe the glands.

____

____

____

2. Use the flow diagram below to show how a negative feedback system in the endocrine system works to control the amount of hormones that are released into the body.

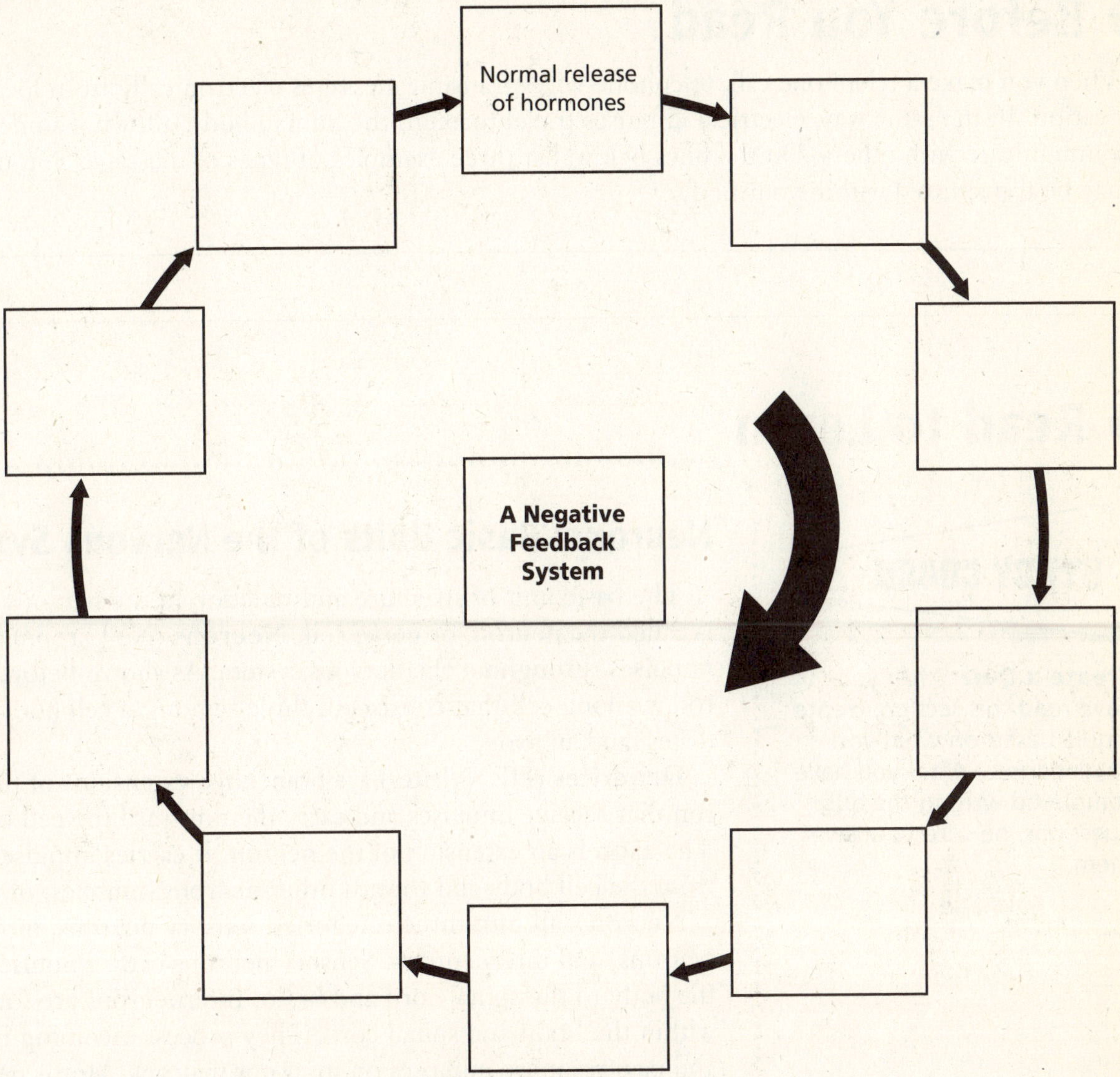

Normal release of hormones

A Negative Feedback System

Visit the Glencoe Science Web site at **science.glencoe.com** to find your biology book and learn more about the endocrine system.

Name Date Class

Section 36.1 The Nervous System

Before You Read

When you make a telephone call, telephone wires transmit messages electronically from location to location. In the same way, electrical impulses travel through the human body, allowing some parts to communicate with others. On the lines below, list three examples of types of messages you think may be transmitted within your body.

Read to Learn

STUDY COACH

Create a Quiz After you have read this section, create a quiz based on what you have learned. After you have completed writing the quiz questions, be sure to answer them.

Neurons: Basic Units of the Nervous System

The basic unit of structure and function in the nervous system is called the neuron, or nerve cell. **Neurons** (NYU ronz) conduct impulses throughout the nervous system. As shown below, a neuron is a long cell that consists of three regions: a cell body, dendrites, and an axon.

Dendrites (DEN drites) are branchlike extensions of the neuron that receive impulses and carry them toward the cell body. The **axon** is an extension of the neuron. It carries impulses away from the cell body and toward other neurons, muscles, or glands.

Neurons fall into three categories: sensory neurons, motor neurons, and interneurons. Sensory neurons carry impulses from the body to the spinal cord and brain. Interneurons are found within the brain and spinal cord. They process incoming impulses and pass response impulses on to motor neurons. Motor neurons carry the response impulses away from the brain and spinal cord to a muscle or gland.

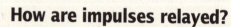

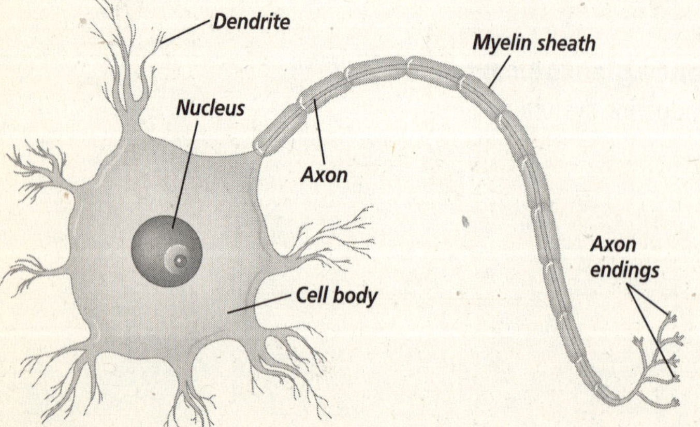

How are impulses relayed?

Imagine that you are in a crowded, noisy store and you feel a tap on your shoulder. You turn your head, and you see a friend standing behind you. What happened in your body to cause the tap to get your attention? First, the touch stimulated sensory receptors located in the skin of your shoulder. This produced a sensory impulse which was carried to the spinal cord and then to your brain. From the brain, an impulse was sent to your motor neurons, which transmitted the impulse to the muscles in your neck. The result? Your neck muscles turned your head in response to the tap.

What occurs when the neuron is at rest?

You have learned that the plasma membrane controls the concentration of ions in a cell. Because the plasma membrane of a neuron is more permeable to potassium ions (K^+) than to sodium ions (Na^+), more potassium ions are inside the cell membrane than outside it. Similarly, more sodium ions are outside the cell membrane than inside it.

The neuron membrane also contains an active transport system, called the sodium/potassium (Na^+/K^+) pump. The pump uses ATP (the cell's energy storing molecules) to pump three sodium ions out of the cell for every two potassium ions it pumps in. This increases the concentration of positive charges on the outside of the membrane. In addition, the presence of many negatively charged proteins and organic phosphates means that the inside of the membrane is more negatively charged than the outside. Under these conditions, which exist when the cell is at rest, the plasma membrane is said to be polarized. A polarized membrane has the potential to transmit an impulse. ✓

How are impulses transmitted?

When a stimulus excites a neuron, gated sodium channels in the membrane open up. That allows sodium ions to enter the cell. As the positive sodium ions build up inside the membrane, the inside of the cell becomes more positively charged than the outside. This is called depolarization. The change in the charge moves down the length of the axon like a wave. As the diagram on page 444 shows, the gated channels and the Na^+/K^+ pump then return the neuron to its resting state.

An impulse can only move down the complete length of an axon when stimulation of the neuron is strong enough. If the threshold level—the level at which depolarization occurs—is not reached, the impulse will die out quickly.

What are white and gray matter?

Most axons are surrounded by a white covering of cells called the myelin sheath. The myelin sheath is like the plastic coating on an electric wire. It insulates the axon, hindering the movement

Think it Over

1. **Analyze** When someone taps you on the shoulder, which neuron goes into action?

______

______

______

Reading Check

2. In order for an impulse to be transmitted, what state must the resting cell's membrane be in?

______

______

______

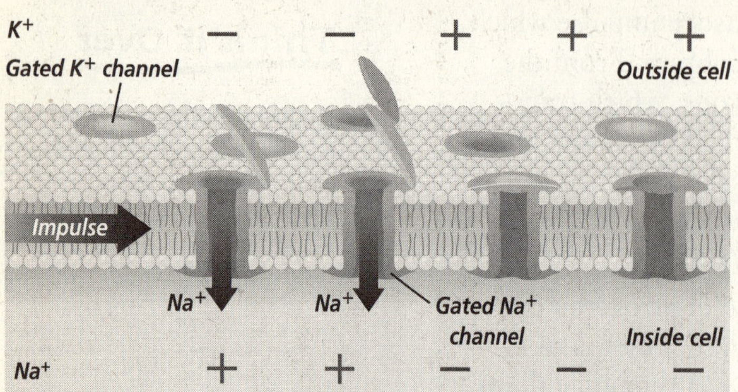

A Gated sodium channels open, allowing sodium ions to enter and make the inside of the cell positively charged and the outside negatively charged.

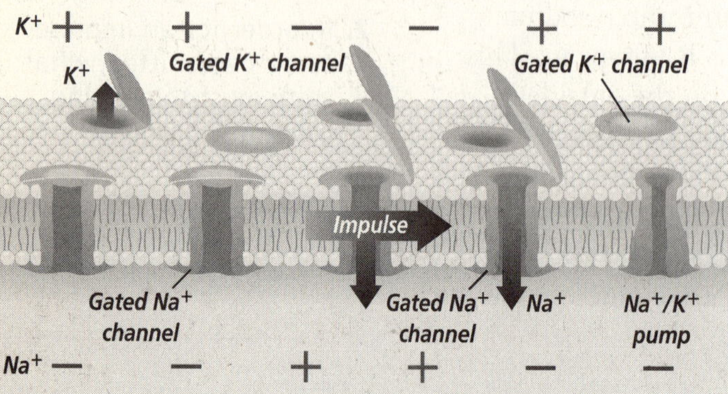

B As the impulse passes, gated sodium channels close, stopping the influx of sodium ions. Gated potassium channels open, letting potassium ions out of the cell. This action repolarizes the cell.

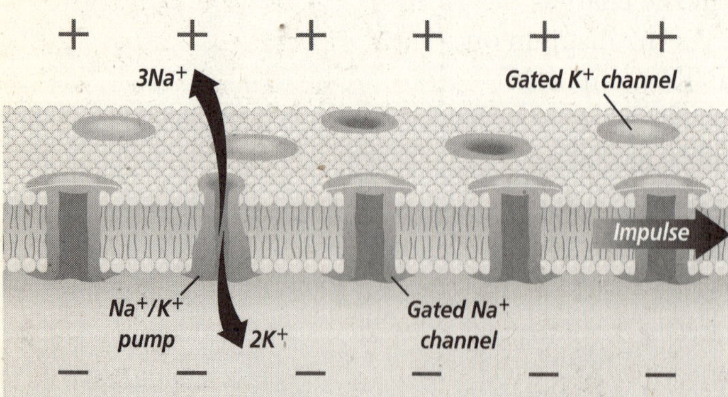

C As gated potassium channels close, the Na^+/K^+ pump restores the ion distribution.

of ions across its plasma membrane. The ions move quickly down the axon until they reach a gap in the sheath. Here, the ions pass through the plasma membrane of the nerve cell and depolarization occurs. As a result, the impulse jumps from gap to gap, greatly increasing the speed at which it travels.

The myelin sheath gives axons a white appearance. In the brain and spinal cord, masses of myelinated axons make up what is called "white matter." The absence of myelin in masses of neurons accounts for the grayish color of "gray matter" in the brain.

What are the spaces between neurons called?

Neurons lie end to end—axons to dendrites—but they do not actually touch. There is a tiny space between one neuron's axon and another neuron's dendrites. This space is called a **synapse.** Impulses traveling to and from the brain must move across this space. How do impulses make this leap?

As an impulse reaches the end of an axon, calcium channels open, allowing calcium to enter the end of the axon. The calcium causes vesicles in the axon to fuse with the plasma membrane, releasing their chemicals into the synaptic space. These chemicals are called **neurotransmitters.** They diffuse across the space to the dendrites of the next neuron. As the neurotransmitters reach the dendrites, they signal receptor sites to open the ion channels. This process is illustrated on page 445.

The open channels change the polarity in the neuron, starting a new impulse. Enzymes in the synapse typically break down the neurotransmitters shortly after transmission. This prevents the continual firing of impulses.

Section **36.1** **The Nervous System,** *continued*

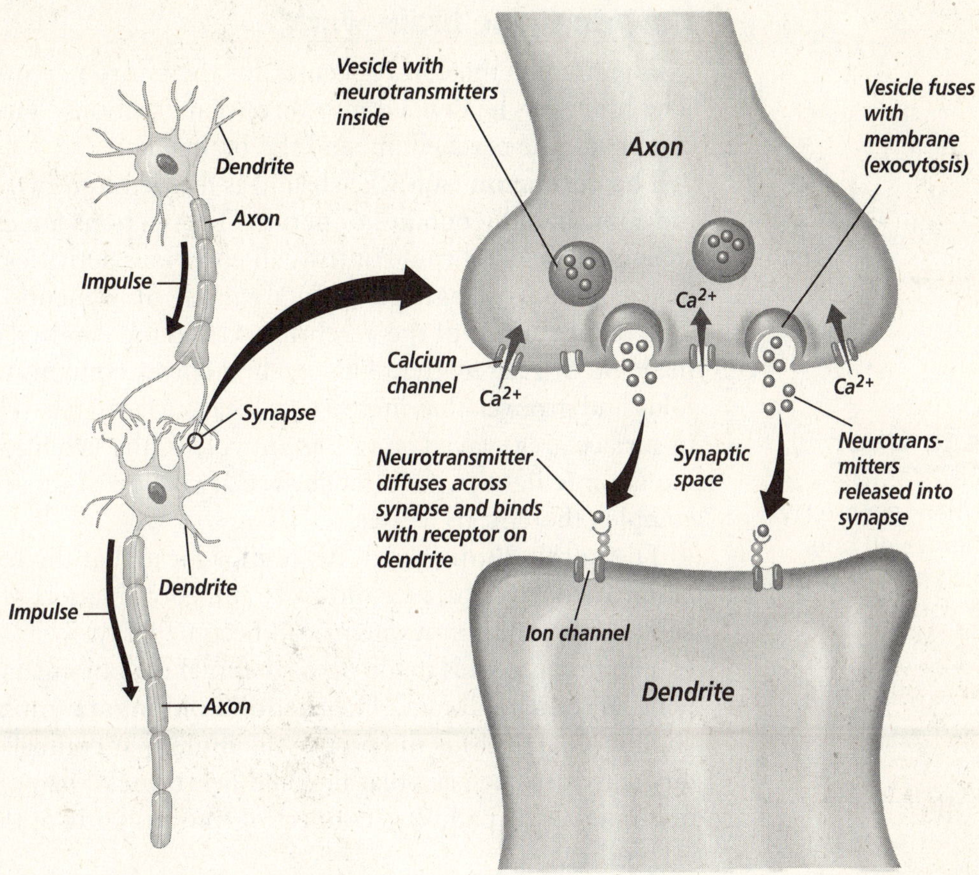

The Central Nervous System

When you make a telephone call to a friend, your call travels through wires to a control center. There it is switched over to wires that connect with a friend's telephone. In the same way, an impulse travels through the neurons in your body. The impulse usually reaches the control center of the nervous system—your brain—before being rerouted. The brain and the spinal cord together make up the **central nervous system,** which coordinates all your body's activities. ✓

Another division of your nervous system is called the **peripheral** (puh RIH frul) **nervous system.** It is made up of all the nerves that carry messages to and from the central nervous system. It is similar to the telephone wires that run between a phone system's control center and the phones in individual homes. Together, the central nervous system (CNS) and the peripheral nervous system (PNS) respond to stimuli from the external environment.

✓Reading Check

3. What two parts of the body make up the central nervous system?

____

____

How does the brain work?

The brain is the control center for the entire nervous system. The brain can be divided into three main sections. They are the cerebrum, the cerebellum, and the brain stem.

The **cerebrum** (suh REE brum) is divided into two halves that are connected by bundles of nerves. The sections are called hemispheres. The cerebrum controls all conscious activities, intelligence, memory, language, skeletal muscle movements, and senses. The outer surface of the cerebrum, called the cerebral cortex, is made up of gray matter. The cerebral cortex contains numerous folds and grooves that increase its total surface area. This increase in surface area played an important role in the evolution of human intelligence. Greater surface area allowed more and more complex thought processes.

The **cerebellum** (ser uh BE lum) is located at the back of the brain. It controls balance, posture, and coordination. If the cerebellum is injured, movements can become jerky.

The brain stem is made up of the medulla oblongata, the pons, and the midbrain. The **medulla oblongata** (muh DU luh • ah blon GAH tuh) is the part of the brain that controls involuntary activities such as breathing and heart rate. The pons and midbrain act as pathways connecting various parts of the brain to each other.

Think it Over

4. **Apply** When you are trying to balance on one foot, which part of your brain are you using?

______

______

______

The Peripheral Nervous System

Remember that the PNS carries impulses between the body and the CNS. For example, when a stimulus is picked up by receptors in your skin, it initiates an impulse in the sensory neurons. The impulse is carried to the CNS. There, the impulse transfers to the motor neurons, which carry the impulse to a muscle.

The PNS can be separated into two divisions—the somatic nervous system and the autonomic nervous system. ✓

Reading Check

5. What two divisions make up the peripheral nervous system?

______

______

What is the somatic nervous system?

The **somatic nervous system** is made up of 12 pairs of cranial nerves from the brain, 31 pairs of spinal nerves from the spinal cord, and all of their branches. These nerves are actually bundles of neuron axons bound together by connective tissue. The cell bodies of the neurons are found in clusters along the spinal column. Most nerves contain both sensory and motor axons.

The nerves of the somatic system relay information mainly between your skin, the CNS, and skeletal muscles. This pathway

Section 36.1 **The Nervous System,** *continued*

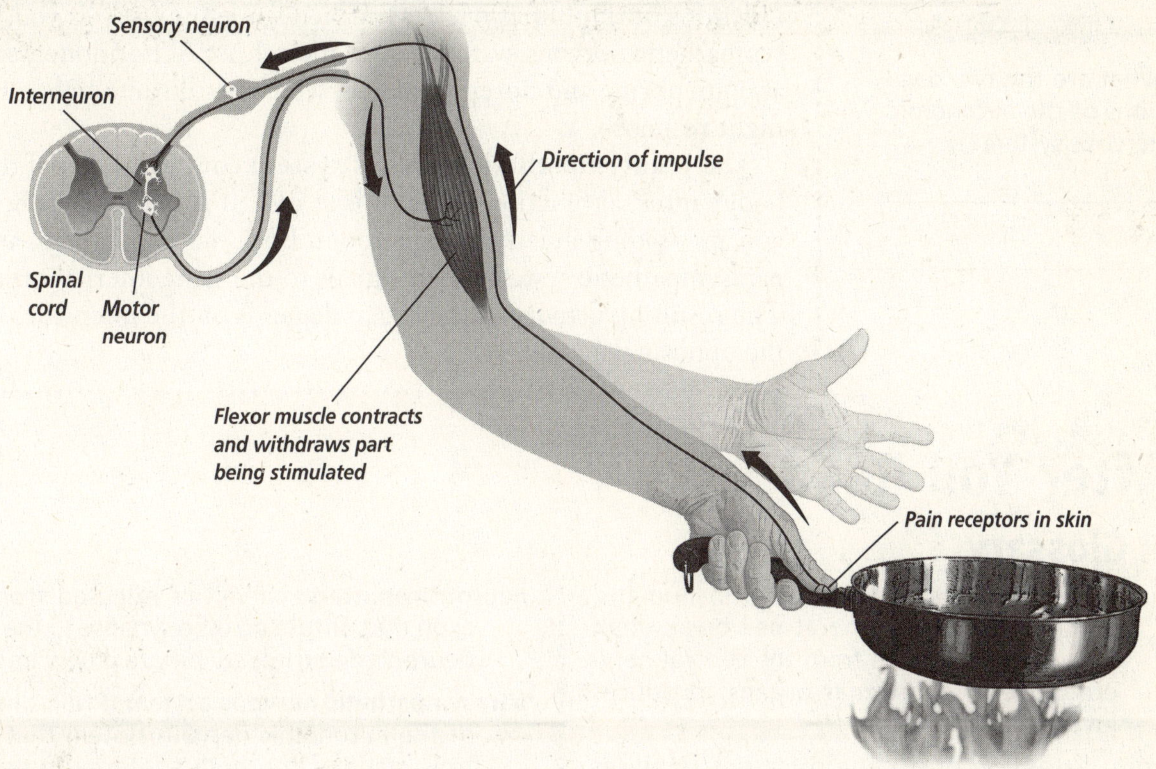

is voluntary, meaning that you can decide whether or not to move body parts under the control of the system.

Sometimes a stimulus results in an automatic, unconscious response within the somatic system. When you touch something hot, you automatically jerk your hand away. Such an action is a **reflex,** an automatic response to a stimulus. As illustrated above, a reflex impulse travels to the spinal column or brain stem where it causes an impulse to be sent directly back to a muscle. It does not go to the brain for interpretation. The brain becomes aware of the reflex only after it occurs.

What is the autonomic nervous system?

Have you ever heard scary sounds in the middle of the night? Maybe your heart began to pound or your palms got sweaty. These internal reactions to being scared are controlled by the autonomic nervous system. The **autonomic nervous system** carries impulses from the CNS to internal organs. These impulses produce responses that are involuntary, meaning they are not under conscious control.

There are two divisions of the autonomic nervous system—the sympathetic nervous system and the parasympathetic nervous system. The **sympathetic nervous system** controls many internal

6. What are the two divisions of the autonomic nervous system?

____

____

functions during times of stress. When something scares you, the sympathetic nervous system causes the release of hormones, such as epinephrine and norepinephrine, which results in a fight-or-flight response.

The **parasympathetic nervous system** controls many of the body's internal functions when it is at rest. It is in control when you are reading quietly in your room. Both the sympathetic and parasympathetic systems send signals to the same internal organs. The resulting activity of the organ depends on the intensities of the opposing signals.

After You Read

Mini Glossary

autonomic nervous system (ANS): in humans, portion of the peripheral nervous system that carries impulses from the central nervous system to internal organs; produces involuntary responses

axon: extension of a neuron; carries impulses away from a nerve cell

central nervous system (CNS): in humans, the central control center of the nervous system made up of the brain and spinal cord

cerebellum (ser uh BE lum): rear portion of the brain; controls balance, posture, and coordination

cerebrum (suh REE brum): largest part of the brain, composed of two hemispheres connected by bundles of nerves; controls conscious activities, intelligence, memory, language, skeletal muscle movements, and the senses

dendrite (DEN drite): branchlike extension of a neuron; transports impulses toward the cell body

medulla oblongata (muh DU luh • ah blon GAH tuh): part of the brain stem that controls involuntary activities such as breathing and heart rate

neuron (NYU ron): basic unit of structure and function in the nervous system; conducts impulses throughout the nervous system; composed of dendrites, a cell body, and an axon

neurotransmitters: chemicals released from an axon that diffuse across a synapse to the next neuron's dendrites to initiate a new impulse

parasympathetic nervous system (PNS): division of the autonomic nervous system that controls many of the body's internal functions when the body is at rest

peripheral (puh RIH frul) nervous system: division of the nervous system made up of all the nerves that carry messages to and from the central nervous system

reflex (REE fleks): simple, automatic response in an animal that involves no conscious control, usually acts to protect an animal from serious injury, automatic response to a stimulus; reflex stimulus travels to the spinal column or brain stem and is sent directly back to the muscle

somatic nervous system: portion of the nervous system composed of cranial nerves, spinal nerves, and all of their branches; voluntary pathway that relays information mainly between the skin, the CNS, and skeletal muscles

sympathetic nervous system: division of the autonomic nervous system that controls many of the body's internal functions during times of stress

synapse: tiny space between one neuron's axon and another neuron's dendrites over which a nerve impulse must pass

1. Read the terms and their definitions in the Mini Glossary on page 448. Use the space below to write a brief paragraph explaining the functions of the cerebrum and the cerebellum.

2. Use the chart below to fill in the missing divisions of the nervous system.

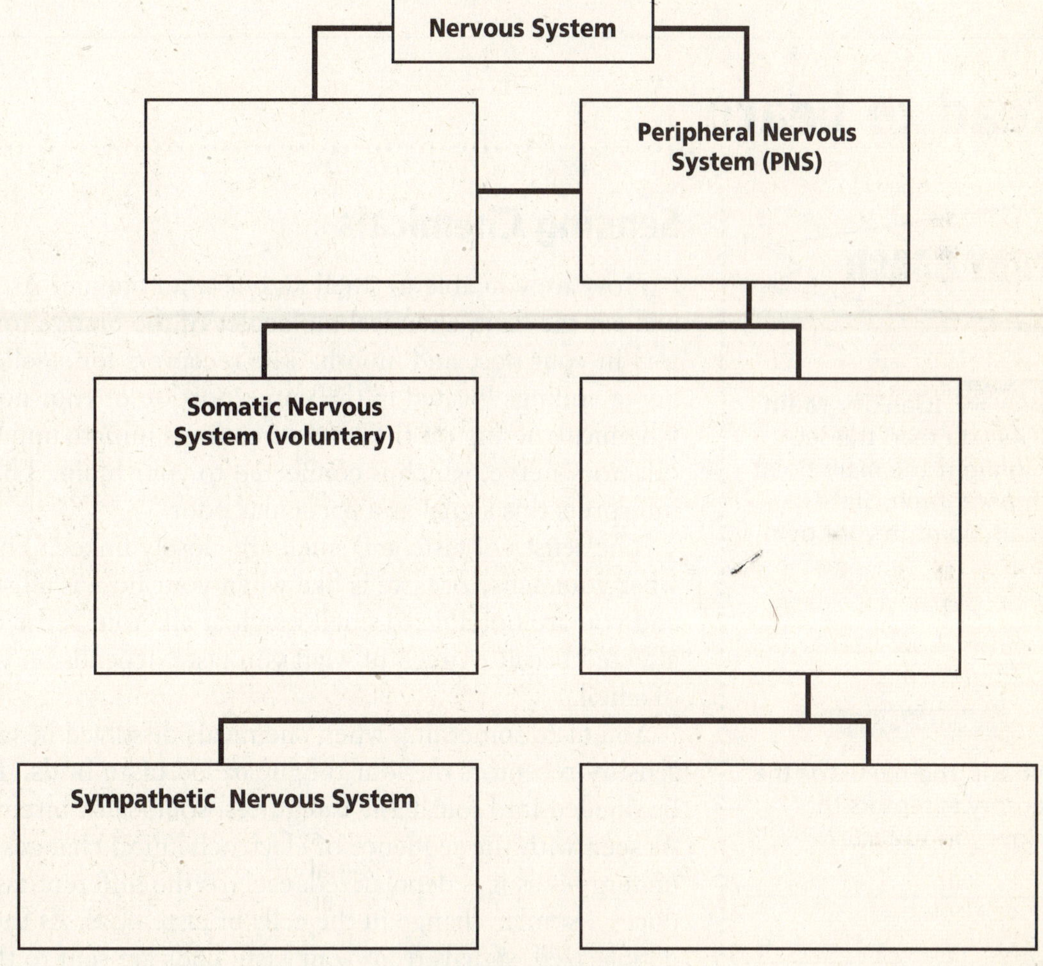

Visit the Glencoe Science Web site at **science.glencoe.com** to find your biology book and learn more about the nervous system.

Section 36.2 The Senses

Before You Read

Before you sit down for a meal, you probably can guess what you are having by sniffing the air coming from the kitchen. During the meal, you may have positive or negative reactions to the way certain foods taste. You might also touch your food or beverage to see if it is too hot or too cold. In this section, you will learn about how the senses work. On the lines below, list the senses you know of and the body parts most closely associated with each sense.

Read to Learn

Mark the Text **Identify Main Ideas** As you read the section, highlight the main point in each paragraph. State each main point in your own words.

1. What is the name for the sensory receptors that allow you to taste?

Sensing Chemicals

How are you able to smell and taste an orange? As you sniff and eat the fruit, chemical molecules of the orange touch receptors in your nose and mouth. The receptors for smell are hairlike nerve endings located in the upper portion of your nose. Chemicals acting on these nerve endings initiate impulses in the olfactory nerve, which is connected to your brain. The brain then interprets this signal as a particular odor.

The senses of taste and smell are closely linked. Think about what your sense of taste is like when your nose is all stuffed up and you are not able to smell much at all. Your sense of taste is affected because much of what you taste depends on your sense of smell.

You taste something when chemicals dissolved in saliva contact sensory receptors on your tongue called **taste buds.** Tastes can be divided into four basic categories: sour, salty, bitter, and sweet. As seen with the sequence of electrochemical changes a neuron undergoes as it is depolarized, each of the different tastes produces a similar change in the cells of taste buds. As these cells are depolarized, signals from your taste buds are sent to the cerebrum. There the signal is interpreted and you become aware of a particular taste. ✓

A young adult has approximately 10 000 taste buds. As a person ages, the sense of smell becomes less sharp and the taste buds may decrease in number or become less sensitive. A reduced sense of taste can result.

Sensing Light

How are you able to see? Your sense of sight depends on receptors in your eyes that respond to light energy. The **retina** is a thin layer of tissue made up of light receptors and sensory neurons. It is found at the back of the eye. Light enters the eye through the pupil and is focused by the lens onto the back of the eye, where it strikes the retina.

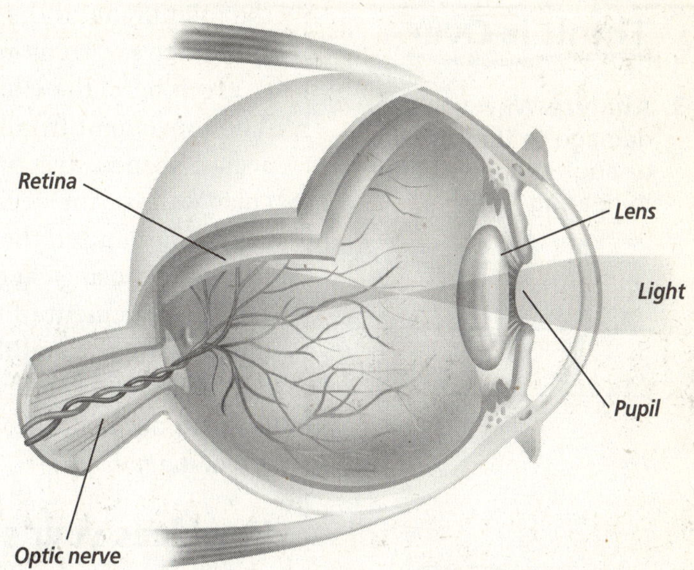

The retina contains two types of light receptor cells—rods and cones. **Rods** are receptor cells adapted for vision in dim light. They help you detect shape and movement. **Cones** are receptor cells adapted for sharp vision in bright light. They also help you detect color.

At the back of the eye, retinal tissue comes together to form the optic nerve. The optic nerve leads to the brain, where images are interpreted.

Reading Check

2. What forms the optic nerve?

____

____

____

Sensing Mechanical Stimulation

When you walk in a park, how are you able to hear the leaves rustle or feel the things you reach out to touch? Both of these senses, hearing and touch, depend on receptors that respond to mechanical stimulation.

How does your sense of hearing work?

Every sound causes the air around it to vibrate. These vibrations travel outward from the sources in sound waves. Sound waves enter your outer ear and travel down to the end of the ear canal. In the ear canal, they strike a membrane called the eardrum and cause it to vibrate. The vibrations then pass to three small bones in the middle ear—the malleus, the incus, and the stapes. As the stapes vibrates, it causes

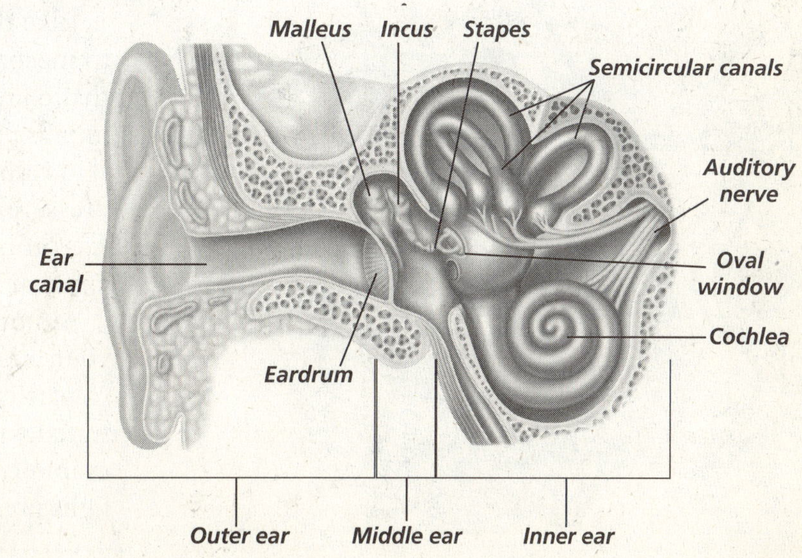

Think it Over

3. **Analyze** Why would damage to the cochlea or auditory nerve result in hearing loss?

______

______

______

Reading Check

4. Where are receptors located that respond to pain, pressure, and temperature?

______

______

______

the membrane of the oval window, a structure between the middle and inner ear, to move back and forth.

From here, the vibrations continue to travel deeper into the ear. The movement from the oval window causes fluid in the **cochlea,** a snail-shaped structure in the inner ear, to move. Inside the circular walls of the cochlea are structures that are lined with hair cells. The fluid in the cochlea moves like a wave against the hair cells causing them to bend.

The movement of the hairs produces electrical impulses, which travel along the auditory nerve to the sides of the cerebrum. Once they reach the cerebrum, they are interpreted as sound. Hearing loss can occur if the auditory nerve or the hair cells in the cochlea are damaged.

How does your sense of balance work?

The inner ear also converts information about the position of your head into nerve impulses, which travel to your brain, informing it about your body's balance.

Maintaining balance is the function of your **semicircular canals.** Like the cochlea, the semicircular canals are filled with a thick fluid and lined with hair cells. When you tilt your head, the fluid moves, causing the hairs to bend. This movement stimulates the hair cells to produce impulses. Then neurons from the semicircular canals carry the impulses to the brain. The brain sends an impulse to stimulate your neck muscles and readjust the position of your head.

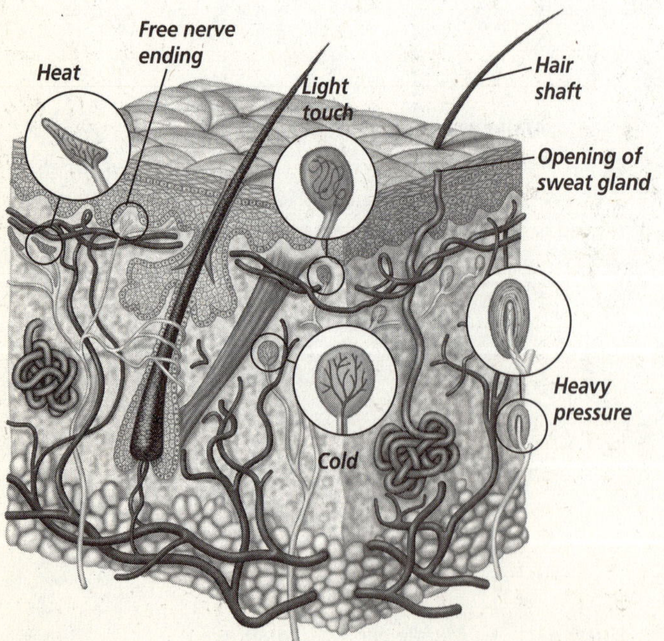

How does your sense of touch work?

Like the ear, your skin also responds to mechanical stimulation with receptors that convert the stimulus into a nerve impulse. Receptors in the dermis of the skin respond to changes in temperature, pressure, and pain. With the help of these receptors, your body can respond to its external environment. ✓

Although receptors are found all over your body, those responsible for responding to particular stimuli are usually concentrated within certain areas of your body. For example, many receptors that respond to light pressure are found in the dermis of your fingertips, eyelids, lips, the tip of your tongue, and the palms of your hands. When these receptors are stimulated, you will feel a light touch.

Receptors that respond to heavier pressure are found inside your joints, in muscle tissue, and in certain organs. They also are abundant on the skin of your palms and fingers and on the soles of your feet. When these receptors are stimulated, you feel heavy pressure.

Free nerve endings extend into the lower layers of the epidermis. Free nerve endings act as receptors for itch, tickle, hot and cold, and pain sensations. Heat receptors are found deep in the dermis, while cold receptors are found closer to the surface of your skin. Pain receptors can be found in all tissues of the body except those in the brain.

After You Read

Mini Glossary

cochlea: snail-shaped structure in the inner ear containing fluid and hairs; produces electrical impulses that the brain interprets as sound

cones: receptor cells in the retina adapted for sharp vision in bright light and color detection

retina: thin layer of tissue found at the back of the eye made up of light receptors and sensory neurons

rods: receptor cells in the retina that are adapted for vision in dim light; also help detect shape and movement

semicircular canals: structures in the inner ear containing fluid and hairs that help the body maintain balance

taste bud: sensory receptors located on the tongue that result in taste perception

1. Read the terms and their definitions in the Mini Glossary above. Circle all of the terms that deal with the sense of sight. Then write a paragraph using the terms.

____

____

____

____

Section 36.2 The Senses, *continued*

2. Match the statements in Column 1 to the term that matches in Column 2.

	Column 1	Column 2
_____	1. Your eyes are squinting from the bright sunlight.	a. olfactory nerves
_____	2. You are dizzy.	b. taste buds
_____	3. You smell something burning.	c. cones
_____	4. The lemon is sour.	d. the cochlea
_____	5. You have hearing loss.	e. semicircular canals

Visit the Glencoe Science Web site at **science.glencoe.com** to find your biology book and learn more about the senses.

Section 36.3 The Effects of Drugs

Before You Read

Have you ever taken an antibiotic for an infection or a pain reliever for an ache or pain? Drugs used as medication can be helpful in treating certain health problems and relieving pain. They should, however, be taken only when they are needed, and never without an adult's or doctor's permission. On the lines below, list a positive and a negative point about the medicinal use of drugs.

____

____

____

____

Read to Learn

Drugs Act on the Body

A **drug** is a chemical that affects the body's functions. Most drugs interact with receptor sites on cells, probably the same ones used by neurotransmitters of the nervous system or hormones of the endocrine system. Some drugs increase the rate at which neurotransmitters are synthesized and released. Drugs also can slow the rate at which neurotransmitters are broken down. Other drugs interfere with a neurotransmitter's ability to interact with its receptor.

Mark the Text **Locate Information** Highlight every heading in the reading that asks a question. Then highlight each answer as you find it.

Medicinal Uses of Drugs

A medicine is a drug that, when taken into the body, helps prevent, cure, or relieve a medical problem. Some of the many kinds of medicines used to relieve medical conditions are discussed here.

How do drugs relieve pain?

Headache, muscle ache, and cramps are common pain sensations. You have just read about how pain receptors in your body send signals to your brain. Pain-relieving medicines manipulate either the receptors that initiate the impulses or the central nervous system that receives them.

Pain relievers that do not cause a loss of consciousness are called analgesics. Some analgesics, such as aspirin, work by inhibiting receptors at the site of pain from producing nerve

Reading Check

1. What is a pain reliever called that does not cause a loss of consciousness?

impulses. Analgesics that work on the central nervous system are called **narcotics.** Many narcotics are made from the opium poppy flower. Opiates, as they are called, can be useful in controlled medical therapy because these drugs are able to relieve severe pain from illness or injury. ✓

How are circulatory problems treated?

Many drugs have been developed to treat heart and circulatory problems such as high blood pressure. These medicines are called cardiovascular drugs. In addition to treating high blood pressure, cardiovascular drugs may be used to normalize an irregular heartbeat, increase the heart's pumping capacity, or enlarge small blood vessels.

How are nervous disorders treated?

Several kinds of medicines are used to help relieve symptoms of nervous system problems. Among these medicines are stimulants and depressants. Drugs that increase the activity of the central and sympathetic nervous systems are called **stimulants.** Amphetamines (am FE tuh meenz) are synthetic stimulants that increase the output of CNS neurotransmitters. Amphetamines are seldom prescribed because they can lead to dependence. However, because they increase wakefulness and alertness, amphetamines are sometimes used to treat patients with sleep disorders.

Drugs that lower, or depress, the activity of the nervous system are called **depressants,** or sedatives. The primary medicinal uses of depressants are to encourage calmness and produce sleep. For some people, symptoms of anxiety interfere with the ability to function effectively. By slowing down the activities of the CNS, a depressant can temporarily relieve some of this anxiety.

The Misuse and Abuse of Drugs

Think it Over

2. **Compare** How does the medicinal use of drugs differ from drug misuse?

The misuse or abuse of drugs can cause serious health problems—even death. Drug misuse occurs when a medicine is taken for an unintended use. Instances of drug misuse include giving your prescription medicine to someone else, not following the prescribed dosage of medication, and mixing medicines.

Drug abuse is the inappropriate use of a drug for a non-medical purpose. Drug abuse may involve use of an illegal drug, such as cocaine; use of an illegally obtained medicine, such as someone else's prescribed drugs; or excessive use of a legal drug, such as alcohol or nicotine. Drugs abused in this way can have powerful effects on the nervous system and other systems of the body.

What is addiction to drugs?

When a person believes he or she needs a drug to feel good or to function normally, that person is psychologically dependent on the drug. When a person's body develops a chemical need for the drug in order to function normally, that person is physiologically dependent. Both are forms of **addiction.**

When a drug user experiences tolerance to or withdrawal from a frequently used drug, that person is addicted to the drug. **Tolerance** occurs when a person needs a larger or more frequent dose of a drug to achieve the same effect. The need for more is related to the body's becoming less responsive to the effects of the drug. When a person stops taking a drug and actually becomes ill, it is called **withdrawal.**

Classes of Commonly Abused Drugs

Each class of drug produces its own effect on the body, and its own particular symptoms of withdrawal.

What are stimulants?

You already know that stimulants increase the activity of the central and sympathetic nervous systems. Increased CNS stimulation can result in mild elevation of alertness, increased nervousness, anxiety, and even convulsions. ✓

Cocaine stimulates the CNS by working the part of the inner brain that governs emotions and basic drives, such as hunger and thirst. When these needs are met under normal circumstances, neurotransmitters—such as dopamine—are released to reward centers and the person experiences pleasure. Cocaine artificially increases levels of these neurotransmitters in the brain. As a result, false messages are sent to reward centers signaling that a basic drive has been satisfied. The user quickly feels a pleasurable high called a rush. This feeling does not last. Soon the effects of the drug change. Physical hyperactivity follows. Often anxiety and depression set in.

Cocaine also disrupts the body's circulatory system by interfering with the sympathetic nervous system. At first cocaine slows the heart rate. However, it soon produces a rapid increase in heart rate and a narrowing of blood vessels, known as vasoconstriction. The result is high blood pressure. Heavy use of this drug weakens the immune system and often leads to heart abnormalities. Cocaine may affect the unborn babies of addicted mothers. Sometimes the babies are born already dependent on the drug.

Think it Over

3. **Compare** How does drug misuse differ from drug abuse?

___

___

___

Reading Check

4. What are some of the side effects caused by stimulants?

___

___

___

Amphetamines are stimulants that increase levels of CNS neurotransmitters. Like cocaine, amphetamines also cause vasoconstriction, a racing heart, and increased blood pressure. Other adverse side effects of amphetamine abuse include irregular heartbeat, chest pain, paranoia, hallucinations, and convulsions.

Not all stimulants are illegal. Caffeine—a substance found in coffee, some carbonated soft drinks, cocoa, and tea—is a CNS stimulant. Its effects include increased alertness and some mood elevation. Caffeine also causes an increase in heart rate and urine production, which can lead to dehydration.

Nicotine, a substance found in tobacco, also is a stimulant. By increasing the release of the hormone epinephrine, nicotine increases heart rate, blood pressure, breathing rate, and stomach acid secretion. Nicotine is the addictive ingredient in tobacco. There are many other harmful chemicals also found in tobacco products. Smoking cigarettes leads to an increased risk of lung cancer and cardiovascular disease. Use of chewing tobacco is associated with oral and throat cancers. ✓

Reading Check

5. What are some of the harmful effects of nicotine?

______

______

______

What are depressants?

Depressants slow down the activities of the CNS. All CNS depressants relieve anxiety, but most produce drowsiness. One of the most widely abused drugs in the world today is alcohol. It is easily produced from various grains and fruits. Alcohol is distributed throughout a person's body via the bloodstream. Like other drugs, alcohol affects cellular communication by influencing the release of or interacting with receptors for several important neurotransmitters in the brain. Alcohol also appears to block the movement of sodium and calcium ions across the cell membrane. That process is important in the transmission of impulses and the release of neurotransmitters.

Tolerance to the effects of alcohol develops as a result of heavy alcohol consumption. Addiction to alcohol—alcoholism—can destroy nerve cells and cause brain damage. Chronic alcohol use contributes to a number of organ diseases. For example, cirrhosis, a hardening of the tissues of the liver, commonly afflicts alcoholics. ✓

Reading Check

6. How does alcohol affect the body's organs?

______

______

______

Barbiturates (bar BIH chuh ruts) are sedatives and anti-anxiety drugs. When barbiturates are used in excess, the user's respiratory and circulatory systems become depressed. Chronic use results in addiction.

What are narcotics?

Most narcotics are opiates, derived from the opium poppy. They act directly on the brain. Heroin is the most abused narcotic in the United States. It depresses the CNS, slows breathing, and lowers heart rate. Addiction develops quickly, and withdrawal from heroin is painful.

What are hallucinogens?

Natural hallucinogens have been known and used for thousands of years, but the abuse of hallucinogenic drugs did not become widespread in the United States until the 1960s, when new synthetic versions became widely available.

Hallucinogens (huh LEW sun uh junz) stimulate the CNS—altering moods, thoughts, and sensory perceptions. The user sees, hears, feels, tastes, or smells things that are not actually there. This disorientation can impair the user's judgment and place him or her in a potentially dangerous situation. Hallucinogens increase heart rate, blood pressure, respiratory rate, and body temperature. They sometimes cause sweating, salivation, nausea, and vomiting. After large enough doses, convulsions may occur.

LSD, also called acid, is a synthetic drug. The mechanism by which LSD produces hallucinations is not certain, but it may involve the blocking of a CNS neurotransmitter.

What are anabolic steroids?

Anabolic steroids are synthetic drugs that are similar to the hormone testosterone. Like testosterone, anabolic steroids stimulate muscles to increase in size. Physicians use anabolic steroids to treat hormone imbalances or diseases that result in a loss of muscle mass. Abuse of anabolic steroids is associated with infertility in men, high cholesterol, and extreme mood swings. ✓

Reading Check

7. What class of drugs has been associated with infertility in men, high cholesterol, and extreme mood swings?

____

____

____

Breaking the Habit

Once a person has become addicted to a drug, breaking the habit can be very difficult. Remember that an addiction can involve both physiological and psychological dependencies. Besides the desire to break the addiction, people usually need both medical and psychological therapy to be successful in their treatment. Support groups such as Alcoholics Anonymous encourage addicts to share their experiences in an effort to maintain sobriety. Often people going through the same recovery are able to offer the best support.

What is nicotine replacement therapy?

Nicotine replacement therapy is one example of a relatively successful drug treatment approach. People who are trying to break their addiction to tobacco often go through stressful withdrawal symptoms when they stop smoking cigarettes. To ease the intensity of the withdrawal symptoms, patients wear adhesive patches that slowly release small amounts of nicotine into their bloodstream. Alternatively, pieces of nicotine-containing gum are chewed periodically to temporarily relieve cravings.

After You Read

Mini Glossary

addiction: psychological and/or physiological drug dependence

depressant: type of drug that lowers or depresses the activity of the nervous system

drug: chemical substance that affects body functions

hallucinogen (huh LEW sun uh jun): drug that stimulates the central nervous system so that the user becomes disoriented and sees, hears, feels, tastes, or smells things that are not there

narcotic: type of pain relief drug that affects the central nervous system

stimulant: drug that increases the activity of the central and sympathetic nervous systems

tolerance: the body becomes less responsive to a drug and an individual needs larger or more frequent doses of the drug to achieve the same effect

withdrawal: psychological response or physiological illness that occurs when a person stops taking a drug

1. Read the terms and their definitions in the Mini Glossary above. Then write a paragraph using at least four of the words.

____

____

____

____

Section 36.3 The Effects of Drugs, *continued*

2. Fill in the chart below by supplying at least one fact in each box.

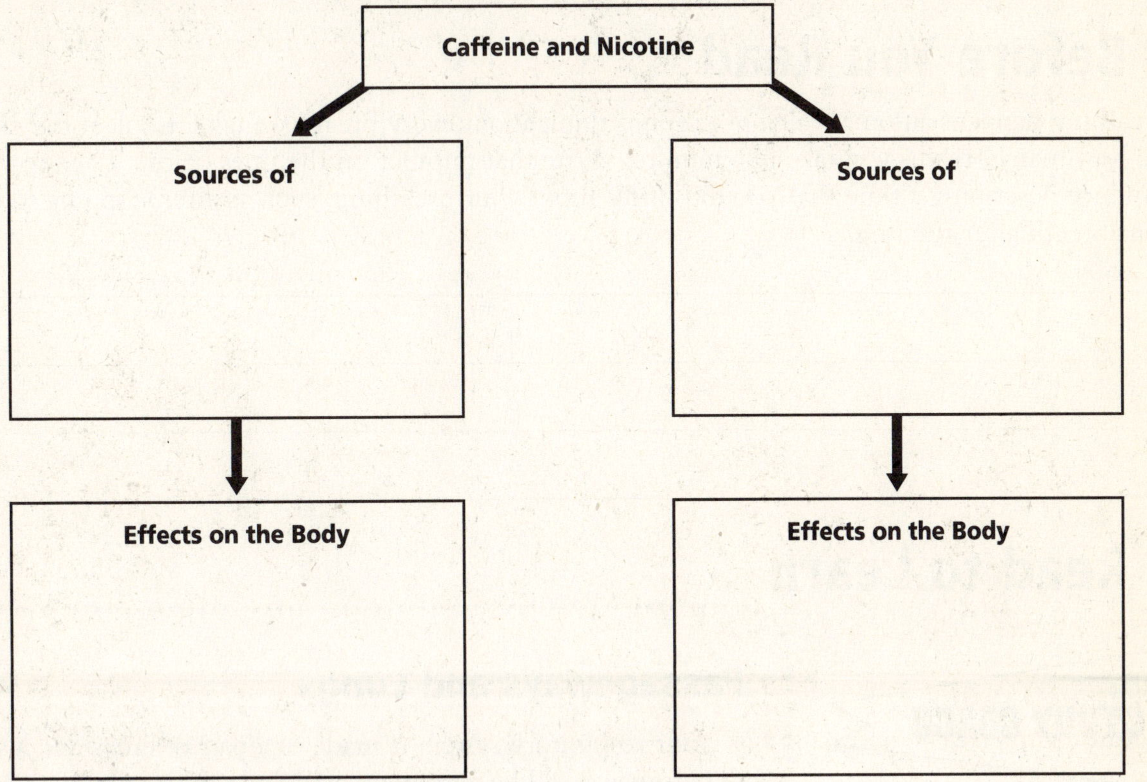

Visit the Glencoe Science Web site at **science.glencoe.com** to find your biology book and learn more about the effects of drugs.

Section 37.1 The Respiratory System

Before You Read

Breathing happens automatically. We do not think about every breath we take. Look at the clock and see how many breaths you take in a minute. Write that number on the lines below. Then write one sentence describing a time that you did think about your breathing, such as after stopping to catch your breath after running.

Read to Learn

Mark the Text **Identify Main Ideas** As you read this section, stop after every few paragraphs and put what you have just read in your own words. Highlight the main idea in each paragraph.

Passageways and Lungs

Your respiratory system is made of a pair of lungs and a series of passageways. The passageways include the nasal passages, the throat, the windpipe, and the bronchi. You probably think of breathing when you hear the term respiratory system. Breathing is just one of the functions that the respiratory system carries out. Gas exchange, or respiration, is another important function performed by the respiratory system. Respiration includes all of the steps involved in getting oxygen to the cells of your body and getting rid of carbon dioxide. Recall that cellular respiration also involves the formation of ATP within the cells.

The first step in the process of respiration involves taking air into your body. Air enters through the nose or mouth. It flows into the pharynx, or throat, passes the epiglottis, and moves through the larynx. The air then travels down the windpipe, or **trachea** (TRAY kee uh), a tubelike passageway that leads to two tubes or bronchi (BRAHN ki) (singular, bronchus), which lead into the lungs. Use the illustration on page 463 to trace the steps. When you swallow food, the epiglottis covers the entrance to the trachea, which prevents food from getting into the air passages.

What happens if the air is not clean?

The air you breathe is far from clean. Depending on where you live, you may breathe in as much as a million particles of foreign matter per day. The nasal cavity, trachea, and bronchi are lined with cells that secrete mucous. These cells also have tiny hairlike projections called cilia. The cilia constantly beat in the direction of the throat. They move the foreign matter to where it can be swallowed or expelled by coughing or sneezing. These cells prevent most of the foreign matter from reaching the lungs. ✓

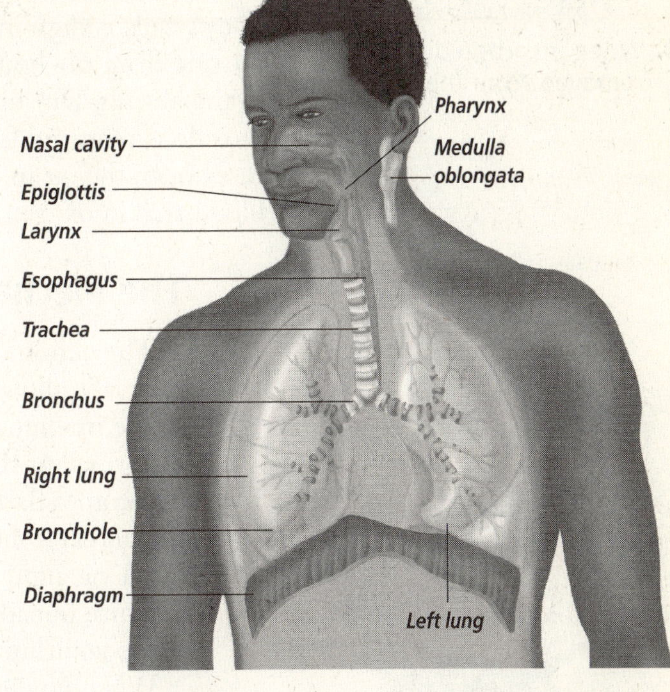

Where does gas exchange happen?

Like the branches of a tree, each bronchus branches into bronchioles. The bronchioles branch into many microscopic tubules that eventually open into thousands of thin-walled sacs called alveoli. **Alveoli** (al VEE uh li) (singular, alveolus) are the sacs of the lungs where oxygen and carbon dioxide are exchanged by diffusion between air and blood. The clusters of alveoli are surrounded by networks of tiny blood vessels, or capillaries. Blood in these vessels has come from the cells of the body and contains wastes from cellular respiration. Diffusion of gases takes place easily because the wall of each alveolus and the wall of each capillary are only one cell thick. External respiration involves the exchange of oxygen or carbon dioxide between the air in the alveoli and the blood that circulates through the walls of the alveoli. ✓

Once oxygen diffuses into the blood vessels surrounding the alveoli, the heart pumps it to the body cells. There it is used for cellular respiration. Remember that cellular respiration is the process by which cells use oxygen to break down glucose and release energy in the form of ATP. Carbon dioxide is a waste product of this process. The carbon dioxide diffuses into the blood, which carries it back to the lungs.

✓ Reading Check

1. How are particles of foreign matter expelled from the respiratory system?

____

____

____

✓ Reading Check

2. Why is diffusion possible between the wall of a capillary and the wall of an alveolus?

____

____

____

Reading Check

3. When is carbon dioxide removed from the body?

That means that the blood that comes to the alveoli from the body's cells is high in carbon dioxide and low in oxygen. Carbon dioxide from the body diffuses from the blood into the air spaces in the alveoli. During exhalation, or breathing out, the carbon dioxide is removed from the body. At the same time, oxygen diffuses from the air in the alveoli into the blood. This makes the blood rich in oxygen.

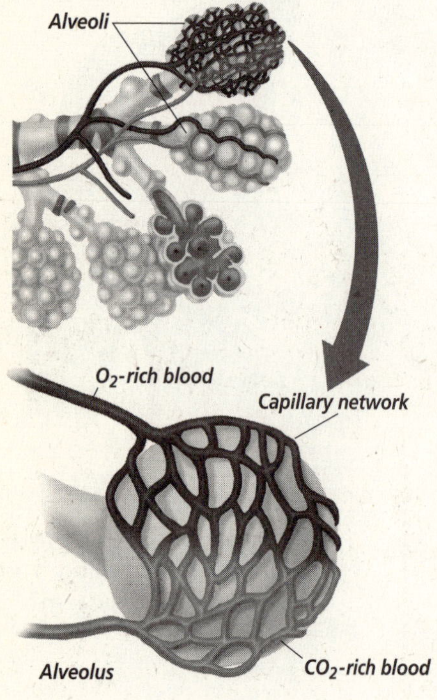

The Mechanics of Breathing

The action of your diaphragm and the muscles between your ribs allows you to breathe in and out. When you inhale, the muscles between your ribs contract and your rib cage rises. At the same time, the diaphragm muscle contracts. It becomes flattened and moves lower in the chest cavity. This creates more space in the chest cavity, which creates a slight vacuum. Air rushes into your lungs because the air pressure outside your body is greater than the air pressure inside your lungs.

When you exhale, the muscles associated with the ribs relax, and your ribs drop down in your chest cavity. Your diaphragm relaxes, returning to its resting position. As the muscles relax, the chest cavity becomes smaller. This forces most of the air out of the alveoli.

In healthy lungs, the alveoli are elastic. They stretch as you inhale and return to their original size as you exhale. Even after you exhale, the alveoli contain a small amount of air.

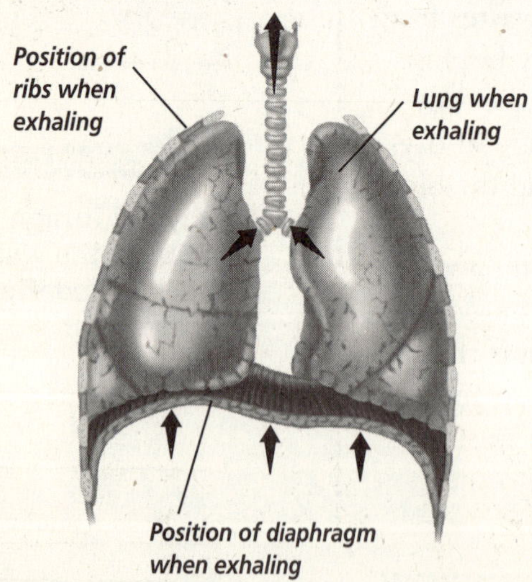

A **When relaxed, your diaphragm is positioned in a dome shape beneath your lungs, decreasing the volume of the chest cavity and forcing air out of the lungs.**

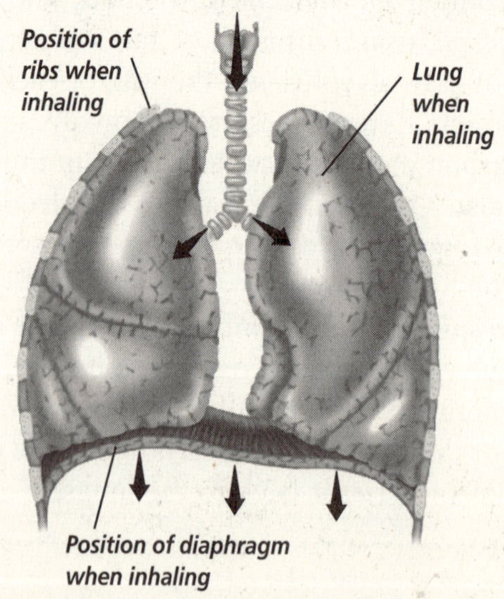

B **When contracting, the diaphragm flattens, enlarging the chest cavity and drawing air into the lungs.**

Control of Respiration

Breathing is usually an involuntary process. It is partially controlled by an internal feedback mechanism. The medulla oblongata receives signals about the chemistry of your blood. It responds to higher levels of carbon dioxide in your blood by sending nerve signals to the rib muscles and diaphragm. The nerve signals cause the muscles to contract and you inhale. During exercise you breathe faster. This causes a more rapid exchange of gases between air and blood.

Reading Check

4. What structure is signaled by blood chemistry to control breathing?

____

____

____

After You Read

Mini Glossary

alveoli (al VEE uh li): sacs in the lungs where oxygen diffuses into the blood and carbon dioxide diffuses into the air

trachea (TRAY kee uh): tubelike passageway for air flow that connects with two bronchi tubes that lead into the lungs

1. Read the terms and their definitions in the Mini Glossary above. On the lines below, write each term in a sentence.

____

____

2. Place the list of respiratory structures below in the order an oxygen molecule would pass them as it moves from the outside air into a blood vessel.

bronchiole	**pharynx**	**nose/mouth**	**capillary**
bronchi	**trachea**	**larynx**	**alveolus**

1. ______________________ 5. ______________________
2. ______________________ 6. ______________________
3. ______________________ 7. ______________________
4. ______________________ 8. ______________________

3. Describe the process by which carbon dioxide leaves the body.

____

____

Visit the Glencoe Science Web site at **science.glencoe.com** to find your biology book and learn more about the respiratory system.

Name Date Class

Section 37.2 The Circulatory System

Before You Read

To donate blood, you need to be in good health. If you pass the health questions and the blood test, and if you are old enough, you can donate blood. Donating blood saves lives. Some companies are trying to develop artificial blood. Artificial blood has not been able to replace human blood in most cases. Why do you think that might be?

Read to Learn

Make Flash Cards Making flash cards is a good way to learn material. Write a quiz question on one side of the flash card, and the answer on the other side. Quiz yourself until you know the answers.

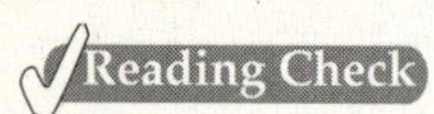

1. What is blood made of?

Your Blood: Fluid Transport

Blood is a tissue made of fluid, cells, and fragments of cells. The fluid portion of the blood is called **plasma.** Plasma is straw colored and makes up about 55 percent of the total volume of blood. Red and white blood cells and cell fragments are suspended in plasma. ✓

What do red blood cells do?

Red blood cells are round, disk-shaped cells. **Red blood cells** carry oxygen to body cells. They make up 44 percent of the total volume of blood. Red blood cells are produced in the red bone marrow of your ribs, humerus, femur, sternum, and other long bones.

The red blood cells in humans have nuclei in an early stage of cell development. The nucleus is lost before the cell enters the bloodstream. Red blood cells remain active in the bloodstream for about 120 days. Then they break down and are removed as waste. Old red blood cells are destroyed in the spleen and in the liver.

How is oxygen carried by the blood?

Red blood cells have an iron-containing protein molecule called **hemoglobin** (HEE muh gloh bun). Oxygen becomes loosely attached to the hemoglobin in blood cells that have entered the lungs. These oxygenated blood cells carry oxygen from the lungs to the body's cells. As blood passes through body tissue with low oxygen concentrations, oxygen is released from the hemoglobin and diffuses into the tissues.

Hemoglobin can also carry some carbon dioxide after it releases the oxygen. Remember that once biological work has been done in a cell, wastes in the form of carbon dioxide diffuse into the blood. The bloodstream carries the carbon dioxide to the lungs. About 70 percent of this carbon dioxide combines with water in the blood plasma to form bicarbonate. The remaining 30 percent travels back to the lungs dissolved in plasma or attached to the hemoglobin molecules.

Reading Check

2. What protein molecule carries oxygen in the blood?

What is the function of white blood cells?

White blood cells play a major role in protecting the body from foreign substances and from microscopic organisms that cause disease. White blood cells make up only one percent of the total volume of blood.

How does blood clot?

What happens if you cut yourself? If the cut is not deep, you bleed until the blood clots. It usually does not take long for the blood to clot. That's because, in addition to red and white blood cells, blood contains small cell fragments called **platelets.** They help blood to clot after an injury. Platelets help connect a sticky network of protein fibers called fibrin. This forms a web over the wound that traps escaping blood cells. Then a dry, leathery scab forms. Platelets are produced from cells in the bone marrow. They have a short life span and are removed from the blood by the spleen and liver about a week after they are produced.

Reading Check

3. Where are platelets produced?

ABO Blood Groups

If a person is injured so severely that a large amount of blood is lost, a transfusion of blood from another person may be required. Whenever blood is transfused from one person to another, it is important to know the blood group of each person. There are four human blood groups, A, B, AB, and O. You inherited the characteristics of one of these blood groups from your parents. Sometimes the term *blood type* is used to describe the blood group to which a person belongs. If your blood group is O, you are said to have type O blood.

Reading Check

4. List the four human blood groups.

What are the differences between the blood groups?

Differences in blood groups are due to the presence or absence of proteins on the membranes of red blood cells. The proteins are called antigens. **Antigens** stimulate an immune response in the body. An immune response defends the body against foreign proteins. The letters A and B stand for the types of blood surface antigens found on human red blood cells.

Blood plasma contains proteins called **antibodies** (AN tih bahd eez). The antibodies are shaped to correspond with the different blood surface antigens. The antibody in the blood plasma reacts with its matching antigen on red blood cells if they are brought into contact with one another. This reaction results in clumped blood cells that can no longer function. That is why a transfusion from the wrong blood group can be so dangerous. Each blood group contains antibodies for the blood surface antigens that are found in other blood groups. A blood group does not contain antibodies for antigens found on its own red blood cells.

For example, if you have type A blood, you have the A antigen on your red blood cells. You have the anti-B antibody in your plasma. Your blood flows smoothly because there isn't anything for the anti-B antibody to react with. What would happen if you received a transfusion of type B blood? This blood group contains anti-A antibodies, and B antigens. The result would be clumped blood cells that cannot carry oxygen or nutrients to body cells.

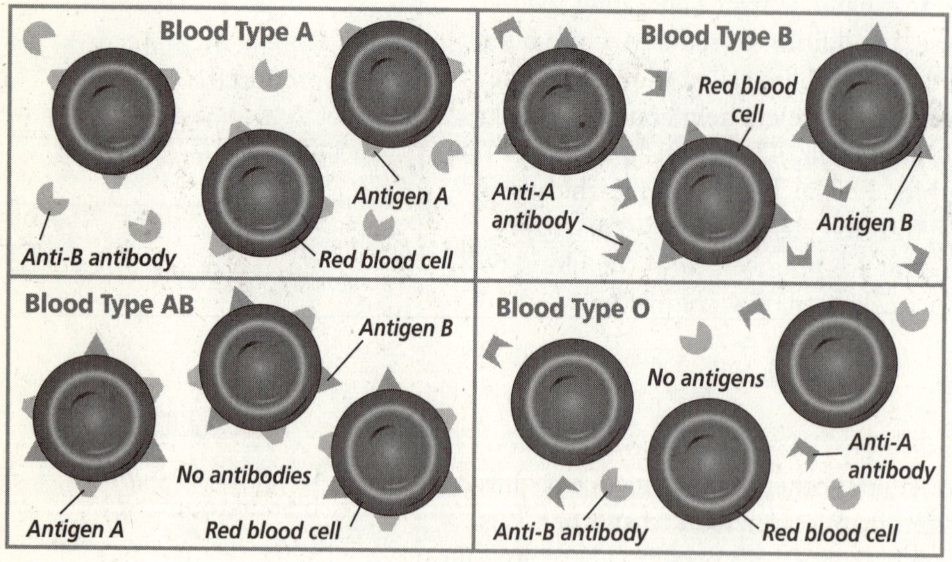

What is Rh factor?

Another characteristic of red blood cells involves the presence or absence of an antigen called Rh or Rhesus factor. Rh factor is an inherited characteristic. People are Rh positive (Rh^+) if they have the Rh antigen on their red blood cells. They are Rh negative (Rh^-) if they do not.

Section
37.2 The Circulatory System, *continued*

Rh factor can cause complications in some pregnancies. The problem begins when an Rh^- mother becomes pregnant with an Rh^+ baby. Sometimes at birth, the blood cells of the baby are mixed with those of the mother. If the Rh^- mother is exposed to the blood of the Rh^+ baby, the mother will make anti-Rh^+ antibodies. If the mother becomes pregnant again, the antibodies can cross the placenta and enter the fetus. If the new fetus is Rh^+, the anti-Rh^+ antibodies from the mother will destroy red blood cells in the fetus.

Prevention of this problem is possible. When the Rh^+ fetus is 28 weeks old and again shortly after the Rh^+ baby is born, the Rh^- mother is given a substance that prevents the production of Rh antibodies in her blood. As a result, the next fetus will not be in danger.

Think it Over

5. Analyze When does Rh factor cause complications?

____

____

____

Your Blood Vessels: Pathways of Circulation

Blood is a fluid channeled through blood vessels. The three main types of blood vessels are arteries, capillaries, and veins. Each is different in structure and function. ✓

Arteries are large, thick-walled, muscular, elastic blood vessels that carry blood away from the heart. The blood that they carry is under great pressure. As the heart contracts, it pushes blood through the arteries. Each artery's elastic walls expand slightly. As the heart relaxes, the artery shrinks a bit, which helps to push the blood forward. As a result, blood surges through the arteries in pulses that correspond with the rhythm of the heartbeat.

The arteries branch off from the heart. They divide into smaller arteries that, in turn, divide into even smaller vessels called arterioles. Arterioles (ar TEER ee ohlz) enter tissues, where they branch into the smallest blood vessels, the capillaries. **Capillaries** (KA puh ler eez) are microscopic blood vessels with walls that are only one cell thick. These vessels are so tiny that red blood cells must move through them in single file. Capillaries form a dense network that reaches almost every cell in the body. Thin capillary walls allow nutrients and gases to diffuse easily between blood cells and surrounding tissue cells.

✓ Reading Check

6. What are the three main types of blood vessels?

____

____

____

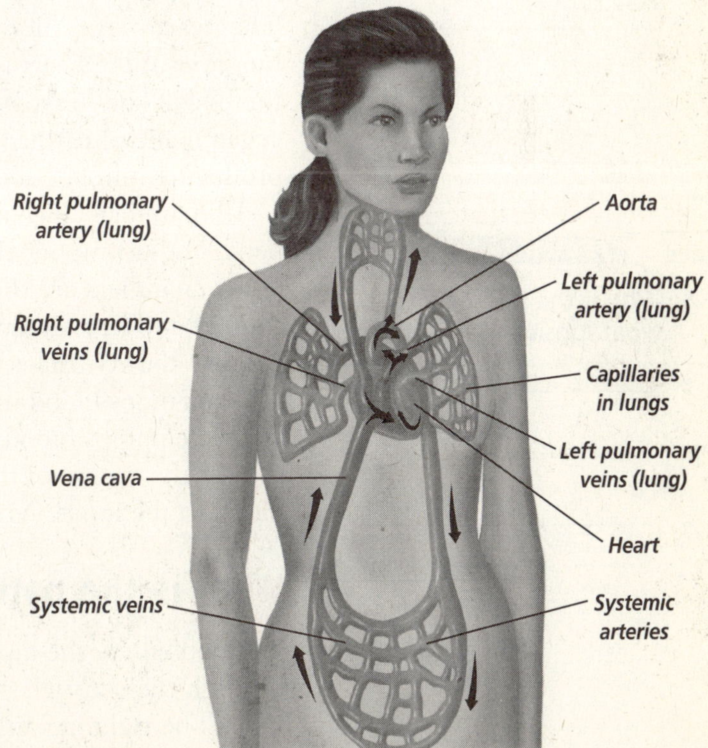

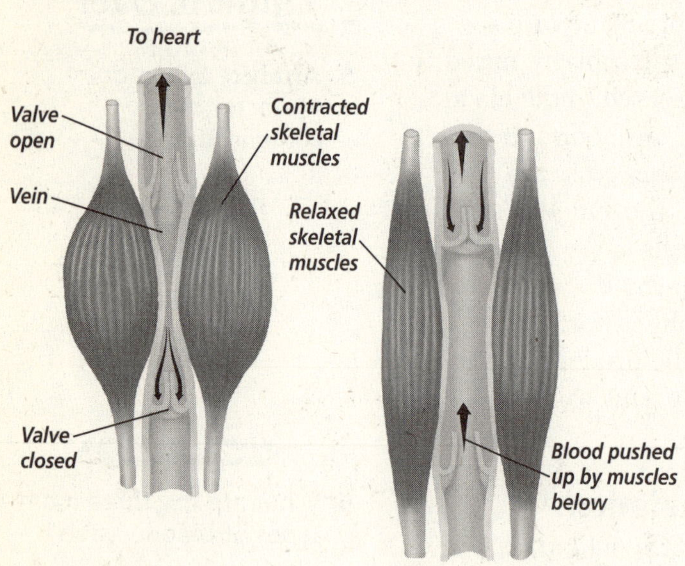

As the blood leaves the tissues, the capillaries join to form slightly larger vessels called venules. The venules merge to form **veins,** the large blood vessels that carry blood from the tissues back toward the heart. Blood in veins is not under pressure as great as blood in arteries. In some veins, especially those in your arms and legs, blood has to travel uphill against gravity. These veins, shown at left, are equipped with valves that prevent the blood from flowing backward. The veins work with skeletal muscles to open and close the valves. When the skeletal muscles contract, the top valves open, and blood is forced toward the heart. When the skeletal muscles relax, the top valves close to prevent blood from flowing backward, away from the heart.

Your Heart: The Vital Pump

The thousands of blood vessels in your body would be of little use if there were not a way to move blood through them. The heart moves blood through the vessels. In fact, the main function of the heart is to keep blood moving constantly through the body. The heart is well adapted for its job. It is a large organ made of cardiac muscle cells that are rich in energy-producing mitochondria. ✓

All mammals, including humans, have hearts with four chambers. The two upper chambers of the heart are the **atria.** The two lower chambers are the **ventricles.** The walls of each atrium are thinner and less muscular than those of each ventricle. That's because the ventricles perform more work than the atria. Each atrium pumps blood into the corresponding ventricle. The left ventricle pumps blood to the entire body. So its muscles are thicker than those of the right ventricle. The right ventricle pumps blood to the lungs. As a result, the heart is somewhat lopsided. ✓

✓ Reading Check

7. What is the main function of the heart?

______

______

______

✓ Reading Check

8. Why is the heart somewhat lopsided?

______

______

______

What is the path of blood through the heart?

Blood enters the heart through the atria and leaves the heart through the ventricles. Both atria fill up with blood at the same time. The right atrium receives oxygen-poor blood from the head and body through two large veins called the **venae cavae** (vee nee • KAY vee) (singular, vena cava). The left atrium receives oxygen-rich blood from the lungs through four pulmonary veins. These veins are the only veins that carry blood rich in oxygen. After the two atria have filled with blood, they contract, pushing the blood down into the two ventricles.

After the ventricles have filled with blood, they contract at the same time. When the right ventricle contracts, it pushes the oxygen-poor blood out of the heart and toward the lungs through the pulmonary arteries. These arteries are the only arteries that carry blood poor in oxygen. At the same time, the left ventricle forcefully pushes oxygen-rich blood out of the heart through the **aorta** to the arteries of the body. The aorta is the largest blood vessel in the body.

Reading Check

9. What is unusual about the pulmonary veins?

Use the illustration at right to trace a drop of blood as it travels through the heart. Begin with the blood coming back from the body through a vena cava. The oxygen-poor drop travels first to the right atrium, then into the right ventricle. The right ventricle pumps it to the lungs through a pulmonary artery. In the lungs, the blood drops off its carbon dioxide and picks up oxygen. Then it moves through the pulmonary veins to the left atrium, into the left ventricle, and finally out to the body through the aorta. Eventually it will return to the heart.

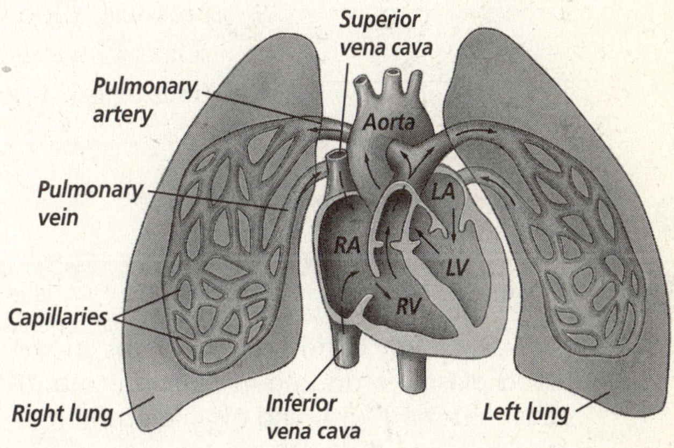

What regulates the heartbeat?

Each time the heart beats, a surge of blood flows from the left ventricle into the aorta and then into the arteries. The surge of blood can be felt in arteries that are close to the surface of the body. This is called a **pulse.**

The pacemaker sets the heart rate. The pacemaker is a bundle of nerve cells located at the top of the right atrium. It generates an electrical impulse that spreads over both atria. The impulse signals the two atria to contract at almost the same time. The impulse also triggers a second set of cells at the base of the right atrium to send the same electrical impulse over the ventricles. This causes the ventricles to contract. The pacemaker causes the atria to contract 70–80 times per minute.

Reading Check

10. How does the pacemaker cause the heart to contract?

What controls the pacemaker?

The pacemaker controls the heartbeat. A portion of the brain called the medulla oblongata regulates the rate of the pacemaker. If the heart beats too fast, sensory cells in arteries near the heart

Reading Check

11. What are the names of the highest and lowest blood pressure?

____

____

____

become stretched. These cells send a signal through the nervous system to the medulla oblongata. The medulla oblongata sends signals that slow the pacemaker. If the heart slows too much, blood pressure drops, signaling the medulla oblongata to speed up the pacemaker and increase the heart rate.

What is blood pressure?

A pulse beat represents the pressure that blood exerts as it pushes against the walls of an artery. **Blood pressure** is the force that the blood exerts on the blood vessels. Blood pressure rises and falls as the heart contracts and then relaxes.

Blood pressure rises sharply when the ventricles contract, pushing blood through the arteries. The high pressure is called systolic pressure. Blood pressure then drops dramatically as the ventricles relax. The lowest pressure occurs just before the ventricles contract again. It is called diastolic pressure.

After You Read

Mini Glossary

antibodies (AN tih bahd eez): proteins in the blood plasma produced in reaction to antigens that react with and disable antigens

antigens: foreign substances that stimulate an immune response in the body

aorta: largest blood vessel in the body; transports oxygen-rich blood from the left ventricle of the heart to the arteries

arteries: large, thick-walled muscular vessels that carry blood away from the heart

atria: two upper chambers of the mammalian heart through which blood enters

blood pressure: force that blood exerts on blood vessels; rises and falls as the heart contracts and relaxes

capillaries (KA puh ler eez): microscopic blood vessels with walls only one cell thick that allow diffusion of gases and nutrients between the blood and surrounding tissues

hemoglobin (HEE muh gloh bun): iron-containing protein molecule in red blood cells that binds to oxygen and carries it from the lungs to the body's cells

plasma: fluid portion of the blood that makes up 55 percent of the total volume of the blood; contains red and white blood cells

platelets: small cell fragments in the blood that help clot blood after an injury

pulse: surge of blood through an artery that can be felt on the surface of the body

red blood cells: round, disk-shaped cells in blood that carry oxygen to body cells; make up 44 percent of the total volume of blood

veins: large blood vessels that carry blood toward the heart

venae cavae (vee nee • KAY vee): two large veins that fill the right atrium of the mammalian heart with oxygen-poor blood from the head and body

ventricles: two lower chambers of the mammalian heart; receive blood from the atria and send it to the lungs and body

white blood cells: blood cells that play a major role in protecting the body from foreign substances and microscopic organisms; make up only one percent of the total volume of the blood

Section

37.2 The Circulatory System, *continued*

1. Read the terms and their definitions in the Mini Glossary on page 472. Circle the key terms that are the components of blood.
2. Fill in the blank boxes to trace the path blood takes in the human body.

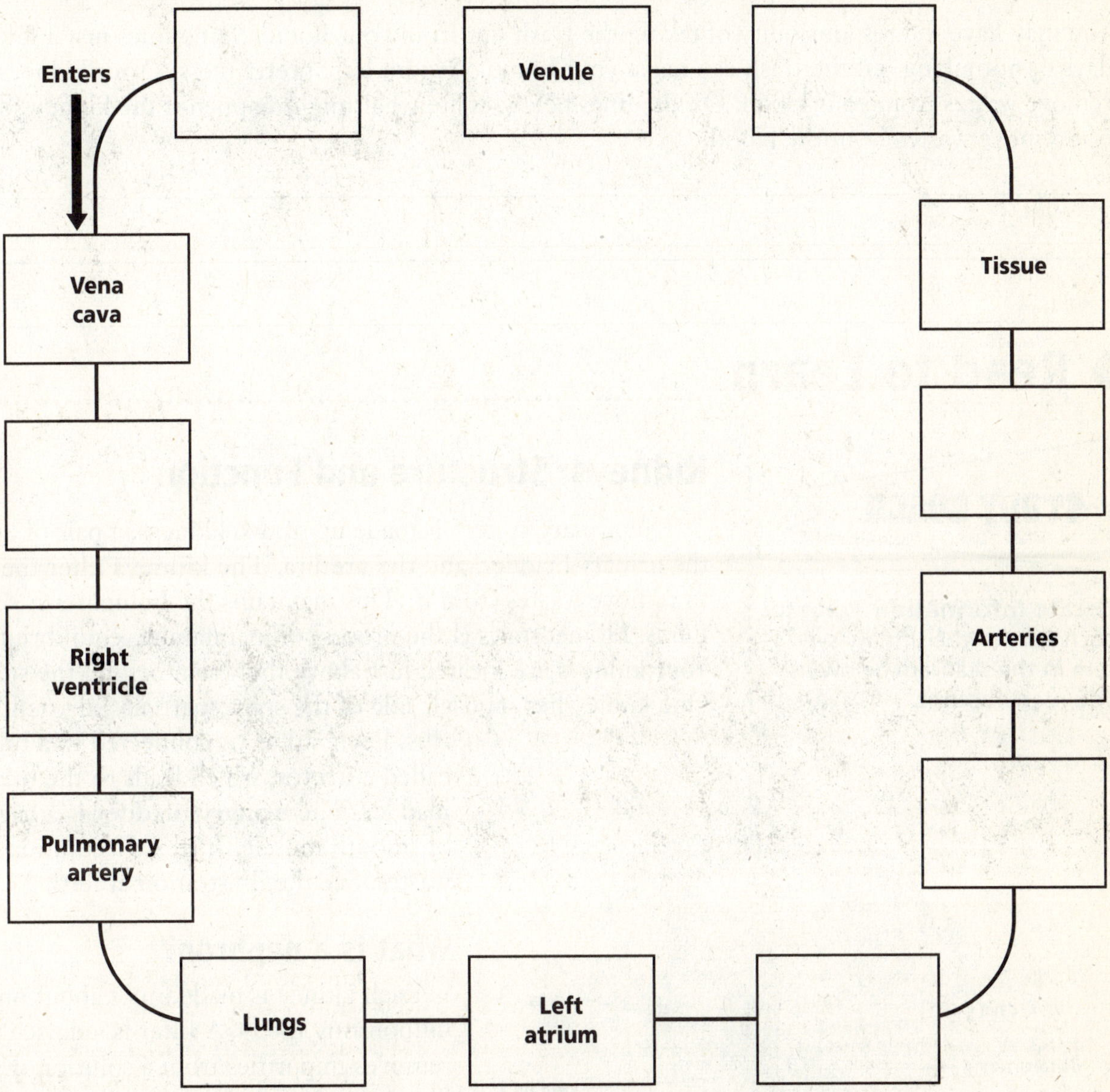

Enters

Vena cava		
[blank]		
Right ventricle		
Pulmonary artery		
Lungs		
Left atrium		
[blank]		
[blank]		
Arteries		
[blank]		
Tissue		
[blank]		
Venule		
[blank]		

Visit the Glencoe Science Web site at **science.glencoe.com** to find your biology book and learn more about the circulatory system.

Name Date Class

Section 37.3 The Urinary System

Before You Read

You may have the responsibility of taking the trash out from your home. Sometimes just a few days of not taking out the trash can make your home or room look pretty messy. Your kidneys remove wastes from your blood. On the lines below, write what might happen if the kidneys did not remove wastes from the blood.

____

____

____

Read to Learn

Locate Information Point to each part of the urinary system in the diagram below as you read about it.

Kidneys: Structure and Function

The urinary system is made up of two kidneys, a pair of ureters, the urinary bladder, and the urethra. The **kidneys** filter the blood to remove wastes from it. This maintains the homeostasis of body fluids. Homeostasis is the process of maintaining equilibrium. Your kidneys are located just above the waist, behind the stomach. One kidney lies on each side of the spine, partially surrounded by ribs. Each kidney is connected to a tube called a **ureter,** which leads to the urinary bladder. The **urinary bladder** is a bag made of smooth muscle. It stores a solution of wastes. See the illustration at left.

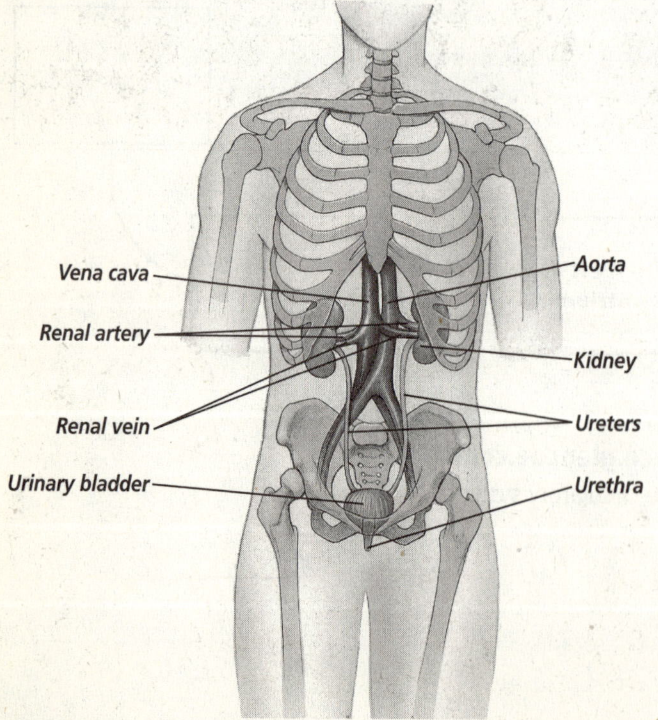

What is a nephron?

Each kidney is made up of about one million tiny filters. A filter is a device that removes impurities from a solution. Each filtering unit of a kidney is called a **nephron.**

Blood entering a nephron carries wastes produced by body cells. The blood entering the nephron is under high pressure. It immediately flows into a bed of capillaries called the glomerulus. Because of the pressure, water, glucose, vitamins, amino acids, protein waste products (called urea), salt, and ions from the blood pass out of the capillaries into a part of the nephron called the Bowman's capsule. Blood cells and most proteins are too large to pass through the walls of a capillary, so these components stay within the blood vessels. ✓

The liquid forced into the Bowman's capsule passes through a narrow, U-shaped tubule. As the liquid moves along the tubule, most of the ions and water and all of the glucose and amino acids are reabsorbed into the bloodstream. This reabsorption of substances is the process by which the body's water is conserved and homeostasis is maintained. Small molecules, including water, move back into the capillaries by diffusion. Other molecules and ions move back into the capillaries by active transport.

The liquid that remains in the tubules, composed of waste molecules, excess water, and ions, is **urine.** Humans produce about 2 L of urine a day. This waste fluid flows out of the kidneys, through the ureter, and into the urinary bladder where it may be stored. Urine passes from the urinary bladder out of the body through a tube called the **urethra** (yoo REE thruh).

Reading Check

1. What components do not pass into the Bowman's capsule?

____

____

____

Nephron
Bowman's capsule
Glomerulus
From renal artery
Renal artery
To renal vein
Tubule
Renal vein
Ureter
Capillaries
To ureter

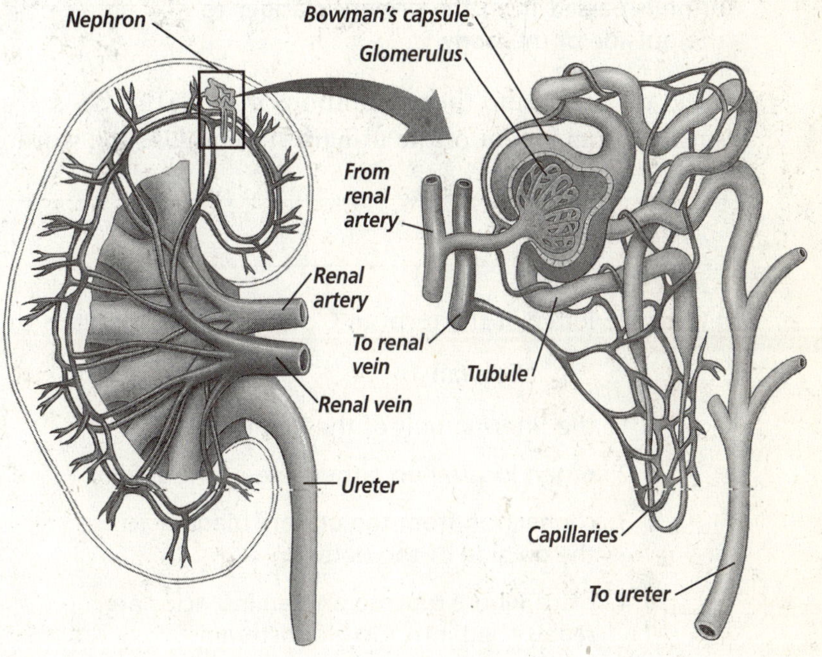

The Urinary System and Homeostasis

The major waste products of cells are nitrogenous wastes, which come from the breakdown of proteins. These wastes include ammonia and urea. Both compounds are toxic to the human body and must be removed from the blood regularly. In addition to removing these wastes, the kidneys control the level of sodium in blood by removing and reabsorbing sodium ions. This helps control the osmotic pressure of the blood. The kidneys also regulate the pH of blood by filtering out hydrogen ions and allowing bicarbonate to be reabsorbed into the blood. Glucose is a sugar that is not usually filtered out of the blood by the kidneys. Individuals with diabetes have too much glucose in their blood. ✓

Reading Check

2. In addition to removing wastes, what else do the kidneys do?

____

____

____

After You Read

Mini Glossary

kidneys: organs of the vertebrate urinary system; remove wastes, control sodium levels of the blood, and regulate blood pH levels

nephron: individual filtering unit of the kidneys

ureter: tube that transports urine from each kidney to the urinary bladder

urethra (yoo REE thruh): tube through which urine is passed from the urinary bladder to the outside of the body

urinary bladder: smooth muscle bag that stores urine until it is expelled from the body

urine: liquid composed of wastes that is filtered from the blood by the kidneys, stored in the urinary bladder, and eliminated through the urethra

1. Read the terms and their definitions in the Mini Glossary above. Circle the key term for the smallest component of the urinary system. Use the word in a sentence on the lines below.

____

____

2. Write the letter of the term in Column 2 that matches the statement in Column 1.

	Column 1	Column 2
____	1. the filtering unit of the kidney	a. urinary bladder
____	2. organ located on either side of the spine	b. nephron
____	3. connection from the urinary bladder to the outside of the body	c. kidney
____	4. place where glucose and amino acids are reabsorbed into the bloodstream	d. ureter
____	5. connection between the kidney and the urinary bladder	e. Bowman's capsule
____	6. made of smooth muscle and store a solution of wastes	f. urethra

Visit the Glencoe Science Web site at **science.glencoe.com** to find your biology book and learn more about the urinary system.

Name Date Class

Section 38.1 Human Reproductive Systems

Before You Read

During puberty, humans mature physically and become able to reproduce. At what ages do you think males and females go through puberty? Write your response on the lines below.

Read to Learn

Mark the Text **Identify Main Ideas** Circle the key terms in this section. Highlight the text that describes how sperm and eggs move through the reproductive system. In a different color, highlight the text that explains the role of the endocrine system in the reproductive process.

Human Male Anatomy

The main function of the organs, glands, and hormones of the male reproductive system is to produce sperm and deliver them to the female. Sperm are the male sex cells.

Where do sperm form?

Sperm are produced in the testes. The male body has two testes. The testes are located in a sac called the **scrotum.** The scrotum is suspended directly behind the base of the penis. Before a male infant is born, the testes form in the embryo's abdomen. The testes then descend into the scrotum. Sperm can only develop in temperatures that are about 2° to 3°C colder than normal body temperature. For this reason, the scrotum is located outside the abdomen. Muscles in the walls of the scrotum help maintain a proper temperature. The muscles contract in response to cold temperatures. This pulls the scrotum closer to the body and keeps the testes warm. The muscles relax in response to warm temperatures. This lowers the scrotum and allows air to circulate around it. This cools the testes and the sperm inside the scrotum.

Inside each testis is a network of tightly coiled tubes. Sperm are produced by the meiosis of the cells that line these tubes. Remember that meiosis produces haploid cells, not diploid cells. When a single cell in a testis divides by meiosis, the cell produces four haploid cells. All four of these haploid cells will develop into mature sperm. The maturation process takes about 74 days. A sexually mature male can produce about 300 million mature sperm per day, each day of his life.

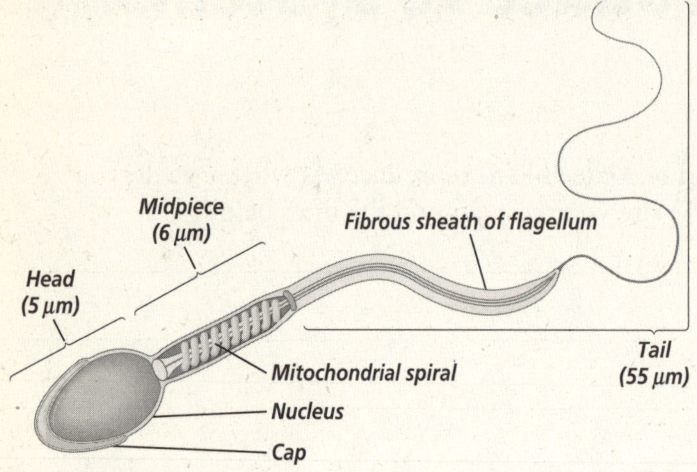

A sperm is well adapted to its job, to reach and then to enter the female egg. As you can see in the illustration to the left, a sperm has a head that contains its nucleus. The head is covered by a cap that contains enzymes. These enzymes help penetrate the female egg. Mitochondria are found in the midpiece of the sperm. The mitochondria provide energy for movement. The tail is a flagellum that whips back and forth to push the sperm along its way. Once inside the female reproductive tract, sperm usually live for about 48 hours. ✓

Reading Check

1. What does the midpiece of the sperm contain?

____

____

____

Reading Check

2. What is the purpose of the vas deferens?

____

____

____

How do sperm leave the testes?

Sperm leave the testes before they mature. They move out through a number of coiled ducts. These ducts empty into one single coiled tube called the **epididymis** (e puh DIH duh mus). The epididymis is located within the scrotum. The sperm mature in this tube. When sperm are released from the epididymis, they are mature. The mature sperm then enter the vas deferens. The **vas deferens** (VAS • DE fuh renz) is a duct that carries sperm from the epididymis toward the ducts that will push the sperm out of the body. Sperm can stay in the vas deferens for two or three months. When sperm leave the vas deferens, they are pushed toward the urethra. Peristaltic contractions, waves of involuntary muscle movement, within the vas deferens, move the sperm through this duct and into the urethra. The urethra is a tube in the penis that has two functions. It carries sperm out of the male body. The urethra also carries urine from the urinary bladder out of the male body. A muscle that is located at the base of the bladder stops urine and sperm from mixing together. ✓

How do fluids help transport sperm?

Sperm mix with fluids that are secreted by several glands. The **seminal vesicles** are a pair of glands located at the base of the urinary bladder. They secrete a fluid into the vas deferens. This secreted fluid is thick, like mucus, and it is rich in fructose, a sugar. Fructose provides energy for the sperm cells.

Section **38.1** **Human Reproductive Systems,** *continued*

The **prostate gland** is a doughnut-shaped gland that lies below the urinary bladder. Notice that the prostate gland surrounds the top portion of the urethra. Remember that the urethra is a tube in the penis that carries sperm and urine out of the male body. The prostate gland secretes an alkaline fluid that helps the sperm move. This fluid is thinner than the fluid secreted by the seminal vesicles. The alkaline fluid secreted by the prostate also helps the sperm survive. Two other tiny glands are located beneath the prostate. These are the bulbourethral glands. The **bulbourethral** (bul boh yoo REE thrul) **glands** secrete a clear, sticky alkaline fluid. This alkaline fluid also helps protect sperm because it neutralizes the acidic environment of the male urethra and the female vagina. **Semen** includes sperm and all of these fluids. ✓

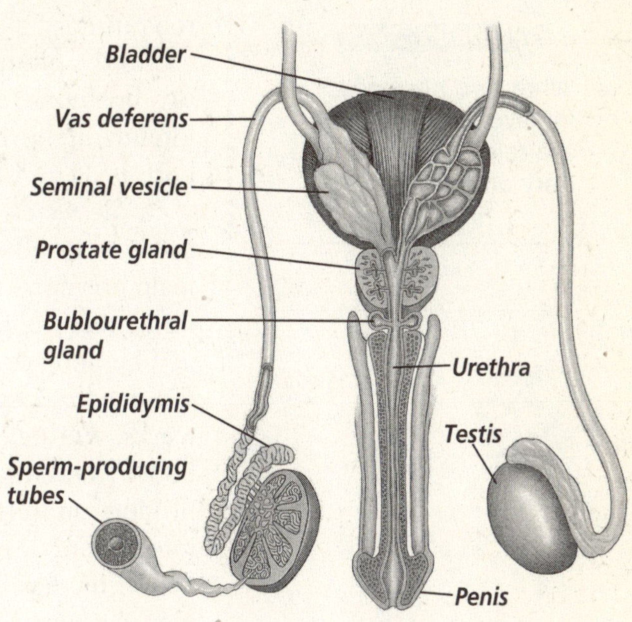

Puberty in Males

Remember that the glands of the endocrine system release hormones. Hormones control the development and activity of the male reproductive system.

What role do hormones play in male puberty?

Puberty begins in the early teen years. **Puberty** is the time when secondary sex characteristics begin to develop. Secondary sex characteristics in males include growth and maintenance of the male sex organs, an increase in body hair, an increase in muscle mass, increased growth of the long bones of the arms and legs, and deepening of the voice. Secondary sex characteristics begin to develop so that sexual maturity is reached. Sexual maturity means that the potential for sexual reproduction exists. The changes that occur during puberty are controlled by sex hormones. These hormones are secreted by the endocrine system.

What hormones are involved in male puberty?

The onset of puberty in males causes the hypothalamus to produce several kinds of hormones. These hormones interact with

✓ Reading Check

3. Which glands secrete alkaline fluids?

____

____

____

Reading Check

4. What two hormones involved in male puberty are released by the pituitary gland?

___

___

the pituitary gland. The hypothalamus secretes a hormone that causes the pituitary gland to release two other hormones. These two hormones are follicle-stimulating hormone (FSH) and luteinizing hormone (LH). When released into the bloodstream, both FSH and LH are carried to the testes. ✓

In the testes, FSH causes the production of sperm cells. LH causes the endocrine cells that are in the testes to produce the male hormone testosterone (teh STAHS tuh rohn). Testosterone influences the production of sperm cells.

The body controls the amounts of these hormones by a negative-feedback system. Remember how a negative feedback system works. As the levels of testosterone in the blood increase, the body decreases the production of FSH and LH. Increased sperm production in the testes also feeds back into the system. This information leads to a block in the production of FSH and LH. When the levels of testosterone in the blood drop, the body increases its production of FSH and LH.

Testosterone is a steroid hormone. A steroid hormone is fat-soluble. As a result, it can diffuse freely into a cell through the cell's plasma membrane. Testosterone is responsible for the growth and development of secondary sex characteristics in the male. As you read earlier in this section, these characteristics include growth and maintenance of male sex organs, as well as the production of sperm.

Reading Check

5. What are the three functions of the female reproductive system?

___

___

___

Human Female Anatomy

The female reproductive system has three main functions. The first is to produce eggs. Eggs are female sex cells. The second function is to receive sperm. The third function is to provide an environment in which a fertilized egg can develop. ✓

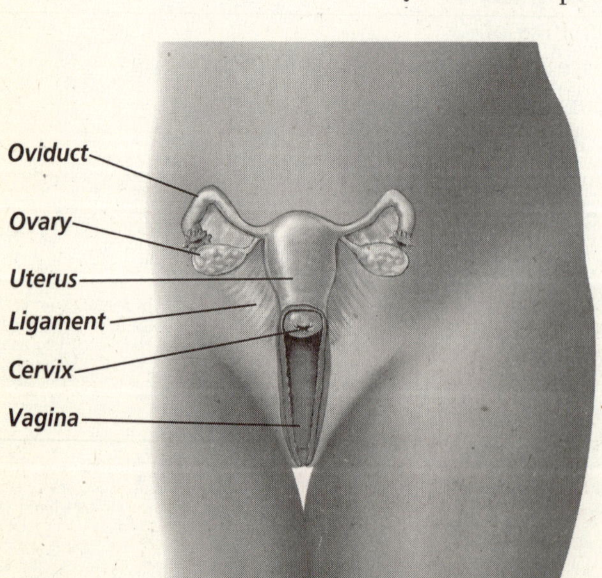

Females have two ovaries in which eggs are produced. Notice that there is one ovary on each side of the lower abdomen. Each ovary is about the size and shape of an almond.

Close to each ovary is the open end of an oviduct. The **oviduct** is a tube that transports eggs from the ovary to the uterus. Remember that during pregnancy in female mammals, the fetus develops in the uterus. The oviduct contains smooth muscle lined with cilia. Peristaltic contractions of the muscles in the wall of the oviduct, along with beating cilia, move eggs through the oviduct. In this way, an egg is moved from an ovary into the uterus.

The human uterus is located between the urinary bladder and the rectum. The uterus is the size of and has the shape of an upside-down pear. The walls of the uterus are composed of three layers. The outer layer is made up of connective tissue. The middle layer is made up of thick muscle. The inner layer is a thin, mucous lining. The inner layer is called the endometrium (en doh MEE tree um). The **cervix** is the lower end of the uterus. At its lower end, the cervix becomes smaller and leads into a narrow opening in the vagina. The vagina leads to the outside of the female body. ✓

Reading Check

6. How many layers do the walls of the uterus have?

______

______

______

Puberty in Females

Puberty in females begins in the early teen years. The hypothalamus signals the pituitary to produce and release FSH and LH. These two hormones, FSH and LH, are the same hormones that are produced in males. In females, LH causes eggs to be released into the oviduct. This process is discussed later in this section. FSH stimulates the development of follicles in the ovary. A **follicle** is a group of epithelial cells. These epithelial cells surround a developing egg cell. FSH also causes a hormone called estrogen to be released from the ovary. Estrogen is a steroid hormone responsible for the secondary sex characteristics of females. These characteristics include the growth and maintenance of female sex organs. As in males, secondary sex characteristics include an increase in growth rates of the long bones of the arms and legs. Females also develop more hair, especially under the arms and in the pubic area. In females, however, and unlike males, the hips broaden, and more fat is deposited in the breasts, buttocks, and thighs. The menstrual cycle begins. ✓

Reading Check

7. What are the secondary sex characteristics in females?

______

______

______

______

______

______

How are eggs produced?

The female body begins to develop eggs before birth. Before a female is born, cells in the ovaries divide until the first stage of meiosis, prophase I, is reached. When prophase I is reached, the cells begin a resting period. At birth, a female's ovaries contain about two million potential eggs. These potential eggs are called primary oocytes. Many oocytes break down, so that at puberty, the female ovaries contain about 40 000 primary oocytes.

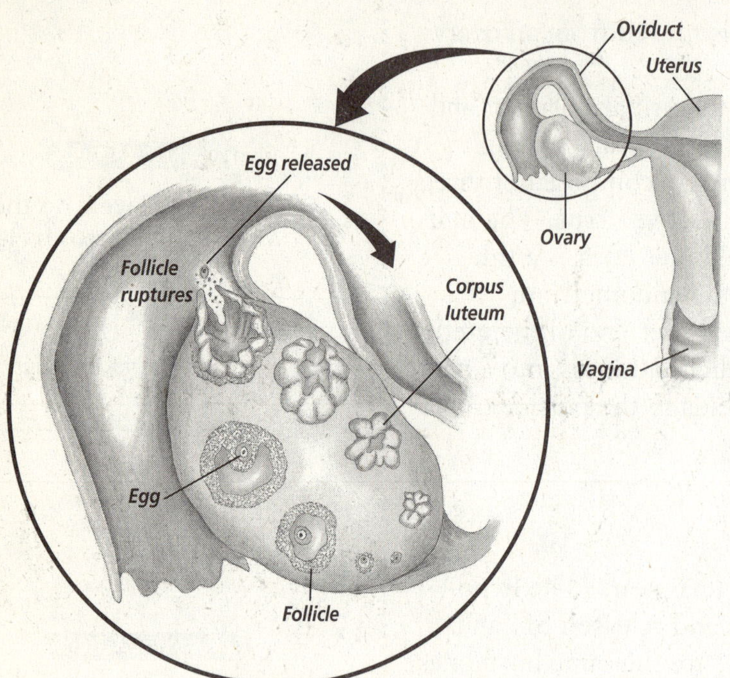

How are eggs released?

Beginning at puberty, meiosis starts up again in several of the prophase I cells. This happens about once a month. Each cell completes meiosis I and begins meiosis II. During meiosis II, one of the egg cells will rupture from the ovary and move into the oviduct as shown in the illustration on the left. **Ovulation** is the term used to describe the process of the egg rupturing through the ovary wall and moving into the oviduct. During the reproductive life of a female, about 400 eggs will ovulate, moving from the ovary into the oviduct.

The Menstrual Cycle

The **menstrual cycle** is the series of changes that the female body experiences each month. These changes include producing an egg and preparing the uterus for receiving the egg. Once an egg has been released during ovulation, the part of the follicle that remains in the ovary develops into the corpus luteum. The **corpus luteum** is a structure that secretes the two female hormones—estrogen and progesterone. Progesterone causes changes to occur in the lining of the uterus. These changes prepare the uterus to receive a fertilized egg. The menstrual cycle begins during puberty. It will continue for 30 to 40 years. It stops at menopause. At menopause, the female stops releasing eggs and the secretion of the female hormones estrogen and progesterone decreases. ✓

8. What two female hormones are secreted by the corpus luteum?

____

____

The length of each menstrual cycle varies from female to female. The average is 28 days. If the egg that is released at ovulation is not fertilized, the lining of the uterus is shed. This causes some bleeding in the female body that lasts for several days. The menstrual cycle can be divided into three phases. The first is the flow phase. The second is the follicular phase. The third phase is the luteal phase. The timing of each of these three phases correlates with hormone output from the pituitary gland, changes in the ovary, and changes in the uterus. Internal feedback controls hormone secretion during the menstrual cycle.

What is the flow phase?

The first day of the menstrual cycle, day 1, is the day that menstrual flow begins. The flow phase generally ends by the fifth day of the cycle. The lining of the uterus is called the endometrium. It is made up of blood, tissue fluid, mucus, and epithelial cells. The endometrium is shed during menstrual flow. This flow passes from the uterus through the cervix. It then passes into the vagina and then to the outside of the body. Contractions of the uterine muscle help to expel the uterine lining. During the flow phase, the level of FSH in the blood begins to rise. Another follicle in one of the ovaries begins to mature as the meiosis of the prophase I cells continues.

9. What is shed during menstrual flow?

____

____

____

What is the follicular phase?

The second phase of the menstrual cycle varies in length. In a 28-day cycle, it lasts about nine days, from day 6 to day 14. As the follicle containing a primary oocyte continues to develop, it secretes the hormone estrogen. This hormonal secretion stimulates the repair of the endometrial lining of the uterus. The endometrial cells undergo mitosis, and the lining of the uterus thickens. Notice that the feedback about the increase in estrogen levels goes to the hypothalamus and the pituitary gland. This feedback signals the pituitary to slow the production of FSH and LH. Just before ovulation, estrogen levels peak. This causes a sudden, sharp increase in the release of LH.

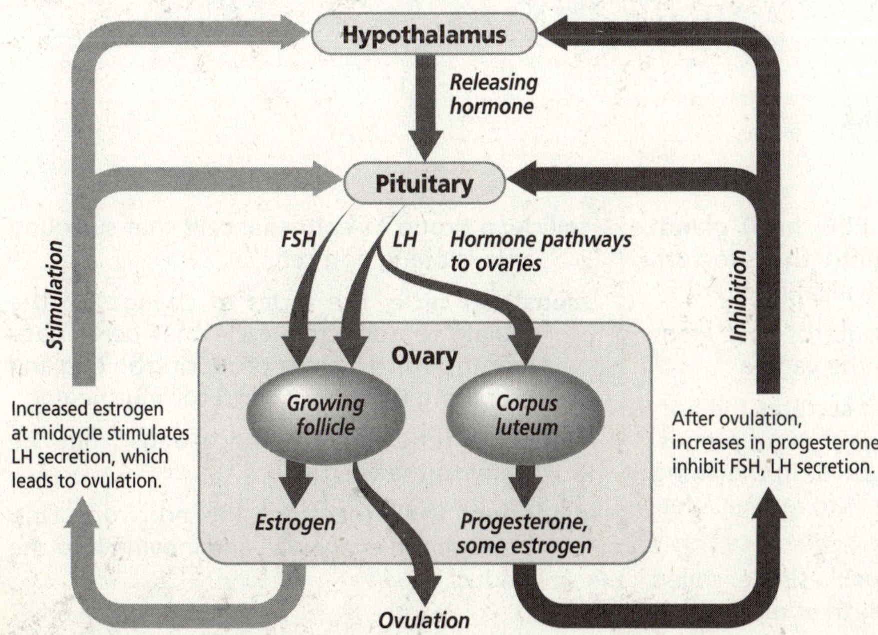

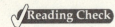

Reading Check

10. What are the three phases of the menstrual cycle?

____

____

____

The increase in LH causes the follicle to rupture, and the egg is released into the oviduct. Ovulation occurs at about day 14 of the menstrual cycle. The body temperature of the female will rise about 0.5°, and the cells of the cervix produce large amounts of mucus during this time.

What is the luteal phase?

The last phase of the menstrual cycle is called the luteal phase. It lasts about fourteen days, from day 15 to day 28. The luteal phase takes its name from the corpus luteum. Remember that the corpus luteum secretes the hormones estrogen and progesterone. Progesterone causes changes to occur in the lining of the uterus that prepare the uterus to receive a fertilized egg. ✓

During the luteal phase, LH stimulates the corpus luteum to develop from the ruptured follicle. Progesterone and estrogen are produced. Progesterone increases the blood supply of the endometrium. This causes it to accumulate lipids and tissue fluid. These changes correspond to the arrival of a fertilized egg. Through a negative feedback system, progesterone prevents the production of LH.

If the egg is not fertilized, the rising levels of progesterone and estrogen from the corpus luteum cause the hypothalamus to block the release of FSH and LH. The corpus luteum breaks up and stops secreting progesterone or estrogen. As hormone levels drop, the lining of the uterus begins to shed. If fertilization occurs, the endometrium will begin to secrete a fluid that is rich in nutrients. These nutrients will enable a fetus to grow and develop.

After You Read

Mini Glossary

bulbourethral (bul boh yoo REE thrul) gland: tiny gland located beneath the prostate that secretes a clear, sticky alkaline fluid

cervix: the lower end of the uterus that leads into a narrow opening in the vagina

corpus luteum: a structure that secretes the hormones estrogen and progesterone; progesterone causes changes to occur in the lining of the uterus that prepare it to receive a fertilized egg

epididymis (e puh DIH duh mus): a single, coiled tube in which sperm finish their maturation process

follicle: a group of epithelial cells that surround a developing egg cell

menstrual cycle: the series of changes in the female reproductive cycle that occur each month, which include producing an egg and preparing the uterus for receiving the egg

oviduct: a tube that transports eggs from the ovary to the uterus

ovulation: the process of the egg rupturing through the ovary wall and moving into the oviduct

Section 38.1 Human Reproductive Systems, *continued*

prostate gland: a doughnut-shaped gland that lies below the urinary bladder and surrounds the top portion of the urethra; secretes an alkaline fluid that helps the sperm move

puberty: in males and females, the time when secondary sex characteristics begin to develop; as secondary sex characteristics begin to develop, so does sexual maturity, the potential for sexual reproduction

scrotum: a sac that contains the testes; suspended directly behind the base of the penis

semen: the combination of sperm and the fluids produced in glands to either protect the sperm or move them through and out of the male body

seminal vesicle: a gland located at the base of the urinary bladder that secretes a fructose-rich fluid into the vas deferens

vas deferens (VAS • DE fuh renz): a duct that carries sperm from the epididymis toward the ducts that will force the sperm out of the body; sperm can stay in the vas deferens for two or three months

1. Read the terms and their definitions in the Mini Glossary. Describe how sperm are moved from the testes out of the male body.

2. Place each of the terms below under the correct reproductive system in the table. Then explain the role that each term plays in the reproductive system.

bulbourethral gland	follicle	semen
cervix	oviduct	seminal vesicle
corpus luteum	prostate gland	vagina
epididymis	scrotum	vas deferens

Male Reproductive System		Female Reproductive System	
Term	**Role**	**Term**	**Role**

Visit the Glencoe Science Web site at **science.glencoe.com** to find your biology book and learn more about human reproductive systems.

Name Date Class

Section 38.2 Development Before Birth

Before You Read

The union of an egg and a sperm results in a fertilized egg. This single cell must undergo many changes before it develops into a fetus. On the lines below, list the changes that you know occur in the developing embryo and fetus.

Read to Learn

Mark the Text **Locate Information** As you read this section, highlight the portions of text that describe the changes to the embryo and fetus during each of the three trimesters of pregnancy.

1. Why do so few sperm survive once they are in the vagina?

Fertilization and Implantation

After an egg ruptures from a follicle, it stays alive for about 24 hours in the oviduct. For this egg to be fertilized, sperm must be in the oviduct during those first hours after ovulation occurs. During sexual intercourse, between 300 and 500 million sperm are forced out of the male penis and into the female's vagina. Sperm can live for 48 hours in the female body. Therefore, fertilization can occur if intercourse occurs at any time from two days before to one day after the egg moves into the oviduct.

How is a zygote formed?

Remember that a zygote is a diploid cell that is formed when a sperm fertilizes an egg. Only one sperm will fertilize the egg. Fluids secreted by the vagina are acidic, and the acid destroys most of the sperm. Some sperm will survive, however, because the semen contains alkaline fluids that neutralize the acid and protect the sperm. The sperm that survive swim through the vagina and into the uterus. A few hundred sperm survive the entire journey and pass into the two oviducts. The egg is present in only one of the oviducts as shown in the illustration on page 487. ✓

Remember that the head of the sperm contains enzymes that help the sperm penetrate the egg. The egg's membrane carries an electrical charge. When one sperm enters the egg, the electrical charge of the cell membrane changes. This stops other sperm from entering the egg. The nucleus of the sperm, which is located in its head, unites with the egg's nucleus to form the zygote. This zygote contains all the genetic information that will be needed for the fetus.

How does the fertilized egg travel to the uterus?

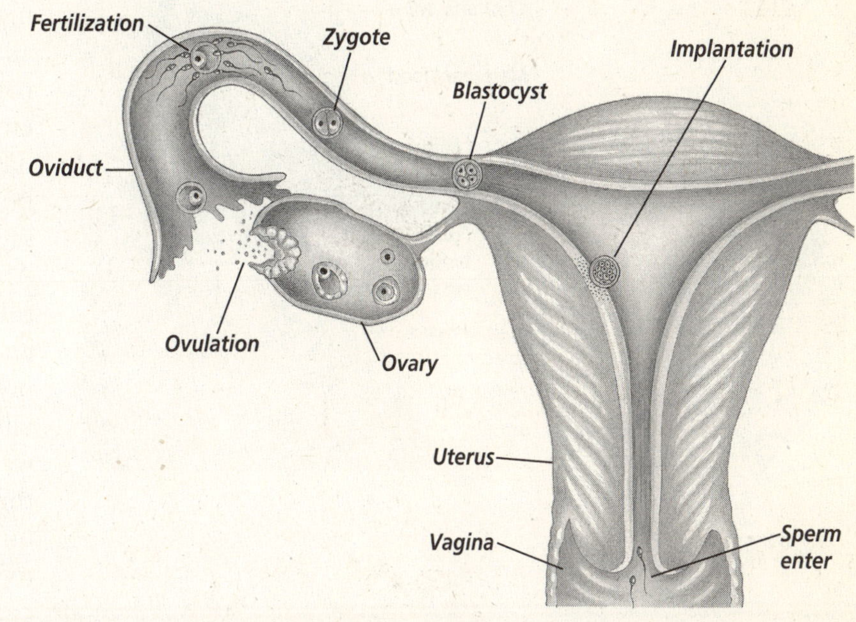

As the zygote begins to move down the oviduct, it begins to divide. The process of division by mitosis is repeated again and again. As the zygote moves along, it gets nutrients from the fluids that are secreted by the mother's body. By the sixth day, the zygote has moved into the uterus. Due to the continuous cell divisions, a hollow ball of cells forms. The hollow ball of cells is called a blastocyst. *Blastocyst* is the term used when talking about human embryonic development. Remember that the term *blastula* is used when talking about the embryonic development of other animals.

The newly formed blastocyst attaches to the uterine lining. This attachment occurs six days after fertilization. The attachment of the blastocyst to the uterine lining is called **implantation.** A small mass of cells that are located within the blastocyst will soon develop into a human embryo.

Embryonic Membranes and the Placenta

When you studied reptiles and birds, you learned about amniotic eggs and their importance to the evolutionary advancement of animals. Membranes that are similar to those of the amniotic egg form around the human embryo. The purpose of these membranes is to protect and nourish the embryo. These membranes are called the amniotic sac. The amnion is a thin, inner membrane. It is filled with a clear, watery amniotic fluid. This fluid serves as a shock absorber. It also helps to regulate the body temperature of the developing embryo. ✓

✓ Reading Check

2. What is the purpose of the amniotic sac?

____

____

____

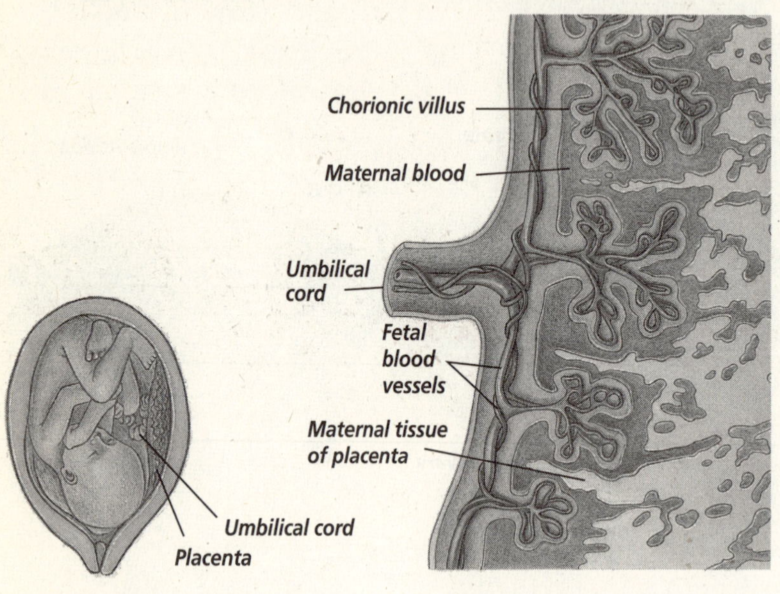

The allantois membrane is an outgrowth of the digestive tract of the embryo. The **umbilical cord** is made up of blood vessels of the allantois. The umbilical cord is a ropelike structure that attaches the embryo to the wall of the uterus. The chorion is the outer membrane that surrounds the amniotic sac and the embryo that is inside. About 12 days after the egg is fertilized, fingerlike projections of the chorion begin to develop. These projections are called chorionic villi, and they grow into the uterine wall as shown in the illustration to the left. The chorionic villi combine with part of the uterine lining to form the placenta.

Reading Check

3. What two roles does the placenta serve?

____

____

How are substances exchanged between the mother and the embryo?

To survive and develop, the embryo needs proper nutrients. The embryo also needs to eliminate its own body wastes. The placenta delivers nutrients to the embryo. It carries wastes away from the embryo. ✓

Blood vessels from the mother's uterine wall lie close to the blood vessels of the embryo's chorionic villi. The mother's blood vessels are not, however, connected to the embryo's blood vessels. Oxygen and nutrients that are transported by the mother's blood diffuse into the blood vessels of the chorionic villi. Remember that the chorionic villi are in the placenta. The blood carries these substances from the chorionic villi into the umbilical cord and into the embryo. To remove wastes, the reverse of this process occurs. Waste products from the embryo move from the blood vessels in the umbilical cord into the placenta. The waste products then diffuse out of the blood vessels in the chorionic villi into the mother's blood. Finally, the mother's excretory system removes these waste products.

What role do hormones play in pregnancy?

The two female hormones estrogen and progesterone play important roles during pregnancy. Remember that both of these hormones, but especially progesterone, cause the uterine lining to thicken. This helps prepare the uterus for implantation of the blastocyst. Once the blastocyst is implanted, the chorionic membrane of the embryo begins to secrete another hormone. The hormone is human chorionic gonadotropin (hCG). This hormone keeps the corpus luteum alive. Because the corpus luteum is alive, it continues to secrete estrogen and progesterone. By the third or fourth month of pregnancy, the placenta takes over the job of the corpus luteum. The placenta will secrete enough estrogen and progesterone to continue and sustain the pregnancy. ✓

Reading Check

4. Which hormones are needed to sustain pregnancy?

____

____

____

Fetal Development

There are three different processes in fetal development. One is growth. Growth means that the actual number of cells increases. The second process is development. The growing number of cells will move and arrange themselves into specific organs. The third process is cellular differentiation. Cellular differentiation means that cells change so that they can perform certain specific tasks. Remember that not all cells do the same things.

In humans, pregnancy normally lasts about 280 days, or about nine months. The time is calculated from the first day of the mother's last menstrual period. The baby develops for about 266 days, or the time of egg's fertilization until birth. This time span is divided into three trimesters. Each trimester lasts about three months. During the first trimester, at the eighth week, the embryo is known as the fetus. During each of the three trimesters, the embryo, later called the fetus, grows and develops. Its cells will also undergo change and differentiate.

What changes occur during the first trimester?

During the first trimester, all the organ systems of the embryo begin to develop. These include the heart and the lungs. Arms and legs are formed, the skeleton begins to harden, and the tissue that forms the eyes develops. Muscles also begin to appear. With muscles, the embryo can move. While all this is happening to the embryo, the mother may not even realize she is pregnant.

The first seven weeks following fertilization are critical to the embryo. During these initial weeks, the embryo is more sensitive to outside influences than at any other time. Substances such as alcohol, tobacco, and other drugs can harm a developing embryo. Exposure to poisons such as pesticides also can cause permanent damage to the developing embryo. If the mother develops certain

Reading Check

5. What major changes occur in the developing embryo and fetus during the first trimester?

infections such as chicken pox or measles, the developing embryo can be harmed. A pregnant woman needs to eat well to sustain the embryo. Folic acid, a vitamin that is found in green leafy vegetables, broccoli, and dried beans, will help prevent neural tube defects. A neural tube defect is a birth defect that occurs in the brain or the spinal cord of the developing fetus. They are the most common of all serious birth defects.

By the eighth week, all the organ systems have been formed. At this time the embryo is known as a fetus. At the end of the first trimester, the fetus weighs about 28 g. From the top of its head to its buttocks, it is about 7.5 cm long. The gender of the fetus can be determined at this time. An ultrasound will show the external sex organs.

What changes occur during the second trimester?

Most of the changes that occur to the fetus during the second trimester involve body growth. During the fourth month, the fetus grows rapidly. During the fifth month, growth slows down. During the fifth month, the mother can feel the fetus move. During the sixth month, the fetus's eyes open, and its eyelashes form. By the end of the second trimester, the fetus weighs about 650 g and is about 34 cm long.

At this time, the fetus may be able to survive outside the uterus. In order to survive, however, the fetus would need constant and significant medical help. The fetus cannot maintain a constant body temperature, and the lungs are not mature enough to work on their own. The death rate of fetuses that are born during the second trimester is high.

What changes occur during the third trimester?

During the last three months, the body mass of the fetus will more than triple. The fetus continues to kick, stretch, and move freely within the amniotic cavity that surrounds it. During the eighth month, fat is deposited under the skin. This fat will help keep the newborn warm.

During the last weeks of pregnancy, the fetus grows large enough so that it fills the space inside the embryonic membranes. By the end of the third trimester, the fetus weighs about 3300 g and is about 51 cm long. The body systems have developed, and the fetus can now survive independently outside the uterus. The time for birth has arrived.

Section
38.2 Development Before Birth, *continued*

Can genetic disorders be predicted?

Our knowledge of human heredity is expanding. Technology in the medical field is advancing rapidly. Because of these two factors, scientists can predict more easily and more accurately if certain genetic disorders will be passed on from parents to their children. Scientists can now identify genes that carry certain genetic disorders. These disorders include cystic fibrosis, Huntington's disease, and Tay-Sachs disease. Early identification or detection of certain genetic disorders such as phenylketonuria (PKU) and cystic fibrosis is essential to the treatment of these disorders. ✓

Reading Check

6. Name three genetic disorders.

____

____

____

What does a genetic counselor do?

As services, such as genetic counseling, become more available, some people may choose to visit a genetic counselor to gain additional information about their own genetic makeup. A genetic counselor has a medical background with additional training in genetics. Sometimes a team will work with potential parents. The team may include geneticists, clinical psychologists, social workers, and other consultants. A genetic counselor records the medical history of both parents and their families. The counselor collects and analyzes all the available information. The information may include pedigrees, biochemical analyses of blood, karyotypes, and DNA. When the analysis of this information is completed, the counselor meets with the potential parents. The counselor will explain the risk factors for their giving birth to children with genetic disorders.

After You Read

Mini Glossary

implantation: name given to the attachment of the blastocyst, or hollow ball of cells, to the uterine lining, which occurs six days after fertilization

umbilical cord: a ropelike structure made up of blood vessels of the allantois that attaches the embryo to the wall of the uterus

1. Read the key terms and the definitions in the Mini Glossary above. In your own words, explain how these two terms are related.

2. Use the sequencing diagram below to show the development of the human embryo into a fetus. Show the changes that occur during each trimester.

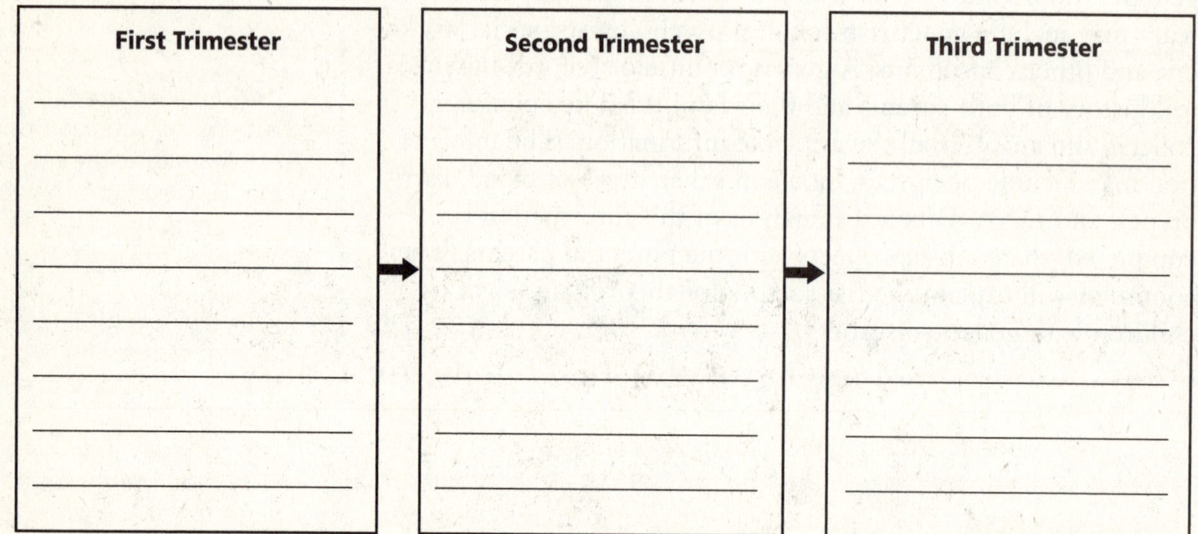

Visit the Glencoe Science Web site at **science.glencoe.com** to find your biology book and learn more about development before birth.

Name Date Class

Section 38.3 Birth, Growth, and Aging

Before You Read

Have you ever looked at your baby pictures? Although you may see similarities between the way you look now and the way you looked then, you may also notice many differences. On the lines below, list the changes that are most noticeable.

Read to Learn

Birth

A mother gives birth to her baby when the fetus is pushed out of the uterus and out of her body. Scientists still do not know what triggers the mother's body to begin the birthing process. Different hormones released from the pituitary gland, the uterus, and the placenta may stimulate the uterus to move the baby out of the womb and into the world. Birth occurs in three recognizable stages. They are dilation, expulsion, and the placental stage.

What happens during the dilation stage?

The changes that a female goes through to give birth are called **labor.** These changes are both physiological and physical. Labor starts when the uterus begins to contract. At first, these involuntary contractions are mild. The mother has no control over them. Oxytocin, a hormone released by the pituitary gland, stimulates these contractions. The contractions of the uterine muscles open, or dilate, the cervix. As labor continues, the contractions begin to occur regularly. The contractions grow stronger as they get closer together. With each contraction, the cervix opens only a tiny bit more. When the opening of the cervix is about 10 cm, it is fully dilated. Usually, the amniotic sac that surrounds the fetus breaks and the amniotic fluid flows through the vagina. The vagina is called the birth canal because the baby will come through this canal. ✓

Mark the Text **Identify Details** As you read this section, highlight the information that explains the changes in the human body as it develops from infancy to old age. On a separate sheet of paper, create four diagrams: one of an infant, one of a child, one of a teenager, and one of an adult. Using the information that you have highlighted, write descriptions on your diagrams of the changes that occur at each stage of development.

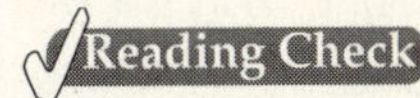

1. How large is the opening of the cervix when it is fully dilated?

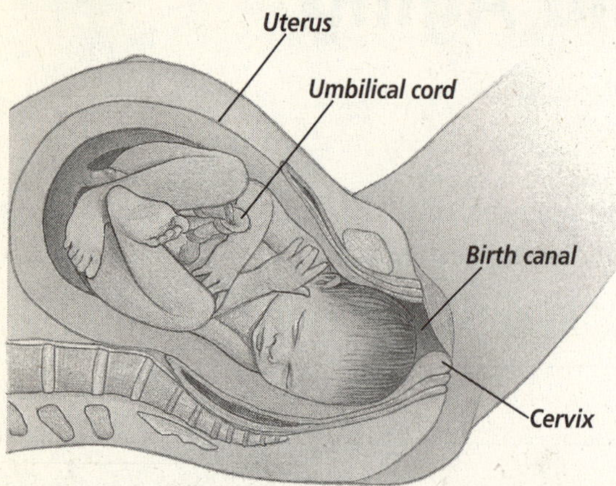

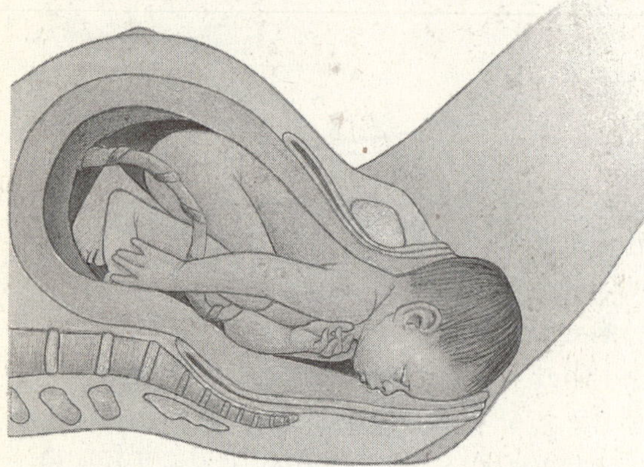

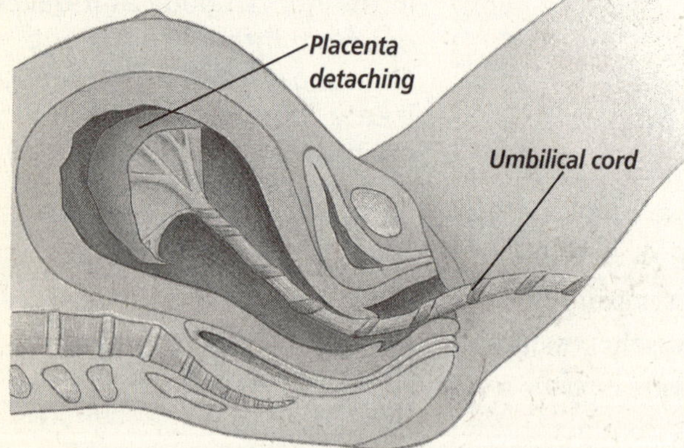

How is a baby pushed out of the mother's body?

Expulsion occurs when the baby is pushed out of the mother's body by the involuntary uterine contractions. These contractions become so powerful that they push the baby from the uterus, through the cervix, and into the birth canal as shown in the illustration to the left. The mother can assist in the expulsion phase. By contracting her own voluntary abdominal muscles as the involuntary uterine muscles contract, the mother helps push the baby out. The expulsion stage usually lasts from about 20 minutes to one hour.

What is the placental stage?

The last stage of the birthing process is called the placental stage. Ten to fifteen minutes after the baby is born, the placenta separates from the wall of the uterus. The placenta is pushed out of the mother's body, along with whatever is left of the embryonic membranes. Because these materials are pushed out after the baby is born, they are commonly called the afterbirth. The uterine muscles continue to contract with a good deal of force to constrict the uterine blood vessels and prevent the mother from bleeding too much.

After the baby is born, its umbilical cord is clamped. The cord is then cut near the baby's abdomen. The bit of cord that is left dries up and falls off. The navel is the abdominal scar that is left after the umbilical cord falls off.

Growth and Aging

Once a baby is born, it continues to grow and it begins to learn. Human growth varies with age and, to a certain extent, with gender. Remember that growth is an increase in the amount of living material in an organism. Growth also includes the formation of new structures within the organism.

What role do hormones play in human growth?

Human growth is regulated by human growth hormone, hGH. It is one of the hormones produced and controlled by the endocrine system. The pituitary gland secretes hGH. This hormone causes all body cells to grow, but it acts mainly on the skeleton and the skeletal muscles. hGH increases the rate of protein synthesis. It also increases the metabolism, or burn rate, of fat molecules. Other hormones that influence growth are throxin, the female hormone estrogen, and the male hormone testosterone.

What is the first stage of growth?

Infancy is the first stage of growth. A baby is considered to be an infant for the first two years of life. During these two years, a child grows tremendously. The infant shows an amazing increase in physical coordination and in mental development. An infant will double its birth weight by the time it is five months old. The infant's birth weight will triple in the first twelve months of life. By the time an infant is two, the child will weigh four times more than it weighed at birth. In the first two years, an infant learns to control its arms and legs, roll over, sit, crawl, and walk. At the end of these two years, the child may begin to talk using simple words.

2. Infer Which takes more coordination? (Circle your choice.)
a. sitting
b. walking
c. crawling

What changes occur during childhood and adolescence?

Childhood is the period of growth and development that goes from the end of infancy, when a child is two years old, to adolescence. Adolescence is the time when puberty begins, usually the early teen years. During childhood, most children experience steady growth. In addition, a child develops the ability to reason and to solve problems.

Adolescence follows childhood. At puberty, which marks the beginning of adolescence, most young teenagers experience a growth spurt. These spurts can be surprisingly large. An increase in height of 5 to 8 cm in one year is not uncommon in teenage boys. During the teen years, adolescents gain their maximum height. Heredity, nutrition, and environment determine a person's height. By the time a young person reaches adulthood, his or her organs have reached maximum mass. Physical growth is complete.

What happens as adults age?

Reading Check

3. What are the three primary stages of growth and development after infancy?

___

___

___

As the body ages, there are distinct, noticeable changes. Metabolism, the rate at which fat is burned, slows down. Digestion becomes slower. The skin begins to lose its elasticity, and wrinkles appear. Less pigment is produced in the hair follicles so the hair begins to turn white. Bones can become thinner and more brittle. They may break or fracture much more easily. The disks between vertebrae compress, so height is lost. People actually become shorter. Vision and hearing may diminish, requiring corrective lenses or hearing aids. Many people, however, are intellectually and physically active as they grow older.

After You Read

Mini Glossary

labor: the physiological and physical changes that a female goes through to give birth

1. Review the key term and definition in the Mini Glossary above. Write a definition of **labor** in your own words.

___

___

2. Use the web diagram below to name and describe the three parts of the birthing process.

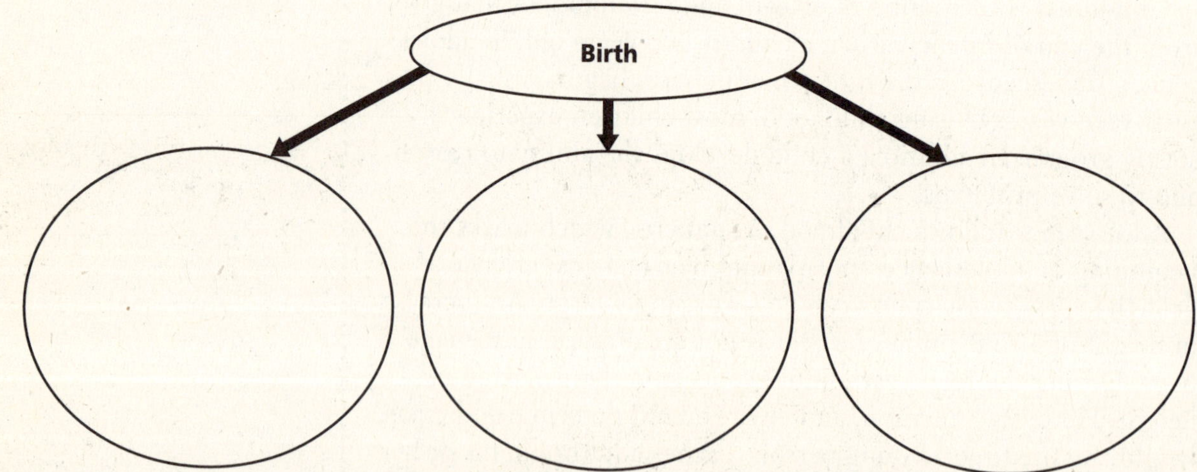

Visit the Glencoe Science Web site at **science.glencoe.com** to find your biology book and learn more about birth, growing, and aging.

Name Date Class

Section 39.1 The Nature of Disease

Before You Read

Think of a disease you know something about. Perhaps you had a cold recently or have read about a disease in the news. Write the name of the disease and answer the following questions on the lines below. Is the disease contagious? How is it spread? What causes the disease? Is there a cure? These are the types of questions that scientists ask about diseases.

____

____

____

Read to Learn

What is an infectious disease?

A disease disrupts the homeostasis in the body. Disease-producing organisms such as bacteria, protozoans, fungi, viruses, and other parasites are called **pathogens.** The main sources of pathogens are soil, contaminated water, and infected animals, including people.

Not all microorganisms are pathogenic, or disease-producing. Some microorganisms that live in the body are helpful. At birth, microorganisms establish themselves on your skin and in your upper respiratory system, urinary tract, reproductive tract, and lower intestinal tract. These microorganisms have a symbiotic relationship with your body. Both you and the microorganisms benefit from the close association.

Microorganisms in the body help maintain a healthy balance called equilibrium. They keep harmful bacteria and other microorganisms from growing. If conditions change and the beneficial organisms are eliminated, pathogens can establish themselves and cause infection and disease. On the other hand, if these beneficial organisms enter areas of the body where they are not normally found or if a person becomes weak or injured, these formerly harmless organisms can become pathogens. ✓

Any disease caused by the presence of pathogens in the body is called an **infectious** (ihn FEK shus) **disease.** The table on page 498 lists some infectious diseases.

Summarize Information As you read, summarize the key information under each heading into one or two sentences.

✓ Reading Check

1. How are some microorganisms beneficial?

____

____

____

The Nature of Disease, *continued*

isease	Cause	Affected Organ System	Transmission
mallpox	Virus	Skin	Droplet
Chicken pox	Virus	Skin	Droplet
Rabies	Virus	Nervous system	Animal bite
Poliomyelitis	Virus	Nervous system	Contaminated water
Colds	Viruses	Respiratory system	Direct contact
Influenza	Viruses	Respiratory system	Direct contact
HIV/AIDS	Virus	Immune system	Exchange of body fluids
Hepatitis B	Virus	Liver	Exchange of body fluids
Tetanus	Bacteria	Nervous system	Puncture wound
Food poisoning	Bacteria	Digestive system	Contaminated food/water
Strep throat	Bacteria	Respiratory system	Droplet
Diphtheria	Bacteria	Respiratory system	Droplet
Tuberculosis	Bacteria	Respiratory system	Droplet
Spinal meningitis	Bacteria	Nervous system	Droplet

Determining What Causes a Disease

One of the first problems scientists face when studying a disease is finding out what caused the disease. Not all diseases are caused by pathogens. Some disorders are inherited. Hemophilia (hee muh FIH lee uh) is an inherited disorder. Osteoarthritis (ahs tee oh ar THRIH tus) may be caused by wear and tear on the body as it ages. Some diseases, such as cirrhosis (suh ROH sihs), are caused by exposure to chemicals or toxins such as alcohol that can destroy liver cells. Other diseases can be caused by malnutrition. Scurvy, which results in poor wound healing, swollen gums, and loose teeth, is caused by a lack of vitamin C. Pathogens cause infectious diseases and some cancers. Scientists follow a standard set of procedures to determine which pathogen causes a disease. ✓

Reading Check

2. Are all diseases caused by pathogens?

____

____

____

How was the first pathogen identified?

The first proof that pathogens actually cause disease came in 1876. Robert Koch (KAHK), a German doctor, was looking for the cause of anthrax. Anthrax is a deadly disease that primarily affects cattle and sheep but also can occur in humans. Koch discovered a rod-shaped bacterium in the blood of cattle that had died of anthrax. He cultured, or grew, the bacteria. Then he injected samples of the culture into healthy animals. When those animals became sick and died, Koch compared the bacteria in their blood with the bacteria he had originally taken from the anthrax victims. He found that the two sets of blood cultures contained the same bacteria. ✓

Reading Check

3. What did Koch discover in the blood of infected cattle?

____

____

____

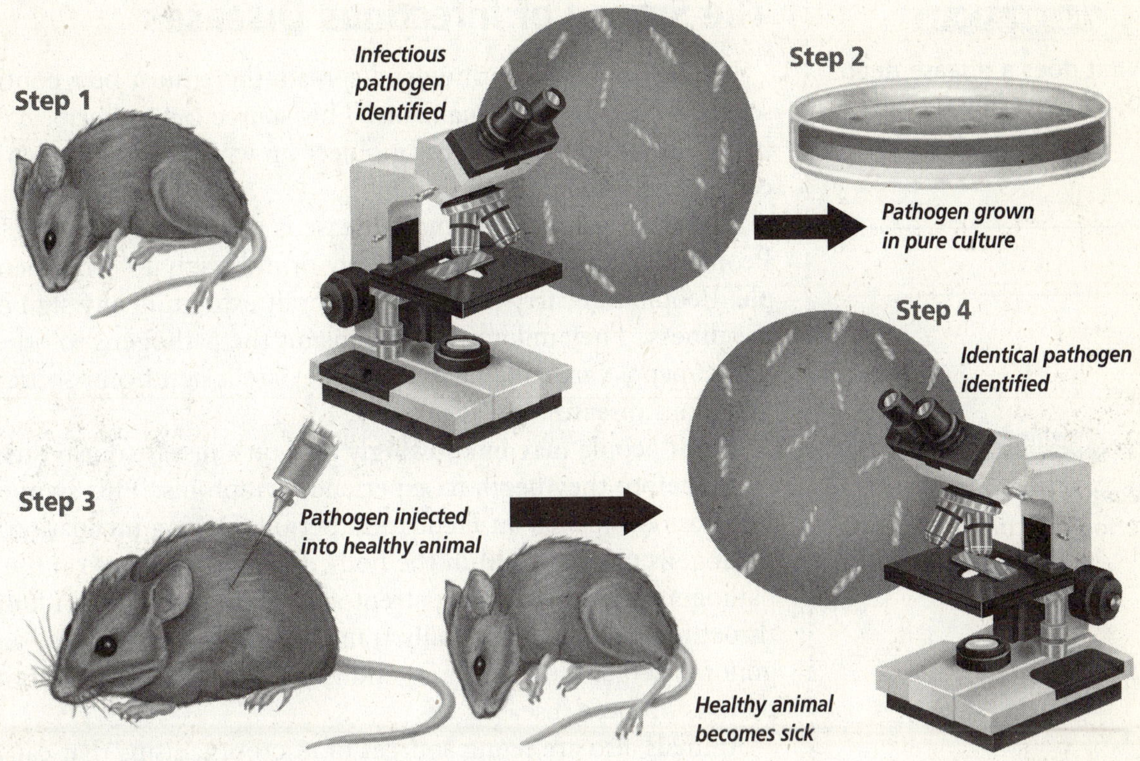

What was the procedure Koch established?

Koch established experimental steps shown in the illustration above for directly linking a specific pathogen to a specific disease. These steps, first published in 1884, are known today as **Koch's postulates.**

1. The pathogen must be found in the host in every case of the disease.
2. The pathogen must be isolated from the host and grown in a pure culture containing no other organisms.
3. When the pathogen from the pure culture is placed in a healthy host, it must cause the disease.
4. The pathogen must be isolated from the new host and be shown to be the original pathogen.

Are there exceptions to Koch's postulates?

Koch's postulates are useful in determining the cause of most but not all diseases. Some organisms, such as the pathogenic bacterium that causes syphilis, have never been grown in a pure culture. Viral pathogens multiply only within cells. As a result, living tissue must be used as a culture medium for viruses.

Think it Over

4. **Conclude** Why can't viruses be identified using Koch's postulates?

____

____

____

Reading Check

5. What does a disease need to continue and spread?

______

______

______

Reading Check

6. What is the symptom-free period of a disease called?

______

______

______

The Spread of Infectious Diseases

For a disease to continue and spread, there must be a continual source of the disease organisms. This source can either be a living organism or a nonliving object on which the pathogen can survive. ✓

The main source of human disease is the human body itself. People may pass pathogens directly or indirectly to other people. People can carry pathogens without exhibiting any sign of the illness. They unknowingly transmit the pathogens to others. These people are called carriers. They are a significant source of infectious diseases.

Other people may unknowingly pass on a disease during its first stage, before they begin to experience symptoms. This symptom-free period is called an incubation period. During incubation, the pathogens multiply within the body. Humans can pass on the pathogens that cause colds, streptococcal (strep tuh KAH kul) throat infections, and sexually transmitted diseases (STDs) such as gonorrhea (gah nuh REE uh) and AIDS during the incubation periods of these diseases. ✓

Animals also are living sources of microorganisms that cause disease in humans. Some types of influenza, commonly known as the flu, are transmitted from animals to humans. Rabies and Lyme disease also are transmitted from animals to humans.

The major nonliving sources for infectious diseases are soil and water. Soil contains pathogens such as fungi and the bacterium that causes botulism, a type of food poisoning. Water contaminated by human and animal feces is a source for several pathogens, especially those responsible for intestinal diseases.

How are diseases transmitted?

Pathogens can be transmitted to a host in four main ways: by direct contact, by an object, through the air, or by a carrier organism called a vector.

The common cold, influenza, and STDs are spread by direct contact. STDs, such as genital herpes and the virus that causes AIDS, are usually transmitted by the exchange of body fluids, especially during sexual intercourse.

Bacteria and other microorganisms can live on objects such as money, toys, or towels. Transmission occurs when people handle contaminated objects. You can prevent this type of transmission by thoroughly cleaning objects such as eating utensils and countertops that can harbor pathogens. Washing your hands often throughout each day can prevent transmission of disease.

Airborne transmission of a disease can occur when a person coughs or sneezes, spreading pathogens contained in droplets of mucus into the air. *Streptococcus*, the bacterium that causes strep throat infections, and the virus that causes measles are two examples of disease-causing organisms that can be spread through the air.

Diseases transmitted by vectors are most commonly spread by insects and other arthropods. Mosquitoes transmit diseases such as malaria and the West Nile virus. Ticks transmit Lyme disease and Rocky Mountain spotted fever. Flies also are significant vectors of disease. They land on infected materials, such as animal wastes, and then land on fresh food that is eaten by humans, transmitting the pathogens. ✓

Reading Check

7. Name a vector and a disease it transmits.

What causes the symptoms of a disease?

When a pathogen invades your body, it encounters your immune system. The immune system consists of individual cells, tissues, and organs that work together to protect the body against organisms that may cause infection or disease. If the pathogen overcomes the defenses of your immune system, it can multiply. This damages the tissues it has invaded, even killing host cells.

How do pathogens damage the host?

Viruses cause damage by taking over a host cell's genetic and metabolic machinery. Many viruses cause the eventual death of the cells they invade.

Toxins do most of the damage to host cells. Toxins are poisonous substances that are sometimes produced by microorganisms. These poisons are transported by the blood and can cause serious effects and sometimes death. Toxins can slow or stop protein synthesis in the host cell, destroy blood cells and blood vessels, produce fever, or cause spasms by disrupting the nervous system.

For example, the toxin produced by tetanus bacteria affects nerve cells and produces uncontrollable muscle contractions. If the condition is left untreated, paralysis and death occur. Tetanus bacteria are normally present in soil. A small amount of this toxin can kill many people. ✓

Reading Check

8. Where is the tetanus bacterium found?

Patterns of Diseases

Diseases can spread rapidly as people fly from one part of the country or world to another. Government health departments try to identify a pathogen, its method of transmission, and the geographic distribution of the disease it causes. The Centers for Disease Control and Prevention in the United States publishes a weekly report about the incidence of specific diseases.

Some diseases, such as typhoid fever, occur only occasionally in the United States. These periodic outbreaks often occur because someone traveling in a foreign country brought the disease back home. On the other hand, many diseases, such as chicken pox, are constantly present in the population. Such a disease is called an **endemic disease.**

Sometimes, an epidemic breaks out. An **epidemic** occurs when many people have the same disease at about the same time. Influenza can become an epidemic, sometimes spreading to many parts of the world. During the early 1950s, a polio epidemic spread across the United States. Victims were paralyzed or died when the polio virus attacked the nerve cells of the brain and spinal cord.

Think it Over

9. What is the difference between an endemic disease and an epidemic?

____

____

____

Treating Diseases

A person who becomes sick may be treated with medicinal drugs, such as antibiotics. An **antibiotic** is a substance produced by a microorganism that will kill or inhibit the growth and reproduction of other microorganisms, especially bacteria. Although antibiotics can be used to cure some bacterial infections, antibiotics do not affect viruses.

Bacteria can become resistant to antibiotics, and the drugs become ineffective. Penicillin, an antibiotic first used in the 1940s, is still one of the most effective antibiotics known. However, more and more bacteria have evolved that are resistant to it. For example, the bacterium that causes pneumonia, ear infections, and meningitis has become resistant to penicillin. This creates a problem because penicillin is the primary drug used to treat these diseases.

Fortunately, the use of antibiotics is only one way to fight infection. Your body has its own built-in defense system, the immune system, which works to keep you healthy.

Think it Over

10. Analyze Which diseases cannot be effectively treated with antibiotics? (Circle your choice.)

a. viruses

b. bacterial infections

After You Read

Mini Glossary

antibiotics: substances produced by a microorganism that, in small amounts, will kill or inhibit growth and reproduction of other microorganisms

endemic disease: disease that is constantly present in a population

epidemic: occurs when many people in a given area are afflicted with the same disease at about the same time

infectious (ihn FEK shus) disease: any disease caused by pathogens in the body

Koch's postulates: experimental steps relating a specific pathogen to a specific disease

pathogens: disease-producing agents such as bacteria, protozoans, fungi, viruses, and other parasites

1. Review the terms and their definitions in the Mini Glossary above. On the lines below explain why antibiotics may lose their effectiveness over time.

____

____

2. Fill in the table below with facts about the nature of disease.

The Nature of Disease	
Pathogens	**1. bacteria** **2.** **3.** **4.** **5.**
Causes of disease other than pathogens	**1. inherited** **2.** **3.** **4.**
Sources of pathogens	**1. human body** **2.** **3.** **4.**
Transmission from sources	**1. direct contact** **2.** **3.** **4.**

Visit the Glencoe Science Web site at **science.glencoe.com** to find your biology book and learn more about the nature of disease.

Name Date Class

Section 39.2 Defense Against Infectious Diseases

Before You Read

Have you ever been around someone who was sneezing and coughing? Did you later get sick, or were you able to fight off the infection? Why do you think that you sometimes catch other people's *bugs* and other times you do not? On the lines below, write a sentence about the last time you got sick. Then explain why you think your body was unable to defend itself.

Read to Learn

Mark the Text **Identify Main Ideas** As you read this section, highlight the main ideas. Then study the ideas and restate them in your own words.

1. What is your body's first barrier against pathogens?

Innate Immunity

Your body produces a variety of white blood cells. These cells defend your body against invasion by pathogens. No matter what pathogens are present, a healthy immune system is always ready. The body's **innate immunity** is always present and defends the body against any and all pathogens.

How do skin and body secretions protect you?

When a potential pathogen comes in contact with your body, often the first barrier it meets is your skin. Skin keeps many microorganisms from entering the body.

In addition to the skin, pathogens also encounter your body's secretions of mucus, oil, sweat, tears, and saliva. The main function of mucus is to prevent various areas of the body from drying out. It also traps many microorganisms and other foreign substances that enter the respiratory and digestive tracts. Mucus is continually swallowed and passed to the stomach. There, acidic gastric juice destroys most bacteria and their toxins. Sweat, tears, and saliva contain the enzyme lysozyme, which is capable of breaking down the cell walls of some bacteria. ✓

What causes inflammation of body tissues?

If a pathogen gets past the skin and body secretions, your body has several other nonspecific defense mechanisms. These can destroy the invader and restore homeostasis. Think about what happens when you get a splinter. If bacteria or other pathogens enter and damage body tissues, inflammation (ihn fluh MAY shun) results. Inflammation has four symptoms—redness, swelling, pain, and heat. The figure to the right shows what happens when inflammation begins. First, damaged tissue cells called mast cells and white blood cells called basophils release histamine (HIHS tuh meen).

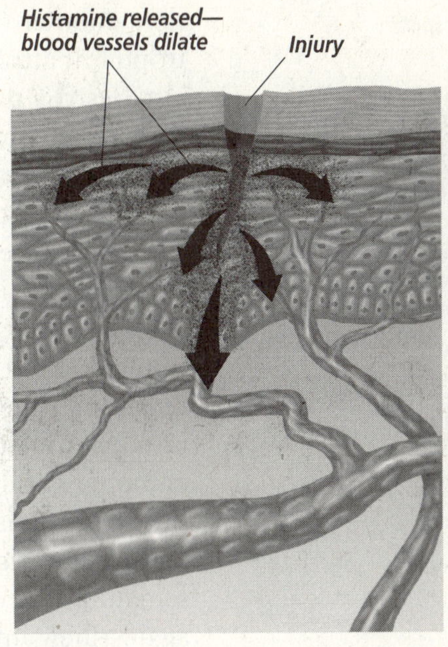

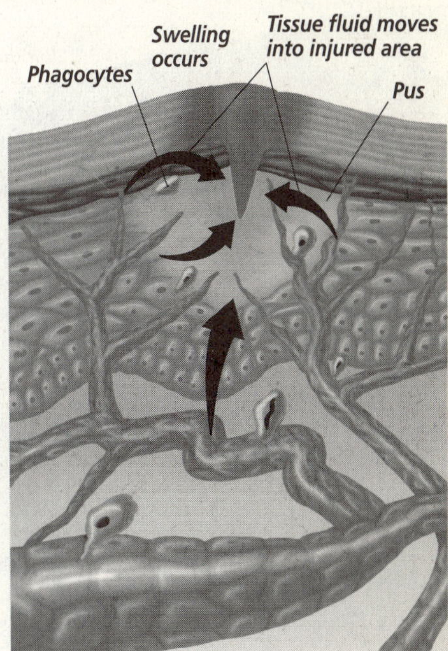

Histamine causes blood vessels in the injured area to dilate, or enlarge. These dilated blood vessels cause the redness of an inflamed area. Fluid that leaks from the vessels into the injured tissue helps the body destroy toxic agents and helps restore homeostasis. This increase in tissue fluid causes swelling and pain, and may also cause the area to become warmer. Inflammation can occur with other types of injuries as well as infections. Physical force, chemical substances, extreme temperatures, and radiation can cause inflammation. ✓

Reading Check

2. What causes blood vessels in an injured area to dilate?

____

____

____

What is phagocytosis of pathogens?

Pathogens that enter your body may encounter cells that engulf and destroy them, a process known as phagocytosis. **Phagocytes** (FA guh sites) are white blood cells that destroy pathogens by surrounding and engulfing them. They are like fighter cells attacking and devouring the invaders they encounter. Phagocytes include monocytes. Monocytes develop into macrophages. Phagocytes also include neutrophils and eosinophils.

Macrophages are white blood cells that provide the first defense against pathogens that have entered the tissues. Macrophages are sometimes called giant scavengers or big eaters because of the manner in which they engulf pathogens or damaged cells. They will attack anything they recognize as foreign. Enzymes inside the macrophage digest the particles it has engulfed.

Section
39.2 Defense Against Infectious Diseases, *continued*

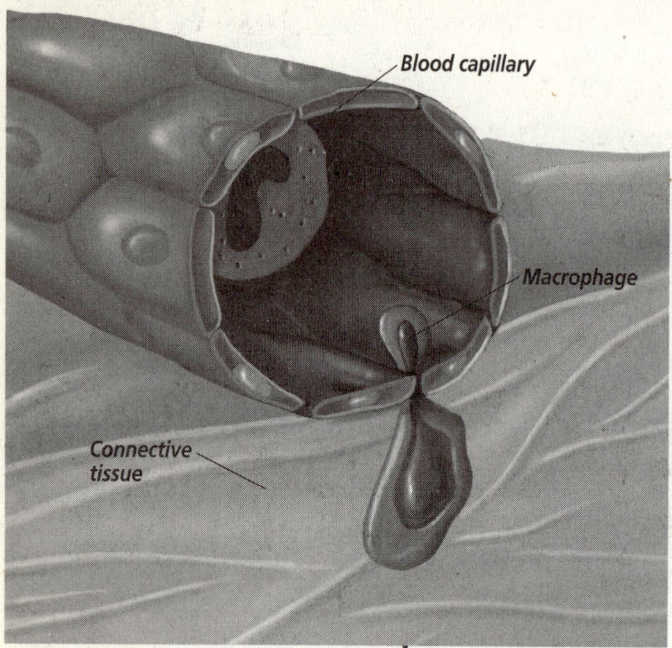

If the macrophages do not stop the infection, another type of phagocyte, called a neutrophil, is attracted to the site. Neutrophils also destroy pathogens by engulfing and digesting them.

If the infection is still not stopped, a third type of phagocyte arrives on the scene. Monocytes are small, immature macrophages that circulate in the bloodstream. These cells squeeze through blood vessel walls to move into the infected area. Once they reach the site of the infection, they mature into macrophages. They begin consuming pathogens and dead neutrophils. Once the infection is over, some monocytes mature into tissue macrophages. They remain in the area and prepare to guard against new infections.

Reading Check

3. What clears away pus?

____

____

____

After a macrophage has destroyed large numbers of pathogens, dead neutrophils, and damaged tissue cells, it eventually dies. After a few days, infected tissue develops a substance called pus. **Pus** consists of living and dead white blood cells, living and dead pathogens, and body fluids. Pus formation usually continues until the infection subsides. Eventually, the pus is cleared away by macrophages.

What are protective proteins?

When an infection is caused by a virus, the body faces a problem. Phagocytes alone cannot destroy viruses. Recall that a virus multiplies within a host cell. A phagocyte that engulfs a virus will be destroyed if the virus multiplies within it. One way your body can counteract viral infections is with interferons. **Interferons** are proteins that protect cells from viruses. Interferons are host-cell specific. This means that human interferons will protect human cells from viruses but cannot protect cells of other species from the same virus.

Acquired Immunity

The cells of your innate immune system continually check your body for foreign invaders. When a pathogen is detected, these cells defend your body. As the infection continues, another type of immune response that fights the invading pathogen is activated. Certain white blood cells gradually develop the ability to recognize a specific foreign substance. This acquired immune response causes these white blood cells to destroy the pathogen. Defending against a specific pathogen by gradually building up a resistance to it is called **acquired immunity.**

Normally, the immune system recognizes components of the body as something that belongs to the body. It recognizes foreign substances, called antigens, as not belonging. An acquired immune response begins when the immune system recognizes an antigen. It responds by producing antibodies against it. Antigens are foreign substances that stimulate an immune response. Antibodies are proteins in the blood that correspond specifically to each antigen. The development of acquired immunity is the job of the lymphatic system. The process of acquiring immunity to a specific disease can take days or weeks. The illustration at right shows the lymphatic system.

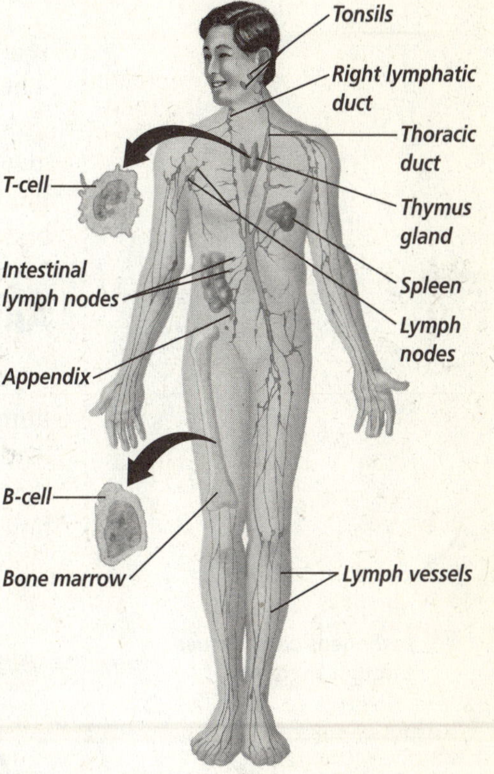

What is the lymphatic system?

Your lymphatic (lihm FA tihk) system not only helps the body defend itself against disease, but also maintains homeostasis by keeping body fluids at a constant level.

Body cells are constantly bathed in fluid. This **tissue fluid** is composed of water and dissolved substances that diffuse from the blood into the spaces between the cells that make up the surrounding tissues. This tissue fluid collects in open-ended lymph capillaries. Once the tissue fluid enters the lymph vessels, it is called **lymph.**

What are the glands of the lymphatic system?

At locations along the lymphatic system, the lymph vessels pass through lymph nodes. A **lymph node** is a small mass of tissue that contains lymphocytes. It filters pathogens from the lymph. A **lymphocyte** (LIHM fuh site) is a type of white blood cell that defends the body against foreign substances. ✓

Tonsils are large clusters of lymph tissue located at the back of the mouth cavity and at the back of the throat. They form a protective ring around the openings of the nasal and oral cavities. Tonsils protect against bacteria and other pathogens that enter your nose and throat.

Reading Check

4. What function do lymph nodes serve?

______

______

______

The spleen is an organ that stores certain types of lymphocytes. It also filters and destroys bacteria and worn-out red blood cells. The spleen does not filter lymph.

Another important part of the lymphatic system is the thymus gland. It is located above the heart. The thymus gland stores immature lymphocytes until they mature and are released into the body's defense system.

Antibody Immunity

Acquired immunity involves the production of two kinds of immune responses: antibody immunity and cellular immunity. Antibody immunity is a type of chemical warfare in your body that involves several types of cells. The illustration below shows how antibody immunity defends your body against pathogens.

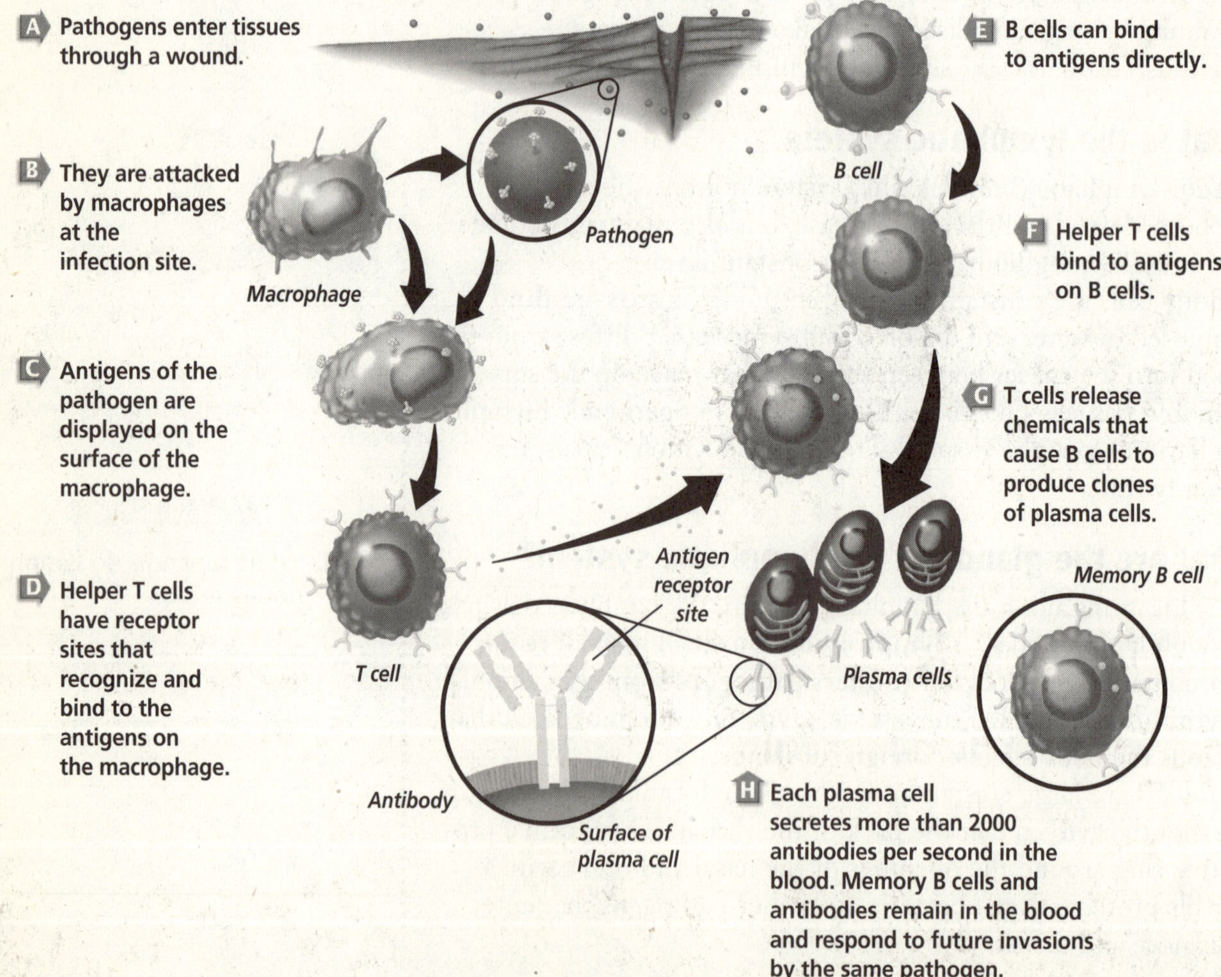

A Pathogens enter tissues through a wound.
B They are attacked by macrophages at the infection site.
Macrophage
Pathogen
C Antigens of the pathogen are displayed on the surface of the macrophage.
D Helper T cells have receptor sites that recognize and bind to the antigens on the macrophage.
T cell
Antigen receptor site
Antibody
Surface of plasma cell
B cell
E B cells can bind to antigens directly.
F Helper T cells bind to antigens on B cells.
G T cells release chemicals that cause B cells to produce clones of plasma cells.
Memory B cell
Plasma cells
H Each plasma cell secretes more than 2000 antibodies per second in the blood. Memory B cells and antibodies remain in the blood and respond to future invasions by the same pathogen.

When a pathogen invades the body, it is attacked by the cells of your innate immune system. If the infection is not controlled, then your body builds up acquired immunity. It produces antibodies to use against the antigen. A type of lymphocyte called a T cell becomes involved. A **T cell** is a lymphocyte that is produced in bone marrow and processed in the thymus gland. Two kinds of T cells play different roles in immunity.

One kind of T cell, a helper T cell, interacts with B cells. A **B cell** is a lymphocyte that becomes a plasma cell and makes antibodies when activated by a T cell. B cells are made in the bone marrow. Plasma cells release antibodies into the bloodstream and tissue spaces. Some activated B cells do not become plasma cells but remain in the bloodstream as memory B cells. Memory B cells are ready to respond if the same pathogen invades the body again.

Cellular Immunity

Cellular immunity also involves T cells with antigens on their surfaces. The T cells involved in cellular immunity are cytotoxic, or killer, T cells. T cells stored in the lymph nodes, spleen, and tonsils, transform into cytotoxic T cells. They are specific for a single antigen. However, unlike B cells, they do not form antibodies. Cytotoxic T cells produce identical clones. They travel to the infection site and release enzymes directly into the pathogens, which die.

The cells that protect the body against pathogens sometimes can cause problems within the body. The immune system may overreact to a harmless substance such as pollen. Mast cells release histamines in large amounts. This causes the symptoms of an allergic reaction: sneezing, increased mucus production in the nasal passages, and redness. The immune system also can attack its own cells. This attack of the body's own tissue is called an autoimmune disorder. Lupus and rheumatoid arthritis are autoimmune disorders. T cells and antibodies also can attack transplanted tissue, such as a kidney or heart, which comes from outside the body. ✓

Reading Check

5. What happens when the immune system overreacts to a harmless substance such as pollen?

____

____

____

Passive and Active Immunity

Acquired immunity to a disease may be passive or active. Passive acquired immunity develops by acquiring antibodies that are generated in another host. Active acquired immunity develops when your body produces antibodies in response to being exposed to antigens.

Think it Over

6. **Compare** How do humans develop active immunity by artificial and natural means?

Reading Check

7. Why did dairy workers who had cowpox not get smallpox during smallpox epidemics?

How does passive immunity develop?

Passive immunity may develop in two ways. Natural passive immunity develops when antibodies are transferred from a mother to her unborn baby through the placenta or to a newborn infant through the mother's milk. Artificial passive immunity occurs when a human is injected with antibodies from a person or animal who is already immune.

How does active immunity develop?

Active immunity can be gained naturally. When a person is exposed to particular antigens, the body produces antibodies that correspond to these antigens. When a person recovers from the infection, that person will usually be immune to the pathogen for the rest of his or her life.

Active immunity can be created artificially. This is done through vaccinations, usually a shot or injection, with a particular vaccine. A **vaccine** is a substance that consists of weakened, dead, or incomplete portions of pathogens or antigens. When a vaccine is injected into the body, it causes an immune response. Vaccines produce immunity because the body reacts as if it were infected with the disease.

In the late 1790s an English doctor named Edward Jenner demonstrated the first safe vaccination procedure. Dr. Jenner knew that people who worked with dairy cows sometimes acquired cowpox from cows that had the disease. Cowpox is similar to smallpox but is a much milder disease. Dairy workers who had had cowpox did not catch smallpox.

Jenner decided to test whether immunity to cowpox would create immunity to smallpox. Jenner infected a young boy with cowpox. The boy developed a mild cowpox infection. Six weeks later, Jenner scratched the boy's skin with smallpox viruses. The boy did not get sick. He had acquired active immunity to smallpox from the cowpox vaccination. The viruses for cowpox and smallpox are so similar that the immune system can't tell them apart. ✓

AIDS and the Immune System

In 1981, a number of cases of a rare pneumonia appeared in the San Francisco, California, area. The pneumonia was caused by a protozoan. Medical investigators noticed a relationship between this pneumonia and a rare form of skin cancer called Kaposi's sarcoma. The pneumonia and the skin cancer seemed to be associated with a failure throughout the body's immune system.